Engineering

Engineering Physics

SECOND EDITION

A. MARIKANI

Professor and Head
Department of Physics
Mepco Schlenk Engineering College
Sivakasi, Tamil Nadu

PHI Learning Private Limited
Delhi-110092
2015

₹ 350.00

ENGINEERING PHYSICS, Second Edition
A. Marikani

ISBN-978-81-203-4823-3

The export rights of this book are vested solely with the publisher.

Tenth Printing (Second Edition) ... ... **July, 2015**

Published by Asoke K. Ghosh, PHI Learning Private Limited, Rimjhim House, 111, Patparganj Industrial Estate, Delhi-110092 and Printed by Raj Press, New Delhi-110012.

CONTENTS

PREFACE

The first edition of this book is revised as per the syllabus prescribed by Anna University, Chennai. This book is suitable for first year of B.E./B.Tech. students of affiliated engineering colleges of Anna University, Chennai for their courses in Engineering Physics I and Engineering Physics II. Though it is written as per the syllabi prescribed by Anna University, Chennai, the organization and content of the book are suitable for other universities/institutions as well.

The text emphasizes the basic concepts of physics. In the second edition of this book, the following chapters/topics are included. They are, (i) elasticity, (ii) acoustics, (iii) thermal physics, (iv) non-linear optics, (v) biomaterials and (vi) crystal growth.

The book also contains numerous solved problems, short and descriptive type questions and exercise problems to help students assess their progress and familiarize them with the types of questions set in examinations.

The author is thankful to the Management, Thiru A. Tenzing, Correspondent, and Dr. S. Arivazhagan, Principal, Mepco Schlenk Engineering College, Sivakasi for their constant encouragement in bringing out this book.

The author extends his sincere thanks to Ms. Pushpita Ghosh, Director, Ms. Shivani Garg, Senior Editor, Editorial Department, Ms. Babita Mishra, Editorial Coordinator, PHI Learning, Delhi and Mr. D. Sakthivel, Mr. K. Prakash Babu, and Mr. P. Arivazhagan, PHI Learning, Tamil Nadu region for successfully bringing this book in time.

The comments and suggestions for further improvement of the book from students and professors will be thankfully acknowledged.

A. Marikani

PREFACE TO THE FIRST EDITION

Physics is the basis for most of the engineering principles and concepts. Therefore, a common course in engineering physics is designed by universities to be taught by all the first-year engineering students. This book emphasizes the basic concepts of physics. Though it is written as per the syllabi prescribed by Anna University (Chennai, Coimbatore, Trichy, Madurai and Tirunelveli), the organization and content of the book are suitable for other universities and institutions as well. This book covers all the topics for the first and second semesters. The book also covers the syllabus prescribed by SASTRA University, Tanjavur, Tamil Nadu.

The author has been teaching Engineering Physics since 1986. Based on his experience, the author has written this book in a simple and lucid language.

The book consists of sixteen chapters and appendices. Chapter 1 discusses ultrasonics and their industrial and medical applications. Chapter 2 is devoted to the properties of lasers and their several applications.

Different types of optical fibres and their geometries and use are discussed in Chapter 3. Quantum concepts of radiation have been discussed in Chapter 4. Chapter 5 is devoted to electron optics. Crystallography and crystal imperfections are discussed in Chapters 6 and 7. Properties of conducting, semiconducting, magnetic, dielectric and superconducting materials and their applications are discussed in Chapters 8, 9, 10, 11 and 12. Topics like nanomaterials, carbon nanotubes, and other modern engineering materials like shape memory alloys and metallic glasses are discussed in Chapters 13, 14 and 15. Chapter 16 provides an introduction to the concept of spectroscopy, electromagnetic spectrum and atomic energy levels. The breakdown of classical mechanics, uncertainty principle, quantum mechanical tunneling, perturbation theory, smart materials, classification of nanomaterials, AFM and STM are discussed in appendices.

The book also contains numerous solved problems, short and descriptive type questions and exercises to help students assess their progress and familiarize them with the types of questions set in examinations.

The comments and suggestions for further improvement of the book from students and teachers will be thankfully acknowledged.

A. Marikani

ACKNOWLEDGEMENTS

I wish to express my sincere thanks and gratitude to our respected Correspondent Thiru A. Vairaprakasam, and beloved Principal Dr. S. Balakrishnan, Mepco Schlenk Engineering College, Sivakasi for their constant encouragement and support to bring out this book.

I extend my heartfelt thanks to the staff of PHI Learning who are involved in this project for their support and tireless work to bring out this book in time.

I thank my family members for their encouragement and cooperation during the preparation of this manuscript. I would like to thank the Almighty for His endless blessings and kindness.

A. Marikani

CHAPTER 1

CRYSTAL PHYSICS

1.1 INTRODUCTION

Crystallography is the branch of science that deals with the crystal structures of elements. The crystal structures of elements are studied by means of either X-ray diffraction, electron beam diffraction or neutron beam diffraction. Among these three methods, the X-ray diffraction is mostly used because it is a powerful experimental tool. The X-ray diffraction is used either to determine the structure of new material, or it is used to identify the chemical composition of a common material having known structure.

The crystal structure of a material is studied due to certain reasons. By studying the crystal structure of a material (i) one can know about whether the material is crystalline or non-crystalline and (ii) the structure-dependent properties of the materials.

1.2 CRYSTALLINE AND NON-CRYSTALLINE MATERIALS

In a natural crystal, the atoms or ions are arranged in a regular and periodic manner in three dimensions. Such crystals are said to be **crystalline materials**. The examples of the crystalline materials are the metals like silver, copper, gold, aluminium etc.

The crystalline materials are further classified into single crystals and polycrystalline materials. In a solid, if the periodic arrangement of atoms extends up to the entire specimen without any interruption, then it is said to be **single crystals**. Single crystals may exist in nature or it can be synthesized artificially. It is generally very difficult to grow the single crystals because the environment should be carefully controlled. When a crystalline solid is composed of a large number of small crystals or grains, then it is said to be **polycrystalline material**.

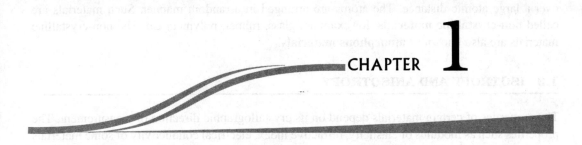

In some materials there is no regular and systematic periodic arrangement of atoms relatively over a large atomic distance. The atoms are arranged in a random manner. Such materials are called **non-crystalline materials**, for example, glass, rubber, polymers etc. The non-crystalline materials are also known as **amorphous materials**.

1.3 ISOTROPY AND ANISOTROPY

The properties of certain materials depend on its crystallographic direction of measurement. The properties such as modulus of elasticity, refractive index, electrical conductivity of some materials vary for [100] and [111] crystallographic directions. The substance whose measured properties depend on its crystallographic direction of measurement is called **anisotropic** and the directionality of properties is called **anisotropy**. The properties of some substances do not depend upon the direction of measurement. They are said to be **isotropic**.

The triclinic structure is normally anisotropic. The magnetic material, such as iron gets more easily magnetized along [100] directions compared to [110] and [111] directions. Similarly, Ni gets more easily magnetized along [111] compared to [100] and [110] directions. This shows the directional properties of magnetic materials.

1.4 LATTICE, BASIS AND CRYSTAL STRUCTURE

1.4.1 Lattice

The atoms or ions are arranged in a natural crystal in a systematic periodic arrangement. In order to study the arrangement of atoms in a natural crystal, the points (dots) are arranged. The arrangement of points is called a **lattice**. If the points are arranged in two dimension, then it is said to be a two-dimensional lattice. If the points are arranged in a three dimension, then it is said to be a three-dimensional lattice. Consider a two-dimensional lattice as shown in Figure 1.1. Each and every point is called a lattice point. The distances between any two lattice points along *X*- and *Y*-directions are represented as $\boldsymbol{a}$ and $\boldsymbol{b}$ respectively. They are called lattice translational vectors. In three-dimensional lattice, one can need three lattice translational vectors $\boldsymbol{a}$, $\boldsymbol{b}$ and $\boldsymbol{c}$.

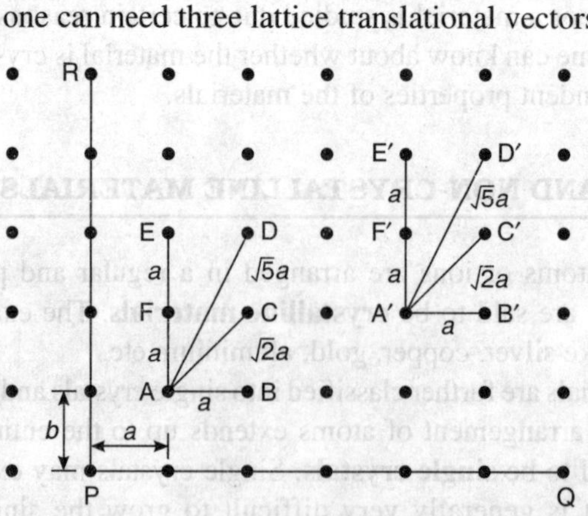

Figure 1.1 Lattice, basis and crystal structure.

The term lattice can be defined in another way. In an arrangement of points, if the arrangement looks the same when it is viewed from different positions, then that arrangement is called a **lattice**. Consider the set of lattice points A, B, C, D, E, & F and A′, B′, C′, D′, E′, & F′. The distances $AB = A'B' = a$, $AC = A'C' = \sqrt{2}a$, $AD = A'D' = \sqrt{5}a$, $AE = A'E' = 2a$ and $AF = A'F' = a$. This shows that this arrangement of points looks the same, when it is viewed either from the set of lattice points A, B, C, D, E, F or A′, B′, C′, D′, E′, F′. Therefore, this arrangement is said to be a lattice.

1.4.2 Basis

The smallest arrangement or a group of atoms to be fixed in each and every lattice point is said to be a **basis**. Consider a two-dimensional lattice as shown in Figure 1.2a. The group of atoms that is associated with every lattice point is called a **basis**. The basis is arranged in each and every lattice points so as to get a structure. In the case of a natural crystal also, a group of atoms is repeatedly arranged so as to form a crystal structure. Consider a sodium chloride unit cell. The Na and Cl ions are repeatedly arranged in NaCl unit cell. The Na and Cl ions combined together are called a basis for NaCl unit cell.

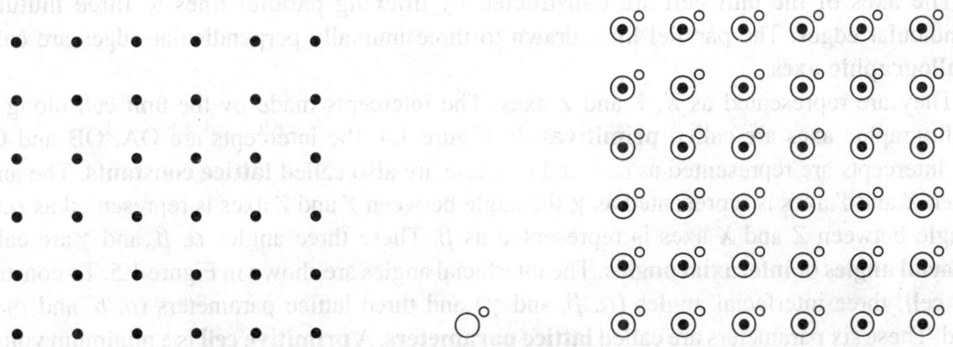

(a) Lattice (b) Basis (c) Crystal structure

Figure 1.2 Lattice, basis and crystal structure.

1.4.3 Crystal Structure

A crystal structure is formed by arranging the basis in each and every lattice points. It can be expressed as

$$\text{Lattice} + \text{basis} \rightarrow \text{crystal structures} \quad (1.1)$$

The equation represents a crystal structure and it is obtained by arranging the basis in each and every lattice points. Consider that the basis, shown in Figure 1.2(b), is arranged in each and every lattice points as in Figure 1.2(a) which results in a structure as shown in Figure 1.2(c). Therefore, the crystal structure is the result of two quantities, namely lattice and basis.

1.5 UNIT CELLS

The regular and periodic arrangement of atoms in a natural crystal shows that a small group of atoms form a repetitive pattern. The repetitive pattern formed by arranging a small group of atoms one above the other is called **unit cell**. It is similar to the arrangement of bricks one above the other while constructing a wall. The unit cell is the basic structural unit or the building block of a crystal structure. The unit cell is chosen to represent the symmetry of a crystal structure and it is shown in Figure 1.3. There are fourteen different types of unit cells in three dimensions.

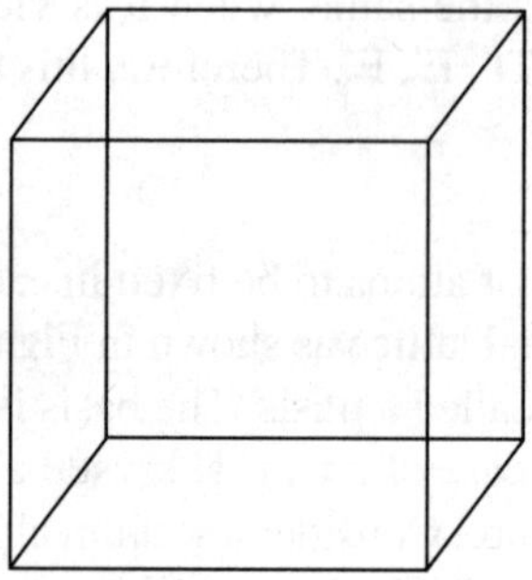

Figure 1.3 Unit cell.

The axes of the unit cell are constructed by drawing parallel lines to three mutually perpendicular edges. The parallel lines drawn to three mutually perpendicular edges are called **crystallographic axes**.

They are represented as *X*, *Y* and *Z* axes. The intercepts made by the unit cell along the crystallographic axes are called **primitives**. In Figure 1.4, the intercepts are OA, OB and OC. These intercepts are represented as *a*, *b* and *c*. These are also called **lattice constants**. The angle between *X* and *Y* axes is represented as γ, the angle between *Y* and *Z* axes is represented as α and the angle between *Z* and *X* axes is represented as β. These three angles α, β, and γ are called **interfacial angles** or **interaxial angles**. The interfacial angles are shown in Figure 1.5. To construct a unit cell, three interfacial angles (α, β, and γ) and three lattice parameters (*a*, *b*, and *c*) are needed. These six parameters are called **lattice parameters**. A **primitive cell** is a minimum volume occupying cell and it contains only one atom per cell and it is shown in Figure 1.6.

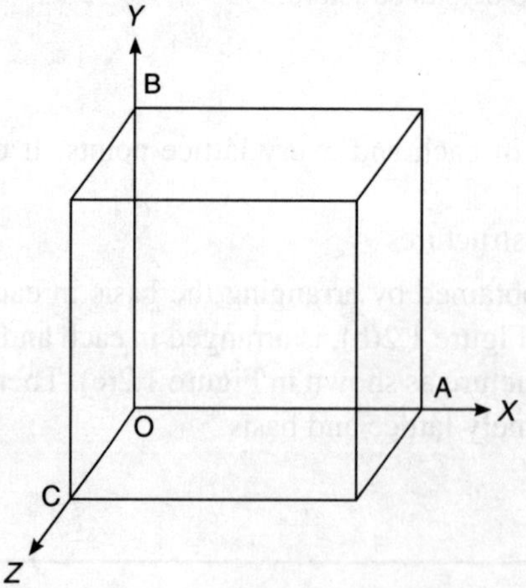

Figure 1.4 Crystallographic axes.

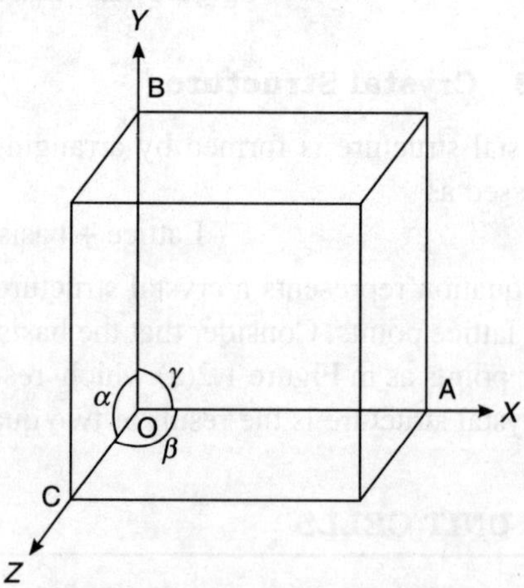

Figure 1.5 Interfacial angles.

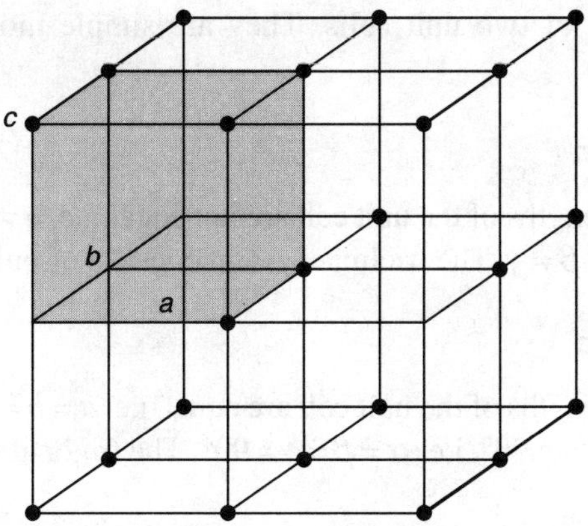

Figure 1.6 Primitive cell.

1.6 CRYSTAL SYSTEMS

The crystals are classified into seven crystal systems on the basis of the unit length of the axis and there exist a relation between the interfacial angles. They are cubic, tetragonal, orthorhombic, monoclinic, triclinic, trigonal and hexagonal. The trigonal crystal system is also called rhombohedron.

1.6.1 Cubic System

In a cubic crystal, all the lengths of the unit cell are equal, i.e. $a = b = c$ and the interfacial angles are equal to 90°, i.e. $\alpha = \beta = \gamma = 90°$. The cubic systems consist of three unit cells. They are simple cubic, body-centred cubic and face-centred cubic unit cells.

1.6.2 Tetragonal System

In a tetragonal crystal, all the lengths of the unit cell along *X*- and *Y*-axes are equal and they are not equal to the length of unit cell along *Z*-axis, i.e. $a = b \neq c$. and the interfacial angles are equal to 90°, i.e., $\alpha = \beta = \gamma = 90°$. The tetragonal system consists of two unit cells. They are simple tetragonal and body-centred tetragonal unit cells.

1.6.3 Orthorhombic System

In an orthorhombic system, all the lengths of the unit cell are not equal, (i.e. $a \neq b \neq c$) and the interfacial angles are equal to 90°, (i.e. $\alpha = \beta = \gamma = 90°$). The orthorhombic system consists of four unit cells. They are simple orthorhombic, body-centred orthorhombic, face-centred orthorhombic and base-centred orthorhombic unit cells.

1.6.4 Monoclinic System

In a monoclinic system, all the lengths of the unit cell are not equal, i.e. $a \neq b \neq c$ and the interfacial angles α and β are equal to 90°, whereas γ is not equal to 90°, i.e. $\alpha = \beta = 90°$, $\gamma \neq 90°$. The monoclinic system consists of two unit cells. They are simple monoclinic and base-centred monoclinic unit cells.

1.6.5 Triclinic System

In a triclinic system, all the lengths of the unit cell are not equal, i.e. $a \neq b \neq c$ and all the interfacial angles are not equal, i.e. $\alpha \neq \beta \neq \gamma$. The triclinic system consists of only one unit cell.

1.6.6 Trigonal System

In a trigonal system, all the lengths of the unit cell are equal, i.e. $a = b = c$ and the interfacial angles are equal, but they are other than 90°, i.e. $\alpha = \beta = \gamma \neq 90°$. The trigonal system consists of only one unit cell.

1.6.7 Hexagonal System

In a hexagonal system, the lengths of the unit cell $a = b$, but they are not equal to c, i.e. $a = b \neq c$ and the interfacial angles $\alpha = \beta = 90°$ but $\gamma = 120°$. The hexagonal system consists of only one unit cell.

1.7 BRAVAIS LATTICES

In 1848, Bravais arranged the points in different ways in three dimensions so that the environment looks the same from each point. This arrangement provides 14 different types of unit cells in three dimensions. These fourteen different types of unit cells in three dimensions are known as Bravais lattices. The fourteen different types of Bravais lattices (unit cells) are shown in Figure 1.7. The crystal systems, lattice parameters, lattice symbols and number of unit cells are displayed in Table 1.1.

Table 1.1 Crystal systems and number of unit cells

Crystal systems	*Lattice parameters*	*Lattice symbols*	*Number of unit cells*
Cubic	$a = b = c, \alpha = \beta = \gamma = 90°$	P, I, F	3
Tetragonal	$a = b \neq c, \alpha = \beta = \gamma = 90°$	P, I	2
Orthorhombic	$a \neq b \neq c, \alpha = \beta = \gamma = 90°$	P, I, F, C	4
Monoclinic	$a \neq b \neq c, \alpha = \beta = 90° \neq \gamma$	P, C	2
Triclinic	$a \neq b \neq c, \alpha \neq \beta \neq \gamma$	P	1
Trigonal	$a = b = c, \alpha = \beta = \gamma \neq 90°$	P	1
Hexagonal	$a = b \neq c, \alpha = \beta = 90°, \gamma = 120°$	P	1

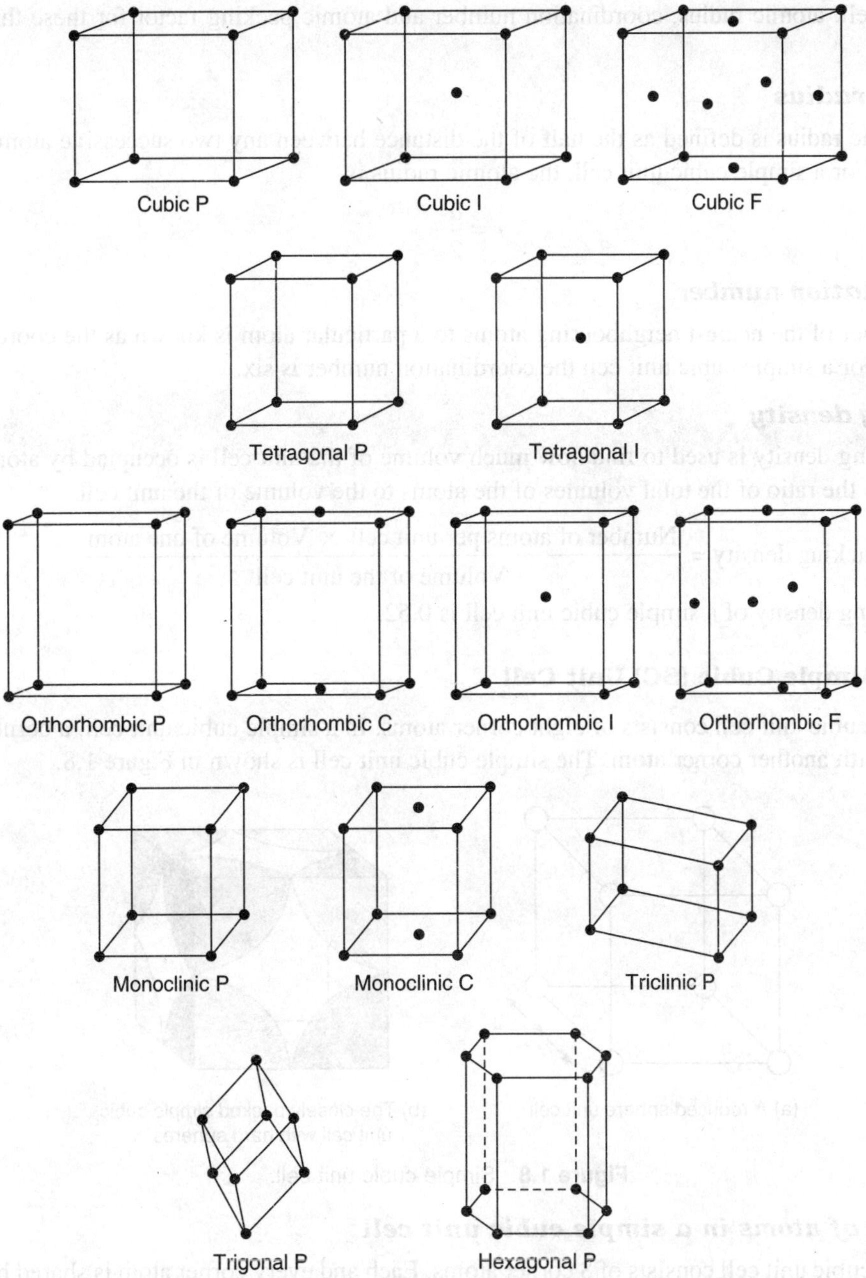

Figrue 1.7 Bravais lattices.

1.8 CUBIC UNIT CELLS

There are three unit cells in cubic crystal structures. They are simple cubic unit cell, body-centred cubic unit cell and face-centred cubic unit cell. In this section let us discuss the number of atoms in the unit cell, atomic radius, coordination number and atomic packing factor for these three unit cells.

Atomic radius

The atomic radius is defined as the half of the distance between any two successive atoms in the unit cell. For a simple cubic unit cell, the atomic radius is

$$r = \frac{a}{2}$$

Coordination number

The number of the nearest neighbouring atoms to a particular atom is known as the coordination number. For a simple cubic unit cell the coordination number is six.

Packing density

The packing density is used to find how much volume of the unit cell is occupied by atoms. It is defined as the ratio of the total volumes of the atoms to the volume of the unit cell.

$$\text{Packing density} = \frac{\text{Number of atoms per unit cell} \times \text{Volume of one atom}}{\text{Volume of the unit cell}} \tag{1.2}$$

The packing density of a simple cubic unit cell is 0.52.

1.8.1 Simple Cubic (SC) Unit Cell

A simple cubic unit cell consists of eight corner atoms. In a simple cubic unit cell, a corner atom touches with another corner atom. The simple cubic unit cell is shown in Figure 1.8.

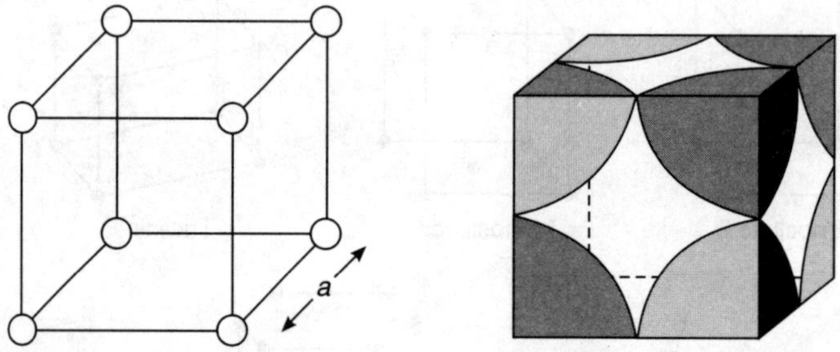

(a) A reduced sphere unit cell

(b) The closely packed simple cubic unit cell with hard spheres

Figure 1.8 Simple cubic unit cell.

Number of atoms in a simple cubic unit cell

A simple cubic unit cell consists of 8 corner atoms. Each and every corner atom is shared by eight adjacent unit cells. Therefore, one corner atom contributes $\frac{1}{8}$th of its parts to one unit cell. Since, there are eight corner atoms in a unit cell, the total number of atoms is $\frac{1}{8} \times 8 = 1$. Therefore, the number of atoms in a simple cubic unit cell is one.

Atomic radius

In a simple cubic lattice, a corner atom touches another corner atom. Therefore, $2r = a$. So, the atomic radius of an atom in a simple cubic unit cell is $\frac{a}{2}$.

Coordination number

Consider a corner atom in a simple cubic unit cell. It has four nearest neighbours in its own plane. In a lower plane, it has one more nearest neighbour and in an upper plane, it has one more nearest neighbour. Therefore, the total number of nearest neighbour is six.

Packing density

The packing density of a simple cubic unit cell is calculated as follows:

$$\text{Packing density} = \frac{\text{Number of atoms per unit cell} \times \text{Volume of one atom}}{\text{Volume of the unit cell}}$$

$$= \frac{1 \times \frac{4}{3}\pi r^3}{a^3}$$

Substituting $r = \frac{a}{2}$, we get

$$\text{Packing density} = \frac{4}{3}\pi\left(\frac{a}{2}\right)^3 \frac{1}{a^3} = \frac{\pi}{6} = 0.52$$

The packing density of a simple cubic unit cell is 0.52. It means that 52% of the volume of the unit cell is occupied by atoms and the remaining 48% of the volume is vacant.

1.8.2 Body-centred Cubic (BCC) Unit Cell

A body-centred cubic unit cell has eight corner atoms and one-body centred atom. In a body-centred cubic unit cell, the atoms touch along the body diagonal. The body-centred cubic unit cell is shown in Figure 1.9.

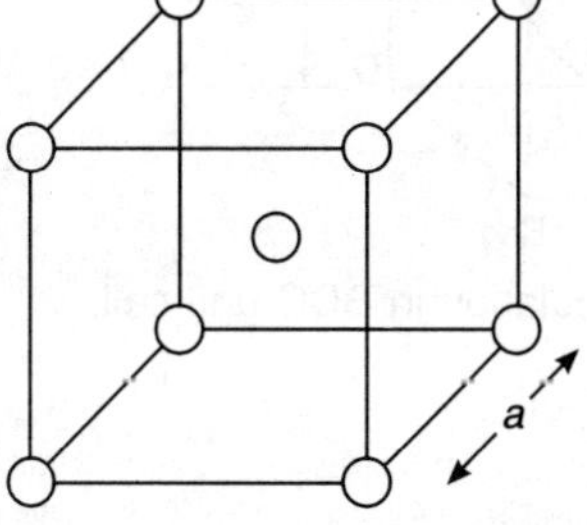

(a) A reduced sphere unit cell

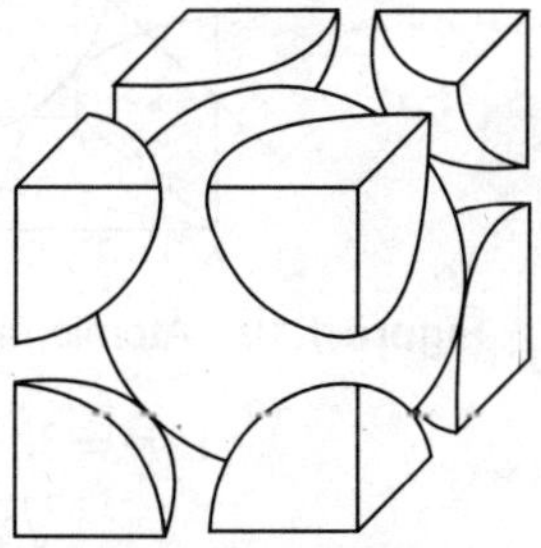

(b) The closely-packed BCC unit cell with hard spheres

Figure 1.9 Body-centred cubic unit cell.

Number of atoms in a body-centred cubic unit cell

A corner atom in a body-centred cubic unit cell is shared by eight adjacent unit cells. Therefore, one corner atom contributes $\frac{1}{8}$th of its parts to one unit cell. Since, there are eight corner atoms in a unit cell, the total number of atoms contributed by the corner atoms is $\frac{1}{8} \times 8 = 1$. In addition, a body-centred cubic unit cell has a body-centred atom at the centre of the unit cell. Therefore, the total number of atoms present in a body-centred cubic unit cell is two.

Atomic radius

In a body-centred cubic unit cell, the corner atoms touches along the body diagonal. From Figure 1.10, the length of the body diagonal is $4r$.

From triangle ABD,

$$AD^2 = AB^2 + BD^2 \qquad (1.3)$$

For a cubic unit cell, the cube edge is a. Therefore, AB = BD = a. Substituting, the values of AB, and BD we get,

$$AD^2 = a^2 + a^2 = 2a^2$$

$$AD = \sqrt{2}a$$

Consider the triangle ADH,

$$AH^2 = AD^2 + DH^2 \qquad (1.4)$$

Substituting the values of AH = $4r$, AD = $\sqrt{2}a$, and DH = a

$$(4r)^2 = (\sqrt{2}a)^2 + a^2$$

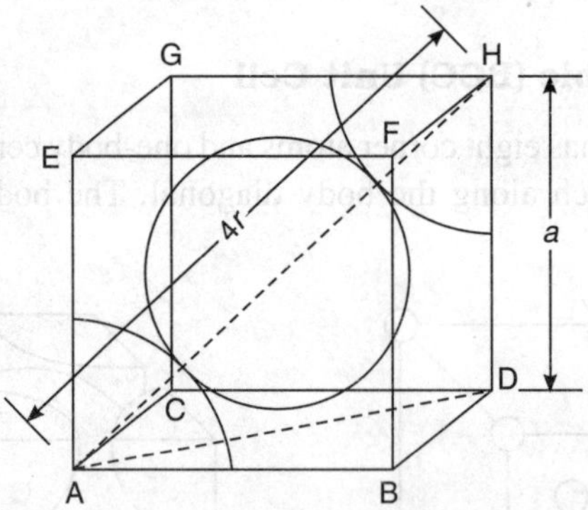

Figure 1.10 Atomic radius calculation in BCC unit cell.

$$16r^2 = 3a^2$$

$$r = \frac{\sqrt{3}}{4}a \qquad (1.5)$$

The atomic radius of an atom in a BCC unit cell is $r = \frac{\sqrt{3}}{4}a$

Coordination number

In a BCC unit cell, a body-centred atom is surrounded by eight corner atoms. For a body-centred atom, a corner atom is the nearest neighbour. Therefore, the number of the nearest neighbour is eight.

Packing density

The packing density of a body-centred cubic unit cell is calculated as follows:

$$\text{Packing density} = \frac{\text{Total volume of the atoms}}{\text{Volume of the unit cell}}$$

$$\text{Packing density} = \frac{\text{Number of atoms per unit cell} \times \text{Volume of one atom}}{\text{Volume of the unit cell}}$$

$$= \frac{2 \times \frac{4}{3}\pi r^3}{a^3}$$

Substituting $r = \frac{\sqrt{3}a}{4}$, we get

$$\text{Packing density} = 2 \times \frac{4}{3}\pi \left(\frac{\sqrt{3}a}{4}\right)^3 \frac{1}{a^3} = \frac{\sqrt{3}\pi}{8} = 0.68$$

The packing density of a body-centred cubic unit cell is 0.68. It means that 68% of the volume of the unit cell is occupied by atoms and the remaining 32% of the volume is vacant.

1.8.3 Face-centred Cubic (FCC) Unit Cell

A face-centred cubic unit cell consists of eight corner atoms and six face-centred atoms. In a face-centred cubic unit cell, an atom touches another atom along the face diagonal. The face-centred cubic unit cell is shown in Figure 1.11.

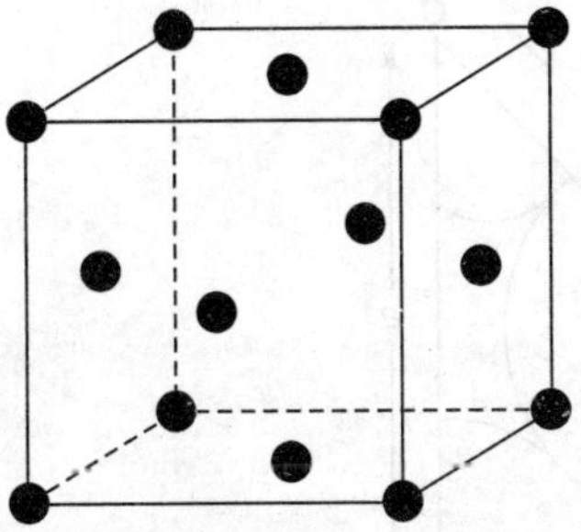

(a) A reduced sphere unit cell

(b) The closely-packed FCC unit cell with hard spheres

Figure 1.11 Face-centred cubic unit cell.

Number of atoms in a face-centred cubic unit cell

In a FCC unit cell, there are eight corner atoms and six face-centred atoms. A corner atom is shared by eight adjacent unit cells. Therefore, one corner atom contributes $\frac{1}{8}$th of its parts to one unit cell. Since, there are eight corner atoms in a unit cell, the total number of atoms contributed by the corner atoms is $\frac{1}{8} \times 8 = 1$. There are six face-centred atoms. A face-centred atom is shared by two unit cells. Therefore, one face-centred atom contributes half of its parts to one unit cell. So, the total number of atoms contributed by the face-centred atoms is $\frac{1}{2} \times 6 = 3$ and the total number of atoms present in a face-centred cubic unit cell is four.

Atomic radius

In a face-centred cubic unit cell, the atom touches along the face diagonal. Therefore the face diagonal is equal to $4r$. To find the atomic radius of the atom in face-centred cubic unit cell, consider the triangle ABC (Figure 1.12).

From triangle ABC,

$$AC^2 = AB^2 + BC^2$$

For a cubic unit cell, the cube edge is a. Therefore, AB = BC = a. Substituting the values of AB, BC and AC, we get

$$(4r)^2 = a^2 + a^2$$

$$16r^2 = 2a^2$$

$$r = \frac{a}{\sqrt{8}} = \frac{a}{2\sqrt{2}}$$

The atomic radius of an atom in FCC unit cell is

$$r = \frac{a}{2\sqrt{2}} \qquad (1.6)$$

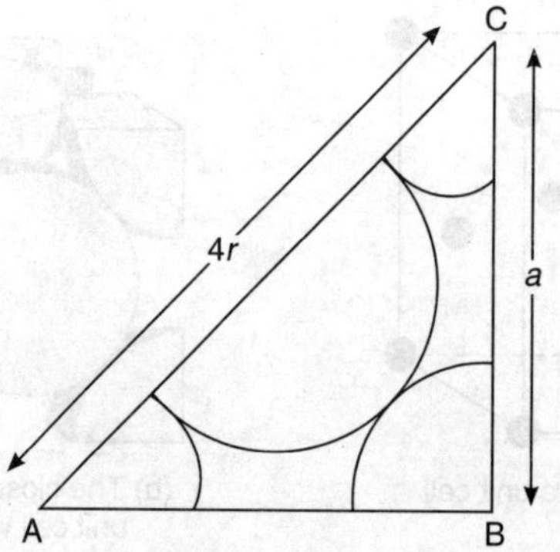

Figure 1.12 Atomic radius calculation for FCC unit cell.

Coordination number

Consider a face-centred atom in the upper plane of a FCC unit cell. It is surrounded by four corner atoms. These corner atoms are the nearest neighbours for this face-centred atom. There are four more face-centred atoms that are nearest neighbours for this reference face-centred atom in a lower plane. Similarly this reference face-centred atom has four more face-centred atoms as its nearest neighbours in an upper plane. Therefore, the total number of the nearest neighbours is 12.

Packing density

The packing density of a face-centred cubic unit cell is calculated as follows:

$$\text{Packing density} = \frac{\text{Total volume of the atoms}}{\text{Volume of the unit cell}}$$

$$\text{Packing density} = \frac{\text{Number of atoms per unit cell} \times \text{Volume of one atom}}{\text{Volume of the unit cell}}$$

$$= \frac{4 \times \frac{4}{3}\pi r^3}{a^3}$$

Substituting $r = \dfrac{a}{2\sqrt{2}}$, we get

$$\text{Packing density} = 4 \times \frac{4}{3}\pi\left(\frac{a}{2\sqrt{2}}\right)^3 \frac{1}{a^3} = \frac{\pi}{3\sqrt{2}} = 0.74$$

The packing density of a face-centred cubic unit cell is 0.74. It means that 74% of the volume of the unit cell is occupied by atoms and the remaining 26% of the volume is vacant. The packing density 0.74 is the maximum value and hence this unit cell is said to be cubic close-packed (CCP) structure.

1.9 HEXAGONALLY CLOSE-PACKED (HCP) STRUCTURE

A hexagonally close-packed structure consists of three layers of atoms, namely bottom layer, middle layer and upper layer. The middle layer lies just above the bottom layer at a distance of $c/2$. The upper layer lies at a distance of c from the bottom layer. The bottom layer consists of six corner atoms and one face-centred atom. The middle layer has three atoms. The upper layer has six corner atoms and one face-centred atom. The lattice parameters for a HCP structure is: $a = b \neq c$ and $\alpha = \beta = 90°$ and $\gamma = 120°$. A hexagonally close-packed unit cell is shown in Figure 1.13.

The arrangement of atoms in a hexagonally close-packed structure is explained using a simple figure shown in Figure 1.14. The bottom layer has an arrangement of seven atoms having equal radius in such a way that one atom is surrounded by six other atoms. The centre of these atoms are marked as 'A' and the centre of the surrounding atoms are joined, it will constitute a hexagon. These six atoms are the corner atoms for the bottom layer and the middle atom is the face-centred atom. The second layer is arranged over the first layer of atoms. There are six possible places to arrange the second layer. Among these six places, three alternate places are marked as 'B' and the remaining three places are marked as 'C'. Consider that the second layer is placed over the places marked as 'B'. There are two possible ways to arrange the third layer of atom. Firstly, the third layer is arranged directly over the first layer. This will constitute a stacking sequence of ABABAB... and it will form a HCP unit cell. Secondly, the third layer is placed over the places marked as 'C'. The stacking sequence goes on as ABCABCABC... and it will constitute a FCC structure.

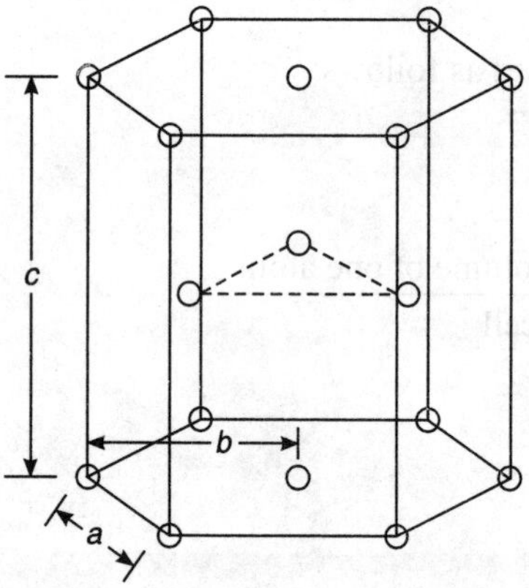

Figure 1.13 HCP unit cell.

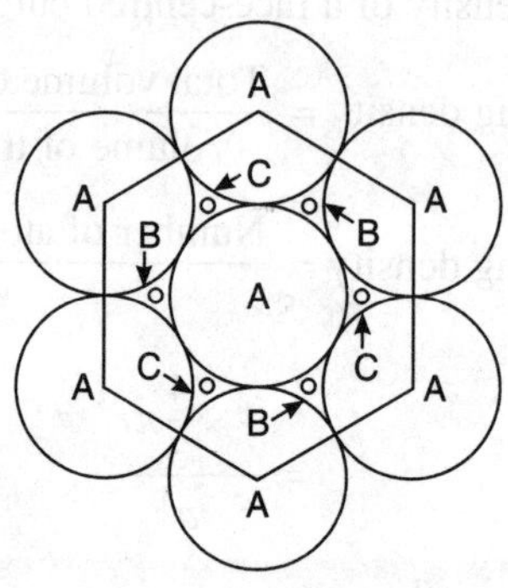

Figure 1.14 Arrangement of atoms in a HCP unit cell.

Number of atoms in a HCP unit cell

The bottom layer of a HCP unit cell consists of six corner atoms and one face-centred atom. Each and every corner atom in the bottom layer is shared by six unit cells. Therefore, a corner atom contributes $\frac{1}{6}$th of its parts to one unit cell. Since there are six corner atoms in the bottom layer, the total number of atoms contributed by the corner atoms is $\frac{1}{6} \times 6 = 1$. The face-centred atom in the bottom layer is shared by two unit cells. Therefore, it contributes $\frac{1}{2}$ of its parts to one unit cell. The total number of atoms present in the bottom layer is $1 + \frac{1}{2} = \frac{3}{2}$. The upper layer also has $\frac{3}{2}$ atoms. The middle layer has three atoms. Therefore, the total number of atoms present in the unit cell is $\frac{3}{2} + 3 + \frac{3}{2} = 6$.

Atomic radius

Consider the bottom layer of atoms. A corner atom touches another corner atom as shown in Figure 1.15. Therefore, $2r = a$ and hence $r = \frac{a}{2}$.

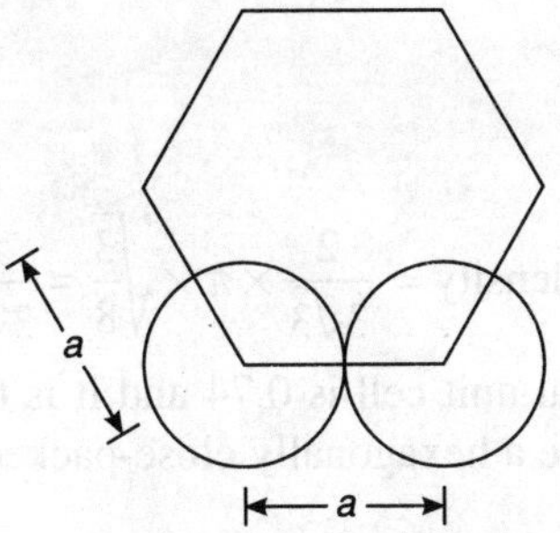

Figure 1.15 HCP structure-atomic radius.

Coordination number

Consider the face-centred atom in the bottom layer. It is surrounded by six corner atoms. So, there are six nearest neighbours for a face-centred atom in its own plane. The middle layer of atoms also touches the bottom layer. So, the three middle layers of atoms are the nearest neighbours lying in the upper plane. The unit cell that lies below this unit cell also has three more atoms in the middle layer and these three atoms are the nearest neighbours to the face-centred atom. Therefore, the total number of the nearest neighbour is 6 + 3 + 3 = 12.

Packing density

The packing density of a HCP unit cell is calculated as follows:

$$\text{Packing density} = \frac{\text{Number of atoms per unit cell} \times \text{Volume of one atom}}{\text{Volume of the unit cell}}$$

The volume of the unit cell is calculated using the relation,

$$\text{Volume of a hexagon} = \text{area of the bottom surface} \times \text{height}$$

$$= 6 \times \text{area of an equilateral triangle} \times \text{height}$$

$$= 6 \times \frac{\sqrt{3}}{4} a^2 \times c$$

$$\text{Packing density of a HCP unit cell} = \frac{6 \times \frac{4}{3}\pi r^3}{6 \times \frac{\sqrt{3}}{4} a^2 c}$$

Substituting $r = \frac{a}{2}$, we get

$$\text{Packing density} = \frac{6 \times \frac{4}{3}\pi \left(\frac{a}{2}\right)^3}{6 \times \frac{\sqrt{3}}{4} a^2 c}$$

$$= \frac{2}{3\sqrt{3}} \times \pi \times \frac{a}{c}$$

Substituting $\frac{c}{a}=\sqrt{\frac{8}{3}}$, we get

$$\text{Packing density} = \frac{2}{3\sqrt{3}} \times \pi \times \sqrt{\frac{3}{8}} = \frac{\pi}{3\sqrt{2}} = 0.74$$

The packing density of a hexagonal unit cell is 0.74 and it is the maximum value. Therefore, a hexagonal unit cell is also said to be a hexagonally close-packed (HCP) structure.

1.10 RELATION BETWEEN *c* AND *a*

Consider the bottom surface of a hexagonal unit cell. It has six corner atoms and one face-centred atom. Let A, B, C, D, E and F be the corner atoms and O be the face-centred atom as shown in Figure 1.16. The middle layer of atoms is placed over the first layer. Let I, G, and H be the second layer of atoms. Consider the triangle AFO. It is an equilateral triangle. In the triangle AFO, A and F are corner atoms and O is the face-centred atom. Let us bisect the faces AF and AO. Let OE′ and FJ be the perpendicular bisectors drawn to the lines AF and AO. X is the centroid. The second layer of atoms lies exactly at a distance of $\frac{c}{2}$ from the bottom layer.

In Figure 1.17, OG = a and XG = $\frac{c}{2}$. Consider the triangle XJO. The angles of the triangle XJO are 30°, 60° and 90°. In the triangle XJO

$$\sin 60° = \frac{OJ}{OX}$$

$$OX \sin 60° = \frac{a}{2}$$

$$OX \times \frac{\sqrt{3}}{2} = \frac{a}{2}$$

$$OX = \frac{a}{\sqrt{3}}$$

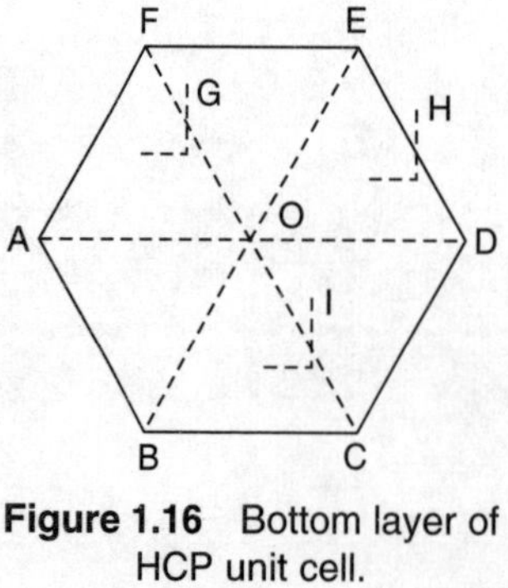

Figure 1.16 Bottom layer of HCP unit cell.

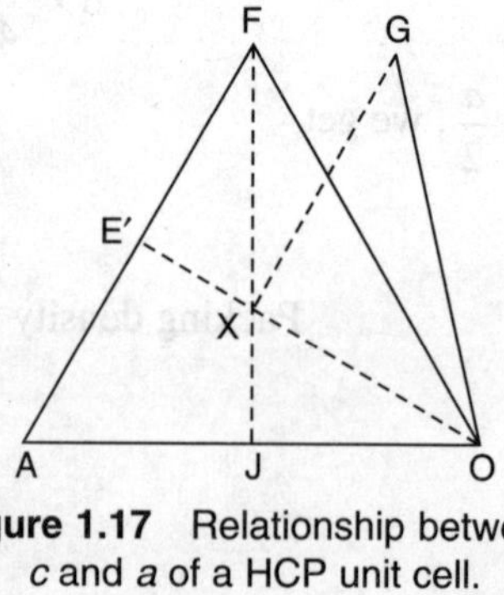

Figure 1.17 Relationship between *c* and *a* of a HCP unit cell.

Consider the triangle, XOG. Applying Pythagoras theorem, we get

$$OG^2 = XG^2 + OX^2 \tag{1.7}$$

$$a^2 = \left(\frac{c}{2}\right)^2 + \left(\frac{a}{\sqrt{3}}\right)^2$$

$$a^2 - \frac{a^2}{3} = \frac{c^2}{4}$$

$$\frac{c^2}{a^2} = \frac{8}{3}$$

$$\frac{c}{a} = \sqrt{\frac{8}{3}} = 1.633 \tag{1.8}$$

The ratio of c to a in a hexagonal unit cell is, $\frac{c}{a} = \sqrt{\frac{8}{3}}$.

1.11 COMPARISON OF ATOMIC RADIUS, COORDINATION NUMBER AND PACKING DENSITY OF SC, BCC, FCC AND HCP UNIT CELLS

The number of atoms in a unit cell, atomic radius, coordination number and packing density, for a simple cubic (SC), body-centred cubic (BCC), face-centred cubic (FCC) and hexagonally close-packed unit cells are listed in Table 1.2.

Table 1.2 Packing density, atomic radius, coordination number of SC, BCC, FCC and HCP unit cells

Crystal systems	*Number of atoms per unit cell*	*Atomic radius*	*Coordination number*	*Packing density*
Simple cubic	1	$\frac{a}{2}$	6	0.52
Body-centred cubic	2	$\frac{\sqrt{3}}{4}a$	8	0.68
Face-centred cubic	4	$\frac{a}{2\sqrt{2}}$	12	0.74
Hexagonally close-packed structure	6	$\frac{a}{2}$	12	0.74

From Table 1.2, one can infer that the packing density increases with the increase in coordination number.

1.12 RELATION BETWEEN THE ATOMIC WEIGHT *A* AND THE INTERATOMIC DISTANCE *a*

The mass of the unit cell is given by

$$\text{Mass} = \text{volume of the unit cell} \times \text{density}$$

$$m = V \times \rho \tag{1.9}$$

Let A be the atomic weight. The atomic weight of a substance represents the mass of one atom of that substance. Let n be the number of atoms in a unit cell. The mass of the unit cell is nA. Usually, the atomic weight is given in atomic mass unit (amu). To convert it into kilogram, it should be divided by the Avogadro's constant, N_A. Therefore, the mass of the unit cell is

$$m = \frac{nA}{N_A} \tag{1.10}$$

From Eq. (1.9) and Eq. (1.10), we get

$$\rho V = \frac{nA}{N_A}$$

For cubic unit cell, $V = a^3$. Therefore, the above equation becomes

$$a = \left(\frac{nA}{\rho N_A}\right)^{1/3} \tag{1.11}$$

Equation (1.11) gives the relationship between the interatomic distance and atomic weight.

1.13 CRYSTAL PLANES AND MILLER INDICES

In natural crystals, the atoms or ions are arranged in a regular and periodic arrangement. This periodic arrangement of atoms or ions in a natural crystal produces parallel equidistant planes. These parallel and equidistant planes formed by the periodic arrangement of atoms or ions in a natural crystal are said to be crystal planes. The crystal planes with lattice spacings d_1, d_2 and d_3 are shown in Figure 1.18.

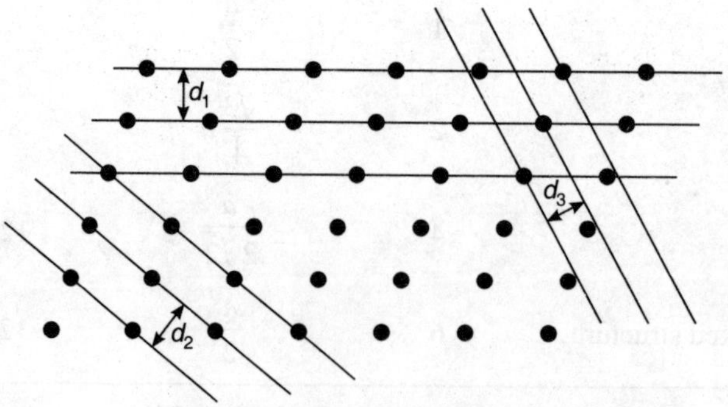

Figure 1.18 Crystal planes.

William Hallowes Miller (1801–1880) devised a method to represent a crystal planes. According to Miller the crystallographic planes are specified in terms of indexing schemes. The planes are indexed using the reciprocals of the axial intercepts. The Miller indices are the set of numbers, used to represent a crystal plane, obtained from the reciprocals of the intercepts made by the crystal planes.

1.13.1 Procedure Used to Find the Miller Indices of a Plane

In order to find the Miller indices of a plane, the following steps are to be followed:

(i) The intercepts made by the plane are noted and they should be written in terms of the lattice constants, *a*, *b* and *c*.
(ii) The coefficients of the intercepts are noted.
(iii) Find the inverse of these coefficients.
(iv) Find LCM and then multiply the fractions by LCM.
(v) Write the integers within the parenthesis.
(vi) The integers written within the parenthesis represent the Miller indices of the given plane.

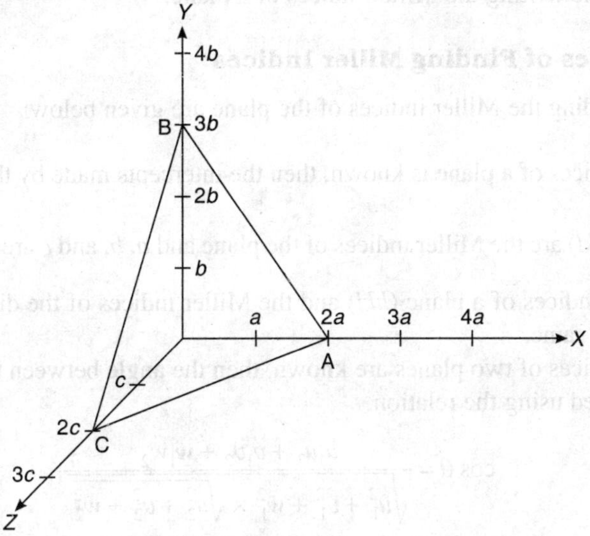

Figure 1.19 Miller indices of a plane.

Consider a plane ABC as shown in Figure 1.19. It makes intercepts along the *X*, *Y* and *Z* axes. The intercepts are 2*a*, 3*b* and 2*c*. The coefficients of the intercepts are 2, 3 and 2. The inverses are $\frac{1}{2}$, $\frac{1}{3}$ and $\frac{1}{2}$. The least common multiplier (LCM) is 6. Multiplying these fractions by LCM, we get 3, 2 and 3. These three values can be written within the parenthesis as (323). It represents the Miller indices of the plane ABC.

A set of structurally equivalent planes is called **family of planes.** They are represented by braces { }. The {110} planes are (110), (101), (011), $(1\bar{1}0)$, $(\bar{1}10)$, $(10\bar{1})$, $(\bar{1}01)$, $(\bar{1}\bar{1}0)$, $(\bar{1}0\bar{1})$, $(0\bar{1}1)$, $(01\bar{1})$ and $(0\bar{1}\bar{1})$.

1.13.2 Salient Features of Miller Indices

The important features of the Miller indices are:

(i) Miller indices represent a set of parallel planes. It does not represent a single plane.

(ii) If a plane lies along an axis, it is not possible to find the intercepts made by the plane correctly. To find the Miller indices of a plane that lies along an axis, consider a parallel plane and find the Miller indices of that plane. The Miller indices of these two parallel planes are the same.

(iii) If a plane does not make any intercept along a particular axis, then it is assumed that it will meet that axis at infinity. The intercepts for that axis is taken as ∞.

(iv) The negative intercept is also taken into account. For example, the Miller index notation $(0\bar{1}0)$ represents the plane has negative intercept at *Y*-axis.

(v) The *X*, *Y* and *Z* axes are represented as (100), (010) and (001) respectively.

(vi) The Miller indices of a plane (102) is read as one zero two and it should not be read as one hundred and two.

(vii) There is no comma, or any other special characters should be introduced in between the integers while writing the Miller indices of a plane.

1.13.3 Advantages of Finding Miller Indices

The advantages of finding the Miller indices of the plane are given below:

(i) If Miller indices of a plane is known, then the intercepts made by the plane is $\frac{a}{h}$, $\frac{b}{k}$ and $\frac{c}{l}$, where (*hkl*) are the Miller indices of the plane and *a*, *b*, and *c* are the lattice constants.

(ii) The Miller indices of a plane (*hkl*) and the Miller indices of the direction of that plane [*hkl*] are the same.

(iii) If Miller indices of two planes are known, then the angle between these two planes can be determined using the relation,

$$\cos\theta = \frac{u_1u_2 + v_1v_2 + w_1w_2}{\sqrt{u_1^2 + v_1^2 + w_1^2} \times \sqrt{u_2^2 + v_2^2 + w_2^2}} \tag{1.12}$$

where $(u_1v_1w_1)$ and $(u_2v_2w_2)$ are the Miller indices of the planes.

(iv) Miller indices of a plane is used to find the relationship between interplanar distance, *d* and interatomic distance, *a*.
For cubic unit cell

$$d = \frac{a}{\sqrt{h^2 + k^2 + l^2}} \tag{1.13}$$

1.13.4 Miller Indices Determination for the Planes in Cubic Unit Cells

Consider a cubic unit cell. Let us find the Miller indices of the planes (i) BCFG, (ii) ADFG, and (iii) AFH as shown in Figure 1.20. To find the Miller indices of the above planes, the crystallographic axes X, Y and Z are drawn by drawing parallel lines to the edges EF, EH and EA. They are represented as X, Y, and Z axes respectively. The Miller indices of the planes are determined by finding the intercepts made by the planes as follows:

(i) The plane BCFG

The intercepts made by the plane BCFG are a, ∞b, ∞c. The coefficients of the intercepts are 1, ∞, ∞. The inverse of these coefficients are 1, $\frac{1}{\infty}$, $\frac{1}{\infty}$. The Miller indices are (100).

(ii) The plane ADFG

The intercepts made by the plane ADFG are a, ∞b, c. The coefficients of the intercepts are 1, ∞, 1. The inverse of these coefficients are 1, 0, 1. The Miller indices are (101).

(iii) The plane AFH

The intercepts made by the plane AFH are a, b, c. The coefficients of the intercepts are 1, 1, 1. The inverse of these coefficients are, 1, 1, 1. The Miller indices are (111).

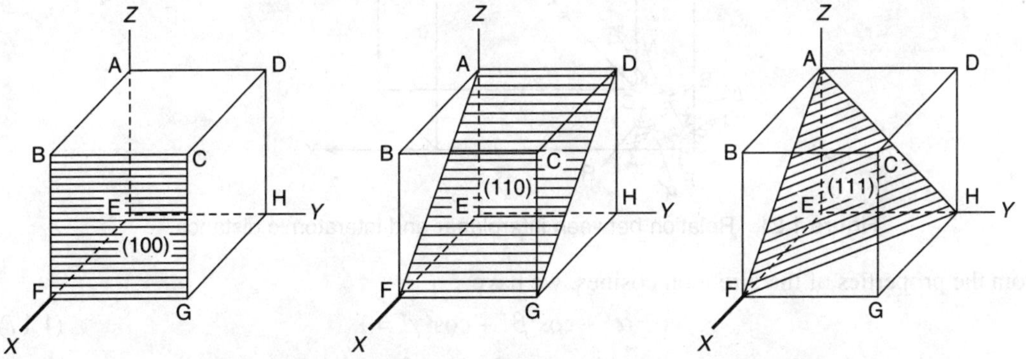

Figure 1.20 Miller indices determination for a cubic unit cell.

1.14 RELATION BETWEEN THE INTERPLANAR DISTANCE AND INTERATOMIC DISTANCE

The distance between any two adjacent atoms of the same kind is known as interatomic distance. It is represented by the letter a. The distance between any two successive parallel planes is known as interplanar distance. It is represented by the letter d.

Consider a cubic unit cell. Consider a plane ABC in the cubic unit cell. Let (hkl) be the Miller indices of the plane ABC. The intercepts made by the plane along X, Y and Z axes are $\frac{a}{h}$, $\frac{b}{k}$ and $\frac{c}{l}$ respectively. Consider that another parallel plane OPQ is passing through the origin, O. Let ON be the perpendicular line drawn between O and the plane ABC. It represents the interplanar distance, d. Let the angle between ON and X axis be α', the angle between ON and Y axis be β' and the angle between ON and Z axis be γ'.

From Figure 1.21

$$\cos\alpha' = \frac{ON}{OA} = \frac{d}{a/h} = \frac{dh}{a} \tag{1.14}$$

$$\cos\beta' = \frac{ON}{OB} = \frac{d}{b/k} = \frac{dk}{b} \tag{1.15}$$

and

$$\cos\gamma' = \frac{ON}{OC} = \frac{d}{c/l} = \frac{dl}{c} \tag{1.16}$$

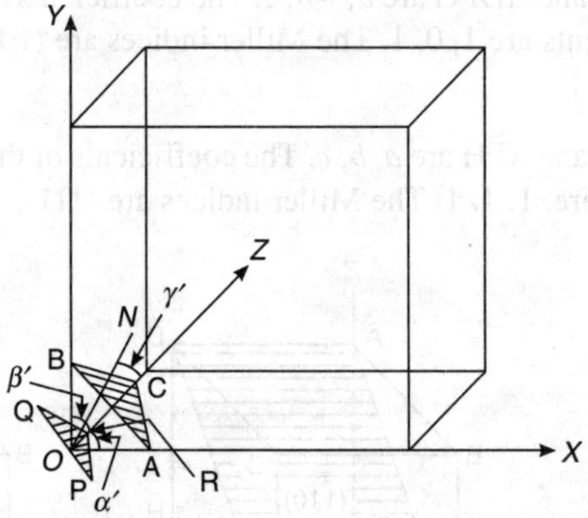

Figure 1.21 Relation between interplanar and interatomic distance.

From the properties of the direction cosines, we have

$$\cos^2\alpha' + \cos^2\beta' + \cos^2\gamma' = 1 \tag{1.17}$$

$$\left(\frac{dh}{a}\right)^2 + \left(\frac{dk}{b}\right)^2 + \left(\frac{dl}{c}\right)^2 = 1$$

$$\frac{1}{d^2} = \left(\frac{h}{a}\right)^2 + \left(\frac{k}{b}\right)^2 + \left(\frac{l}{c}\right)^2 \tag{1.18}$$

For cubic unit cell, $a = b = c$. Equation (1.18) can be rewritten as

$$\frac{1}{d^2} = \frac{1}{a^2}(h^2 + k^2 + l^2)$$

$$d = \frac{a}{\sqrt{h^2 + k^2 + l^2}} \tag{1.19}$$

Equation (1.19) gives the relation between the interplanar distance and interatomic distance.

1.15 DIRECTION OF A PLANE AND MILLER INDICES

A perpendicular line drawn to a plane is called its direction. The direction of a plane is represented by a set of three numbers written within square brackets, which are obtained by identifying the smallest integer position intercepted by the line from the origin of the crystallographic axes. The direction $[1\bar{1}1]$ represents the line starts from the origin passes through the points, 1, –1, 1. The direction of a plane passing through a negative axis is represented by a bar over the index as $[1\bar{1}1]$. A set of structurally equivalent directions is called **family of directions**. They are represented by angular brackets < >. The <110> planes are [110], [101], [011], $[1\bar{1}0]$, $[\bar{1}10]$, $[\bar{1}01]$, $[10\bar{1}]$, $[\bar{1}\bar{1}0]$, $[\bar{1}0\bar{1}]$, $[0\bar{1}1]$ and $[0\bar{1}\bar{1}]$.

1.15.1 Procedure to Draw the Direction of a Plane in a Cubic Unit Cell

To draw the direction of a plane in a cubic unit cell, say [111], mark the point 1,1,1, in the cubic unit cell. Draw a line between the origin 0,0,0 and the point 1,1,1. This line represents the [111] direction of the cubic unit cell.

For example, to draw [100], [110] and [111] directions in a cubic unit cell, draw a cubic unit cell. Mark *X*, *Y* and *Z*-axes for the cubic unit cell. Mark the points 1,0,0. Draw a line between the origin and the point 1,0,0. This line represents the direction [100]. Then to draw the direction [110] mark the points 1,1,0. Draw a line between the origin and the point 1,1,0. This line represents the direction [110]. The directions [100], [001], [010], [011], [101], [110] and [111] are shown in Figure 1.22.

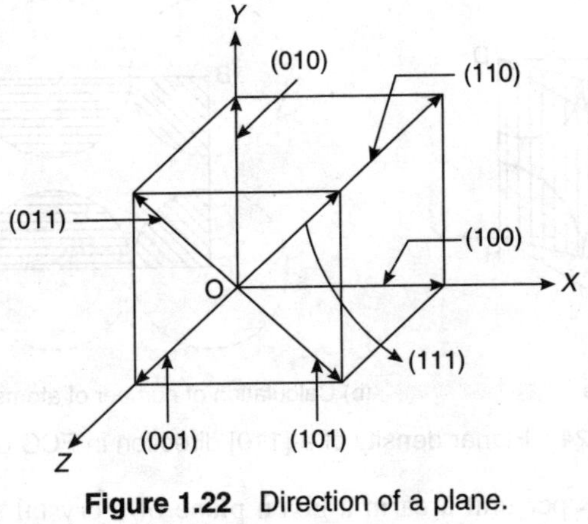

Figure 1.22 Direction of a plane.

1.16 LINEAR DENSITY AND PLANAR DENSITY

The number of atoms per unit length along a given direction in a crystal structure is known as linear density [LD]. It is represented in m^{-1}.

$$\text{Linear density} = \frac{\text{Number of atoms centred on a direction vector}}{\text{Length of the direction vector}} \quad (1.20)$$

Let us determine the linear density along the [110] direction of a FCC unit cell. The [110] direction is shown in Figure 1.23. The atoms *X*, *Y*, and *Z* lies along the [110] plane. The atom *X* contains only half of the atom, whereas the remaining half is shared by another unit cell, *Y* has full atom and Z has only half of the atom. In total, [110] direction has 2 atoms. The length of [110] direction is $\sqrt{2}a$. Therefore,

$$\text{Linear density} = \frac{\text{Number of atoms lying on a direction vector}}{\text{Length of the direction vector}}$$

$$= \frac{2}{\sqrt{2}a} = \frac{\sqrt{2}}{a}$$

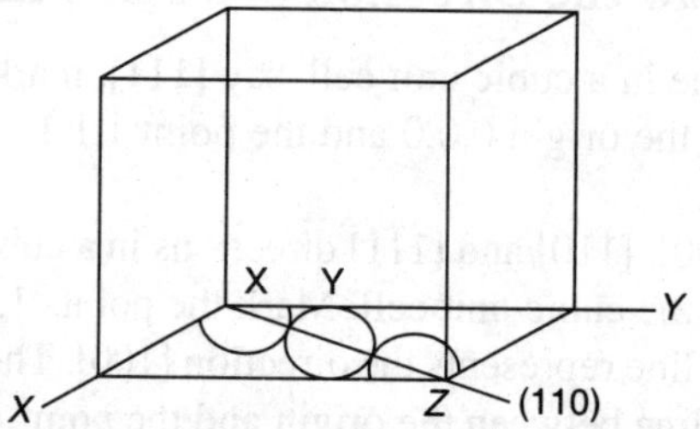

Figure 1.23 [110] direction in the FCC unit cell for linear density calculation.

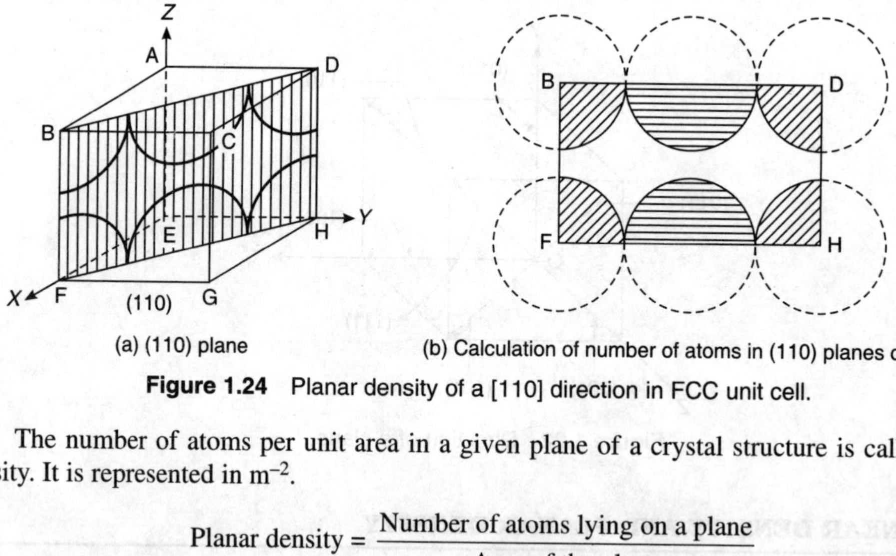

(a) (110) plane

(b) Calculation of number of atoms in (110) planes of FCC

Figure 1.24 Planar density of a [110] direction in FCC unit cell.

The number of atoms per unit area in a given plane of a crystal structure is called planar density. It is represented in m^{-2}.

$$\text{Planar density} = \frac{\text{Number of atoms lying on a plane}}{\text{Area of the plane}} \tag{1.21}$$

Consider a (110) plane of a FCC unit cell as shown in Figure 1.24. The (110) plane has 2 atoms. Its area is $\sqrt{2}a^2$. Therefore, the planar density is

$$\text{Planar density} = \frac{\text{Number of atoms lying on a plane}}{\text{Area of the plane}}$$

$$= \frac{2}{\sqrt{2}a^2}$$

1.17 SOME SPECIAL CRYSTAL STRUCTURES

1.17.1 Diamond Unit Cell

Diamond crystallizes in cubic crystal structure. In diamond unit cell all the atoms are carbon. The diamond unit cell consists of two interpenetrating FCC unit cells. The diamond lattice may be considered as two identical atoms at 000, $\frac{1}{4}\frac{1}{4}\frac{1}{4}$ positions associated with each point of the FCC lattice. The unit cell has eight corner atoms, six face-centred atoms and four atoms on the body diagonals. The body diagonal atom lies at a distance of one-fourth of the body diagonal. The body diagonal atoms are tetrahedrally bonded with corner atoms and hence touch the corner atoms. The diamond unit cell is shown in Figure 1.25.

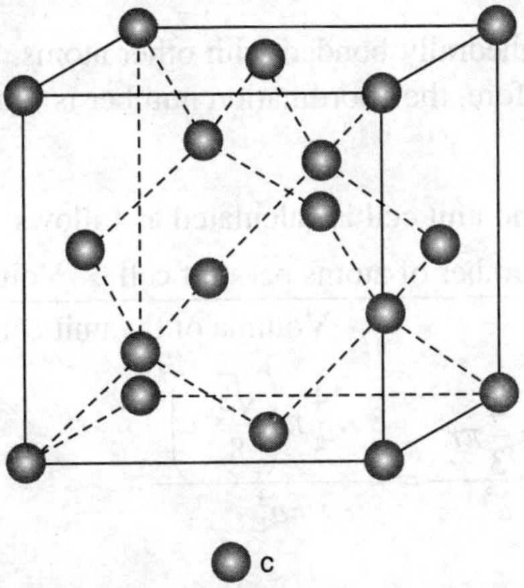

Figure 1.25 Diamond unit cell.

Number of atoms in a unit cell

There are eight corner atoms. Each and every corner atoms are shared by eight adjacent unit cells. Therefore, one corner atom contributes $\frac{1}{8}$th of its parts to one unit cell. Since there are eight corner atoms, the number of atoms contributed by the corner atoms is $\left(\frac{1}{8}\times 8=1\right)$ one.

There are six face-centred atoms. Each and every face-centred atom is shared by two adjacent unit cells. So, a face-centred atom contributes $\frac{1}{2}$ of its parts to one unit cell Therefore, the total number of atoms contributed by the face-centred atoms is $\left(\frac{1}{2}\times 6=3\right)$ three.

The body diagonal atoms lies within the unit cell. There are four body diagonal atoms. Therefore, the total number of atoms in a diamond unit cell is (1 + 3 + 4 = 8) eight.

Atomic radius

The atoms that lie along the body diagonal of the unit cell are tetrahedrally bonded with corner atoms and hence touch the corner atoms. The body diagonal atom lies at a distance of one-fourth of the body diagonal. Therefore,

$$2r = \frac{1}{4} \times \text{body diagonal of the unit cell}$$

$$2r = \frac{1}{4} \times \sqrt{3}a$$

$$r = \frac{\sqrt{3}}{8}a \qquad (1.22)$$

Coordination number

The body diagonal atom is tetrahedrally bonded with other atoms. Therefore, the number of the nearest neighbour is four. Therefore, the coordination number is four.

Packing density

The packing density of a diamond unit cell is calculated as follows:

$$\text{Packing density} = \frac{\text{Number of atoms per unit cell} \times \text{Volume of one atom}}{\text{Volume of the unit cell}}$$

$$= \frac{8 \times \frac{4}{3}\pi r^3}{a^3} = \frac{8 \times \frac{4}{3}\pi\left(\frac{\sqrt{3}}{8}a\right)^3}{a^3}$$

$$= \frac{\sqrt{3}}{16}\pi$$

$$= 0.34$$

The packing density of diamond unit cell is 0.34. This shows that nearly 66% of the diamond unit cell is vacant. Diamond is one of the hardest materials. The high value of hardness of diamond is due to its crystal structure and the strong interatomic covalent bonds. Some of the elements that crystallize in diamond unit cell are C, Si, Ge and tin.

1.17.2 Cesium Chloride Crystal Structure

Cesium chloride crystallizes in cubic unit cell. The Cs^+ ion lies in the body-centred position, whereas the Cl^- ions occupy the corner atom position. Each and every Cl^- ion at the corner is shared by eight unit cells. Therefore, one Cl^- ion contributes $\frac{1}{8}$th of its parts to one unit cell. The total number of Cl^- ions at the corner atom position is $\left(\frac{1}{8} \times 8 = 1\right)$ one. The number of Cs^+ ion present in the unit cell is one. Therefore, in total the unit cell has only one CsCl atom. For a Cs^+ ion, the Cl^- ions are the nearest neighbours. Therefore the coordination number is eight. This unit cell is not a true body-centred cubic unit cell, because atoms of two different kinds are involved. Some compounds exhibiting this crystal structure is CsCl, CsBr, CsI, TiI, BeCu, AlNi. The CsCl unit cell is shown in Figure 1.26.

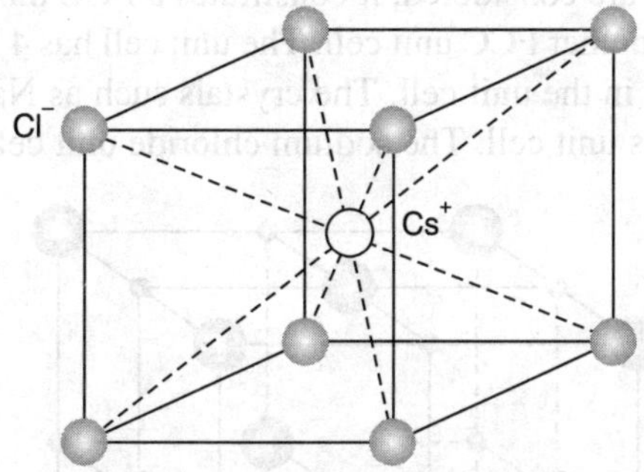

Figure 1.26 CsCl unit cell.

1.17.3 Zinc Blende Structure

The ZnS crystal structure is presented in Figure 1.27. This crystal structure is called zinc blende or *sphalerite* structure, after the mineralogical term for zinc sulphide. The S atom occupies the corner position and the face-centred positions. The Zn atom occupies the interior tetrahedral position. If the Zn atom and S atom are reversed and vice versa, the same type of crystal structure is obtained. This structure is also the interpenetration of two FCC lattices. If only Zn atoms are considered, it constitutes the FCC lattice. Similarly, if only the S atoms are considered it constitutes a FCC lattice. There is a tetrahedral bonding between the Zn and S atoms. The coordination number is four due to tetrahedral bonding as Zn atom is surrounded by four S atoms and vice versa. In a unit cell, there are 4 ZnS atoms. Some of the common materials, which crystallizes in this crystal structure are ZnS, CdS, ZnSe, GaAs, CuCl, SiC, and ZnTe.

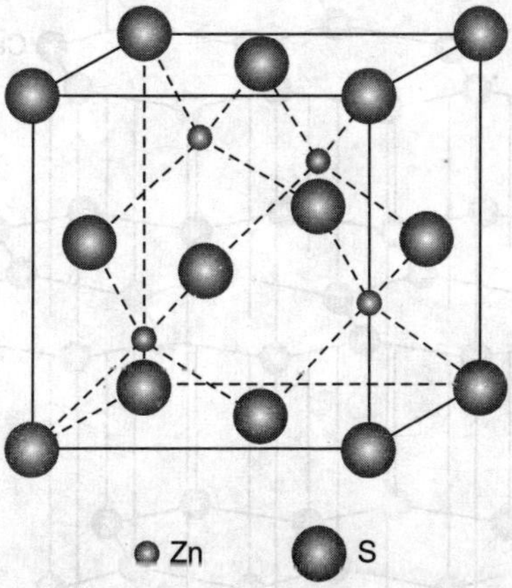

Figure 1.27 ZnS unit cell.

1.17.4 Sodium Chloride Unit Cell

Sodium chloride crystallizes in cubic unit cell. The sodium and chlorine ions are alternatively arranged in the sodium chloride unit cell. The unit cell looks like the interpenetration of two FCC unit cells. If only the Na^+ ions are considered, it constitutes a FCC unit cell. Similarly, if Cl^- ions are considered, it constitutes another FCC unit cell. The unit cell has 4 Na^+ ions and 4 Cl^- ions and hence 4 NaCl ions are present in the unit cell. The crystals such as NaCl, MnS, LiF, MgO, CaO, FeO and NiO crystallize in this unit cell. The sodium chloride unit cell is shown in Figure 1.28.

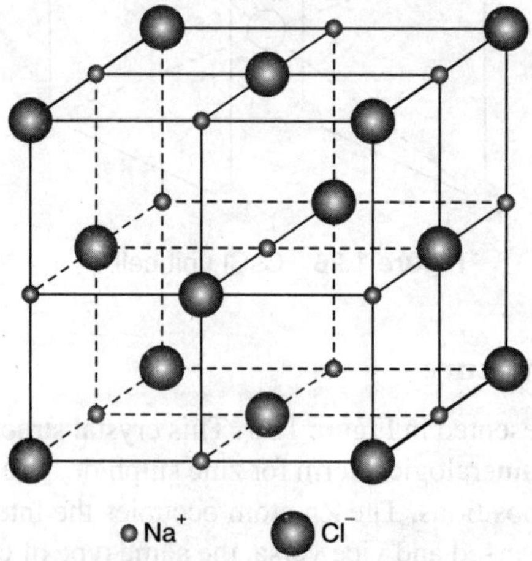

Figure 1.28 NaCl unit cell.

1.17.5 Graphite Structure

The graphite structure is shown in Figure 1.29. In graphite, the carbon atoms are connected together in a hexagonal pattern by covalent bonding in a two-dimensional pattern. That is, the carbon atoms

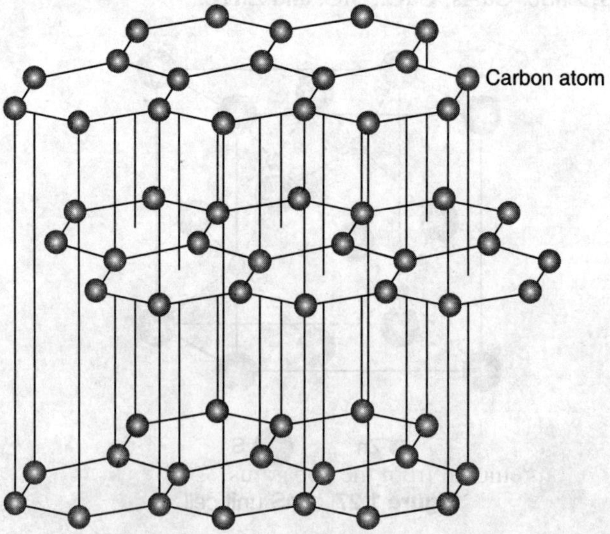

Figure 1.29 Graphite structure.

are arranged in the form of layers or sheet structures. Each carbon atom is connected by three other carbon atoms in a hexagonal pattern by sharing of electrons. Carbon has four valence electrons. Three electrons are used for forming covalent bond in hexagonal pattern and the remaining one electron is free. The fourth electron is delocalized and resonates between the three covalent bonds. Therefore, the graphite has very good electrical conductivity. The bond length between each carbon atom is 1.42 Å. In graphite the sheets are held together by van der Waals bond. The distance between each sheet is 3.4 Å. This weak inter-sheet bonding is the reason for soft characteristics of graphite.

1.18 CRYSTAL GROWTH

Semiconductor industry mostly uses silicon crystal for the fabrication of devices. Device fabrication needs Si in the form of nearly perfect single crystals. The semiconductor grade Si is generally polycrystalline compound, and it consists of a number of small crystals called grains. These grains are oriented in different directions, and they contain several defects. Impure Si contains one unwanted impurity atom in about 10^9 atoms of silicon. In order to use Si for the fabrication of devices, it must be perfect single crystalline form. Single crystals are prepared using crystal growth technique. Crystal growth is the process of converting a random oriented or a polycrystalline material into an orderly arranged crystalline material.

There are a number of methods used for the crystal growth technique. This section deals with the single crystal growth from solution, melt and vapour phase techniques.

1.18.1 Solution Growth Technique

It is a low temperature process of growing single crystals. In the solution growth, a supersaturated solution of the given substance is prepared, and it is kept without any disturbance at a constant low temperature. Single crystals begin to grow from supersaturated solution after some days or week or month. Then the entire supersaturated solution gets converted into a large size single crystal. The supersaturated copper sulphate solution, kept in a beaker at room temperature without any disturbance, produces single crystals by the next day.

Advantages of solution growth

The advantages of the solution growth technique are given as under:

1. It is a low temperature process.
2. It is suitable for those materials that suffer decomposition in the melt or in the solid at high temperature.
3. Since it is a low temperature process, it requires simple and straightforward equipment that gives good degree of control to an accuracy of ± 0.01°C.
4. Low temperature crystal growth produces a variety of different morphologies and polymorphic form of same substances.
5. The use of low temperature for the crystal growth avoids the thermal shock during growth of the crystal or on removal from the apparatus.

Disadvantages of solution growth

1. Since it is a low temperature process, the growth rate is very low, and it may include solvent into the growing crystal.

1.18.2 Melt Growth

In this method, the given polycrystalline material is melted by heating it above its melting point and resolidified by cooling the liquid below its freezing point. The melt growth method is the most important method of crystal growth. There are a number of melt growth methods. Let us discuss about (i) Bridgman method and (ii) Czochralski method.

Bridgman–Stockbarger method

Bridgman–Stockbarger method is one of the methods used to prepare single crystals. It is also called Bridgman method. The diagrammatic representation of this method is given in Figure 1.30. In this method, the materials are taken in a polycrystalline form in a crucible that contains a tip at the bottom. The crucible is initially kept in the upper portion of the furnace. The upper portion of the furnace is kept slightly above the melting point of the polycrystalline material. Due to high temperature, the polycrystalline material gets melted. The crucible with the molten liquid is slowly lowered, and it is brought into the lower portion of the furnace. The lower portion of the furnace is kept at a temperature slightly lower than the melting point of the polycrystalline material. When the crucible is lowered initially, single crystal is formed at the tip of the crucible because the pointed end tends to cool slightly faster than the bulk material. The crucible is continuously lowered slowly at the rate of 1 mm/hour to 30 mm/hour into the lower portion of the furnace. The single crystal formed at the tip acts as a seed crystal and hence the single crystal grows continuously.

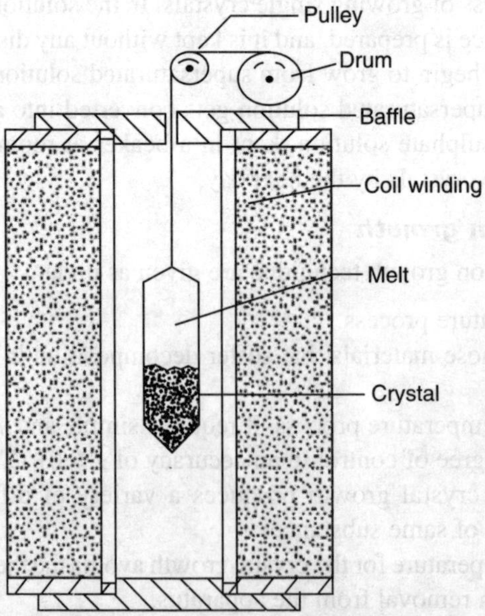

Figure 1.30 Bridgman technique.

The advantages and drawbacks of this method are given as under:

Advantages

1. The Bridgman technique is fairly simple, and cost is low.
2. This method is best suited for low melting point materials.

Drawbacks

1. This method has low growth rate, typically 1 mm/hour to 30 mm/hour. Since the melt is in contact with the container for a longer time, there is a possibility to dissolve some of the oxygen from the wall of the crucible, which is made up of SiO_2.
2. During cooling there will be a contraction in the container, and it compresses the solid material. This compression will induce some stresses in the crystal and this will lead to dislocation in the material.
3. The container wall acts as a preferential nucleation site and this results in the formation of the polycrystalline material rather than single crystal.
4. This method cannot be used for the materials which decompose before melting.

Czochralski method

Czochralski method is one of the most popular and widely used industrial methods for growing single crystals of semiconductors. It is named after its inventor Czochralski. The Czochralski method is shown in Figure 1.31. In this method, the material to be grown is taken in the form of powder in a crucible. The crucible is kept inside a furnace, and it is heated slightly above the melting point of the material. The material is converted into a molten liquid. A small single crystal of the material, known as seed crystal, is made to contact with the molten liquid. The seed crystal is slowly pulled in the upward direction at the rate of few mm/hour. The molten liquids that are in contact with the seed crystal, crystallizes when the seed crystal is slowly moved in the upward direction. Thus slowly pulling the seed crystal produces a large size single crystal.

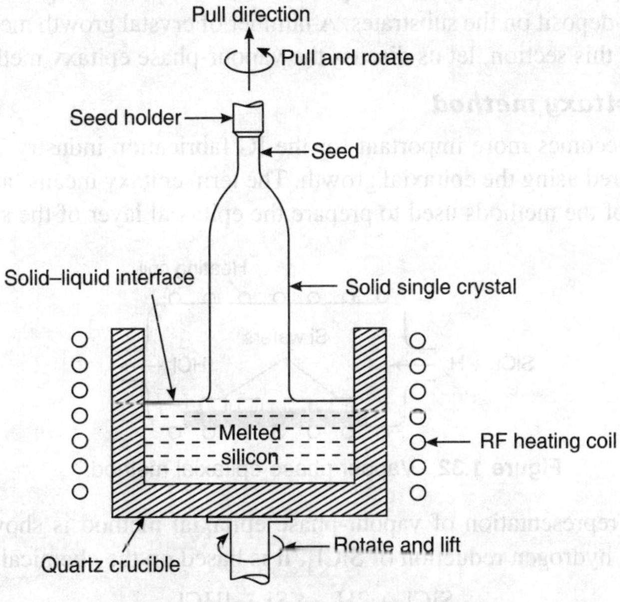

Figure 1.31 Czochralski method.

The rate of pulling the seed crystal is determined by the trial and error method. Once the correct value of the rate of pulling of the seed crystal is determined, the single crystals can be readily reproduced. In semiconductor industry, this method is used to produce silicon crystal of size 200-mm to 300-mm diameter and length more than 1 m.

Advantages of Czochralski method

Czochralski method has several advantages in comparison with other methods of crystallization:

1. The advantage of the Czochralski method is that large single crystals can be grown, thus it is used extensively in the semiconductor industry.
2. A high level of automation in the process of crystallization allows us to conduct the crystallization process with minimal participation of the operator.
3. It is used to obtain the single crystals of a given crystallographic orientation, with small deviations of the growth axis from this direction.
4. Single crystals of various chemical and stoichiometric compositions can be prepared using this method.
5. Thermal units of Czochralski method is simple, reliable, low cost and efficiency in the operation.

Drawbacks of Czochralski method

1. This method is not suitable for incongruently melting compounds.
2. It needs a seed crystal of the same composition. This will limits the use of this method for exploratory synthetic research.

1.18.3 Vapour Growth Technique

The process of growing the single crystals from the vapour phase is called vapour growth technique. In this method, the vapours of the required compound is made to pass through the reaction chamber, and then it is made to deposit on the substrates. A number of crystal growth methods using vapour phase is available. In this section, let us discuss the vapour-phase epitaxy method.

Vapour-phase epitaxy method

The crystal growth becomes more important for the IC fabrication industry. The IC fabrication uses the crystal prepared using the epitaxial growth. The term epitaxy means 'arranged upon'. The vapour phase is one of the methods used to prepare the epitaxial layer of the silicon crystal.

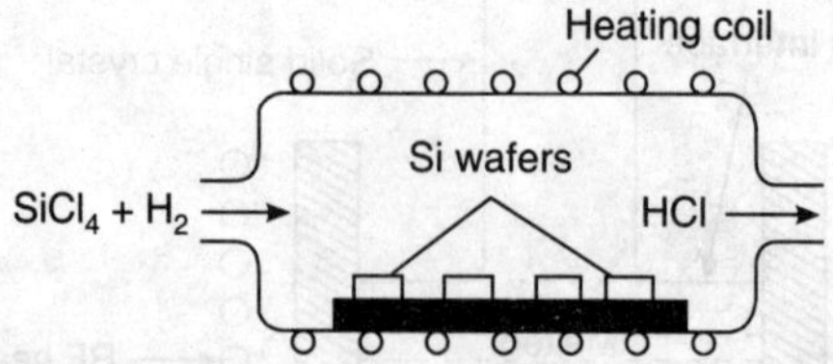

Figure 1.32 Vapour-phase epitaxial method.

The schematic representation of vapour-phase epitaxial method is shown in Figure 1.32. This method uses the hydrogen reduction of $SiCl_4$. It is based on the chemical reaction,

$$SiCl_4 + 2H_2 \rightarrow Si + 4HCl$$

In this method, the single crystals of silicon (Si wafer) are used as substrates. The vapours of $SiCl_4$ and H_2 are passed through the reaction chamber. The reaction chamber is called reactor. The reactor is kept at 1250°C. At this temperature $SiCl_4$ decomposes, and it produces Si and HCl. Silicon gets deposited over the silicon substrate. The substrate serves as the seed crystal. The crystal orientation and the crystal structure are same as that of the substrate. The first layer of atom gets coated over the substrate serves as the substrate for the second layer and so on. The epitaxial layer is made as either p-type or n-type by passing the hydrogen gas through solution containing boron or phosphorous atoms (for example, boron trichloride or phosphorous trichloride), before it is introduced into the reactor.

Advantages of this method

The advantages of this method are as follows:

1. The vapour-phase epitaxy method is an economic method and it is easy to operate.
2. This method has higher growth rate.
3. It is used to produce thin layers.
4. The composition of the substance is easily controlled.

Drawbacks of this method

1. This method requires high temperatures in the order of 800°C to 1250°C.

SOLVED PROBLEMS

1.1 From the knowledge of crystal structure, calculate the density of diamond ($r = 0.071$ nm).

Solution

$$\text{Density} = \frac{\text{Mass}}{\text{Volume}}$$

$$\text{Density} = \frac{8 \times \text{Mass of one carbon atom}}{\text{Volume of the unit cell}}$$

$$\text{Density} = \frac{8 \times \text{Mass of one carbon atom}}{\left(\frac{8r}{\sqrt{3}}\right)^3}$$

$$\text{Density} = \frac{8 \times 12}{6.022 \times 10^{26} \times \left(\frac{8 \times 0.071 \times 10^{-9}}{\sqrt{3}}\right)^3}$$

$$\text{Density} = 4520.3 \text{ kg m}^{-3}$$

The density of diamond = 4520.3 kg m^{-3}

1.2 Determine the percentage volume change that occurs when Ti changes from a BCC structure to HCP structure. With the BCC structure the lattice parameter $a = 0.332$ nm and with the HCP structure $a = 0.296$ nm and $c = 0.468$ nm.

Solution

Volume of the BCC unit cell = $a^3 = (0.332 \times 10^{-9})^3 = 3.659 \times 10^{-29}$ m^{-3}

Volume of the HCP unit cell = $6 \times \frac{\sqrt{3}}{4} a^2 \times c = 6 \times \frac{\sqrt{3}}{4} (0.296 \times 10^{-9})^2 \times (0.468 \times 10^{-9})$

$= 1.065 \times 10^{-28}$ m^3

Change in volume = $1.065 \times 10^{-28} - 3.659 \times 10^{-29}$

$= 6.991 \times 10^{-29}$

$$\text{Volume change in percentage} = \frac{6.991 \times 10^{-29}}{3.659 \times 10^{-29}} \times 100\% = 191.06\%$$

The percentage volume change is 191.06%.

1.3 Copper has FCC structure and its atomic radius is 1.278 Å. Calculate its density. The atomic weight of copper is 63.54.

Given data

Atomic radius of copper = 1.278 Å

Atomic weight of copper = 63.54

Solution

The relation between the density and the atomic weight is given by the equation

$$\rho V = \frac{nA}{N_A}$$

Since, copper crystallizes in cubic unit cell, substituting $V = a^3$, we get

$$\rho a^3 = \frac{nA}{N_A}$$

The density is given by

$$\rho = \frac{nA}{N_A a^3}$$

For FCC

$$r = \frac{a}{2\sqrt{2}}$$

i.e. $a = 2\sqrt{2}\,r = 2 \times \sqrt{2} \times 1.278 \times 10^{-10} = 3.6147 \times 10^{-10}$

For FCC, $n = 4$. Substituting the values of n, A, N_A, and a, we get

$$\rho = \frac{nA}{N_A a^3} = \frac{4 \times 63.54}{6.022 \times 10^{26} \times (3.6147 \times 10^{-10})^3} = 8936 \text{ kg m}^{-3}$$

The density of copper is 8936 kg m^{-3}.

1.4 NaCl crystallizes in cubic structure. The density of NaCl is 2180 kg m^{-3}. Calculate the distance between adjacent atoms.

Given data

Density of NaCl = 2180 kg m^{-3}

Solution

Molecular weight of NaCl = atomic weight of Na + atomic weight of Cl

= 23 + 35.5 = 58.5

The density and the atomic weight are related using the equation,

$$\rho V = \frac{nA}{N_A}$$

Since NaCl is a cubic unit cell, substituting $V = a^3$, we get

$$\rho a^3 = \frac{nA}{N_A}$$

The interatomic distance is given by,

$$a = \left(\frac{nA}{N_A \rho}\right)^{1/3}$$

Substituting the values of n, A, N_A, ρ, we get

$$a = \left(\frac{4 \times 58.5}{6.022 \times 10^{26} \times 2180}\right)^{1/3}$$

$$a = 5.628 \times 10^{-10} \text{ m}$$

The interatomic distance in NaCl crystal is 5.628×10^{-10} m.

1.5 Calculate the interplanar spacing for (101) and (221) planes in a simple cubic lattice whose lattice constant is 0.42 nm.

Given data

Lattice constant = 0.42 nm

Solution

The relation between interplanar and interatomic distance is given by

$$d = \frac{a}{\sqrt{h^2 + k^2 + l^2}}$$

For (101) plane

$$d = \frac{0.42 \times 10^{-9}}{\sqrt{1^2 + 0^2 + 1^2}} = \frac{0.42}{\sqrt{2}} \times 10^{-9} = 0.2969 \text{ nm}$$

For (221) plane

$$d = \frac{0.42 \times 10^{-9}}{\sqrt{2^2 + 2^2 + 1^2}} = \frac{0.42}{3} \times 10^{-9} = 0.14 \text{ nm}$$

The interplanar spacing for (101) plane is 0.2969 nm and for (221) plane is 0.14 nm.

1.6 Identify the axial intercepts made by the following planes: (i) (102), (ii) (231) and (iii) $(3\bar{1}2)$.

Solution

If the Miller indices of the plane is (*hkl*), then the intercepts made by the plane is $\frac{a}{h}, \frac{b}{k}$ and $\frac{c}{l}$. For the plane (102), the intercepts are $\frac{a}{1} = a, \frac{b}{0} = \infty$ and $\frac{c}{2}$.

For the plane (231), the intercepts are $\frac{a}{2}, \frac{b}{3}$ and $\frac{c}{1} = c$.

For the plane $(3\bar{1}2)$, the intercepts are $\frac{a}{3}, \frac{b}{-1} = -b$ and $\frac{c}{2}$.

1.7 Find the angle between two planes (111) and (212) in a cubic lattice.

Solution

The angle between any two planes having Miller indices $(u_1v_1w_1)$ and $(u_2v_2w_2)$ is

$$\cos\theta = \frac{u_1u_2 + v_1v_2 + w_1w_2}{\sqrt{u_1^2 + v_1^2 + w_1^2} \times \sqrt{u_2^2 + v_2^2 + w_2^2}}$$

$$\cos\theta = \frac{1\times 2 + 1\times 1 + 1\times 2}{\sqrt{1^2 + 1^2 + 1^2} \times \sqrt{2^2 + 1^2 + 2^2}}$$

$$\cos\theta = \frac{5}{\sqrt{3}\times 3} = 0.962$$

$$\theta = \cos^{-1}(0.962) = 15.846° = 15°50'44.91''$$

The angle between the planes (111) and (212) is 15.846°.

1.8 Sketch the following crystallographic planes for the cubic systems (100), (110), (111), (200), (120) and (211).

Solution

For a plane with Miller indices (*hkl*), the intercepts are $\frac{a}{h}, \frac{b}{k}, \frac{c}{l}$.

(i) The intercepts of the plane (100) are a, ∞, ∞. The plane (100) is shown in Figure 1.33(i).
(ii) The intercepts of the cubic plane (110) are a, a, ∞. The plane (110) is shown in Figure 1.33(ii).
(iii) The intercepts of the plane (111) are a, a, a. The plane (111) is shown in Figure 1.33(iii).
(iv) The intercepts of the plane (200) are $a/2$, ∞, ∞. The plane (200) is shown in Figure 1.33(iv).
(v) The intercepts of the plane (120) are a, $a/2$, ∞. The plane (120) is shown in Figure 1.33(v).
(vi) The intercepts of the plane (211) are $a/2$, a, a. The plane (211) is shown in Figure 1.33(vi).

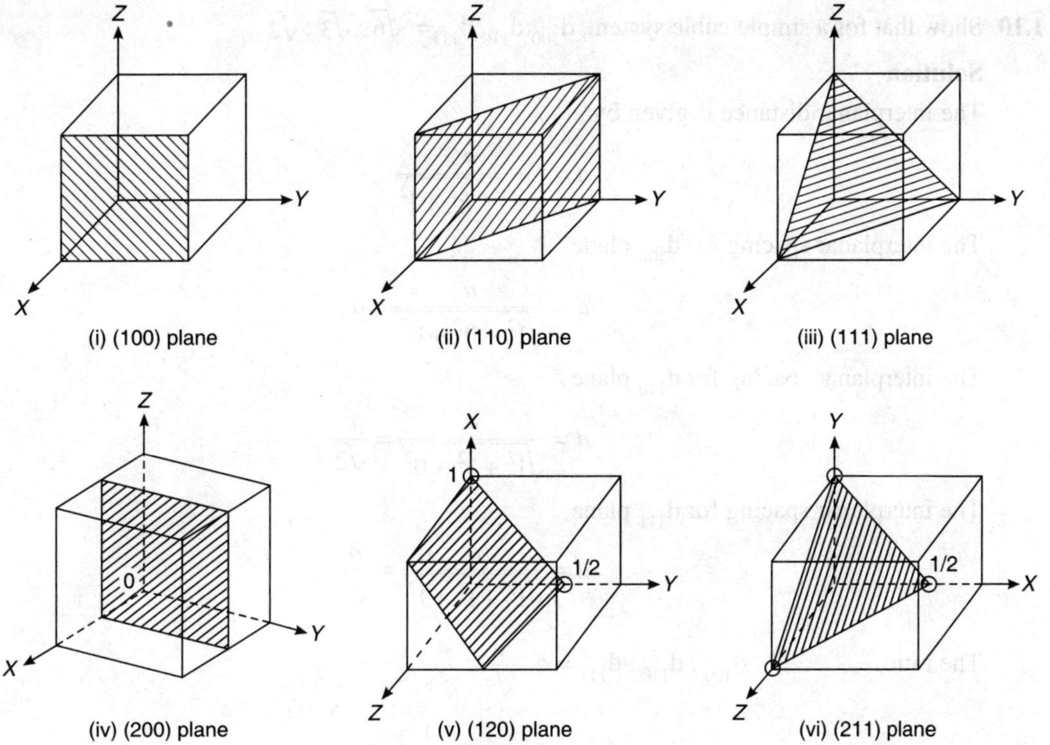

Figure 1.33 Crystal planes.

1.9 The interplanar distance between the planes $(\bar{1}11)$ in aluminium (FCC structure) is 0.2338 nm. What is the lattice constant?

Given data

Interplanar distance, $d = 0.2338$ nm

Miller indices of the crystal plane = $(\bar{1}11)$

Solution

The interplanar distance is given by

$$d = \frac{a}{\sqrt{h^2 + k^2 + l^2}}$$

$$a = d \times \sqrt{h^2 + k^2 + l^2}$$

$$a = 0.2338 \times 10^{-9} \times \sqrt{(-1)^2 + 1^2 + 1^2}$$

$$a = 0.2338 \times 10^{-9} \times \sqrt{3}$$

$$a = 0.4049 \times 10^{-9} \text{ m}$$

The lattice constant is 0.4049 nm.

1.10 Show that for a simple cubic system, $d_{100}:d_{110}:d_{111} = \sqrt{6}:\sqrt{3}:\sqrt{2}$.

Solution

The interplanar distance is given by

$$d = \frac{a}{\sqrt{h^2 + k^2 + l^2}}$$

The interplanar spacing for d_{100} plane

$$d = \frac{a}{\sqrt{1^2 + 0^2 + 0^2}} = a$$

The interplanar spacing for d_{110} plane

$$d = \frac{a}{\sqrt{1^2 + 1^2 + 0^2}} = \frac{a}{\sqrt{2}}$$

The interplanar spacing for d_{111} plane

$$d = \frac{a}{\sqrt{1^2 + 1^2 + 1^2}} = \frac{a}{\sqrt{3}}$$

The ratio, $$d_{100} : d_{110} : d_{111} = a : \frac{a}{\sqrt{2}} : \frac{a}{\sqrt{3}}$$

$$d_{100} : d_{110} : d_{111} = 1 : \frac{1}{\sqrt{2}} : \frac{1}{\sqrt{3}}$$

Multiplying the RHS by $\sqrt{6}$, we get

$$d_{100} : d_{110} : d_{111} = \sqrt{6} : \sqrt{3} : \sqrt{2}$$

1.11 Find the ratio of the intercepts made by (231) plane in a simple cubic crystal.

Solution

The intercepts made by the plane (*hkl*) is $\frac{a}{h}, \frac{b}{k}, \frac{c}{l}$. For the plane (231), the intercepts are $\frac{a}{2}, \frac{b}{3}, \frac{c}{1}$. For a cubic unit cell, $a = b = c$. Therefore, the intercepts are $\frac{a}{2}, \frac{a}{3}, \frac{a}{1}$. The ratio of the intercepts,

$$l_1 : l_2 : l_3 = \frac{a}{2} : \frac{a}{3} : \frac{a}{1} = \frac{1}{2} : \frac{1}{3} : 1$$

The LCM is 6. Multiplying by the LCM, we get

$$l_1 : l_2 : l_3 = 3 : 2 : 6$$

The ratio of the intercepts made by (231) plane in a simple cubic crystal is $l_1 : l_2 : l_3 = 3 : 2 : 6$.

1.12 Consider a crystal with primitives 0.8Å, 1.2Å, 1.5Å, a plane (123) cuts an intercepts 0.8Å along *X*-axis. Find the lengths of the intercepts along *Y* and *Z* axes.

Solution

The intercepts made by the plane (*hkl*) is $\frac{a}{h}, \frac{b}{k}, \frac{c}{l}$. For the plane (123), the intercepts are $\frac{a}{1}, \frac{b}{2}, \frac{c}{3}$.

The ratio of the intercepts

$$l_1 : l_2 : l_3 = a : \frac{b}{2} : \frac{c}{3}$$

It is given that, $l_1 = 0.8$Å. Therefore,

$$0.8\text{Å} : l_2 : l_3 = a : \frac{b}{2} : \frac{c}{3}$$

Substituting the values of *a*, *b* and *c*, we get

$$0.8\text{Å} : l_2 : l_3 = 0.8\text{Å} : \frac{1.2}{2}\text{Å} : \frac{1.5}{3}\text{Å}$$

Solving, we get

$$l_2 = 0.6\text{Å and } l_3 = 0.5\text{Å}.$$

1.13 Find the nearest neighbour distance in simple cubic, body-centred cubic and face-centred cubic unit cells.

Solution

(i) Simple cubic unit cell

In simple cubic unit cells, a corner atom is the nearest neighbour to another corner atom. The distance between any two corner atoms is *a*, i.e. the nearest neighbour distance is *a*.

(ii) Body-centred cubic unit cell

In a BCC unit cell, the body-centred atom is the nearest neighbour to a corner atom. The distance between a body-centred atom and a corner atom is

$$2r = 2\frac{\sqrt{3}}{4}a = \frac{\sqrt{3}}{2}a$$

(iii) Face-centred cubic unit cell

In a FCC unit cell, the face-centred atom is the nearest neighbour to a corner atom. The distance between a face-centred atom and a corner atom is

$$2r = 2\frac{a}{\sqrt{8}} = \frac{a}{\sqrt{2}}$$

1.14 For a simple cubic lattice of lattice parameters 2.04 Å, calculate the spacing of the lattice plane (212).

Solution

The interplanar distance is given by

$$d = \frac{a}{\sqrt{h^2 + k^2 + l^2}}$$

$$= \frac{2.04 \times 10^{-10}}{\sqrt{2^2 + 1^2 + 2^2}}$$
$$= 0.68 \text{ Å}$$

1.15 The radius of copper is 1.278 Å. It crystallizes in FCC unit cell. The atomic weight of copper is 63.54. Calculate the number of atoms per unit cell. Its density is 8980 kg m^{-3}.

Given data

Radius of copper = 1.278 Å

Atomic weight of copper = 63.54

Solution

The atomic radius of a FCC crystal

$$r = \frac{a}{\sqrt{8}}$$

The interatomic distance

$$a = \sqrt{8}r$$
$$= \sqrt{8} \times 1.278 \times 10^{-10}$$
$$= 3.614 \times 10^{-10} \text{ m}$$

The density and atomic weight are related by

$$\rho a^3 = \frac{nM}{N_A}$$
$$n = \frac{\rho a^3 N_A}{M}$$
$$= \frac{8980 \times (3.614 \times 10^{-10})^3 \times 6.022 \times 10^{26}}{63.54} = 4.0173$$

The number of atoms per Cu unit cell is 4.

1.16 The ratio of the intercepts of an orthorhombic crystal are $a : b : c = 0.429 : 1 : 0.379$. What are the Miller indices of the faces with the following intercepts?

(i) 0.214 : 1 : 0.188, (ii) 0.858 : 1 : 0.754,

(iii) 0.429 : ∞ : 0.126

Solution

(i) Given that $a : b : c = 0.429 : 1 : 0.379$

The ratio of the intercepts are $0.214 : 1 : 0.188 = \frac{a}{0.429} \times 0.214 : 1 : \frac{c}{0.379} \times 0.188$

$$= \frac{a}{2} : b : \frac{c}{2}$$

The coefficients are $\frac{1}{2} : 1 : \frac{1}{2}$. The inverses are 2, 1, 2.

Miller indices for the given plane is (212).

(ii) Given that $a : b : c = 0.858 : 1 : 0.754$

The ratio of the intercepts are $0.858 : 1 : 0.754 = \frac{a}{0.429} \times 0.858 : 1 : \frac{c}{0.379} \times 0.754$

$= 2a : b : 2c$

The coefficients are 2, 1, 2. The inverses are $\frac{1}{2} : 1 : \frac{1}{2}$ LCM is 2. Multiplying by the LCM, we get, 1, 2, 1

Miller indices for the given plane is (121).

(iii) Given that $a : b : c = 0.429 : \infty : 0.126$

The ratio of the intercepts are $0.429 : 1 : 0.126 = \frac{a}{0.429} \times 0.429 : \infty : \frac{c}{0.379} : \times 0.126$

$= a : \infty b : \frac{c}{3}$

The coefficients are $1 : \infty : \frac{1}{3}$. The inverses are 1, 0, 3.

Miller indices for the given plane is (103).

1.17 How many atoms per unit area are there in (i) (100), (ii) (110) and (iii) (111) plane in a material that crystallizes in a face-centred cubic (FCC) unit cell.

(i) (100) plane

The number of atoms presents per unit cell in (100) plane is $\frac{1}{4} \times 4 + 1 = 2$

Number of atoms per m^2 $= \frac{2}{a^2} = \frac{2}{(\sqrt{8}r)^2} = \frac{1}{4r^2}$

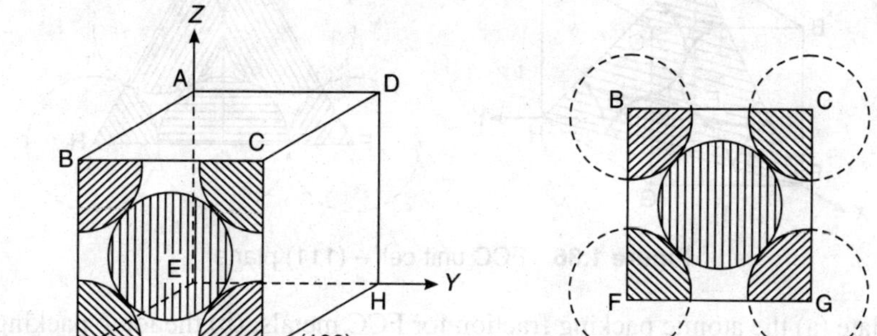

Figure 1.34 FCC unit cell – (100) plane.

(ii) (110) plane

The number of atoms presents per unit cell in (110) plane is $\frac{1}{4} \times 4 + 2 \times \frac{1}{2} = 2$

$$\text{Number of atoms per m}^2 = \frac{2}{a\sqrt{2}a} = \frac{2}{\sqrt{2}(\sqrt{8}r)^2} = \frac{1}{4\sqrt{2}r^2}$$

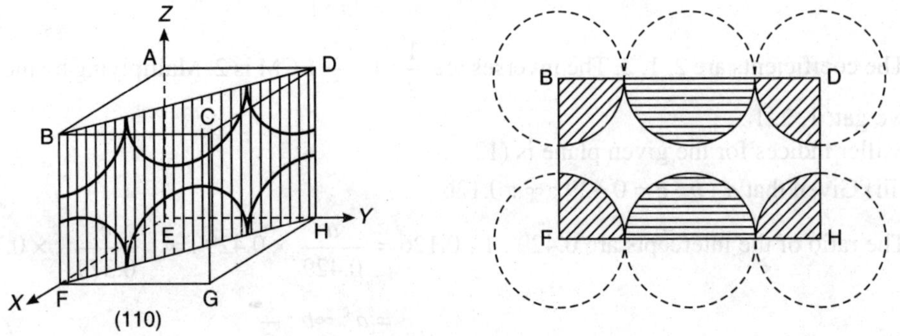

Figure 1.35 FCC unit cell– (110) plane.

(iii) (111) plane

The number of atoms presents per unit cell in (111) plane is $\frac{1}{6} \times 3 + 3 \times \frac{1}{2} = 2$

$$\text{Number of atoms per m}^2 = \frac{2}{\frac{\sqrt{3}}{4}a^2} = \frac{2}{\frac{\sqrt{3}}{4}(4r)^2} = \frac{1}{2\sqrt{3}r^2}.$$

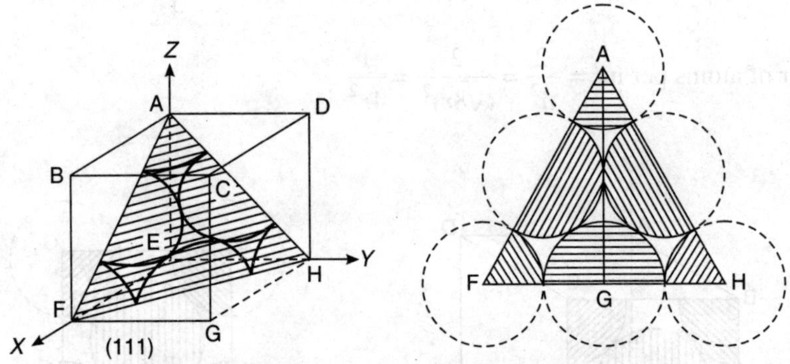

Figure 1.36 FCC unit cell – (111) plane.

1.18 Calculate (a) the atomic packing fraction for FCC metals, (b) the ionic packing fraction of FCC NaCl.

Solution

(a) Atomic packing factor (APF) = The packing density of a face-centred cubic unit cell. This can be calculated using the following formula

$$\text{Packing density} = \frac{\text{Total volume of the atoms}}{\text{Volume of the unit cell}}$$

$$\text{Packing density} = \frac{\text{Number of atoms per unit cell} \times \text{Volume of one atom}}{\text{Volume of the unit cell}}$$

$$= \frac{4 \times \frac{4}{3}\pi r^3}{a^3}$$

Substituting, $r = \dfrac{a}{\sqrt{8}}$, we get

$$\text{Packing density} = 4 \times \frac{4}{3}\pi \left(\frac{a}{2\sqrt{2}}\right)^3 \frac{1}{a^3} = \frac{\pi}{3\sqrt{2}} = 0.74$$

(b) Ionic packing factor of NaCl

$$= \frac{\text{Volume of Na atoms} + \text{Volume of Cl atoms}}{\text{Volume of the unit cell}}$$

$$= \frac{4 \times \frac{4}{3}\pi r^3 + 4 \times \frac{4}{3}\pi R^3}{(2 \times (r + R))^3}$$

where r and R are the radii of Na^+ and Cl^- ions respectively.

$$= \frac{4 \times \frac{4}{3}\pi (r^3 + R^3)}{(2 \times (r + R))^3}$$

Substituting the values of r and R, we get

$$= \frac{4 \times \frac{4}{3}\pi (0.97^3 + 1.81^3) \times (10^{-10})^3}{(2^3 \times (0.97 + 1.81))^3 \times (10^{-10})^3}$$

$$= 0.667.$$

The ionic packing factor of NaCl crystal is 0.667.

SHORT QUESTIONS

1. What is meant by crystallography?
2. What are crystalline materials?
3. What are polycrystalline materials?
4. Define the term non-crystalline materials.
5. What are isotropic substances?
6. Define the term anisotropy.
7. Define lattice.
8. Define basis.

9. Define the term crystal structure.
10. Define the term unit cell.
11. What do you mean by crystallographic axes?
12. What are primitives?
13. Define interfacial angles.
14. What is a primitive cell?
15. What are lattice parameters?
16. Write the names of seven crystal systems.
17. What are Bravais lattices?
18. Define the term atomic radius.
19. What do you mean by coordination number?
20. What is packing density?
21. Determine the packing density of a simple cubic unit cell.
22. Obtain the atomic radius of the atom in a BCC unit cell.
23. Calculate the atomic radius of the atom in a FCC lattice.
24. Deduce the packing density of a simple cubic lattice.
25. Derive the packing density of a body-centred cubic unit cell.
26. Obtain the packing density of a face-centred cubic unit cell.
27. Explain the arrangement of atoms in a HCP structure.
28. Calculate the number of atoms in a HCP unit cell.
29. Deduce the packing density of a HCP unit cell.
30. Deduce the relation between interatomic distance and the atomic weight of a substance.
31. What are crystal planes?
32. What are Miller indices?
33. List out the procedure used to find the Miller indices of a crystal plane.
34. Mention any four salient features of Miller indices.
35. Mention any four advantages of finding Miller indices.
36. What is family of planes? Give examples.
37. What is meant by direction of a plane?
38. What is family of directions? Give examples.
39. What are the steps to be followed to draw the direction of a plane?
40. Define the term linear density.
41. Define the term planar density.
42. Determine the packing density of the diamond unit cell.
43. Determine the atomic radius of the diamond unit cell.
44. What is allotropy? Give examples.
45. What are polymorphisms? Give examples.

DESCRIPTIVE TYPE QUESTIONS

1. Explain a simple cubic unit cell and hence find the atomic radius, number of atoms present in a simple cubic unit cell, coordination number, and packing density of a simple cubic unit cell.
2. Explain a body-centred cubic unit cell. Determine the atomic radius, number of atoms in unit cell, packing density and coordination number for a body-centred cubic unit cell.
3. Explain a face-centred cubic unit cell with a neat sketch. Obtain the number of atoms in a unit cell, atomic radius, coordination number, packing density of a face-centred cubic unit cell.
4. Explain the arrangement of atoms in a hexagonal unit cell. Determine the coordination number, packing density, atomic radius and number of atoms in a HCP structure.
5. Determine the packing density of a BCC, FCC and HCP unit cells.
6. Determine the atomic radius, and coordination number of SC, BCC, FCC and HCP unit cells.
7. Show that a hexagonal unit cell demands an axial ratio of $\frac{c}{a} = \sqrt{\frac{8}{3}}$. Determine the packing density of a HCP unit cell.
8. Describe a diamond unit cell and hence determine the number of atoms in a unit cell, atomic radius, coordination number and packing density of diamond unit cells.
9. Describe with neat sketch the following unit cells: (i) CsCl, (ii) ZnS and (iii) NaCl and (iv) graphite.
10. What is meant by allotropy? Describe the allotropy of carbon.

PROBLEMS

1. Sketch the following crystallographic planes for the cubic systems (210), (211), (220), (310), (311), (321).
2. Calculate the lattice constant of iron if the atomic radii of the iron atoms in BCC and FCC are 1.258 Å and 1.292 Å.
3. Determine the d spacing between (100) planes having lattice constants $a = 5.64$ Å.
4. Copper has an FCC structure and its atomic radius is 0.1278 nm. Calculate the interplanar spacing for (100) and (231) planes.
5. Iron has a BCC structure with atomic radius 0.103 Å. Find the lattice constant and also the volume of the unit cell.
6. Zinc has an HCP structure. The height of unit cell is 0.494 nm. The nearest neighbour distance is 0.27 nm. The atomic weight of Zinc is 65.37. Calculate the volume of the unit cell.
7. A unit cell has the dimensions $a = b = 4$Å; $c = 8$ Å; $\alpha = \beta = 90°$ and $\gamma = 120°$. What is the crystal structure? Give two examples.

8. Compare the packing efficiency of sphere of crystal size in a hexagonal close packing with that of face-centred cubic packing.
9. Draw the crystal planes with Miller indices (110) and (111).
10. Draw (110) and (111) planes in a cubic unit cell.
11. Calculate the interplanar spacing for (101) and (221) planes in a simple cubic lattice where lattice constant is 0.42 nm.
12. Calculate the interplanar spacing between (111) and (220) planes in FCC crystal. Given the atomic radius = 1.246 Å.
13. Calculate the interplanar spacing for (321) plane in simple cubic lattice with interatomic spacing of 4.12 Å.
14. The interplanar distance for (110) plane of a cubic crystal is 0.286 nm. Find its unit cell parameter.
15. How many atoms per unit area are there in (i) (100), (ii) (110) and (iii) (111) planes in a material that crystallizes in a simple cubic unit cell?
16. How many atoms per unit area are there in (i) (100), (ii) (110) and (iii) (111) planes in a material that crystallizes in a body-centred cubic (BCC) unit cell?

CHAPTER 2

PROPERTIES OF MATTER

2.1 INTRODUCTION

Consider a force is applied to an object. The object will move due to the force applied on it. If the object is not moving, there is a change in the dimension of the object. The dimensional change takes place in the following three ways: (i) change in the length, (ii) change in volume of the object without change in the shape, and (iii) change in the shape of the object without change in volume. The force that produces deformation to the object is called deforming force. An equal and opposite force is also acting on the object that tries to bring its original dimension. This force is called the restoring force. If the deforming force is removed from the object, the restoring force acts on the object. Hence the object regains its original length or shape or size. This property of recovering the original length or shape or size of the object after the removal of the deforming force is called elastic property. Elasticity is the branch of science that deals with the elastic property of the material.

The materials are classified into two types on the basis of elastic property: (i) elastic materials and (ii) plastic materials. The materials that regains their original length or shape or size after the removal of the deformation force is called elastic materials. The materials which are not able to regain their original length or shape or size after the removal of the deforming force are called plastic materials. In general, no material is perfectly plastic or perfectly elastic. A material is called elastic material or plastic material by comparing it with other materials. A quartz fiber is said to be more elastic, whereas putty is said to be more plastic.

This chapter describes about the elastic properties such as stress, strain, stress–strain diagram, Young's modulus, bulk modulus, rigidity modulus and the experimental determination of Young's modulus and rigidity modulus.

2.2 CONCEPT OF LOAD, STRESS AND STRAIN

Load

The external force acting on a body that produces change in the dimension of the body is called load.

Stress

Consider one end of a wire or rod of length l is fixed and the other end of the wire or rod is subjected to a load of mass m kg. Due to the application of the load, a downward force is acting on the wire or rod and hence there is an elongation in the length Δl of the material. A restoring force is also acting on the wire or rod that tries to bring the material into its original state. If the load is removed, the rod or wire regains its original length l. The force acting per unit area of cross-section of the wire or rod is called as the stress.

$$\text{Stress} = \frac{\text{Force}}{\text{Area of cross-section}} = \frac{F}{A} \tag{2.1}$$

The unit for stress is N m^{-2}. The term stress is also defined as the restoring force acting per unit area of cross-section. Stress is represented by the symbol σ.

Types of stress

There are three types of stresses.

1. *Longitudinal stress* If the restoring force is acting perpendicular to the area of cross-section and along the length of the wire, the stress is called longitudinal stress. During longitudinal stress, the body undergoes changes in length, but not in shape and volume.
2. *Bulk stress* If the body is subjected to equal forces normally on all the faces, the stress is called bulk stress. The bulk stress produces changes in volume, but not in shape of the body.
3. *Shear stress or tangential stress* If the restoring forces are parallel to the surface, the stress is called shearing stress. The shearing stress produces changes in the shape of the body, but not in volume.

Strain

Consider one end of the wire or rod is fixed and the other end of the wire or rod is subjected to a load of mass m kg. Due to the application of the load, there is a change in the dimension of the wire or rod. The term strain is defined as the ratio of the change in dimension of the material to the original dimension.

$$\text{Strain} = \frac{\text{Change in dimension}}{\text{Actual dimension}} \tag{2.2}$$

Strain is a dimensionless quantity, and it has no unit. The strain is represented by the letter ε.

There are three types of strain, namely longitudinal strain, volume strain and shear strain depending upon the changes in the dimension. If the change is in the dimension of length, volume, or shape, then the strain is respectively called longitudinal strain, volume strain and shear strain. The longitudinal strain, bulk strain and shearing strain are defined as follows:

$$\text{Longitudinal strain} = \frac{\text{Change in length}}{\text{Original length}} = \frac{\Delta l}{l} \tag{2.3}$$

$$\text{Bulk strain} = \frac{\text{Change in volume}}{\text{Original volume}} = \frac{-\Delta V}{V} \tag{2.4}$$

$$\text{Shearing strain} = \frac{\text{Lateral displacement between two layers}}{\text{Perpendicular distance between two layers}} = \frac{\Delta l}{l} \tag{2.5}$$

2.3 HOOKE'S LAW

In 1679, Robert Hooke stated a fundamental law of elasticity known as Hooke's law. According to Hooke's law, within the elastic limit of a body, the stress is directly proportional to the strain, i.e.,

$$\text{Stress} \propto \text{Strain}$$

$$\frac{\text{Stress}}{\text{Strain}} = \text{Constant}, E \tag{2.6}$$

The constant E is known as modulus of elasticity or coefficient of elasticity. Its unit is Nm^{-2} or Pa. For example, a spring balance works on the principle of Hooke's law. Within the elastic limit, the stress–strain graph is a straight line passing through the origin.

2.4 STRESS–STRAIN DIAGRAM

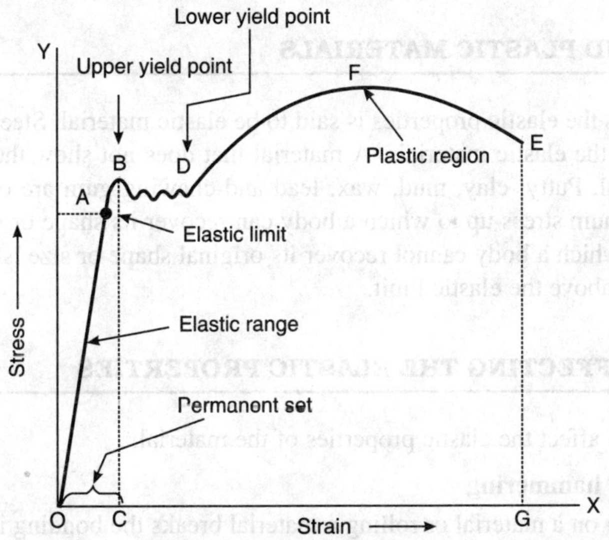

Figure 2.1 Stress–strain diagram.

The relationship between stress and strain is studied by drawing a graph between stress and strain. This plot is called stress–strain diagram. Consider a material is subjected to a gradually varying load and hence stress–strain diagram is obtained as shown in Figure 2.1. In region OA, if stress is increased, the strain is also linearly increasing. In this region OA, if the load is removed, then the material regains its original dimension, and it takes the same path OA. This region OA is called elastic region. Stress is proportional to strain up to a limit, which is called **elastic limit**. Upto this limit, Hooke's law is obeyed. In Figure 2.1, point A is called elastic limit.

If the stress is increased beyond elastic limit A, the strain increases and it reaches point B. Now, the material is partly elastic and partly plastic. If the stress is removed at this stage, it takes a new route BC and return to the original state. The region OC is the residual strain acquired by the material, and it is called **permanent set**.

The stress increases from point B to point D without further increase in the stress. Region BD is in irregular shape. The increase of strain takes place from point B, and it is called **yield point**. The stress corresponding to the yield point is called **yield stress**. The sudden increase of strain gets stopped at D. Point B is called upper yield point, and point D is called lower yield point.

If the stress is gradually increased beyond point D, the strain increases and it takes path DF. This region is called plastic region. In this region, the thickness of the material decreases and the volume of the material remains constant. Point F is the maximum stress a material can withstand. This maximum deforming force is called **ultimate strength** or **tensile strength.** The stress corresponding to this force is called **breaking stress.**

In region FE, the strain increases without further increase of the stress. In this region, a neck is formed in the material (wire). Due to the formation of the neck, the material breaks even though the stress is decreased and the strain is decreasing. The stress corresponding to point E that breaks the material is called **breaking stress**. The area under curve OABDFEG gives the work done per unit volume.

2.5 ELASTIC AND PLASTIC MATERIALS

A material that shows the elastic properties is said to be elastic material. Steel, rubber and quartz are the examples for the elastic materials. A material that does not show the elastic property is called plastic material. Putty, clay, mud, wax, lead and chewing gum are examples for plastic materials. The maximum stress up to which a body can recover its shape or size is called elastic limit. The region in which a body cannot recover its original shape or size is called plastic limit. The plastic limit lies above the elastic limit.

2.6 FACTORS AFFECTING THE ELASTIC PROPERTIES

The following factors affect the elastic properties of the material.

Effect of rolling and hammering

Generally hammering on a material or rolling a material breaks the bonding in the material. This will increase the elastic property of the material.

Effect of annealing

Annealing is the process of heating the material at a particular temperature and gradually cooling. Due to annealing, the particle size gets increased and the material becomes hardened. So, the elastic property gets reduced due to hardening.

Effect of temperature

The increase in temperature generally decreases the elastic property of the materials. However, the elastic property of *invar steel* is not affected by the temperature.

Effect of impurities

The addition of impurities to metals binds the crystal grains better. The addition of carbon to iron and potassium to gold in minute quantities affect their elastic properties. The impurities either increase or decrease the elastic properties of the concerned metals. If the impurity has more elasticity than the material to which it is added, it increases the elasticity. If the impurity is less elastic than the material, it decreases the elasticity.

2.7 YOUNG'S MODULUS

Consider a stress is applied to a material along only one direction, then the stress is called linear stress or longitudinal stress. The corresponding strain is called longitudinal strain or linear strain. The ratio between the longitudinal stress and the longitudinal strain within the elastic limit of the material is called Young's modulus.

Consider a wire of length l is fixed at one end, and it is loaded at another end as shown in Figure 2.2. Let F ($F=mg$) be the force applied, and A is the area of cross-section of the wire, then

$$\text{Longitudinal stress} = \frac{\text{Force}}{\text{Area}} = \frac{F}{A} \tag{2.7}$$

Let l be the length of the wire, and Δl is the increase in the length of the wire, then

$$\text{Longitudinal strain} = \frac{\text{Change in length}}{\text{Actaul length}} = \frac{\Delta l}{l} \tag{2.8}$$

Young's modulus is defined as:

$$\text{Young's modulus} = \frac{\text{Longitudinal stress}}{\text{Longitudinal strain}} \tag{2.9}$$

$$Y = \frac{F/A}{\Delta l/l} = \frac{Fl}{\Delta lA} \tag{2.10}$$

The unit for the Young's modulus is N m^{-2}. The Young's modulus is represented by the letter Y or E.

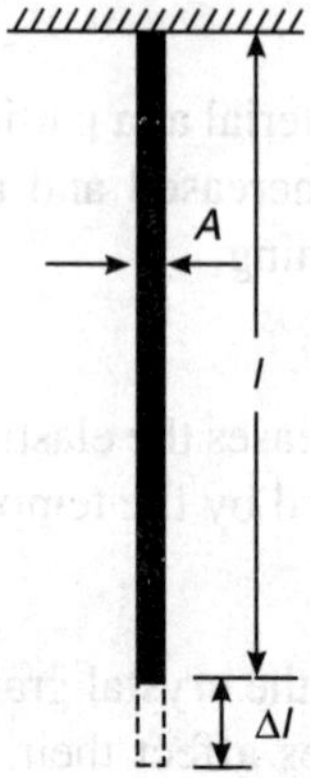

Figure 2.2 Young's modulus.

2.8 BULK MODULUS

Consider a force is applied normally and uniformly to the surfaces of a body, then the volume of the body changes, but there is no change in the shape of the body. The ratio between the uniform force applied normal to the surface of the body and the area of the surface is called bulk stress. Let F be the force applied, and A be the area of cross-section of the surfaces, then

$$\text{Bulk stress} = \frac{\text{Force}}{\text{Area}} = \frac{F}{A} \quad (2.11)$$

Let ΔV be the change in volume, and V be the volume of the body, then

$$\text{Bulk strain} = \frac{\text{Change in volume}}{\text{Actaul volume}} = \frac{-\Delta V}{V} \quad (2.12)$$

The negative sign shows that the volume decreases due to the applied force. The ratio between the bulk stress and bulk strain is called bulk modulus.

$$\text{Bulk modulus} = \frac{\text{Bulk stress}}{\text{Bulk strain}} \quad (2.13)$$

$$K = \frac{F/A}{-\Delta V/V} = \frac{-FV}{\Delta V \cdot A} = \frac{-PV}{\Delta V} \quad (2.14)$$

where P is the bulk pressure. The unit for the bulk modulus is N m^{-2}. It is represented by the letter K. It represents the incompressibility. The reciprocal of the bulk modulus is known as compressibility.

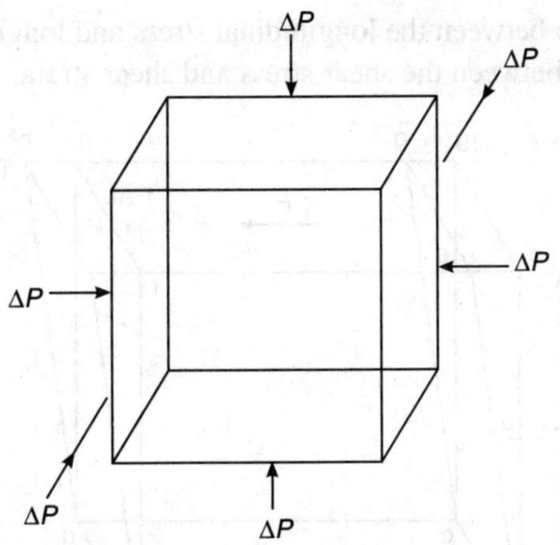

Figure 2.3 Bulk modulus.

2.9 RIGIDITY MODULUS

Consider a tangential force is applied to a body. Assume that there is a displacement of the upper layer of the body and there is no displacement in the bottom portion of the body. The ratio of the tangential force applied to the area of cross-section is called shearing stress.

$$\text{Shearing stress} = \frac{\text{Tangential force}}{\text{Area}} = \frac{F}{A} \tag{2.15}$$

Consider a force F is applied along, the upper surface of the object abcdefgh. Hence the upper layer efgh gets displaced into e′f′g′h′ (Figure 2.4). Let θ be the angle between ae and ae′. The angle θ is called as the shearing angle. Let ΔL be the displacement of the upper surface efgh.

$$\text{Shearing strain} = \frac{\text{Relative displacement of the layers along the force direction}}{\text{Perpendicular distance to the stress}} \tag{2.16}$$

$$\text{Shearing strain} = \frac{\text{ee}'}{\text{ae}} = \frac{\Delta L}{L} \tag{2.17}$$

From Figure 2.4, $\tan\theta = \frac{\text{ee}'}{\text{ae}}$. For smaller values of θ, $\tan\theta$ is nearly equal to θ (i.e., $\tan\theta \approx \theta$).

Therefore, shearing strain $\theta = \Delta L/L$.

The ratio between the shearing stress and the shearing strain is called shear modulus or rigidity modulus.

$$\text{Rigidity modulus} = \frac{\text{Shear stress}}{\text{Shear strain}} = \frac{F/A}{\theta} = \frac{F/A}{\Delta L/l} \tag{2.18}$$

The unit of shear modulus is Nm^{-2}. It is represented by the letter n or η. The equation for the Young's modulus [Eq.(2.10)] and the equation for the shear modulus [Eq.(2.18)] are same. The

Young's modulus is the ratio between the longitudinal stress and longitudinal strain, whereas the rigidity modulus is the ratio between the shear stress and shear strain.

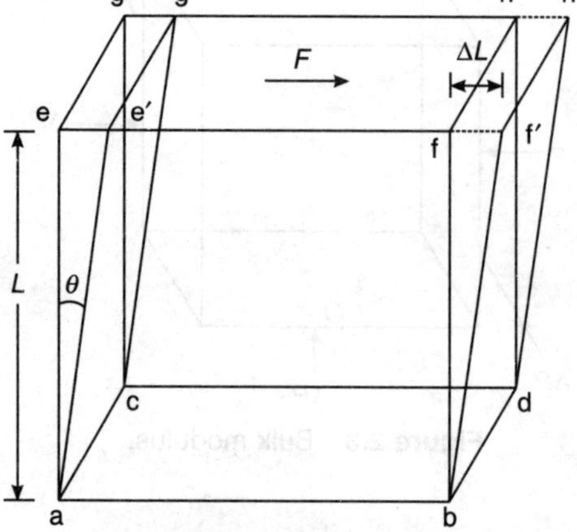

Figure 2.4 Rigidity modulus.

2.10 POISSON'S RATIO

Consider a force or a system of force is applied to a material. The application of the force produces some deformation. The deformation is produced not only in one direction, but also in all other directions. Consider a wire of length l is subjected to a force. Let Δl be the increase in length.

$$\text{Longitudinal strain} = \frac{\Delta l}{l} \tag{2.19}$$

The force applied to the wire not only increases the length of the wire, but also decreases the diameter of the wire. The strain produced along the perpendicular direction of the force is called lateral strain.

$$\text{Lateral strain} = -\frac{\Delta d}{d} \tag{2.20}$$

where Δd is the decrease in diameter, and d is the diameter of the wire. The negative sign shows that the diameter decreases.

The ratio of the lateral strain to the longitudinal strain is called Poisson's ratio.

$$\text{Poisson's ratio} = \frac{\text{Lateral strain}}{\text{Longitudinal strain}}$$

$$\sigma = \frac{-\Delta d/d}{\Delta l/l} = \frac{-\Delta d \times l}{d \times \Delta l} \tag{2.21}$$

The negative sign indicates that the lateral strain and longitudinal strain are having opposite sense.

The Poisson's ratio is represented by the letter σ. The longitudinal strain per unit stress is denoted by α and the lateral strain per unit stress is denoted by β. Therefore, the Poisson's ratio can be written as,

$$\sigma = \frac{\beta}{\alpha} \tag{2.22}$$

where β is the lateral strain per unit stress and α is the longitudinal strain per unit stress.

Limiting values of σ

The relation between K, n and σ is given by Eq.(2.49) and Eq.(2.50) [see Section 2.11.5]. Equating these two equations, we get

$$3K(1 - 2\sigma) = 2n(1 + \sigma) \tag{2.23}$$

Equation (2.23) shows that both positive and negative values for σ are possible.

Case I: *σ is a positive quantity*

If σ is a positive quantity, then $(1 - 2\sigma) > 0$

i.e., $\sigma < 0.5$

Case II: *σ is a negative quantity*

Then, $(1 + \sigma) > 0$ or $\sigma > -1$

These two cases show that the value of σ lies between 0.5 and –1. Substituting $\sigma = 0.5$ in Eq. (2.23), we get $K = \infty$. This value for K is not practically possible. Substituting $\sigma = -1$ in Eq. (2.23), we get, $n = \infty$. This value for σ is also not practically possible. Poisson derived a mathematical relation and showed that the value of $\sigma = 0.25$, but the experimental value for σ lies between 0.2 and 0.4.

2.11 RELATION BETWEEN YOUNG'S MODULUS, BULK MODULUS AND RIGIDITY MODULUS

2.11.1 Relation between Y and α

Consider a cube of unit length and unit volume. Consider a unit tension is acting along one edge of the cube. Let α be the elongation per unit length per unit tension along the direction of force. Then

$$\text{Young's modulus} = \frac{\text{Linear stress}}{\text{Linear strain}} \tag{2.24}$$

Since, stress = 1 (unit tension is acting) and linear strain = $\dfrac{\alpha}{1}\left(= \dfrac{\text{Elongation in length}}{\text{Length}}\right)$.

Therefore,

$$\text{Young's modulus, } Y = \frac{1}{\alpha} \tag{2.25}$$

2.11.2 Relation between K, α and β

Consider a cube ABCDEFGH of unit length as shown in Figure 2.5. Consider a tension T_X, is acting along the sides ABEF and DCGH, the tension T_Y is acting along the sides ABCD and EFGH and the tension T_Z is acting along the sides ACGE and BFHD respectively.

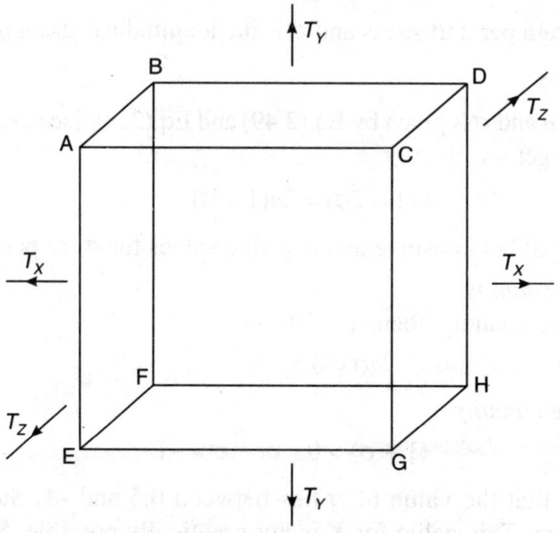

Figure 2.5 Bulk modulus.

There is an elongation of length AC due to tension T_X and the length AC gets contracted due to tensions T_Y and T_Z. Similarly, the side AE gets elongated due to tension T_Y, and it gets shortened due to tensions T_X and T_Z. The side AB gets elongated due to tension T_Z, and it gets shortened due to tensions T_X and T_Y. Let α and β be the elongation per unit length per unit stress and the contraction per unit length per unit stress respectively. Then the changes in the length of sides AC, AE, and AB are written as:

$$AC = 1 + T_X\alpha - T_Y\beta - T_Z\beta \quad (2.26)$$

$$AE = 1 + T_Y\alpha - T_Z\beta - T_X\beta \quad (2.27)$$

$$AB = 1 + T_Z\alpha - T_X\beta - T_Y\beta \quad (2.28)$$

The new volume of the cube, $V = AC \times AE \times AB$

$$V = (1 + T_X\alpha - T_Y\beta - T_Z\beta) \times (1 + T_Y\alpha - T_Z\beta - T_X\beta) \times (1 + T_Z\alpha - T_X\beta - T_Y\beta)$$

Neglecting the higher order terms, we get

$$V = 1 + (\alpha - 2\beta)(T_X + T_Y + T_Z) \quad (2.29)$$

Assuming, $T_X = T_Y = T_Z = T$, we get, $V = 1 + 3T(\alpha - 2\beta)$ (2.30)

Actual volume of the cube $V = 1\ m^3$ (since the cube has unit lengths)

Changes in volume, $\Delta V = 3T(\alpha - 2\beta)$

$$\text{Bulk strain} = \frac{\text{Change in volume}}{\text{Actaul volume}} = \frac{\Delta V}{V} = \frac{3T(\alpha - 2\beta)}{1} \tag{2.31}$$

$$\text{Bulk modulus} = \frac{\text{Bulk stress}}{\text{Bulk strain}} = \frac{T}{3T(\alpha - 2\beta)}$$

$$K = \frac{1}{3(\alpha - 2\beta)} \tag{2.32}$$

$$K = \frac{1}{3\alpha(1 - 2\beta/\alpha)} = \frac{Y}{3(1 - 2\sigma)} \tag{2.33}$$

Equation (2.32) gives the relation between K, α and β. Equation (2.33) gives the relation between K, Y and σ.

2.11.3 Relation between n, α and β

Consider a tangential force F is acting on the upper face ABCD of the cube. Let the upper face is slightly displaced through a distance l. Consider A′B′C′D′ is the new upper face of the cube.

$$\text{Shearing stress applied} = \frac{\text{Force}}{\text{Area}} = \frac{F}{L^2} \tag{2.34}$$

where L is the length of the cube.

$$\text{Shearing strain, } \theta = \frac{l}{L} \tag{2.35}$$

where l is the displacement of the upper side.

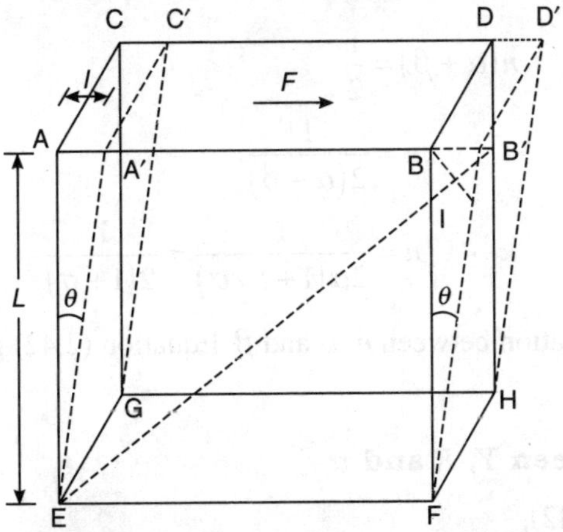

Figure 2.6 Relation between n, α and β.

$$\text{Rigidity modulus, } n = \frac{\text{Shear stress}}{\text{Shear strain}} = \frac{T}{\theta} \tag{2.36}$$

where T is the shear stress applied ($T = F/L^2$).

Due to the application of the tangential force, there is a tensile stress acting along diagonal EB, and there is a compressive stress acting along diagonal AF. Let α and β be the elongation per unit length per unit stress and the contraction per unit length per unit stress. Then,

Extension of the diagonal EB due to tensile stress $= \text{EB}\cdot T\alpha$

Extension of the diagonal EB due to the compressive stress $= \text{EB}\cdot T\beta$

Total extension in length of EB $= \text{EB}\cdot T(\alpha + \beta)$ (2.37)

The increase in length is also determined by drawing a perpendicular line from B to line EB′. The actual increase in length EB is IB′.

Consider the triangle, BIB′

$$\text{IB}' = \text{BB}' \times \cos(\text{BB}'\text{I}) = \text{BB}' \cos 45° = \frac{l}{\sqrt{2}} \tag{2.38}$$

Equating Eq. (2.37) and Eq. (2.38), we get

$$\text{EB}\cdot T(\alpha + \beta) = \frac{l}{\sqrt{2}} \tag{2.39}$$

Substituting, $\text{EB} = \sqrt{2}L$, Eq. (2.39) can be written as

$$\sqrt{2}L\cdot T(\alpha + \beta) = \frac{l}{\sqrt{2}} \tag{2.40}$$

Substituting $\frac{l}{L} = \theta$ and $n = \frac{T}{\theta}$ in Eq. (2.40), we get

$$n(\alpha + \beta) = \frac{1}{2}$$

$$n = \frac{1}{2(\alpha + \beta)} \tag{2.41}$$

$$n = \frac{1}{2\alpha(1 + \beta/\alpha)} = \frac{Y}{2(1+\sigma)} \tag{2.42}$$

Equation (2.41) gives the relation between n, α and β. Equation (2.42) gives the relation between n, Y and σ.

2.11.4 Relation between Y, K and n

From Eq. (2.33) and Eq. (2.42),

$$K = \frac{Y}{3(1 - 2\sigma)} \tag{2.43}$$

$$n = \frac{Y}{2(1+\sigma)} \tag{2.44}$$

Equation (2.43) can be written as:

$$\sigma = \frac{1}{2} - \frac{Y}{6K} \tag{2.45}$$

Equation (2.44) can be written as

$$\sigma = \frac{Y}{2n} - 1 \tag{2.46}$$

Equating Eq. (2.45) and Eq. (2.46), we get,

$$\frac{1}{2} - \frac{Y}{6K} = \frac{Y}{2n} - 1 \tag{2.47}$$

Dividing both sides by Y and multiplying by 6, we get

$$\frac{3}{n} = \frac{9}{Y} - \frac{1}{K} \tag{2.48}$$

Equation (2.48) gives the relation between Y, n and K.

2.11.5 Relation between K and n in terms of Poisson's ratio

Equation (2.33) can be written as:

$$Y = 3K(1 - 2\sigma) \tag{2.49}$$

Equation (2.42) can be written as

$$Y = 2n(1 + \sigma) \tag{2.50}$$

Equating Eq. (2.49) and Eq. (2.50), we get

$$3K(1 - 2\sigma) = 2n(1 + \sigma)$$

Rearranging the above equation, we get

$$\sigma = \frac{3K - 2n}{2n + 6K} \tag{2.51}$$

Equation (2.51) gives the relation between σ, n and K.

2.11.6 Relation between Y, K and n (qualitative)

The relation between the Young's modulus Y, bulk modulus K, rigidity modulus n and Poisson's ratio σ are given by

1. $Y = \dfrac{1}{\alpha}$ (2.52)

2. $K = \dfrac{1}{3\alpha(1 - 2\beta/\alpha)} = \dfrac{Y}{3(1 - 2\sigma)}$ (2.53)

3. $n = \dfrac{1}{2\alpha(1+\beta/\alpha)} = \dfrac{Y}{2(1+\sigma)}$ (2.54)

4. $\dfrac{3}{n} = \dfrac{9}{Y} - \dfrac{1}{K}$ (2.55)

5. $\sigma = \dfrac{3K - 2n}{2n + 6K}$ (2.56)

where α and β are the elongation per unit length per unit stress and the contraction per unit length per unit stress respectively.

2.12 TWISTING COUPLE IN A WIRE

Consider a cylindrical wire of radius r and length l. Assume the wire has a number of concentric rings. One such ring is shown in Figure 2.7(a). Consider one end of the wire is fixed, whereas the other end is twisted [Figure 2.7(b)]. Let OO_1 be the centre of the wire. Line AB gets twisted to AB_1 due to the twisting of the wire. Let θ be the angle of twisting, then from the sector O_1BB_1,

$$BB_1 = r\theta \tag{2.57}$$

Consider the flat surface ABCD [Figure 2.7(c)]. It gets sheared into AB_1C_1D due to the application of the twisting force. Let ϕ be the shearing angle of the wire. From ABB_1, we get

$$BB_1 = l\phi \tag{2.58}$$

Equating Eq. (2.57) and Eq. (2.58), we get

$$1\phi = r\theta \tag{2.59}$$

$$\text{Shearing strain, } \phi = \frac{r\theta}{l} \tag{2.60}$$

The shearing stress σ is given by

$$\text{Shearing stress} = n \times \text{shearing strain}$$

$$\sigma = n\frac{r\theta}{l} \tag{2.61}$$

Let a be the area of cross-section of the wire. The shearing force acting on the wire is

$$F = \text{Shearing stress} \times \text{Area}$$

$$= n\frac{r\theta}{l}a \tag{2.62}$$

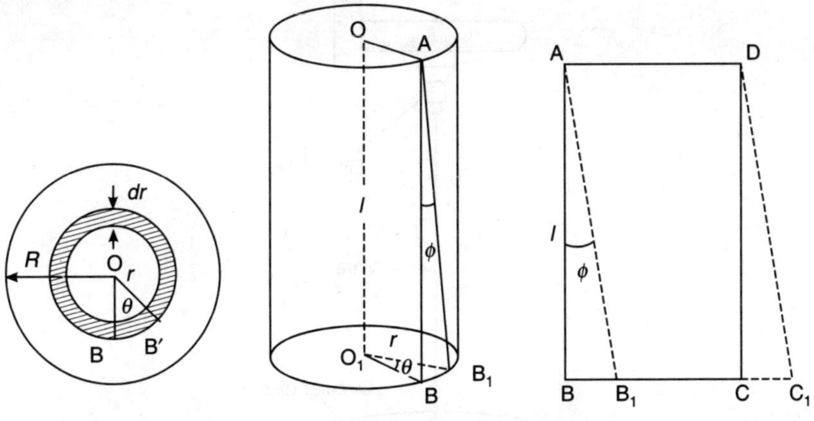

Figure 2.7 Twisting couple of a wire (a) hollow cylinder of radius r and radial thickness dr, (b) twisting of a wire, and (iii) twisting of the rectangular plane ABCD.

Area of the wire is given by

$$A = \text{Circumference} \times \text{Thickness} = 2\pi r\, dr$$

Shearing force acting on the wire,

$$F = n\frac{r\theta}{l} \times 2\pi r dr \tag{2.63}$$

$$\text{Moment of the force} = \text{Force} \times \text{distance}$$

$$\text{Moment of the force acting on the wire} = n\frac{\theta}{l} \times 2\pi r^3 dr \tag{2.64}$$

Total moment of the force acting on the wire is determined by integrating from 0 to r,

$$\text{Total moment of the force acting on the wire} = \int_0^r n\frac{\theta}{l} \times 2\pi r^3 dr \tag{2.65}$$

$$= \frac{n\pi r^4 \theta}{2l} \tag{2.66}$$

The moment of the force is the couple acting on the wire. The couple per angular twist C is given by

$$C = \frac{n\pi r^4}{2l} \tag{2.67}$$

Equation (2.67) gives the twisting couple acting on a wire.

2.13 TORSIONAL PENDULUM

A circular metallic disc suspended using a thin wire that executes torsional oscillation is called torsional pendulum. It executes torsional oscillations, whereas a simple pendulum executes linear oscillations.

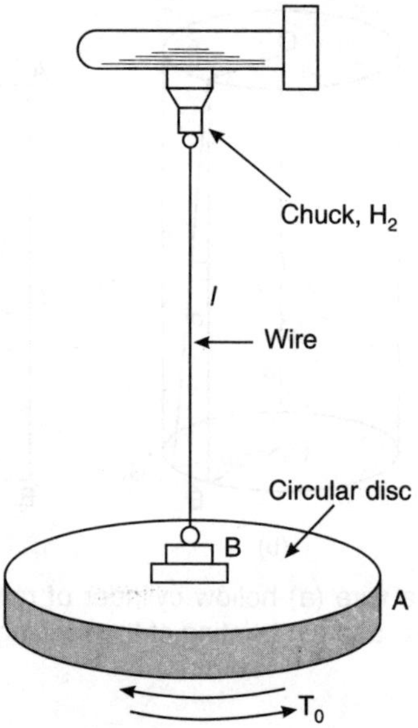

Figure 2.8 Torsional pendulum.

Consider a metallic disc A is suspended symmetrically using a wire as shown in Figure 2.8. The length of the wire can be varied using the chuck (torsion head) H_2. Consider the metallic disc is slightly rotated along the horizontal plane. When the disc is rotating, the wire gets twisted. Let θ be the angle of twisting. A restoring force is also acting on the wire. The restoring force tries to bring the wire into its original position. Therefore, the wire and the disc will rotate in the opposite direction due to the restoring force, and hence the wire will experience a restoring couple.

Let C be the couple per unit twist and θ be the angle of the twist, then the restoring couple of the wire is given by $C\theta$.

Let $d^2\theta/dt^2$ be the acceleration experienced by the wire and I is the moment of inertia of the disc about the wire, then the couple acting on the disc is $I(d^2\theta/dt^2)$.

Equating these two equations, we get,

$$I\frac{d^2\theta}{dt^2} = C\theta \tag{2.68}$$

From Eq. (2.68), we infer that the acceleration is directly proportional to the displacement, and the displacement is directed towards the mean position. Therefore, the disc is said to be executing simple harmonic motion (SHM).

For the simple harmonic motion, the time period is given by

$$T = 2\pi \times \sqrt{\frac{\text{Displacement}}{\text{Accleration}}} \tag{2.69}$$

From Eq. (2.68) and Eq. (2.69), we get

$$T = 2\pi \times \sqrt{\frac{I}{C}} \tag{2.70}$$

Equation (2.70) gives the time period of the torsional pendulum.

2.13.1 Uses of Torsional Pendulum

The torsional pendulum is used to determine:

1. The rigidity modulus of the wire
2. The moment of inertia of the disc and
3. The moment of inertia of an irregular body.

Rigidity modulus of the wire

Consider a torsional pendulum consisting of a circular disc. It is suspended using a wire of length l as shown in Figure 2.9. Assume that the disc is rotated without any additional weights and the time period is noted. Let it be T_0 second. The time period T_0 of the pendulum is given by

$$T_0 = 2\pi \times \sqrt{\frac{I}{C}} \tag{2.71}$$

where C is the couple per unit angular twist and I is the moment of inertia of the disc.

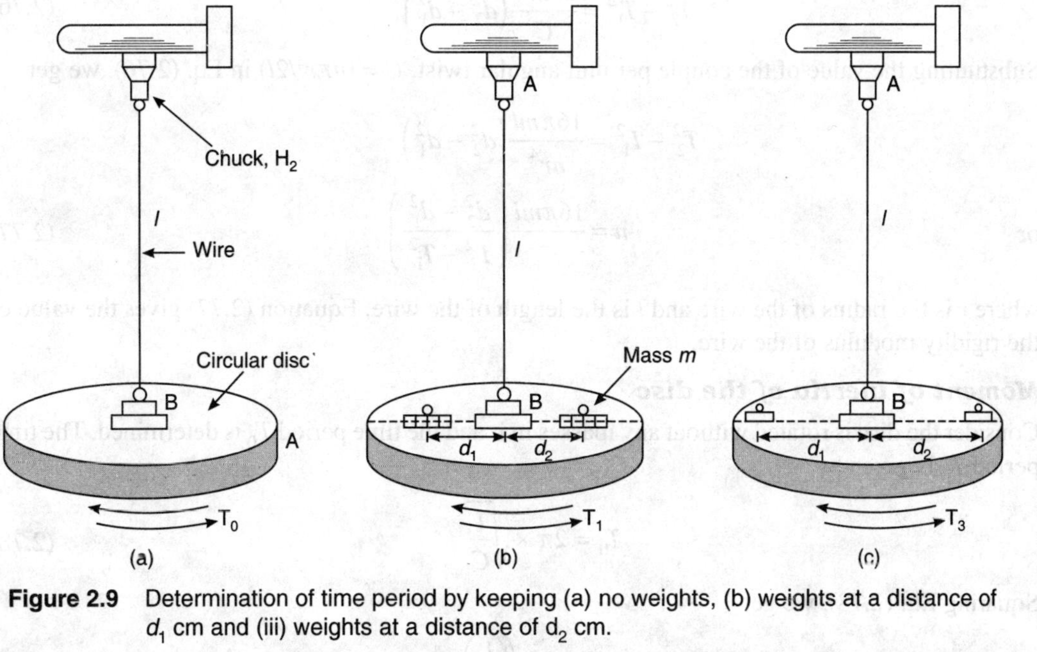

Figure 2.9 Determination of time period by keeping (a) no weights, (b) weights at a distance of d_1 cm and (iii) weights at a distance of d_2 cm.

Consider two equal weights of masses m kg that are placed over the disc at a distance d_1 from the centre of the disc. Consider the disc is rotated and the time period T_1 is noted. The time period is given by

$$T_1 = 2\pi \times \sqrt{\frac{I + 2md_1^2}{C}} \tag{2.72}$$

where m is the mass of the equal weight and d_1 is the distance between the centre of the disc and the equal weights.

The equal weights are placed at a distance d_2 from the centre of the disc and again the time period of the torsional pendulum is noted. Let it be T_2 second. The time period T_2 is given by

$$T_2 = 2\pi \times \sqrt{\frac{I + 2md_2^2}{C}} \tag{2.73}$$

where d_2 is the distance between the centre of the disc and the equal weights. Squaring Eq. (2.72) and Eq. (2.73), we get

$$T_1^2 = 4\pi^2 \times \left(\frac{I + 2md_1^2}{C}\right) \tag{2.74}$$

$$T_2^2 = 4\pi^2 \times \left(\frac{I + 2md_2^2}{C}\right) \tag{2.75}$$

Subtracting Eq. (2.74) from Eq. (2.75), we get

$$T_2^2 - T_1^2 = \frac{8m\pi^2}{C}\left(d_2^2 - d_1^2\right) \tag{2.76}$$

Substituting the value of the couple per unit angular twist, $C = (n\pi r^4/2l)$ in Eq. (2.76), we get

$$T_2^2 - T_1^2 = \frac{16\pi ml}{nr^4}\left(d_2^2 - d_1^2\right)$$

or

$$n = \frac{16\pi ml}{r^4}\left(\frac{d_2^2 - d_1^2}{T_2^2 - T_1^2}\right) \tag{2.77}$$

where r is the radius of the wire and l is the length of the wire. Equation (2.77) gives the value of the rigidity modulus of the wire.

Moment of inertia of the disc

Consider the disc is rotated without any masses in it and the time period T_0 is determined. The time period T_0 is given by

$$T_0 = 2\pi \times \sqrt{\frac{I}{C}} \tag{2.78}$$

Squaring Eq. (2.78), we get

$$T_0^2 = 4\pi^2\left(\frac{I}{C}\right) \tag{2.79}$$

i.e.,
$$I = \frac{CT_0^2}{4\pi^2} \tag{2.80}$$

Substituting the value of C from Eq. (2.67), we get

$$I = \frac{nr^4T_0^2}{8\pi l} \tag{2.81}$$

Substituting the value of n from Eq. (2.77) in Eq. (2.81), we get

$$I = 2mT_0^2\left(\frac{d_2^2 - d_1^2}{T_2^2 - T_1^2}\right) \tag{2.82}$$

Equation (2.82) is used to find the moment of inertia of the disc.

Moment of inertia of an irregular object

The torsional pendulum is rotated without any masses in the disc and the time period is noted. Let T_0 be the time period. The time period is given by

$$T_0 = 2\pi \times \sqrt{\frac{I}{C}} \tag{2.83}$$

Two equal masses, whose moment of inertia is known, are placed over the disc at a distance d from the centre of the disc. The disc is rotated and hence the time period is noted. Let it be T_1. The time period is given by

$$T_1 = 2\pi \times \sqrt{\frac{I + 2I_1}{C}} \tag{2.84}$$

where I_1 is the moment of inertia of the equal masses.

The given irregular object is placed over the disc at the same distance d from the centre of the disc. The time period is noted, and let it be T_2. The time period is given by

$$T_2 = 2\pi \times \sqrt{\frac{I + 2I_2}{C}} \tag{2.85}$$

Squaring Eq. (2.83), Eq. (2.84) and Eq. (2.85), we get

$$T_0^2 = 4\pi^2\left(\frac{I}{C}\right) \tag{2.86}$$

$$T_1^2 = 4\pi^2 \times \left(\frac{I + 2I_1}{C}\right) \tag{2.87}$$

$$T_2^2 = 4\pi^2 \times \left(\frac{I + 2I_2}{C}\right) \tag{2.88}$$

Subtracting Eq. (2.86) from Eq. (2.87), we get

$$T_1^2 - T_0^2 = \frac{8\pi^2 I_1}{C} \tag{2.89}$$

Subtracting Eq. (2.86) from Eq. (2.88), we get

$$T_2^2 - T_0^2 = \frac{8\pi^2 I_2}{C} \tag{2.90}$$

Dividing Eq. (2.89) by Eq. (2.90), we get

$$\frac{T_1^2 - T_0^2}{T_2^2 - T_0^2} = \frac{I_1}{I_2} \tag{2.91}$$

From Eq. (2.91), we get

$$I_2 = I_1\left(\frac{T_2^2 - T_0^2}{T_1^2 - T_0^2}\right) \tag{2.92}$$

Equation (2.92) is used to find the moment of inertia of the irregular object

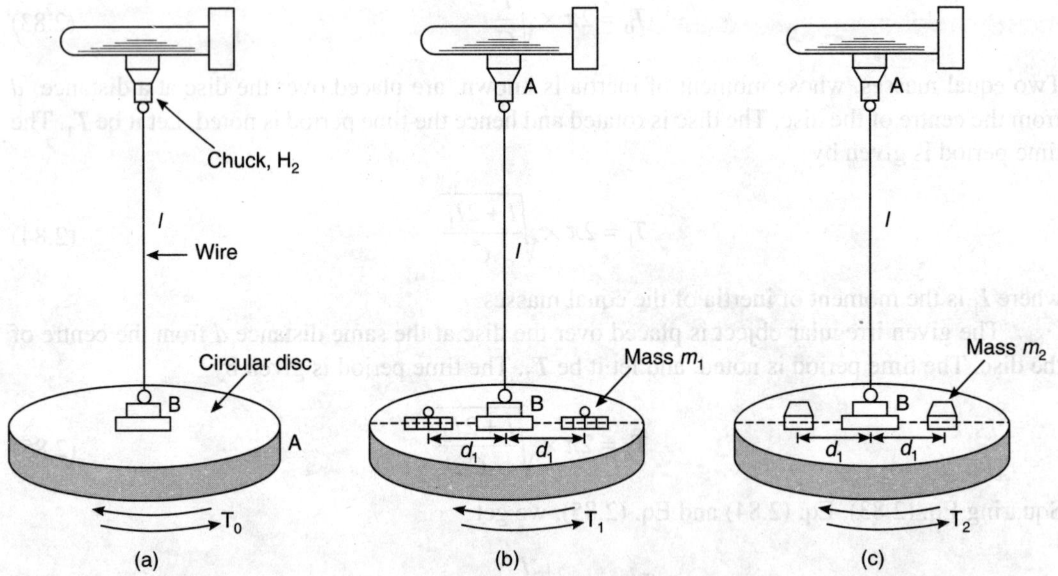

Figure 2.10 Determination of time period by keeping (a) no weights, (b) known weights at a distance *d* cm and (iii) unknown weights at a distance *d* cm.

2.14 BENDING OF BEAM

A material having long length so that its thickness is negligible compared to its length is called a beam. A beam may be a rectangular bar or a circular rod. A beam can be used for the construction of the roof of a building and the construction of bridges over which a heavy load can pass through it. A simple theory was developed for the bending moment of the beam with the following assumptions:

1. The weight of the beam is negligible compared to the weight of the load.
2. No tangential force is applied.
3. The geometrical moment of inertia is not altered due to the application of the load.
4. The application of the load to the beam produces a small curvature.

2.14.1 Bending Moment of a Beam

Consider a beam ABCD of length l. Consider one end of beam AD is fixed and a load is applied to the other end BC of the beam as shown in Figure 2.11. The upper portion AB of the beam gets elongated, and it experiences an inward force. The lower portion DC of the beam gets shortened, and it experiences an outward force. The centre of the beam EF experiences no increase or decrease in length, and it is called neutral filament. The two opposite forces, force due to the applied load and the restoring force, acting respectively downward and upward at end BC of the beam and hence set up a couple. The total moment of the force acting on the beam due to the compressive force on the lower filament and the elongation force on the upper filament is called the bending moment of the beam. A restoring couple is also acting on the beam to bring it into the original position. At equilibrium the moment of the restoring couple is equal to the bending moment of the beam.

Consider a small portion of beam PBCQ as shown in Figure 2.12 that gets slightly curved due to the applications of the load. Let PB, QC and GF be the outer filament, inner filament and neutral filament respectively. Consider O is the centre of the curvature, R is the distance between the centre of the curvature to the neutral filament, and r is the distance between the neutral filaments to the outer filament PB. From Figure 2.12,

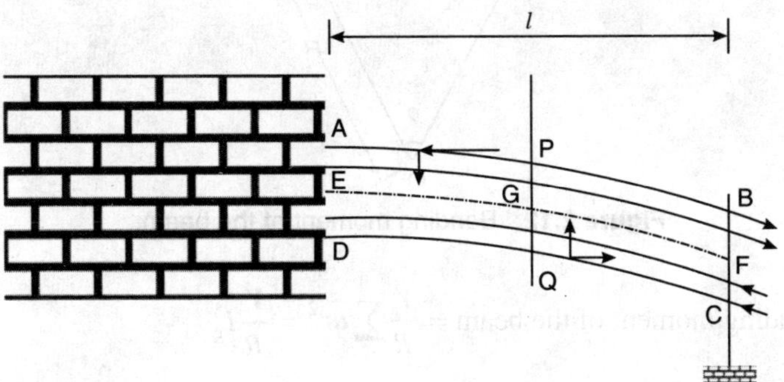

Figure 2.11 Bending of beam.

The length of the neutral filament = $R\phi$, where ϕ is the angle subtended by the neutral filament at the centre. The length of the outer filament, PB = $(R + r)\phi$

$$\text{Linear strain} = \frac{\text{Increase in length}}{\text{Actual length}} \tag{2.93}$$

$$\text{Linear strain} = \frac{(R+r)\phi - R\phi}{R\phi} = \frac{r}{R} \tag{2.94}$$

$$\text{Young's modulus } Y = \frac{\text{Linear stress}}{\text{Linear strain}} \tag{2.95}$$

$$\text{Linear stress} = Y \times \frac{r}{R} \tag{2.96}$$

$$\text{Force} = \text{Stress} \times \text{Area} = Y \times \frac{r}{R} \times a \tag{2.97}$$

where a is the area of cross-section.

$$\text{Moment of the force} = \text{Force} \times \text{Distance} = Y \times \frac{r^2}{R} \times a \tag{2.98}$$

The elongation force is acting in the upper filament, and the compressive force is acting in the lower filament. Therefore, the total moment of the force acting on the upper and the lower filaments of the beam is given by

$$= \frac{Y}{R}\sum ar^2 \tag{2.99}$$

The total moment of the force acting on the beam due to the compressive force on the lower filament and the elongation force on the upper filament is called the bending moment of the beam.

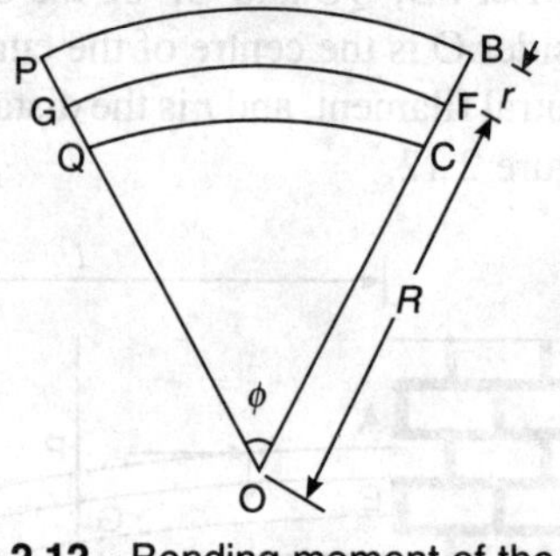

Figure 2.12 Bending moment of the beam.

$$\text{Therefore, the bending moment of the beam} = \frac{Y}{R}\sum ar^2 = \frac{Y}{R}I_g \tag{2.100}$$

where I_g is the geometrical moment of inertia of the beam. The geometrical moment of inertia of the beam is equal to $I_g = Ak^2$, where A is the area of cross-section and k is the radius of gyration.

At equilibrium the bending moment of the beam is equal to the restoring couple acting on the beam. At equilibrium,

$$\text{Bending moment of the beam} = \text{Restoring couple} = \frac{Y}{R}I_g$$

where I_g is the geometrical moment of inertia of the beam.

2.15 CANTILEVER LOADED AT THE FREE END

A beam with long length compared to its thickness, which is fixed at one end and the other end is left free is called a cantilever. Consider a cantilever of length l is fixed at one end, whereas the other end is loaded. Let AB be the neutral axis of the cantilever (Figure 2.13). Due to application of the load, the neutral axis AB gets deflected to AB′. Assume that the weight of the beam does not produce any bending. Consider a section P at a distance of x from the fixed end. The distance PB $\approx$ PB′ = $(1 - x)$.

The bending moment produced at P = $W(l - x)$

The restoring force acting at P = YI_g/R,

where I_g is the geometrical moment of inertia.

At equilibrium, the bending moment and the restoring force are equal. Therefore,

$$W(l-x)=\frac{YI_g}{R} \tag{2.101}$$

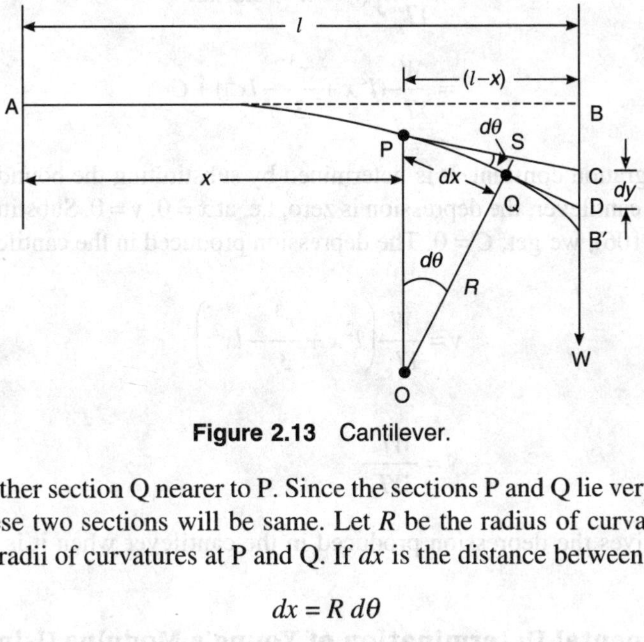

Figure 2.13 Cantilever.

Consider another section Q nearer to P. Since the sections P and Q lie very nearer, the radius of curvature of these two sections will be same. Let R be the radius of curvature and $d\theta$ be the angle between the radii of curvatures at P and Q. If dx is the distance between P and Q, then

$$dx = R\,d\theta$$

Therefore,

$$R=\frac{dx}{d\theta} \tag{2.102}$$

From Eq. (2.101) and Eq. (2.102),

$$W(l-x)=YI_g\frac{d\theta}{dx}$$

$$d\theta=\frac{W(l-x)}{YI_g}dx \tag{2.103}$$

Draw tangents at the points P and Q. Consider these tangents are PC and QD. Let $d\theta$ be the angle between these two tangents. Draw perpendicular line from Q to the line PC. Let CD be the depression dy between the points P and Q. Then,

$$d\theta = \frac{dy}{l - x} \tag{2.104}$$

From Eq. (2.103) and Eq. (2.104), the depression between the points P and Q can be written as

$$dy = \frac{W(l - x)^2}{YI_g} dx \tag{2.105}$$

The depression in the cantilever can be obtained by integrating Eq. (2.105) with respect to x.

$$y = \frac{W}{YI_g} \int (l - x)^2 \, dx$$

$$= \frac{W}{YI_g} \int (l^2 + x^2 - 2lx) dx$$

$$= \frac{W}{YI_g} (l^2 x + \frac{x^3}{3} - lx^2) + C \tag{2.106}$$

where C is the integration constant. It is determined by substituting the boundary conditions. At the fixed end of the cantilever, the depression is zero, i.e, at $x = 0$, $y = 0$. Substituting the boundary condition in Eq. (2.106), we get, $C = 0$. The depression produced in the cantilever is

$$y = \frac{W}{YI_g} \left(l^2 x + \frac{x^3}{3} - lx^2 \right) \tag{2.107}$$

$$y = \frac{Wl^3}{3YI_g} \tag{2.108}$$

Equation (2.108) gives the depression produced in the cantilever when it is loaded at the free end.

2.15.1 Experimental Determination of Young's Modulus Using a Cantilever

One end of the given cantilever is fixed in a wall and the other end is left as free. A pin is fixed at the free end. A load of W kg is suspended at the free end of the cantilever. The position of the tip of the pin is noted using a travelling microscope (Table 2.1). The load is increased in steps of 50 g and the corresponding position of the pin is noted and tabulated. The load is unloaded step by step, and the corresponding positions of the pin are noted using the travelling microscope.

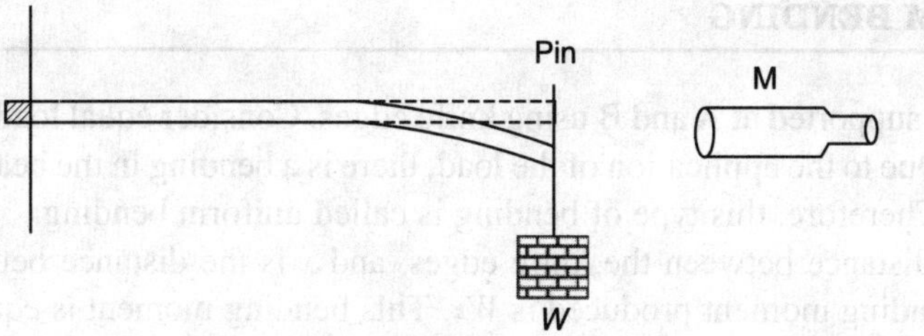

Figure 2.14 Experimental determination of Young's modulus using cantilever.

The Young's modulus of the given cantilever is determined using the relation:

$$Y = \frac{Wl^3}{3yI_g} \tag{2.109}$$

Substituting, $W = mg$, we get

$$Y = \frac{mgl^3}{3yI_g} \tag{2.110}$$

where y is the depression produced, I_g is the geometrical moment of inertia, m is the mass and l is the length of the cantilever. From Table 2.1, the depression y for m kg (say m = 50, g = 0.05 kg) is determined.

The geometrical moment of inertia of the rectangular bar is given by

$$I_g = \frac{bd^3}{12} \tag{2.111}$$

where b is the breadth and d is the thickness of the bar. For a circular rod, the geometrical moment of inertia is given by

$$I_g = \frac{\pi r^4}{4} \tag{2.112}$$

where r is the radius of the rod. By substituting all the values, the Young's modulus of the given cantilever is determined. The Young's modulus of a brass metallic bar or a cylindrical rod is determined using this method.

Table 2.1 Determination of depression of the cantilever

S.No.	*Load in gram*	*Microscope reading during loading*			*Microscope reading during unloading*			*Mean*	*Depression for m kg (m = 50 g)*
		MSR	*VSC*	*TR*	*MSR*	*VSC*	*TR*		
1.	*W*								
2.	*W*+50								
3.	*W*+100								
…	…								
…	…								
…	*W*+400								

2.16 UNIFORM BENDING

Consider a beam is supported at A and B using knife edges. Consider equal loads are applied at the ends of the beam. Due to the application of the load, there is a bending in the beam, and it describes an arc of a circle. Therefore, this type of bending is called uniform bending.

Let l be the distance between the knife edges, and x is the distance between the load and knife edge. The bending moment produced is Wx. This bending moment is equal to the restoring couple, YI_g/R. Therefore,

$$\frac{YI_g}{R} = Wx \tag{2.113}$$

where R is the radius of curvature and I_g is the geometrical moment of inertia. The radius of curvature is determined by drawing a circle. Let AB represents the beam of length l. Let y be the elongation produced and R be the radius of the circle, then from the properties of the circle,

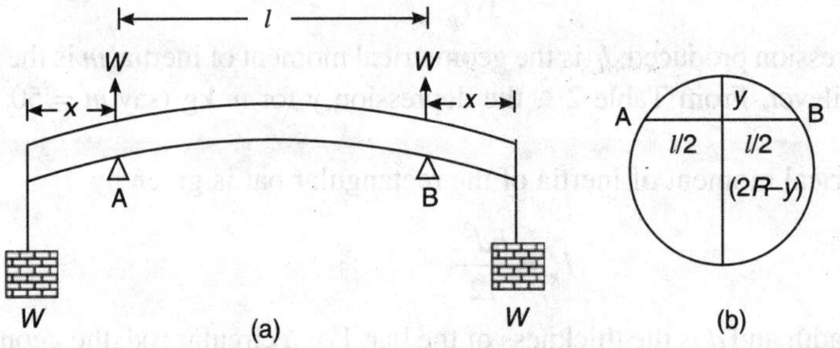

Figure 2.15 Bending of beam (a) uniform bending and (b) determination of elongation.

$$\frac{l}{2} \times \frac{l}{2} = y(2R - y) \tag{2.114}$$

$$\frac{l^2}{4} = 2Ry - y^2$$

Since the depression y is very small, neglecting the values of y^2, we get

$$\frac{l^2}{4} = 2Ry \tag{2.115}$$

i.e.,

$$R = \frac{l^2}{8y} \tag{2.116}$$

Substituting the value of R from Eq. (2.116) in Eq. (2.113), we get

$$y = \frac{Wxl^2}{8YI_g} \tag{2.117}$$

Equation (2.117) gives the elongation produced in the uniform bending. By substituting $I_g = \frac{bd^3}{12}$ in Eq. (2.117) for a rectangular bar, the Young's modulus can be written as

$$Y = \frac{3Wxl^2}{2ybd^3} \tag{2.118}$$

By substituting, $I_g = \frac{\pi r^2}{4}$, for a circular rod, the Young's modulus of a circular rod is given by

$$Y = \frac{Wxl^2}{2y\pi r^4} \tag{2.119}$$

where W ($W = mg$) is the load applied, x is the distance between the knife edge and the end of the circular beam, l is the distance between the knife edges, y is the elongation produced, b is the breadth, d is the thickness of the rectangular bar and r is the radius of the circular rod.

2.16.1 Experimental Determination of Young's Modulus by Uniform Bending

The experimental arrangement is made as shown in Figure 2.16. The given rectangular bar is placed over the knife edges A and B. Let l be the distance of separation between the knife edges. Equal loads (say hanger only) are applied at ends C and D of the rectangular bar. A pin is fixed at the middle of the rectangular bar. The tip of the pin is coincided with the cross wire of the microscope. The position of the pin is noted and is tabulated in Table 2.2. Equal loads (say 50 g) are added to the hangers at ends C and D. The position of the pin tip is again noted. Similarly, by adding equal loads in steps of 50 g to the hangers at ends C and D, the positions of the tip is noted using the travelling microscope. Then the loads are removed in steps of 50 g, and then the positions of the tip during unloading are also noted. The values are tabulated in Table 2.2. From the observation, the elongation produced in the beam for a load of m kg (say 50 g = 0.05 kg) is determined. The breadth, thickness, and the distance between the knife edge l are measured. The Young's modulus of the given beam is determined using the formula:

$$Y = \frac{3Wxl^2}{2ybd^3} \tag{2.120}$$

where W ($W = mg$) is the weight, l is the length, y is the elongation, b is the breadth of the beam, d is the thickness of the beam and x is the distance between the knife edge and the load.

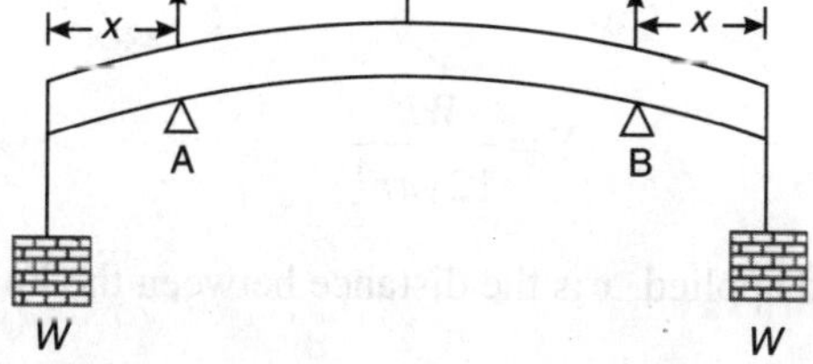

Figure 2.16 Experimental determination of Young's modulus using uniform bending method.

Table 2.2 Determination of elongation of the beam using uniform bending method

S.No.	*Microscope reading during loading*			*Microscope reading during unloading*			*Mean*	*Elongation for m kg (m =50 g)*
	MSR	*VSC*	*TR*	*MSR*	*VSC*	*TR*		
1.								
2.								
...								

2.17 NON-UNIFORM BENDING

Consider a beam is supported at A and B and loaded at the middle. Let l be the distance between A and B. A load of W kg is applied at the centre of beam. This arrangement is similar to two cantilevers fixed at the middle and loaded at the ends. The length of one cantilever is $l/2$ and the reactions at A and B are $W/2$. There is a bending of the cantilever between A and B due to the application of the load at the centre. Since there is no load between the knife edges and the free ends of the cantilever, there is no bending in these regions. Therefore, the bending produced in the entire beam does not describe a circle, and hence this bending is called non-uniform bending.

Substituting, $l = l/2$ and $W = W/2$ in Eq. (2.108), we get

$$\text{Depression produced } y = \frac{\left(\frac{W}{2}\right)\times\left(\frac{l}{2}\right)^3}{3YI_g} \tag{2.121}$$

i.e.,

$$y = \frac{Wl^3}{48YI_g} \tag{2.122}$$

Equation (2.122) gives the depression produced in the non-uniform bending.

By substituting $I_g = \frac{bd^3}{12}$ in Eq. (2.122) for a rectangular bar, the Young's modulus can be written as:

$$Y = \frac{Wl^3}{4ybd^3} \tag{2.123}$$

By substituting, $I_g = \frac{\pi r^4}{4}$ in Eq. (2.124), for a circular rod, the Young's modulus of a circular rod is given by,

$$Y = \frac{Wl^3}{12y\pi r^4} \tag{2.124}$$

where W ($W = mg$) is the load applied, x is the distance between the knife edge and the end of the circular beam, l is the distance between the knife edges, y is the depression produced, b is the breadth, d is the thickness of the rectangular bar and r is the radius of the circular rod.

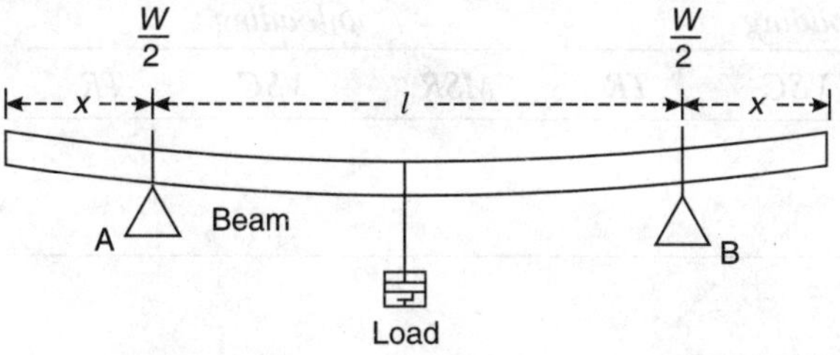

Figure 2.17 Non-uniform bending.

2.17.1 Experimental Determination of Young's Modulus by Non-uniform Bending

The experimental arrangement is shown in Figure 2.18. The given rectangular bar is supported at ends A and B. A hanger W is suspended at the middle of the rectangular bar. A pin is fixed at the middle of the rectangular bar. The tip of the pin is focussed, and it is made to coincide with the vertical cross wire of the microscope. The readings in the main scale and vernier scale of the travelling microscope are noted (Table 2.3). A 50-g weight is added to the hanger. The tip of the pin is again coincided with the vertical cross wire and the readings are noted. Similarly, the weights are added to the hanger in steps of 50 g, and the corresponding readings are noted. Then the

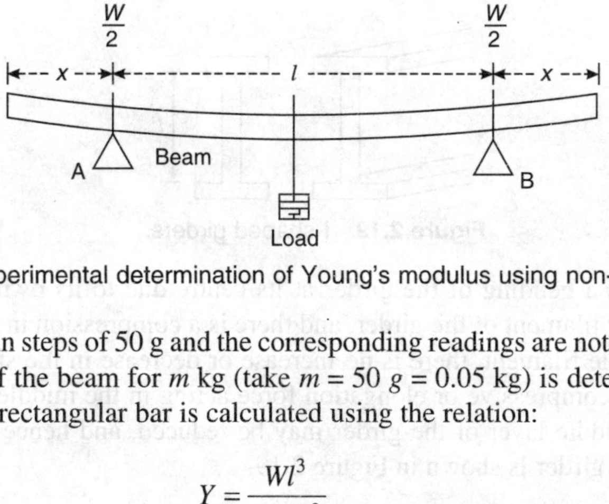

Figure 2.18 Experimental determination of Young's modulus using non-uniform bending.

weights are unloaded in steps of 50 g and the corresponding readings are noted. From the observed data, the depression of the beam for m kg (take m = 50 g = 0.05 kg) is determined. The Young's modulus of the given rectangular bar is calculated using the relation:

$$Y = \frac{Wl^3}{4ybd^3} \tag{2.125}$$

where W ($W = mg$) is the load applied, l is the distance between the knife edges, y is the depression produced, b is the breadth and d is the thickness of the rectangular bar.

Table 2.3 Determination of depression of the beam by non-uniform bending method

S.No.	*Microscope reading during loading*			*Microscope reading during unloading*			*Mean*	*Depression for m* kg (m = 50 g)
	MSR	*VSC*	*TR*	*MSR*	*VSC*	*TR*		
1.								
2.								
...								

2.18 I-SHAPED GIRDERS

Consider a girder is supported at its ends. There may be a bending at the centre of the girder due to its own weight. It is similar to a non-uniform bending. The depression produced in the case of the non-uniform bending for a rectangular bar is

$$y = \frac{Wl^3}{4Ybd^3} \tag{2.126}$$

where W ($W = mg$) is the load applied, l is the length of the girder, y is the depression produced, b is the breadth and d is the thickness of the girder. Equation (2.126) shows that the depression is directly proportional to l^3, inversely proportional to the breadth b and inversely proportional to the cubic power of the thickness, d. This shows that the depression can be minimised by reducing the length l and by increasing the thickness and the breadth of the girder.

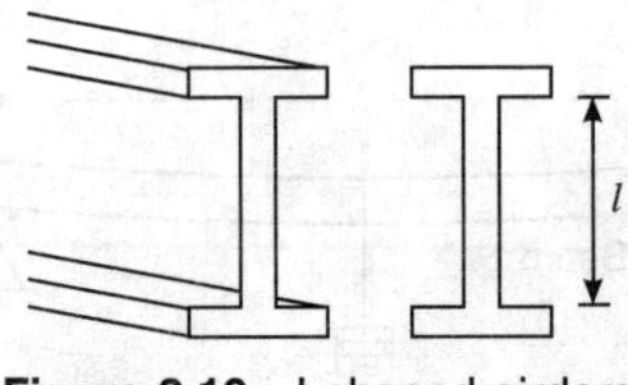

Figure 2.19 I-shaped girders.

Consider there is a bending of the girder at its centre due to its own weight. There is an elongation in the lower filament of the girder, and there is a compression in the upper filament of the girder. In the middle filament, there is no increase or decrease in the size of the filament. It means that there is no compressive or elongation force acting in the middle filament. Therefore, the thickness of the middle layer of the girder may be reduced, and hence the girder looks like I-shaped one. I-shaped girder is shown in Figure 2.19.

Advantages of I-shaped girders

1. I-shaped girder is nearly as strong as a solid steel beam of identical dimensions.
2. I-shaped beam configuration has light weight compared to solid beam configuration of same dimension.
3. The material cost of I-shaped girder is low as compared to the solid steel beam of identical dimensions.

SOLVED PROBLEMS

2.1 Calculate the couple to be applied to twist a wire of length 80 cm fixed at one end and free at the other through 1°. (Radius of wire = 0.03 cm and rigidity modulus of material of wire $=200 \times 10^9$ Pa).

Given data

Length of wire = 80 cm = 0.8 m

Radius of the wire = 0.03 cm = 0.03×10^{-2} m

Angle of twisting, $\theta = 1° = \left(\frac{2\pi}{360} \times 1\right)$ radian

Rigidity modulus, $n = 200 \times 10^9$ Pa

Solution

Couple required to twist the wire, $C = \dfrac{\pi n r^4}{2l}\theta$

$$C = \frac{\pi \times 200 \times 10^9 \times (0.03 \times 10^{-2})^4}{2 \times 0.8} \times \left(\frac{2\pi}{360} \times 1\right)$$

$= 5.55165 \times 10^{-5}$ N m^{-1}

Couple required $C = 5.55165 \times 10^{-5}$ N m^{-1}.

2.2 A 3-m long wire with the area of cross-section 6.25×10^{-5} m^2 is found to stretch 0.3 cm under a load of 1200 kg. What is the Young's modulus of the material of the wire?

Given data

Length of wire $l = 3$ m

Area of cross-section of the wire $A = 6.25 \times 10^{-5}$ m^2

Change in length, $\Delta l = 0.3$ cm $= 0.3 \times 10^{-2}$ m

Load applied $m = 1200$ kg

Solution

Young's modulus $Y = \dfrac{F/A}{\Delta l / l} = \dfrac{mgl}{A \Delta l}$

$$Y = \frac{1200 \times 9.8 \times 3}{6.25 \times 10^{-5} \times 0.3 \times 10^{-2}} = 18.816 \times 10^{10} \text{ N m}^{-2}$$

Young's modulus of the given material $Y = 18.816 \times 10^{10}$ N m^{-2}

2.3 Calculate Poisson's ratio and the rigidity modulus of copper using the following data: Young's modulus of copper is 10.5×10^{10} Nm^{-2} and the bulk modulus of copper is 14.3×10^{10} Nm^{-2}.

Given data

Young's modulus of copper $Y = 10.5 \times 10^{10}$ N m^{-2}

Bulk modulus of copper $K = 14.3 \times 10^{10}$ Nm^{-2}

Solution

Young's modulus, bulk modulus and Poisson's ratio are related as:

$$Y = 3K(1-2\sigma)$$

$$\sigma = \frac{3K - Y}{6K} = \frac{1}{2} - \frac{Y}{6K}$$

Substituting the values of Y and K, we get

$$\sigma = \frac{1}{2} - \frac{10.5\times10^{10}}{6\times14.3\times10^{10}} = 0.3776$$

Poisson's ratio for copper is 0.3776.

2.4 What couple must be applied to a wire, 1-m long, 1-mm diameter, in order to twist one end of it through 90°, the other end remaining fixed? The rigidity modulus is 28×10^{10} N m^{-2}.

Given data

Length of wire = 1 m

Diameter of the wire = 1 mm = 0.001 m

Angle of twisting $\theta = 90° = \frac{\pi}{2}$ radian

Rigidity modulus $n = 28 \times 10^{10}$ N m^{-2}

Solution

Couple required to twist the wire $C = \frac{\pi n r^4}{2l}\theta$

$$C = \frac{\pi\times28\times10^{10}\times(0.5\times10^{-3})^4}{2\times1}\times\frac{\pi}{2}$$

$$= 0.0431795 \text{ N m}^{-1}.$$

Couple required $C = 0.0431795$ N m^{-1}

2.5 A stainless steel wire of length 2 m, area of cross-section 1×10^{-6} m^2 is fixed at one end and a load of 5 kg is applied to the other end of the wire. If the Young's modulus of stainless steel is 18×10^{10} N m^{-2}, calculate the increase in length of the wire.

Given data

Length of the wire $l = 2$ m

Load applied $m = 5$ kg

Area of cross-section $A = 1 \times 10^{-6}$ m^2

Young's modulus $Y = 18 \times 10^{10}$ N m^{-2}

Solution

Young's modulus $Y = \frac{FL}{\Delta l \times A}$

$$\Delta l = \frac{FL}{Y\times A} = \frac{5\times9.8\times2}{18\times10^{10}\times1\times10^{-6}} = 5.444\times10^{-4}\text{ m}$$

Increase in length $\Delta l = 5.444 \times 10^{-4}$ m

2.6 A uniform rigid rod of 1.2 m long is clamped horizontally at one end. A weight of 1 kg is attached to the free end. Calculate the depression of a point 0.9 m distance apart from the clamped end. The diameter of the rod is 2 cm, Y for the material of the rod is 1.013×10^{10} N m^{-2}.

Given data

Length of the rod, $l = 1.2$ m

Load applied, $m = 1$ kg

Depression point = 1.2 – 0.9 = 0.3 m

Diameter of the rod, $d = 2$ cm

Young's modulus, $Y = 1.013 \times 10^{10}$ N-m^{-2}

Solution

Geometrical moment of inertia, $$I_g = \frac{\pi r^4}{4} = \frac{\pi \times [1 \times 10^{-2}]^4}{4} = 7.854 \times 10^{-9} \text{ m}^4$$

Depression of the rod, $$y = \frac{W}{YI_g}\left(l^2 x + \frac{x^3}{3} - lx^2\right)$$

Substituting $l = 1.2$ m and $x = 0.9$ m in the above equation, we get

Depression of the rod, $$y = \frac{0.567W}{YI_g}$$

$$= \frac{0.567 \times 1 \times 9.8}{1.013 \times 10^{10} \times 7.854 \times 10^{-9}} = 0.0698 \text{ m}$$

The depression at a distance of 0.9 m from the fixed end = 6.98 cm

2.7 A solid ball of 2.30 m in diameter is submerged in a lake at such a depth that the pressure exerted by water is 10^4 Nm^{-2}. Find the change in volume of the ball (K for the material = 10×10^{10} N m^{-2}).

Given data

Diameter of the ball, $d = 2.30$ m

Pressure exerted $P = 10^4$ N m^{-2}

Bulk modulus $K = 10 \times 10^{10}$ N m^{-2}

Solution

Bulk modulus $$K = \frac{-PV}{\Delta V}$$

Change in volume $$\Delta V = \frac{-PV}{K} = \frac{-10^4 \times \frac{4}{3}\pi \times 1.15^3}{10 \times 10^{10}} = -6.37 \times 10^{-7} \text{ m}^{-3}$$

The change in the volume of the ball, $\Delta V = -6.37 \times 10^{-7}$m^{-3}. The negative sign shows that the volume of the ball decreases.

2.8 **A rectangular metal bar of 2 cm side, 40 cm long and mass 1 kg is suspended by a wire of 30 cm long and 0.05 cm radius. It makes 50 oscillations in 320 second. Find the rigidity modulus of the wire.**

Given data

Thickness of the metal bar = 2 cm = 2×10^{-2} m
Length of the metal bar = 40 cm = 0.4 m
Mass of the metal bar, M = 1 kg
Length of the wire, l = 30 cm = 0.3 m
Radius of the wire, r = 0.05 cm = 5×10^{-4} m
Time taken by the metal bar, t = 320 s
Number of oscillation = 50

Solution

$$\text{Time period of SHM, } T = 2\pi\sqrt{\frac{I}{C}} \tag{i}$$

$$\text{Squaring the above equation, } T_2 = 4\pi^2 \frac{I}{C} \tag{ii}$$

$$\text{From the above equation, } C = \frac{4\pi^2 I}{T^2} \tag{iii}$$

Substituting, $C = \dfrac{n\pi r^4}{2l}$ in the above equation, we get,

$$n = \frac{8\pi I l}{r^4 T^2} \tag{iv}$$

$$\text{Moment of inertia of metal bar } I = M\left(\frac{l^2 + b^2}{12}\right)$$

$$I = 1 \times \left(\frac{0.4^2 + 0.02^2}{12}\right) = 1.3366 \times 10^{-2} \text{ kg m}^2$$

$$\text{Time period of the pendulum, } T = \frac{\text{Time taken}}{\text{Total number of oscillations}} = \frac{320}{50} = 6.4 \text{ s}$$

From Eq. (iv),
$$n = \frac{8\pi I l}{r^4 T^2}$$

Substituting the values of I, l, r and T, we get,

$$n = \frac{8\pi \times 1.3366 \times 10^{-2} \times 0.3}{(5 \times 10^{-4})^4 \times 6.4^2} = 3.936 \times 10^{10} \text{ Nm}^{-2}$$

The rigidity modulus of the metal bar $n = 3.936 \times 10^{10}$ Nm^{-2}

SHORT QUESTIONS

1. What is meant by elasticity?
2. What is stress? What is its unit?
3. Define the term strain.
4. State Hooke's law.
5. What is yield stress?
6. What is breaking stress?
7. What are the factors affecting the elastic properties of a material?
8. Define Young's modulus.
9. Define bulk modulus.
10. Define rigidity modulus.
11. What is Poisson's ratio? What are its limiting values?
12. Write the relation between Young's modulus and bulk modulus.
13. Write the relation between rigidity modulus and Young's modulus.
14. Write the relations between Young's modulus, bulk modulus and rigidity modulus.
15. Write the relation between bulk modulus and rigidity modulus in terms of Poisson's ratio.
16. What is torsional pendulum?
17. Mention the uses of the torsional pendulum.
18. What is meant by bending of beam?
19. What is a cantilever?
20. What is meant by uniform bending?
21. What is meant by non-uniform bending?
22. What is an I-shaped girder? What are its uses?

DESCRIPTIVE TYPE QUESTIONS

1. (i) Derive the relation between bulk modulus and rigidity modulus
   (ii) Deduce the relation between rigidity modulus and Young's modulus.
2. Derive a mathematical expression for the twisting couple in a wire.
3. Derive an expression for the time period of a torsional pendulum, and hence derive the mathematical expression for the rigidity modulus of the wire.
4. Derive an expression for the determination of (i) the moment of inertia of the disc and (ii) moment of inertia of an irregular object using torsional pendulum.
5. Derive an expression for the depression produced in a cantilever.

6. Describe an experimental method for the determination of Young's modulus using cantilever.
7. (i) What is uniform bending? Derive an expression for the Young's modulus of a beam using uniform bending.
   (ii) Describe an experimental method for the determination of Young's modulus of a rectangular beam using uniform bending.
8. (i) What is non-uniform bending? Derive an expression for the Young's modulus of a beam using non-uniform bending.
   (ii) Describe an experimental method for the determination of Young's modulus of a rectangular beam using non-uniform bending.

PROBLEMS

1. A wire of length 100 cm and diameter 1 mm is clamped at one of its ends. Calculate the couple required to twist the other end of wire by an angle of 9°. Given that the rigidity modulus of the wire is 5.6×10^{10} N m^{-2}. **(Ans** $C = 8.6359 \times 10^{-4}$ N m^{-1})
2. A bar, 1-m long, 5-mm square in cross-section, supported horizontally at its ends and loaded at the middle is depressed 1.96 mm by a load of 100 g. Calculate Young's modulus for the material of the bar (Take $g = 980$ cm/s^2). **(Ans** $Y = 20 \times 10^{10}$ Nm^{-2})
3. A rectangular bar 2 cm in breadth and 1 cm in depth and 1 m in length is supported at the ends and a load of 2 kg is applied at the middle. Calculate the depression of the beam. Young's modulus of the material of the beam is 20×10^{10} Nm^{-2}. **(Ans** $y = 1.225 \times 10^{-6}$ m)
4. A wire of length 1.6 m and diameter 0.32 mm elongates by 1 mm when stretched by a force of 0.33 kg. wt. and twists through 1 radian when equal and opposite torques of 1.45×10^{-5} N m are applied at its end. Find Poisson's ratio for the wire. Given, rigidity modulus of the wire, $n = 2.36 \times 10^{10}$ N m^{-2}. **(Ans** $\sigma = 0.39$)
5. An aluminum wire of length 3 m, and radius 1.76 mm is subjected to a load of 2.3 kg. It produces an elongation of 0.1 mm in length. Calculate the Young's modulus of aluminum. **(Ans** $Y = 6.948 \times 10^{10}$ Nm^{-2})
6. In an experiment, a rod of diameter 1.26 cm was supported on two knife edges, placed 0.7 m apart. On applying a load of 900 g exactly midway between the knife edges, the depression on the middle point was observed to be 0.025 cm. Calculate the Young's modulus of the material of the rod. **(Ans** $Y = 20.376 \times 10^{10}$ Nm^{-2})
7. What couple must be applied to a wire of 1-m long and 2-mm in diameter in order to twist one end of its end through 45°, when the other remains fixed. Given, $n = 5 \times 10^{10}$ N m^{-2}. **(Ans** $C = 6.1685 \times 10^{-2}$ N m^{-1})
8. Calculate Poisson's ratio and the rigidity modulus of silver from the following data:
   Young's modulus of silver = 7.25×10^{10} N/m^2
   Bulk modulus of silver wire = 11×10^{10} N/m^2 **(Ans** $\sigma = 0.39$)

CHAPTER 3

THERMAL PHYSICS

3.1 INTRODUCTION

Thermal physics is the branch of science that deals with the transmission of heat. Whenever there is a temperature difference between two bodies, heat is transmitted from the body with higher temperature to the body with lower temperature. The transmission of heat takes place in three different processes, namely, conduction, convection and radiation. This chapter deals about the basic concepts of physics such as Newton's law of cooling, experimental methods used to find thermal conductivity and transmission of heat through compound media.

3.2 MODES OF HEAT TRANSFER

Heat can be transmitted from one place to another by the following three methods: (i) conduction (ii) convection and (iii) radiation.

Conduction is the process of transfer of heat energy without actual movement of the particles. Consider a rod is heated at one end and the other end is left free. The heat is transmitted from the hot end to the cold end. Whenever the rod is heated, the molecules of the rod acquire some energy from heat and hence vibrate about their equilibrium position. During vibration, the molecules collide with the nearby molecules and hence transfers heat energy. This type of transmission of heat is called conduction. The conduction of heat mostly occurs in solids.

Convection is the process of transmission of heat by the movement of molecules. In the case of convection, there is a transfer of molecules from one place to another and hence heat is transmitted. The convection takes place in liquids and gases. Consider water in a beaker is heated. The water that lies at the bottom of the beaker gets heated easily and hence its density decreases. Therefore, the hot water moves up and the cold water moves down and it gets heated. This process of transmission of heat is known as convection. Sea breeze, land breeze and trade winds are produced by convection.

Radiation is the process of the transmission of heat from one place to another place without any medium. The transmission of heat by sunlight is one of the examples for radiation. The radiation can flow even through vacuum.

3.3 THERMAL CONDUCTIVITY

Consider a cube as shown in Figure 3.1. Let A be the area of cross-section of the face ABCD. Let surfaces ABCD and EFGH are at θ_1°C and θ_2°C. Let θ_1°C > θ_2°C. Let the thickness of the cube is x. Heat flows from the hot surface to the cold surface.

The quantity of heat energy Q flowing in one second is:

- directly proportional to the area of cross-section A
- directly proportional to the temperature difference $(\theta_1 - \theta_2)$
- inversely proportional to the thickness x

Combining these three, the quantity of heat flowing in one second is given by

$$Q \propto \frac{A(\theta_1 - \theta_2)}{x} \tag{3.1}$$

By introducing a proportionality constant, Eq. (3.1) can be written as:

$$Q = \frac{KA(\theta_1 - \theta_2)}{x} = -KA\frac{d\theta}{dx} \tag{3.2}$$

where A is the area of cross-section, K is the proportionality constant and it is known as thermal

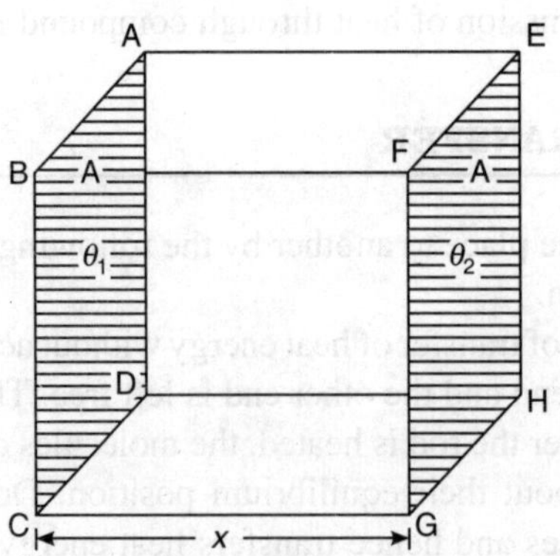

Figure 3.1 Thermal conductivity.

conductivity of the material. The negative sign shows that the temperature decreases with the increase of distance. The thermal conductivity of the material is written as:

$$K = \frac{Q}{A\left(\frac{\theta_1 - \theta_2}{x}\right)} = -\frac{Q}{A\frac{d\theta}{dx}} \qquad (3.3)$$

where $(\theta_1-\theta_2)/x$ is the ratio between the changes in temperature per unit thickness. It is represented as $d\theta/dx$ and is called temperature gradient, A is the area of cross-section and Q is the quantity of heat energy flowing in one second.

Consider, $A = 1\ \text{m}^2$, $d\theta/dx = 1\ \text{K m}^{-1}$, then $K = Q$. The term thermal conductivity is defined as the ratio of the quantity of heat energy flowing in one second through unit area of cross-section to the temperature gradient. The unit for the thermal conductivity is W m^{-1} K^{-1}.

Temperature gradient

The ratio between the change in temperature and thickness is called the temperature gradient. Its unit is Km^{-1}.

$$\text{Temperature gradient} = \frac{\theta_1 - \theta_2}{x} = \frac{d\theta}{dx}$$

Thermal capacity

It is defined as the quantity of heat required to raise the temperature of the whole body through 1 °C.

$$\text{Thermal capacity} = \text{Specific heat} \times \text{mass of the substance} = S \times m$$

where S is the specific heat and m is the mass of the substance. Its unit is J K^{-1}.

Thermal diffusivity

It is defined as the ratio of thermal conductivity of the material to thermal capacity per unit volume of the material. It is given by

$$\text{Thermal diffusivity } h = \frac{K}{\rho S}$$

where K is the thermal conductivity, S is the specific heat and ρ is the density of the material. Its unit is $\text{m}^2\ \text{s}^{-1}$.

Specific heat capacity

It is the quantity of heat energy required to raise the temperature of 1 kg of a substance through 1 K. It is also called as specific heat. Its unit is J kg^{-1} K^{-1}.

3.4 NEWTON'S LAW OF COOLING

Newton's law of cooling states that the rate of loss of heat of a substance is directly proportional to the difference between the temperature of the substance and that of the surroundings.

Consider θ_1 and θ_2 are the temperatures of the substance and the surroundings respectively. Let dQ/dt is the rate change of heat. According to the Newton's law of cooling,

$$-\frac{dQ}{dt} \propto (\theta_1 - \theta_2) \tag{3.4}$$

The negative sign indicates the loss of heat of the substance. Equation (3.4) can be written as:

$$-\frac{dQ}{dt} = K(\theta_1 - \theta_2) \tag{3.5}$$

where K is the proportionality constant.

Let m be the mass of the substance and S is the specific heat of the substance, then the quantity of heat available in the substance is:

$$Q = mSd\theta \tag{3.6}$$

Substituting Eq. (3.6) in Eq. (3.5), we get

$$-mS\frac{d\theta}{dt} = K(\theta_1 - \theta_2) \tag{3.7}$$

$$-\frac{d\theta}{dt} = \frac{K}{mS}(\theta_1 - \theta_2) \tag{3.8}$$

$$-\frac{d\theta}{dt} = C(\theta_1 - \theta_2) \tag{3.9}$$

where C is a constant and it is equal to $C = \dfrac{K}{mS}$.

Equation (3.9) can be written as:

$$-\frac{d\theta}{\theta_1 - \theta_2} = Cdt$$

Integrating the above equation, we get

$$\ln(\theta_1 - \theta_2) = -Ct + \ln C_1$$

where $\ln C_1$ is an integration constant. If a graph is plotted between $\ln(\theta_1 - \theta_2)$ and time, a straight line, as shown in Figure 3.2, is obtained. This straight line behaviour verifies the Newton's law of cooling.

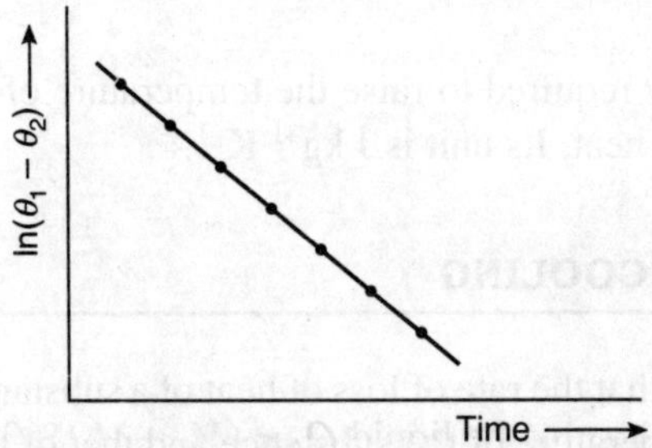

Figure 3.2 Newton's law of cooling.

3.4.1 Specific Heat Capacity Determination

Suppose an empty copper calorimeter with stirrer is taken, and it is weighed. Let M is the mass of the calorimeter. The calorimeter is filled up to nearly 70% of its volume by water at a temperature of 80°C (θ_1°C). The calorimeter is kept inside a constant temperature enclosure. The water is stirred continuously using the stirrer and the time taken by water for every successive one degree fall in temperature is noted until the temperature of water reaches 50°C (θ_2°C). The calorimeter is weighed with water and stirrer. Let it be M_2. Consider t_1 is the time taken by the water to cool from 80°C to 50°C.

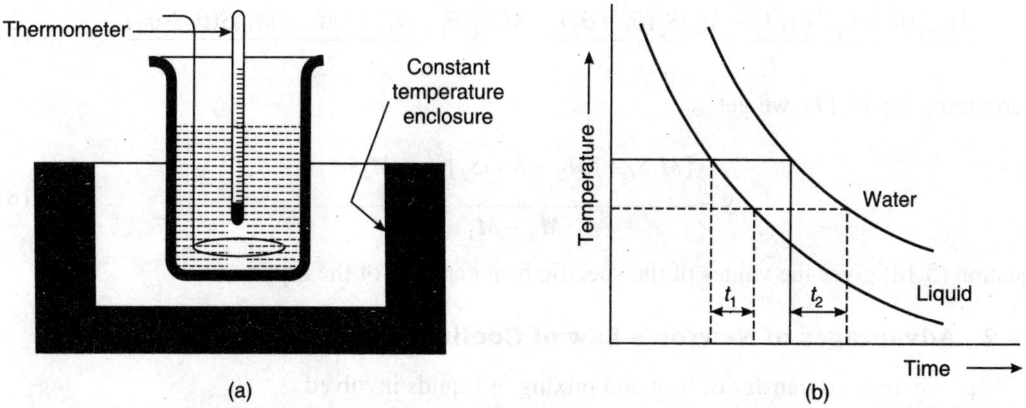

Figure 3.3 Specific heat capacity, (a) calorimeter and constant enclosure and (b) cooling curve for water and liquid.

The calorimeter is emptied, dried, and it is filled nearly 70% of its volume with the given liquid at 80°C. The calorimeter is again kept inside a constant temperature enclosure. The liquid is stirred using the stirrer, and time taken by the liquid for every successive one degree fall in temperature is noted until 50°C. The calorimeter with stirrer and liquid is weighed, and it is taken as M_3. Consider t_2 is the time taken by the liquid to cool from 80°C to 50°C.

$$\text{Quantity of heat lost by the calorimeter } Q_1 = M_1S_1(\theta_1 - \theta_2) \tag{3.10}$$

where S_1 is the specific heat capacity of the calorimeter.

$$\text{Quantity of heat lost by the water } Q_2 = (M_2 - M_1)S_2(\theta_1 - \theta_2) \tag{3.11}$$

where S_2 is the specific heat capacity of the water.
Total quantity of heat lost by the water and calorimeter

$$Q_3 = M_1S_1(\theta_1 - \theta_2) + (M_2 - M_1)S_2(\theta_1 - \theta_2) \tag{3.12}$$

Rate of heat lost by water and calorimeter

$$Q = \frac{Q_3}{t_1} = \frac{M_1S_1(\theta_1 - \theta_2) + (M_2 - M_1)S_2(\theta_1 - \theta_2)}{t_1} \tag{3.13}$$

$$\text{Quantity of heat lost by the liquid } Q_4 = (M_3 - M_1)S(\theta_1 - \theta_2) \tag{3.14}$$

where S is the specific heat capacity of the liquid.

Total quantity of heat lost by the liquid and calorimeter,

$$Q_5 = M_1S_1(\theta_1 - \theta_2) + (M_3 - M_1)S(\theta_1 - \theta_2) \tag{3.15}$$

Rate of heat lost by water and calorimeter

$$Q = \frac{Q_5}{t_2} = \frac{M_1S_1(\theta_1 - \theta_2) + (M_3 - M_1)S(\theta_1 - \theta_2)}{t_2} \tag{3.16}$$

Since the liquid and water are cooled in the same environment, the rate of cooling for liquid and water is same. Equating, Eq. (3.13) and Eq. (3.16), we get

$$\frac{M_1S_1(\theta_1 - \theta_2) + (M_2 - M_1)S_2(\theta_1 - \theta_2)}{t_1} = \frac{M_1S_1(\theta_1 - \theta_2) + (M_3 - M_1)S(\theta_1 - \theta_2)}{t_2} \tag{3.17}$$

Rearranging Eq. (3.17), we get

$$S = \frac{[M_1S_1 + (M_2 - M_1)S_2]\frac{t_2}{t_1} - M_1S_1}{M_3 - M_1} \tag{3.18}$$

Equation (3.18) gives the values of the specific heat capacity of the liquid.

3.4.2 Advantages of Newton's Law of Cooling Method

1. There is no transfer of heat and mixing of liquids involved.
2. No cooling correction is required.
3. The specific heat of the given liquid can be determined at various ranges of temperatures.

3.4.3 Disadvantages of Newton's Law of Cooling Method

1. It is difficult to maintain the surrounding temperature constant.
2. This method cannot be used for solids, since the temperatures of the inner and outer portions of the solids are different.
3. This method is not accurate, and it gives approximate value.
4. It is difficult to find the temperature of the rapidly cooling liquid, when its temperature is rapidly decreasing.
5. A slight variation in conditions between water and liquid produces enormous error.

3.4.4 Verification of Newton's Law of Cooling

Cooling curves are drawn between the temperature and time for water and the liquid. A plot, as shown in Figure 3.3(b), can be obtained. The temperature of the surrounding is noted. Let it be θ_0°C. From the graph, the time is noted for a particular temperature θ_1 for water (or liquid) from Figure 3.3(b). Then the value of $\ln(\theta_1 - \theta_0)$ is noted. Similarly, the time corresponding to different temperatures are noted and the respective values of $\ln(\theta - \theta_0)$ is noted. A graph is plotted between $\ln(\theta - \theta_0)$ and time. A straight line curve, as shown in Figure 3.4, is obtained. The straight line behaviour verifies the Newton's law of cooling.

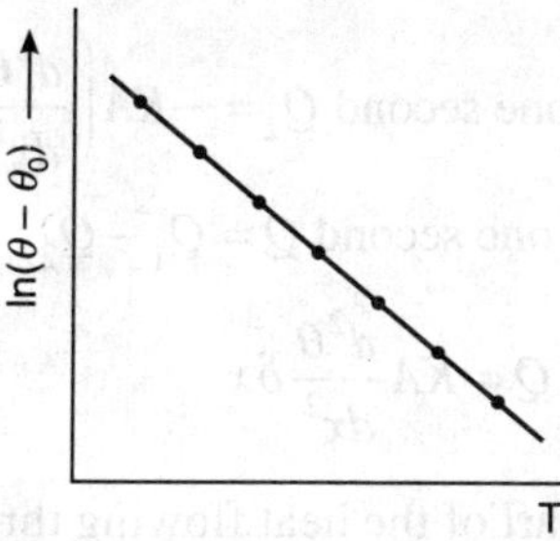

Figure 3.4 Verification Newton's law of cooling.

3.5 RECTILINEAR FLOW OF HEAT (LINEAR FLOW OF HEAT)

A bar AB of uniform area of cross-section is heated at one end A as shown in Figure 3.5. The heat flows from end A to end B. Consider two parallel planes P and Q, such that P is at a distance x from end A, and Q is at a distance $x + \delta x$ from end A. Consider the excess temperature more than the surroundings at plane P is θ. Let $\frac{d\theta}{dx}$ is the temperature gradient at P. $\theta + \frac{d\theta}{dx}\delta x$

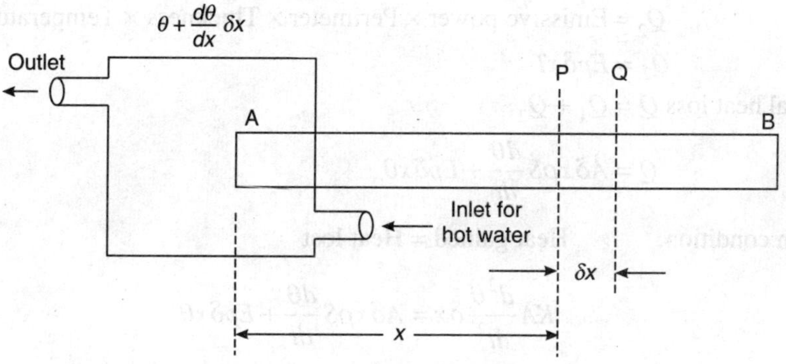

Figure 3.5 Rectilinear flow of heat.

$$\text{Excess temperature at } P = \theta$$

$$\text{Temperature gradient at the plane } P = \frac{d\theta}{dx} \tag{3.19}$$

$$\text{Excess of temperature at } Q = \frac{d\theta}{dx}\delta x + \theta \tag{3.20}$$

$$\text{Temperature gradient at } Q = \frac{d}{dx}\left(\frac{d\theta}{dx}\delta x + \theta\right) \tag{3.21}$$

$$\text{Heat flowing through P in one second } Q_1 = -KA\frac{d\theta}{dx} \tag{3.22}$$

Heat flowing through Q in one second $Q_2 = -KA\left(\frac{d^2\theta}{dx^2}\delta x + \frac{d\theta}{dx}\right)$ (3.23)

Heat gained by portion PQ in one second $Q = Q_1 - Q_2$ (3.24)

$$Q = KA\frac{d^2\theta}{dx^2}\delta x \quad (3.25)$$

Before reaching the steady state, some part of the heat flowing through PQ is radiated, and the remaining part is used to heat portion PQ. The quantity of heat energy used to heat the portion PQ is

$$Q_1 = \text{Mass} \times \text{Specific heat} \times \text{Rate of change of heat}$$

$$Q_1 = A\delta x\rho \times S \times \frac{d\theta}{dx} \quad (3.26)$$

where A is the area of cross-section, δx is the distance between P and Q, ρ is the density, S is the specific heat of the material and $\frac{d\theta}{dx}$ is the rate of change of temperature.

Consider E is the emissive power of the material, p is the perimeter, then the quantity of heat radiated is given by

$$Q_2 = \text{Emissive power} \times \text{Perimeter} \times \text{Thickness} \times \text{Temperature}$$

$$Q_2 = Ep\delta xT \quad (3.27)$$

Total heat loss $Q = Q_1 + Q_2$

$$Q = A\delta x\rho S\frac{d\theta}{dt} + Ep\delta x\theta \quad (3.28)$$

At equilibrium condition, Heat gained = Heat lost

$$KA\frac{d^2\theta}{dt^2}\delta x = A\delta x\rho S\frac{d\theta}{dt} + Ep\delta x\theta \quad (3.29)$$

Divide both sides of Eq. (3.29) by $KA\delta x$

$$\frac{d^2\theta}{dx^2} = \frac{\rho s}{K}\frac{d\theta}{dt} + \frac{Ep}{KA}\theta \quad (3.30)$$

This is the general equation for the rectilinear flow of heat through a bar of uniform area of cross-section.

Case I: When the heat loss due to the radiation is negligible

When the radiation loss is negligible, $E = 0$ Eq. (3.30) can be written as

$$\frac{d^2\theta}{dx^2} = \frac{\rho S}{K}\frac{d\theta}{dt} \quad (3.31)$$

$$\frac{d^2\theta}{dx^2} = \frac{1}{h}\frac{d\theta}{dt} \tag{3.32}$$

where $h = \dfrac{K}{\rho S}$ is known as thermal diffusivity.

Case II: After the steady state is reached

Consider the rod AB has reached the steady state and hence there is no further change of temperature, i.e., $\dfrac{d\theta}{dt} = 0$ in Eq. (3.30), we get

$$\frac{d^2\theta}{dx^2} = \frac{EP}{KA}\theta \tag{3.33}$$

By taking, $\mu^2 = \dfrac{EP}{KA}$, Eq. (3.33) can be written as

$$\frac{d^2\theta}{dx^2} - \mu^2\theta = 0 \tag{3.34}$$

The solution for Eq. (3.34) can be written as,

$$\theta = Ae^{\mu x} + Be^{-\mu x} \tag{3.35}$$

where A and B are constants.

(c) *When an infinite length bar is used*

Consider the length of the bar is infinity. The excess temperature over the surrounding at end B is zero. The boundary conditions are (i) when $x = 0$, $\theta = \theta_0$ and (ii) $x = \infty$, $\theta = 0$. Substituting the boundary condition $x = 0$, $\theta = \theta_0$ in Eq. (3.35), we get $\theta_0 = A + B$. When $x = \infty$, $\theta = 0$, we get $Ae^{\infty} + Be^{-\infty} = 0$ and $A = 0$ and $B = \theta_0$. Equation (3.35) can be written as:

$$\theta = \theta_0 e^{-\mu x} \tag{3.36}$$

Equation (3.36) indicates that the temperature of the bar is exponentially decreasing.

(d) *When the length of the bar is finite*

Consider a bar with finite length L. The boundary conditions are (i) at $x = 0$, $\theta = \theta_0$ and (ii) at $x = L$, $\dfrac{d\theta}{dx} = 0$. Substituting $x = 0$ and $\theta = \theta_0$ in Eq. (3.35), we get

$$A + B = \theta_0 \tag{3.37}$$

Differentiating Eq. (3.35), with respect to x, we get

$$\frac{d\theta}{dx} = A\mu e^{\mu x} - B\mu e^{-\mu x} \tag{3.38}$$

Substitute, $x = L$ and $\dfrac{d\theta}{dx} = 0$ in Eq. (3.38), we get

$$Ae^{\mu L} = Be^{-\mu L} \tag{3.39}$$

From the above equation, the value of B can be written as:

$$B = Ae^{2\mu L} \tag{3.40}$$

From Eq. (3.37), $A = \theta_0 - B$. Substitute the value of A in Eq. (3.40), we get

$$B = \frac{\theta_0}{1 + e^{-2\mu L}} \tag{3.41}$$

Substituting the value of B from Eq. (3.41) in Eq. (3.37), we get

$$A = \frac{\theta_0}{1 + e^{2\mu L}} \tag{3.42}$$

The solution for Eq. (3.30) can be written as:

$$\theta = \theta_0 \left[\frac{e^{\mu x}}{1 + e^{2\mu L}} + \frac{e^{-\mu x}}{1 + e^{-2\mu L}} \right] \tag{3.43}$$

Equation (3.43) gives the temperature of the bar, when the length of the bar is finite.

3.6 LEE'S DISC METHOD

This experimental method is also called Lee's and Charlton's disc method. It is one of the experimental methods used to determine the thermal conductivity of bad conductors such as cardboard, glass plate, etc. This method is based on the principle that at equilibrium the quantity of heat energy conducted through the bad conductor is equal to the quantity of heat radiated by the disc to the atmosphere.

The bad conductor is placed in between the Lee's disc and the steam chamber. The thermometers are inserted into the disc and steam chamber respectively into the holes provided. The entire set up is suspended in a stand as shown in Figure 3.6.

The steam is passed through the steam chamber until the temperature of the disc and the steam chamber becomes constant. These temperatures are called steady state temperatures. The steady state temperatures of the steam chamber and disc are noted as θ_1 and θ_2 respectively. The cardboard is removed and hence the steam chamber is directly placed over the disc. The disc is heated 5°C above its steady temperature [i.e., up to $(\theta_2 + 5)$°C]. Then the disc is suspended in air, and it is allowed to cool. The time for every one degree fall in temperature is noted until the temperature reaches $(\theta_2 - 5)$°C. A graph is plotted between the temperatures θ and time t, and a graph, as shown in Figure 3.7, is obtained. From the graph, the slope at θ_2°C is determined.

The thermal conductivity of the bad conductor is determined using the relation,

$$K = mS\frac{d\theta}{dt} \times \frac{r + 2l}{2r + 2l} \times \frac{d}{\pi r^2} \times \frac{1}{\theta_1 - \theta_2} \tag{3.44}$$

where m, r, l and S are the mass, radius, thickness and specific heat capacity of the disc respectively, θ_1 and θ_2 are the steady state temperatures of the steam chamber and the disc respectively, d is the thickness of the bad conductor and $\frac{d\theta}{dt}$ is the rate of cooling.

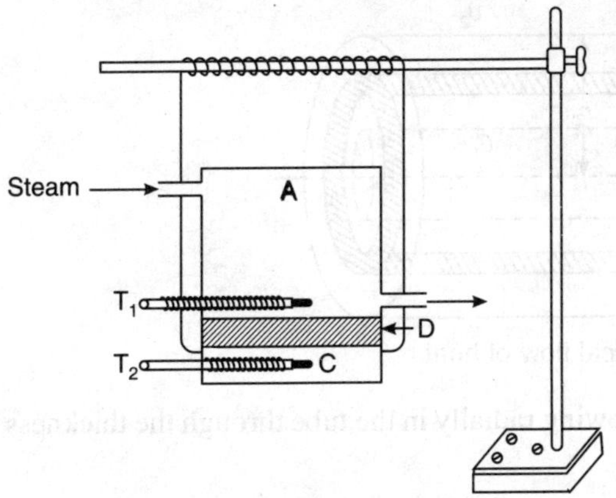

Figure 3.6 Lee's disc experiments.

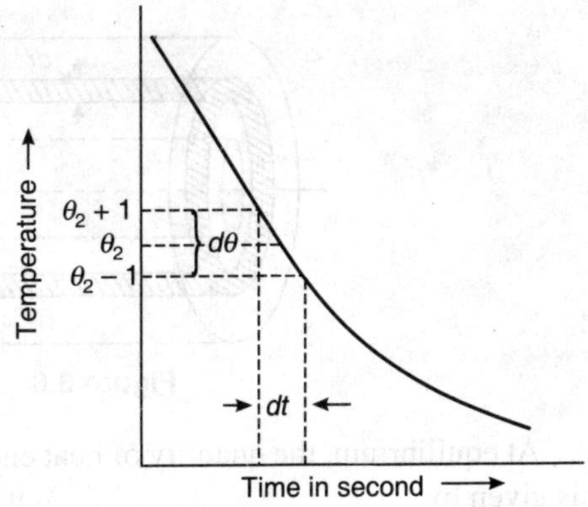

Figure 3.7 Determination of $\frac{d\theta}{dt}$.

At equilibrium,

Quantity of heat conducted per second by the bad conductor
= Quantity of heat radiated by the disc to the surroundings

$$\frac{KA(\theta_1-\theta_2)}{d}=mS\frac{d\theta}{dt}\times\frac{\text{Area of the disc exposed to the surroundings}}{\text{Area of the disc}}$$

$$\frac{KA(\theta_1-\theta_2)}{d}=mS\frac{d\theta}{dt}\times\frac{A+S}{2A+S} \qquad (3.45)$$

where A is the area of the upper or bottom surface of the disc ($A = \pi r^2$) and S is the area of the curved surface ($S = 2\pi rl$).

$$\frac{K\pi r^2(\theta_1-\theta_2)}{d}=mS\frac{d\theta}{dt}\times\frac{\pi r^2+2\pi rl}{2\pi r^2+2\pi rl}$$

$$K=mS\frac{d\theta}{dt}\times\frac{r+2l}{2r+2l}\times\frac{d}{\pi r^2}\times\frac{1}{\theta_1-\theta_2} \qquad (3.46)$$

By substituting the values of m, S, r, d, l, θ_1, θ_2, and $\frac{d\theta}{dt}$, the value of K is determined using Eq. (3.46).

3.7 CYLINDRICAL FLOW OF HEAT

Consider a cylindrical tube of length l. Let r_1 and r_2 be the inner and outer radii of the tube as shown in Figure 3.8. Consider an element of thickness dr at a distance r. Consider steam is passed through the tube for a long time. Let θ_1 be the temperature of the steam. Let θ_1 and θ_2 be the temperatures of the inner and outer portions of the tube respectively.

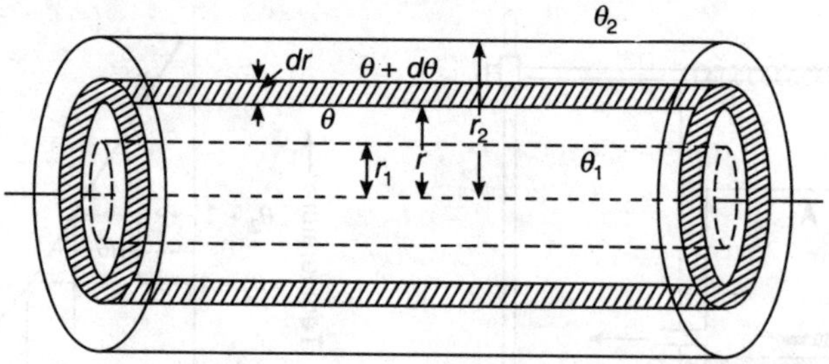

Figure 3.8 Cylindrical flow of heat.

At equilibrium, the quantity of heat energy flowing radially in the tube through the thickness *dr* is given by

$$Q = -KA\frac{d\theta}{dr} \tag{3.47}$$

Substituting, $A = 2\pi rl$, where r is the radius and l is the length, we get

$$Q = -K \times 2\pi rl\frac{d\theta}{dr} \tag{3.48}$$

Rearranging the above equation, we get,

$$Q\frac{dr}{r} = -K \times 2\pi l d\theta$$

Integrating, we get

$$Q\int_{r_1}^{r_2}\frac{dr}{r} = -K \times 2\pi l\int_{\theta_1}^{\theta_2} d\theta$$

$$K = \frac{Q\ln\left(\frac{r_2}{r_1}\right)}{2\pi l(\theta_1 - \theta_2)} \tag{3.49}$$

or

$$K = \frac{Q \times 2.3026 \times \log\left(\frac{r_2}{r_1}\right)}{2\pi l(\theta_1 - \theta_2)} \tag{3.50}$$

where r_1 and r_2 are the inner and outer radii of the rubber tube, Q is the quantity of heat flowing per second radially through the tube, θ_1 and θ_2 are the temperatures of the inner and outer portions of the tube and l is the length of the tube. By substituting the values of r_1, r_2, Q, θ_1, θ_2 and l the thermal conductivity of the cylindrical tube that conducts the heat radially is determined.

3.8 THERMAL CONDUCTIVITY OF RUBBER

The thermal conductivity of rubber tube is determined using the principle of the cylindrical flow of heat.

Consider a calorimeter of mass m_1 and specific heat capacity S_1. Nearly sixty per cent volume of the calorimeter is filled with water, and then the mass of the calorimeter with water is measured. From this, the mass of the water is determined. Let m_2 be the mass of the water and S_2 be its specific heat capacity. Let r_1 and r_2 be the inner and outer radii of the given rubber tube. A known length l (say, 50 cm) of the rubber tube is immersed in water as shown in Figure 3.9. A thermometer T_1 is inserted to measure the temperature of water. Let the initial temperature of water is θ_3°C.

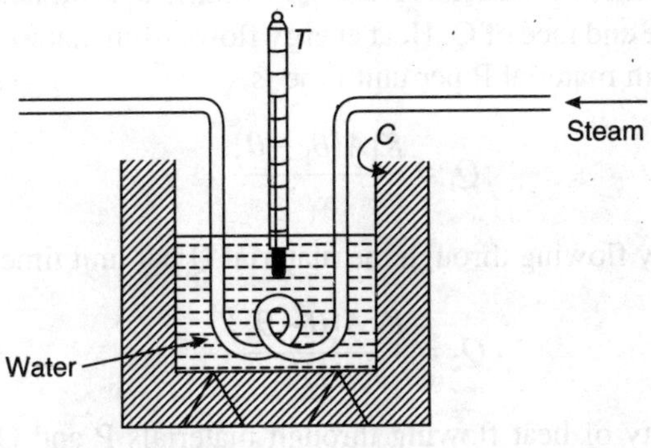

Figure 3.9 Thermal conductivity of rubber.

Consider steam is passed through the rubber tube for a known period of time (say, 15 minutes, i.e., 900 seconds). Let it be t second. The temperature of water is noted. Let it be θ_4°C. Let the temperature of the steam is θ_1°C. The average temperature of the outer portion of the rubber tube is

$$\theta_2 = \frac{\theta_3 + \theta_4}{2}$$

Quantity of heat gained by water $Q = (m_1S_1 + m_2S_2)(\theta_4 - \theta_3)$

$$\text{Thermal conductivity of the rubber tube } K = \frac{Q \times 2.3026 \times \log\left(\frac{r_2}{r_1}\right)}{2\pi l(\theta_1 - \theta_2)}$$

Substituting the values of Q and θ_2, we get,

$$K = \frac{(m_1S_1 + m_2S_2)(\theta_4 - \theta_3) \times 2.3026 \times \log\left(\frac{r_2}{r_1}\right)}{2\pi l\left(\theta_1 - \frac{\theta_3 + \theta_4}{2}\right)} \tag{3.51}$$

Substituting the values of m_1, m_2, S_1, S_2, θ_1, θ_3, θ_4, l, r_1, and r_2, the thermal conductivity of the rubber tube is determined using Eq. (3.51).

3.9 CONDUCTION OF HEAT THROUGH A COMPOUND MEDIA

3.9.1 Bodies in Series

Consider a compound wall is made up of two different materials (or slabs) P and Q of thermal conductivities K_1 and K_2 respectively as shown in Figure 3.10. Let x_1 and x_2 be the thicknesses of these two materials. Consider A is the area of cross-section of the end faces of these two materials P and Q. Let θ_1 and θ_2 be the temperatures of the end faces of the materials P and Q. Consider θ is the temperature of the interface of materials P and Q. Assume $\theta_1 > \theta_2$, i.e., the end face of P is at a higher temperature than the end face of Q. Heat energy flows from material P to Q. The quantity of heat energy flowing through material P per unit time is

$$Q_1 = \frac{K_1 A(\theta_1 - \theta)}{x_1} \tag{3.52}$$

The quantity of heat energy flowing through the material Q per unit time is:

$$Q_2 = \frac{K_2 A(\theta - \theta_2)}{x_2} \tag{3.53}$$

At equilibrium, the quantity of heat flowing through materials P and Q per unit time is equal. Equating Eq. (3.52) and Eq. (3.53), we get

$$\frac{K_1 A(\theta_1 - \theta)}{x_1} = \frac{K_2 A(\theta - \theta_2)}{x_2} \tag{3.54}$$

Simplifying Eq. (3.54), we get

$$\frac{K_1}{x_1}\theta_1 - \frac{K_1}{x_1}\theta = \frac{K_2}{x_2}\theta - \frac{K_2}{x_2}\theta_2$$

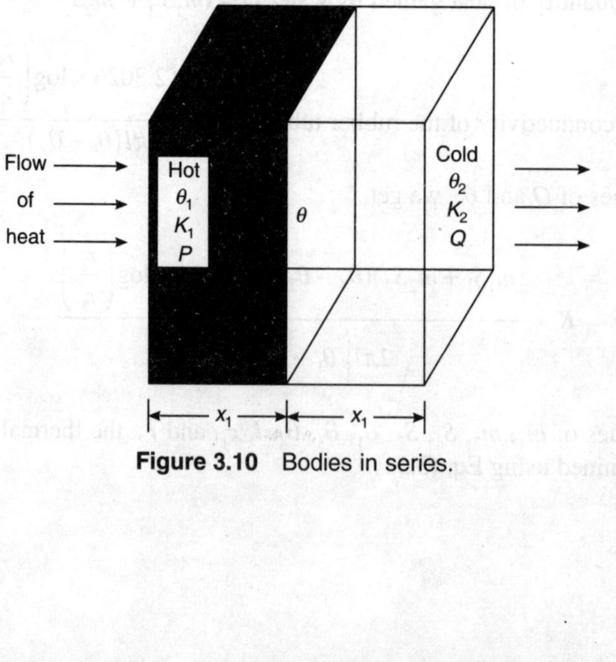

Figure 3.10 Bodies in series.

The temperature of the interface is given by

$$\theta = \frac{\frac{K_1}{x_1}\theta_1 + \frac{K_2}{x_2}\theta_2}{\frac{K_1}{x_1} + \frac{K_2}{x_2}} \tag{3.55}$$

$$\theta = \frac{\sum_{i=1}^{n} \frac{K_i}{x_i}\theta_i}{\sum_{i=1}^{n} \frac{K_i}{x_i}} \tag{3.56}$$

The quantity of heat flowing through P and Q per unit time is obtained by substituting Eq. (3.55) in Eq. (3.52) or in Eq. (3.53), we get

$$Q = \frac{K_1 A}{x_1}\left(\theta_1 - \frac{\frac{K_1\theta_1}{x_1} + \frac{K_2\theta_2}{x_2}}{\frac{K_1}{x_1} + \frac{K_2}{x_2}}\right) = \frac{K_1 A}{x_1}\left(\frac{\frac{K_1\theta_1}{x_1} + \frac{K_2\theta_1}{x_2} - \frac{K_1\theta_1}{x_1} - \frac{K_2\theta_2}{x_2}}{\frac{K_1}{x_1} + \frac{K_2}{x_2}}\right) = \frac{K_1 A}{x_1}\left(\frac{\frac{K_2}{x_2}(\theta_1 - \theta_2)}{\frac{K_1}{x_1} + \frac{K_2}{x_2}}\right)$$

$$Q = \frac{K_1 K_2 A}{x_1 x_2}\left(\frac{\theta_1 - \theta_2}{\frac{K_1}{x_1} + \frac{K_2}{x_2}}\right)$$

Multiplying and dividing the above equation by $\frac{x_1 x_2}{K_1 K_2}$, we get

$$Q = \frac{A(\theta_1 - \theta_2)}{\frac{x_1}{K_1} + \frac{x_2}{K_2}} \tag{3.57}$$

$$Q = \frac{A(\theta_1 - \theta_2)}{\sum_{i=1}^{n} \frac{x_i}{K_i}} \tag{3.58}$$

Equation (3.58) gives the quantity of heat flowing through a compound media arranged in series.

3.9.2 Bodies in Parallel

Consider two different materials (or slabs) P and Q of area of cross-sections A_1 and A_2 are arranged parallel as shown in Figure 3.11. Assume that the thermal conductivity of these two materials as K_1 and K_2 respectively. Let x_1 and x_2 be the thicknesses of materials P and Q respectively. Consider θ_1 and θ_2 be the temperatures of the end faces of P and Q respectively. Consider the temperature of surface P is greater than that of Q, i.e., $\theta_1 > \theta_2$. Therefore, heat flows from surface P to surface Q.

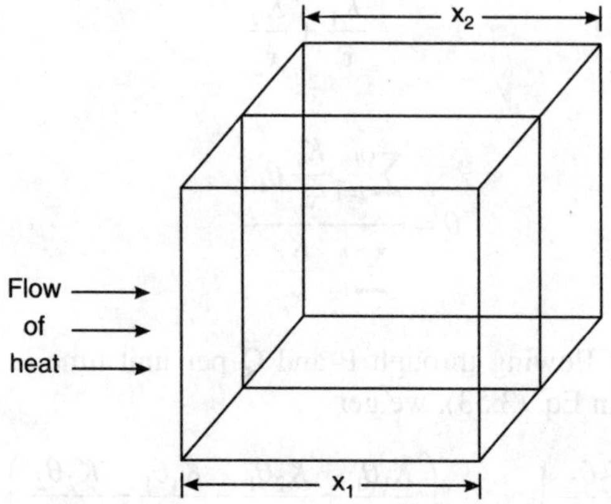

Figure 3.11 Bodies in parallel.

The quantity of heat flowing through material P from one end to another end in time t second is

$$Q_1 = \frac{K_1 A_1 (\theta_1 - \theta_2)}{x_1} t \tag{3.59}$$

The quantity of heat energy flowing through material Q from one end to another end at t second is

$$Q_2 = \frac{K_2 A_2 (\theta_1 - \theta_2)}{x_2} t \tag{3.60}$$

The total amount of heat flow is given by

$$Q = Q_1 + Q_2$$

$$Q = \frac{K_1 A_1}{x_1}(\theta_1 - \theta_2)t + \frac{K_2 A_2}{x_2}(\theta_1 - \theta_2)t \tag{3.61}$$

Then Rate of heat flow $= \dfrac{Q}{t} = \dfrac{K_1 A_1}{x_1}(\theta_1 - \theta_2) + \dfrac{K_2 A_2}{x_2}(\theta_1 - \theta_2)$ (3.62)

$$= \sum_{i=1}^{n} \frac{K_i A_i}{x_i}(\theta_1 - \theta_2) \tag{3.63}$$

Equation (3.63) gives the rate of heat flow through the bodies (i.e., compound wall or slab) arranged in parallel, whenever the heat flow takes place in a similar condition.

SOLVED PROBLEMS

3.1 A Cu rod 0.19-m long and 785×10^{-7} m^2 area of cross-section, which is thermally insulated, is heated at one end through 100°C, while the other end is kept at 30°C. Calculate the amount of heat flow in 10 minutes along the way if the thermal conductivity of Cu is 380 W m^{-1}K^{-1}.

Given data

Length of Cu rod x = 0.19 m

Area of cross-section $A = 785 \times 10^{-7}$ m^2

Temperature of one end T_1 = 100°C = 373 K

Temperature of another end T_2 = 30°C = 303 K

Time t = 10 minutes = 600 s

Thermal conductivity of Cu = 380 W m^{-1} K^{-1}

Solution

Quantity of heat flowing in t s, $Q = \dfrac{KA(T_1 - T_2)}{x} t$

$$Q = \frac{380 \times 785 \times 10^{-7} \times (373 - 303)}{0.19} \times 600 = 6594 \text{ J}$$

Quantity of heat flowing through the Cu rod in 10 minutes, Q = 6594 J.

3.2 In an experiment performed to find the thermal conductivity of cardboard by Lee's disc method, the following observations are made.

Mass of the disc = 800 g

Steady state temperature of the upper disc = 99.5°C

Steady state temperature of the lower disc = 83.5°C

Time taken for the upper disc to cool from 86°C to 81°C = 4 min

Thickness of the cardboard disc = 4.8 mm

Thickness of the upper disc = 1 cm

Radius of the disc = 6 cm

Material of the disc is copper. Find the thermal conductivity of cardboard.

Given data

Mass of the disc = 800 g = 0.8 kg

Steady state temperature of the upper disc = 99.5°C = 372.5 K

Steady state temperature of the lower disc = 83.5°C = 356.5 K

Time taken for the upper disc to cool from 86°C to 81°C = 4 min

Rate of cooling $\dfrac{(359 - 354)}{4 \times 60} = 0.0208 \text{ K s}^{-1} = \text{K s}^{-1}$

Thickness of the cardboard disc = 4.8 mm = 4.8×10^{-3} m

Thickness of the upper disc = 1 cm = 1×10^{-2} m

Radius of the disc = 6 cm = 6×10^{-2} m

Solution

Thermal conductivity $K = mS\dfrac{d\theta}{dt}\times\dfrac{r+2l}{2r+2l}\times\dfrac{d}{\pi r^2}\times\dfrac{1}{\theta_1-\theta_2}$

$$K = 0.8\times 385\times 0.0208\times\frac{6\times10^{-2}+2\times1\times10^{-2}}{2\times6\times10^{-2}+2\times1\times10^{-2}}\times\frac{4.8\times10^{-3}}{\pi\times(6\times10^{-2})^2}\times\frac{1}{372.5-356.5}$$

$K = 0.0971$ W m^{-1} K^{-1}.

Thermal conductivity of the given bad conductor = 0.0971 W m^{-1} K^{-1}

3.3 A liquid takes 5 minutes to cool from 80°C to 50°C. How much time will it take when it cools from 50°C to 40°C? The temperature of the surrounding is 30°C.

Given data

In the first case, initial temperature θ_1 = 80°C = 353 K
In the first case, final temperature θ_2 = 50°C = 323 K
In the second case, initial temperature θ_1 = 50°C = 323 K
In the second case, final temperature θ_2 = 40°C = 313 K
Temperature of the surrounding θ_0 = 30°C = 303 K

Solution

According to Newton's law of cooling

$$\int_{\theta_1}^{\theta_2}\frac{d\theta}{\theta-\theta_0} = -C\int dt$$

$$\ln\frac{\theta_2-\theta_0}{\theta_1-\theta_0} = -Ct$$

$$\ln\frac{\theta_1-\theta_0}{\theta_2-\theta_0} = Ct$$

When the liquid cools from 353 K to 323 K,

$$\ln\frac{353-303}{323-303} = C\times5\times60 \qquad \text{(i)}$$

When the liquid cools from 323 K to 313 K,

$$\ln\frac{323-303}{313-303} = C\times t \qquad \text{(ii)}$$

Solving Eq. (i) and Eq. (ii), we get t = 226.9 s.
The liquid will cool in 226.9 s from 50°C to 40°C.

3.4 A body takes 3 minutes to cool from 70°C to 60°C. What will be its temperature after the next 5 minutes? The temperature of the surrounding is 30°C.

Given data

In the first case, initial temperature θ_1 = 70°C = 343 K
In the first case, final temperature θ_2 = 60°C = 333 K

Temperature of the surroundings, $\theta_0 = 30°C = 303$ K
Time taken, $t = 5$ minutes = 300 s

Solution
According to Newton's law of cooling

$$\int_{\theta_1}^{\theta_2} \frac{d\theta}{\theta - \theta_0} = -C\int dt$$

$$\ln\frac{\theta_1 - \theta_0}{\theta_2 - \theta_0} = Ct$$

When the liquid cools from 343 K to 333 K,

$$\ln\frac{343 - 303}{333 - 303} = C \times 3 \times 60 \quad \text{(i)}$$

When the liquid cools from 333 K to θ K,

$$\ln\frac{333 - 303}{\theta - 303} = C \times 5 \times 60 \quad \text{(ii)}$$

Solving Eq. (i) and Eq. (ii), we get

$$300 \times (\ln 40 - \ln 30) = 180(\ln 30 - \ln(\theta - 303))$$

Solving the above equation, we get $\theta = 321.573$ K
The liquid will cool from 333 K to 321.57 K in the next five minutes.

3.5 A liquid of volume V kept in a vessel takes 120 s to cool from 60°C to 50°C. The same volume of water takes 300 s to cool through the same range of temperature. Determine the specific heat of the liquid. Given the specific heat of water is 4200 J kg^{-1} K^{-1}, mass of water is 100 g, and mass of the calorimeter is 100 g mass of the liquid is 85 g and specific heat of the calorimeter is 385 J kg^{-1} K^{-1}.

Given data
Initial temperature $\theta_1 = 60°C = 333$ K
Final temperature $\theta_2 = 50°C = 323$ K
Time taken by the water and calorimeter $t_1 = 300$ s
Time taken by the liquid and calorimeter $t_2 = 120$ s
Mass of calorimeter $M = 100$ g = 0.1 kg
Mass of water $M_1 = 100$ g = 0.1 kg
Mass of the liquid $M_2 = 100$ g = 0.1 kg
Specific heat of calorimeter $S = 385$ J kg^{-1} K^{-1}
Specific heat of water $S = 4200$ J kg^{-1} K^{-1}

Solution

$$\text{Rate of cooling for calorimeter and water} = \frac{MS(T_1 - T_2) + M_1S_1(T_1 - T_2)}{t_2}$$

$$\text{Rate of cooling for calorimeter and liquid} = \frac{MS(T_1 - T_2) + M_2S_2(T_1 - T_2)}{t_2}$$

At equilibrium,

$$\frac{MS(T_1 - T_2) + M_1S_1(T_1 - T_2)}{t_1} = \frac{MS(T_1 - T_2) + M_2S_2(T_1 - T_2)}{t_2}$$

$$S_2 = \frac{(MS + M_1S_1)(T_1 - T_2)\frac{t_2}{t_1} - MS(T_1 - T_2)}{M_2(T_1 - T_2)}$$

$$S_2 = \frac{(MS + M_1S_1)\frac{t_2}{t_1} - MS}{M_2}$$

$$S_2 = \frac{(0.1 \times 385 + 0.1 \times 4200)\frac{120}{300} - 0.1 \times 385}{0.085}$$

$$= 1704.7 \text{ J kg}^{-1}\text{K}^{-1}$$

Specific heat of the liquid $S_2 = 1704.7$ J kg^{-1} K^{-1}.

3.6 Equal bars of Cu and Al are welded end to end and logged. If the free ends of Cu and Al are maintained at 100°C and 0°C respectively, find the temperature of the welded interface. Assume the thermal conductivities of Cu and Al to be 386 W m^{-1}K^{-1} and 220 W m^{-1} K^{-1} respectively.

Given data

Temperature of one end of Cu θ_1 = 100°C=373 K

Temperature of one end of Al θ_2 = 0°C = 273 K

Thermal conductivity of copper K_1 = 386 W m^{-1} K^{-1}

Thermal conductivity of aluminium K_2 = 220 W m^{-1} K^{-1}

Solution

$$\text{Temperature of the interface } \theta = \frac{\frac{K_1\theta_1}{d_1} + \frac{K_2\theta_2}{d_2}}{\frac{K_1}{d_1} + \frac{K_2}{d_2}}$$

Given, the lengths of the Cu bar and Al bar is equal. Therefore, $d_1 = d_2$.

$$\theta = \frac{K_1\theta_1 + K_2\theta_2}{K_1 + K_2}$$

Substituting the values of K_1, K_2, θ_1 and θ_2, we get

$$\theta = \frac{386 \times 373 + 220 \times 273}{386 + 220}$$

$$\theta = 336.696 \text{ K}$$

Temperature of the interface θ =336.696 K

3.7 **Heat is conducted through a compound plate composed of two parallel plates of different materials A and B of thermal conductivities 0.32 W m^{-1}K^{-1} and 0.11 Wm^{-1}K^{-1} and each of thickness 3.6 cm and 4.2 cm respectively. If the temperatures of the outer faces of the A and B are found to be steady at 96°C and 0°C respectively, find the temperature of interface A/B. Assume the area of cross-section of the materials A and B is same.**

Given data

Thermal conductivity of material A K_1 = 0.32 W m^{-1}K^{-1}
Thermal conductivity of material B K_2 = 0.11 W m^{-1}K^{-1}
Thickness of the material A x_1 = 3.6 cm = 3.6 × 10^{-2} m
Thickness of the material B x_2 = 4.2 cm = 4.2 × 10^{-2} m
Temperature of end A, θ_1= 96°C = 369 K
Temperature of end B, θ_2 = 0°C = 273 K

Solution

Let θ be the temperature of the interface. Then heat conducted through A in one second is:

$$Q_1 = K_1 A_1 \frac{\theta_1 - \theta}{x_1}$$

$$Q_1 = 0.32 \times A \frac{369 - \theta}{0.036}$$

Heat conducted through B in one second $Q_2 = K_2 A_2 \dfrac{\theta - \theta_2}{x_1}$

$$Q_2 = 0.11 \times A \times \frac{\theta - 273}{0.042}$$

As both are same, we have, $Q_1 = Q_2$

$$0.32 \times A \frac{369 - \theta}{0.036} = 0.11 \times A \frac{\theta - 273}{0.042}$$

$$\frac{369 - \theta}{\theta - 273} = \frac{0.11 \times 0.036}{0.32 \times 0.042}$$

$$\theta = 347.15 \text{ K}$$

Temperature of the interface = 347.15 K

SHORT QUESTIONS

1. What is meant by conduction of heat?
2. What is meant by convection?
3. What is radiation of heat?
4. Define the term thermal conductivity.
5. What is temperature gradient?
6. Define the term thermal capacity.
7. Define the term thermal diffusion.
8. State Newton's law of cooling.
9. What are the advantages of Newton's law of cooling method?
10. What are the disadvantages of Newton's law of cooling method?
11. How will you verify Newton's law of cooling?
12. Mention the principle of Lee's disc method.
13. What is meant by cylindrical flow of heat?
14. What is the principle behind the experimental determination of thermal conductivity of rubber?
15. Write the mathematical expression for the quantity of heat conducted through the bodies that are arranged in series and explain the terms.
16. Write the mathematical expression used to find the quantity of heat conducted through two bodies arranged in parallel.

DESCRIPTIVE TYPE QUESTIONS

1. State Newton's law of cooling. Derive a mathematical expression for the specific heat of the given liquid using Newton's law of cooling method. What are the advantages and drawbacks of this method?
2. Derive a mathematical expression for the rectilinear flow of heat.
3. Explain with neat sketch the experimental determination of the thermal conductivity of a bad conductor using Lee's disc method.
4. What is cylindrical flow of heat? Explain the experimental method used to find the thermal conductivity of rubber.
5. Deduce the mathematical expression for the transmission of heat through two bodies arranged in (i) series and (ii) parallel.

PROBLEMS

1. The thermal conductivity of a bad conductor is determined using Lee's disc experiment. Determine the thermal conductivity of the material using the following data:

   Mass of Lee's disc = 900 g
   Radius of disc = 6 cm
   Thickness of the disc = 1.2 cm
   Thickness of the cardboard = 1 mm
   Steady state temperature of the steam chamber = 99°C
   Steady state temperature of the Lee's disc = 80°C
   Rate of cooling = 0.033 K^{-1} s^{-1}
   Specific heat capacity of the disc S = 385 J kg^{-1} K^{-1}

   (**Ans** $K = 0.03104$ W m^{-1} K^{-1})

2. One end of a plate is heated by steam at 100°C and another end is kept at 50°C. If the area of cross-section of the face of the plate is 100 cm^2 and its thickness is 3 cm, calculate the quantity of heat flowing through the plate in 1 second. Given S = 385 J kg^{-1} K^{-1}.

   (**Ans** $Q = 6416.66$ J)

3. Thermal conductivity of Cu is four times that of brass. Two rods of Cu and brass of same length and cross-section are joined end to end. The free end of Cu is at 0°C and that of brass rod is at 100°C. Calculate the temperature of the junction. Neglect radiation losses.

   (**Ans** $\theta = 293$ K)

4. A liquid takes 7 minutes to cool from 60°C to 40°C. What will be its temperature after the next 7 minutes? The temperature of the surroundings is 20°C. (**Ans** $\theta = 294$ K)

5. A body takes 6 minutes to cool from 80°C to 50°C. How much time will it take to cool from 60°C to 30°C? The temperature of the surrounding is 20°C. (**Ans** $t = 720$ s)

6. A slab consists of two parallel layers of different materials (Cu and Al) 4-cm and 2-cm thick and of thermal conductivities 386 W m^{-1} K^{-1} and 220 $Wm^{-1}K^{-1}$ respectively. If the opposite faces of the slab are at 100°C and 0°C, calculate the temperature of the surface dividing the two. (**Ans** $\theta = 319.73$ K)

7. A bar of copper of length 75 cm and a bar of iron of length 125 cm are joined together end to end. Both are of circular cross-section with 2 cm diameter. The free end of the copper and iron are maintained at 100°C and 0°C respectively. The surfaces of the bars are thermally insulated. What is the temperature of the copper–iron junction? What is the heat energy transmitted per unit time across the junction? Thermal conductivity of copper is 386 W m^{-1} K^{-1} and that of iron is 46.2 W m^{-1} K^{-1}. (**Ans** $\theta = 295.59$ K, $Q = 52$ J)

CHAPTER 4

QUANTUM PHYSICS

4.1 INTRODUCTION

The classical theory successfully explained the motion of an object that is visible to our eyes or observable by the microscope. But it fails to explain the behaviour of atomic particles such as electron. It fails to explain the spectral lines emitted by the hydrogen spectrum. It also fails to explain the stability of atoms. In the case of atoms, the electrons are revolving in the extranuclear space. According to the electromagnetic theory of radiation, a revolving charged particle radiates energy in the form of electromagnetic radiations. If a revolving electron emits a radiation, then the energy of the electron decreases continuously and it will collapse with the nucleus. This shows the instability of the atom. But the experimental results show that the atoms are stable.

At the same time, it also fails to explain several new discoveries made during the nineteenth century. The new discoveries like black body radiation, photoelectric effect, Zeeman effect, Compton effect, absorption and emission of light cannot be explained by classical physics. Therefore, in order to explain the spectrum produced by the black body, Max Planck proposed a new theory in 1900 and it is called quantum theory of radiations.

4.2 BLACK BODY

Consider a metal is heated to a high temperature. First the metal becomes red hot, then it becomes yellowish and then it becomes white. The colour of the metal changes with the increase in temperature and at the same time the metal emits a radiation. Similarly, a black body also emits a radiation, from shorter wavelength to a longer wavelength, when it gets heated. At thermal equilibrium, the rate of absorption and the rate of emission of a black body are same. A black body absorbs the entire radiations incident on it regardless of frequency. **A black body is defined as an object that appears black and emits all kinds of radiation when it gets heated.** In the laboratory, a hollow object with a very small hole leading to its interior is used as a black body. Any radiation incident on it enters the cavity through the hole, where it is trapped by reflection back and forth until it is absorbed.

4.2.1 Spectrum of Black Body Radiation

The black body spectra are shown in Figure 4.1. It has the following characteristics:

(i) The black body emits all kinds of radiation ranging from lower wavelength to higher wavelength.

(ii) The black body spectrum shows that the energy density increases with the increase in wavelength and reaches a maximum value and then decreases with the increase in wavelength.

(iii) The wavelength corresponding to the maximum energy density gets shifted towards lower wavelengths, with the increase of temperature.

(iv) If the temperature of the black body is increased, the energy density also increases.

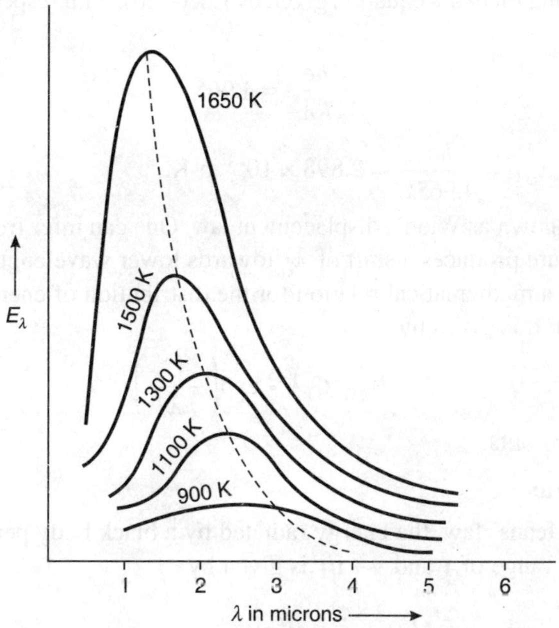

Figure 4.1 Black body spectra.

4.2.2 Laws of Black Body Radiations

Several laws were proposed to explain the properties of black body radiation. They are, namely Kirchhoff's law, Stefan–Boltzmann law, Wien's displacement law and Rayleigh–Jeans' law.

Kirchhoff's law

The Kirchhoff's law states that the ratio between the emissive power and the absorption coefficient of any object at a particular wavelength and at a constant temperature is equal to the emissive power of a perfect black body at that wavelength and temperatures, i.e.

$$\frac{e_\lambda}{a_\lambda} = E_\lambda \tag{4.1}$$

where e_λ and a_λ are the emissive power and the absorption coefficient of any object at a particular wavelength and at a constant temperature and E_λ is the emissive power of a perfect black body at that wavelength and temperatures.

Stefan–Boltzmann law

This law states that the quantity of the energy radiated per second per unit area of cross-section (Q) of a black body is directly proportional to the fourth power of the absolute temperature.

$$Q = \sigma T^4 \tag{4.2}$$

where σ is known as Stefan's constant and it is equal to 5.67×10^{-8} W m^{-2}K^{-4}.

Wien's displacement law

Wien's displacement law states that the wavelength corresponding to the maximum energy density of a black body radiation is inversely proportional to the absolute temperature. The value of λ_m is found out by differentiating Planck's equation given by Eq. (4.20), with respect to λ and substituting $\lambda = \lambda_m$, we get,

$$\frac{hc}{kT\lambda_m} = 4.965 \tag{4.3}$$

From Eq. (4.3), we get, $\lambda_m T = \dfrac{hc}{4.965k} = 2.898 \times 10^{-3}$ m-K.

Equation (4.3) is known as Wien's displacement law. One can infer from Equation (4.3) that the increase in temperature produces a shift of λ_m towards lower wavelengths.

Wien also derived a mathematical relation for the distribution of energy per unit volume of the black body radiation. It is given by

$$\rho_\lambda = C_1 \lambda^{-5} \exp\left(-\frac{C_2}{\lambda T}\right) \tag{4.4}$$

where C_1 and C_2 are constants.

Rayleigh–Jeans' law

According to Rayleigh–Jeans' law, the energy radiated by a black body per unit volume of cross section in the frequency range of ν and $\nu + d\nu$ is given by

$$\rho_\nu d\nu = \frac{8\pi kT}{c^3}\nu^2 d\nu \tag{4.5}$$

In terms of wavelength, Eq. (4.5) can be written as

$$\rho_\lambda d\lambda = \frac{8\pi kT}{c^3}\frac{c^2}{\lambda^2}\left|\frac{c}{\lambda^2}\right| d\lambda$$

i.e. $$\rho_\lambda = \frac{8\pi kT}{\lambda^4} \tag{4.6}$$

Equation (4.5) or Eq. (4.6) is known as Rayleigh–Jeans' law.

4.2.3 Planck's Quantum Theory of Black Body Radiation

In order to explain the black body radiations, Planck proposed some postulates. They are

(i) A chamber that emits the black body radiations contains a number of harmonic oscillators at molecular dimension. The oscillators vibrate in all possible frequency.

(ii) The oscillator emits radiations and the frequency of the radiation emitted is equal to the frequency of vibration of the oscillator.

(iii) The oscillator emits discrete energy, i.e. the energy emitted by the oscillators is not continuous. The energy is in the order of $h\nu$. (i.e. $E = nh\nu$, where $n = 0, 1, 2, 3, 4 \ldots$)

(iv) The oscillators emit or absorb radiations in the order of $h\nu$. This implies that the energy emitted by the oscillator is 0, $h\nu$, $2h\nu$, $3h\nu$, ..., etc.

Derivation

Based on the above postulates Planck derived a mathematical relation for black body radiations. Consider $N_0, N_1, N_2, N_3, \ldots$ are the number of oscillators possessed by a black body. Let $0, E_1, 2E_2, 3E_3, \ldots$ are the energies of these oscillators. The total number of oscillators present in a black body is given by

$$N = N_0 + N_1 + N_2 + N_3 + \cdots \tag{4.7}$$

The number of oscillators present in a black body is obtained from Maxwell–Boltzmann distribution function. The Maxwell–Boltzmann distribution function is given by

$$N = N_0 \exp\left(-\frac{nE}{kT}\right) \tag{4.8}$$

where $n = 0, 1, 2, 3, \ldots$.

Substituting the values of $N_0, N_1, N_2, N_3, \ldots$ in Eq. (4.7), we get

$$N = N_0 + N_0 \exp\left(-\frac{E}{kT}\right) + N_0 \exp\left(-\frac{2E}{kT}\right) + N_0 \exp\left(-\frac{3E}{kT}\right) + \cdots$$

$$N = N_0\left[1 + \exp\left(-\frac{E}{kT}\right) + \exp\left(-\frac{2E}{kT}\right) + \exp\left(-\frac{3E}{kT}\right) + \cdots\right]$$

Equation (4.8) can be simplified using the series

$$1 + x + x^2 + x^3 + \cdots = \frac{1}{(1-x)}$$

$$N = N_0\left[\frac{1}{1 - \exp\left(\frac{-E}{kT}\right)}\right] \tag{4.9}$$

The total energy of the oscillator is given by

$$E = 0 + E_1 + 2E_2 + 3E_3 + 4E_4 + \cdots \tag{4.10}$$

The average energy of the oscillator is given by

$$\bar{\varepsilon} = \frac{\text{total energy}}{\text{total number of oscillators}} = \frac{E}{N} \tag{4.11}$$

From the average energy, one can write, $E_n = \varepsilon N_n$. Substituting the values of n, we get $E_1 = \varepsilon N_1$, $E_2 = \varepsilon N_2$, $E_3 = \varepsilon N_3$, ... and so on. Substituting the values of E_1, E_2, E_3 ... in Eq. (4.10), we get

$$E = 0 + \varepsilon N_0 \exp\left(-\frac{E}{kT}\right) + 2\varepsilon N_0 \exp\left(-\frac{2E}{kT}\right) + 3\varepsilon N_0 \exp\left(-\frac{3E}{kT}\right) + \cdots$$

$$= \varepsilon N_0 \exp\left(-\frac{E}{kT}\right)\left[1 + 2\exp\left(-\frac{E}{kT}\right) + 3\exp\left(-\frac{2E}{kT}\right) + 4\exp\left(-\frac{3E}{kT}\right) + \cdots\right] \tag{4.12}$$

Simplifying Eq. (4.12) using the following expression

$$1 + 2x + 3x^2 + 4x^3 + 5x^4 + \cdots = \frac{1}{(1-x)^2}$$

$$E = \frac{\varepsilon N_0 \exp\left(-\frac{E}{kT}\right)}{\left(1 - \exp\left(-\frac{E}{kT}\right)\right)^2} \tag{4.13}$$

From Eqs. (4.9), (4.11) and (4.13), the average energy of the oscillator is given by

$$\bar{\varepsilon} = \frac{\varepsilon N_0 \exp\left(-\frac{E}{kT}\right)}{\left(1 - \exp\left(-\frac{E}{kT}\right)\right)^2} \times \frac{1 - \exp\left(-\frac{E}{kT}\right)}{N_0}$$

i.e.

$$\bar{\varepsilon} = \frac{\varepsilon \exp\left(-\frac{E}{kT}\right)}{\left(1 - \exp\left(-\frac{E}{kT}\right)\right)} \tag{4.14}$$

Multiplying both the numerator and denominator by $\exp\left(\frac{E}{kT}\right)$, we get

$$\bar{\varepsilon} = \frac{\varepsilon}{\left(\exp\left(\frac{E}{kT}\right) - 1\right)}$$

Substituting $E = \varepsilon = h\nu$, we get

$$\overline{\varepsilon} = \frac{h\nu}{\left(\exp\left(\frac{h\nu}{kT}\right) - 1\right)} \tag{4.15}$$

The numbers of oscillators present per unit volume in the frequency range of ν and $\nu + d\nu$ is given by

$$N = \frac{8\pi\nu^2}{c^3} d\nu \tag{4.16}$$

The energy density of the black body radiations (ρ_ν) is given by

Energy density (ρ_ν)
= Number of oscillators present per unit volume in a frequency interval $d\nu$ × average energy of the oscillators

$$\rho_\nu d\nu = \frac{8\pi h\nu^3}{c^3} \frac{1}{\exp\left(\frac{h\nu}{kT}\right) - 1} d\nu \tag{4.17}$$

Equation (4.17) is known as Planck's equation for black body radiation.

4.2.4 Planck's Radiation Formula in terms of Wavelength

The frequency is given by

$$\nu = \frac{c}{\lambda} \tag{4.18}$$

Differentiating Eq. (4.18), we get

$$d\nu = \frac{c}{-\lambda^2} d\lambda$$

i.e.

$$|d\nu| = \left| \frac{c}{-\lambda^2} d\lambda \right| \tag{4.19}$$

Substituting Eqs. (4.18) and (4.19) in Eq.(4.17), we get

$$\rho_\lambda d\lambda = \frac{8\pi hc^3}{c^3\lambda^3} \frac{1}{\exp\left(\frac{hc}{\lambda kT}\right) - 1} \frac{c}{\lambda^2} d\lambda$$

or

$$\rho_\lambda = \frac{8\pi hc}{\lambda^5} \frac{1}{\exp\left(\frac{hc}{\lambda kT}\right) - 1} \tag{4.20}$$

Equation (4.20) is known as Planck's equation for black body radiation in terms of wavelength.

4.2.5 Deduction of Wien's Displacement Law from Planck's Equation

Wien's displacement formula is applicable to the shorter wavelength region. According to Planck's equation

$$\rho_\lambda = \frac{8\pi hc}{\lambda^5} \frac{1}{\exp\left(\frac{hc}{\lambda kT}\right) - 1} \tag{4.21}$$

When the wavelength is very small, $\exp\left[\frac{hc}{\lambda kT}\right] >> 1$

and hence, by neglecting 1 in Eq. (4.21), we get

$$\rho_\lambda = \frac{8\pi hc}{\lambda^5} \exp\left[-\frac{hc}{\lambda kT}\right] \tag{4.22}$$

Considering $C_1 = 8\pi hc$ and $C_2 = hc/k$, Eq. (4.22) can be rewritten as

$$\rho_\lambda = C_1 \lambda^{-5} \exp\left(-\frac{C_2}{\lambda T}\right) \tag{4.23}$$

Equation (4.23) is Wien's displacement law.

4.2.6 Deduction of Rayleigh and Jeans' Law from Planck's Equation

Rayleigh–Jeans' law is applicable to the longer wavelength region of the black body radiation.

When the wavelength is very high, $\left[\frac{hc}{\lambda kT}\right]$ is low.

In general, if x is small

$$e^x = 1 + x + \frac{x^2}{2!} + \frac{x^3}{3!} + \cdots$$

Since x is small, neglecting the higher order terms, we get

$$e^x \approx 1 + x$$

$$\therefore \qquad \exp\left[\frac{hc}{kT\lambda}\right] = 1 + \frac{hc}{\lambda kT} \tag{4.24}$$

Substituting Eq. (4.24) in Eq. (4.20), we get

$$\rho_\lambda = \frac{8\pi hc}{\lambda^5} \frac{1}{\left(1 + \frac{hc}{\lambda kT} - 1\right)}$$

i.e.

$$\rho_\lambda = \frac{8\pi hc}{\lambda^5} \frac{\lambda kT}{hc}$$

$$= \frac{8\pi kT}{\lambda^4} \tag{4.25}$$

Equation (4.25) represents Rayleigh-Jeans' law.

4.2.7 Planck's Quantum Theory

Based on the postulates of Planck's theory, one can summarize the Planck's quantum theory as follows:

1. The matter is composed of a large number of oscillating particles. The oscillators vibrate with different frequencies.
2. The energy of the oscillating particle is quantized, i.e. the energy is equal to $E = nh\nu$, where ν is the frequency of radiation emitted and h is the Planck's constant.
3. The oscillator emits energy, when it moves from one quantized state to the other quantized state. The oscillator does not emit energy as long as it remains in one energy state.
4. The oscillator absorbs radiation and it gets excited to another quantized state. The oscillator emits or absorbs radiation energy in packets of energy $h\nu$.

According to Planck's theory, the exchange of energy between quantized states is not continuous but discrete. This quantized energy is in small packet or bundle. This bundle of energy or the packet of energy is called quantum (plural quanta). The packet of energy is proportional to ν. These small packets of energy are called photon. The packet of energy propagates like a particle with the speed of light.

4.2.8 Advantages of Planck's Theory

1. The Planck's equation for the energy density explains the entire spectrum of the black body radiations.
2. It is used to deduce Wien's displacement law, and Rayleigh–Jeans' law.
3. It introduces a new concept, i.e. the energy is emitted by a black body in discrete but not continuous. The energy is given by $E = nh\nu$, where h is the Planck's constant.

4.2.9 Properties of Photon

1. The energy of a photon is $E = h\nu = h\dfrac{c}{\lambda}$, where c is the velocity of light ($c = 3 \times 10^8$ m s^{-1}), λ is the wavelength of the radiation and ν is the frequency of the radiation.
2. The speed of the photon is equal to the speed of light.
3. The existence of the photon is of the same general nature as the existence of the electron. The experiments suggest that definite quantities such as h and ν are associated with photon while e and m are associated with electron.
4. The mass of the photon is obtained from Einstein's equation and Plank's equation.

   From Einstein's equation

   $$E = mc^2$$

   From Planck's equation

   $$E = h\nu$$

   Equating these two equations, we get

   $$mc^2 = h\nu$$

   $$m = \frac{h\nu}{c^2} = \frac{hc}{c^2\lambda} = \frac{h}{c\lambda}$$

5. The momentum of the photon is

$$p = mc = \frac{h}{c\lambda} \times c = \frac{h}{\lambda}$$

6. Photons are electrically neutral. They are not affected by both electric and magnetic fields. The photons do not ionize.

4.3 COMPTON EFFECT

When a monochromatic beam of X-rays is made to incident on a scatterer with low atomic weight such as carbon, two scattered beams are observed, one with the same frequency as that of the incident beam and the other with a reduced frequency or increased wavelength. This phenomenon is known as **Compton effect.** The difference between the two scattered wavelengths is known as Compton shift. This phenomenon was observed by Compton in 1921, and he gave an explanation for this observation in 1922, based on the quantum theory of radiation.

4.3.1 Illustration

Compton explained his experimental observation by considering an X-ray photon as the incident beam and an electron at rest as the scatterer. When a monochromatic X-ray photon is made to incident on an electron at rest, some part of the incident photons energy is transferred to the electron at rest. The electron gains some energy from the incident photon and recoils. Due to the transfer of energy, the scattered photon moves with reduced energy or increased wavelength. The difference between the wavelength of the incident photon and the wavelength of the scattered photon is known as Compton shift (Figure 4.2).

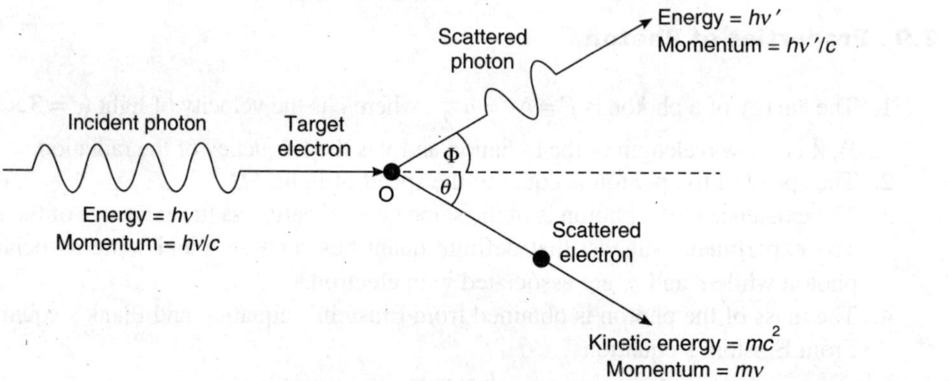

Figure 4.2 Compton effect.

4.3.2 Derivation

Consider a photon of energy $h\nu$ is made to incident on an electron at rest. The collision is considered as inelastic collision, i.e. there is a transfer of energy from incident photon to the electron. The electron gains some amount of energy and recoils. Let the scattered photon moves with reduced energy at an angle ϕ from the direction of incidence. Let the recoiling electron moves along a direction that makes an angle θ from the direction of incidence of photon. Let $h\nu$ be the energy of the incident photon and $h\nu'$ be the energy of the scattered photon. The electron is at rest before collision and $(h\nu - h\nu')$ be the energy gained by the recoiling electron.

Energy before collision

Energy of the photon = $h\nu$

Energy of the electron = m_0c^2

where m_0c^2 is the rest mass energy of the electron.

Total energy before collision = $h\nu + m_0c^2$

Energy after collision

Energy of the photon = $h\nu'$

Energy of the electron = mc^2

where mc^2 is the energy of the recoiling electron.

Total energy after collision = $h\nu' + mc^2$

According to the law of conservation of energy, the total energy before collision is equal to the total energy after collision, i.e.

$$h\nu + m_0c^2 = h\nu' + mc^2 \tag{4.26}$$

Momentum before collision

X-component

Momentum of photon = $\dfrac{h\nu}{c}$

Momentum of the electron = 0

Total momentum = $\dfrac{h\nu}{c}$

Y-component

Momentum of photon = 0

Momentum of the electron = 0

Total momentum = 0

Momentum after collision

X-component

Momentum of photon = $\dfrac{h\nu'}{c}\cos\phi$

Momentum of the electron = $mv\cos\theta$

Total X-component momentum = $\dfrac{h\nu'}{c}\cos\phi + mv\cos\theta$

Y-component

Momentum of photon = $\dfrac{h\nu'}{c}\sin\phi$

Momentum of the electron = $-mv\sin\theta$

Total Y-component momentum = $\dfrac{h\nu'}{c}\sin\phi - mv\sin\theta$

According to the law of conservation of momentum, the momentum before collision is equal to the momentum after collision. The *X*-component of the momentum is given by

$$\frac{h\nu}{c} = \frac{h\nu'}{c}\cos\phi + mv\cos\theta \tag{4.27}$$

The *Y*-component of the momentum can be written as

$$0 = \frac{h\nu'}{c}\sin\phi - mv\sin\theta \tag{4.28}$$

Equations (4.27) and (4.28) can be rewritten as

$$mvc\cos\theta = h(\nu - \nu'\cos\phi) \tag{4.29}$$

$$mvc\sin\theta = h\nu'\sin\phi \tag{4.30}$$

Squaring Eqs. (4.29) and (4.30) and adding we get

$$m^2v^2c^2 = h^2(\nu^2 + \nu'^2\cos^2\phi - 2\nu\nu'\cos\phi) + h^2\nu'^2\sin^2\phi$$

$$= h^2(\nu^2 + \nu'^2 - 2\nu\nu'\cos\phi) \tag{4.31}$$

Equation (4.26) can be written as

$$mc^2 = h(\nu - \nu') + m_0c^2 \tag{4.32}$$

Squaring Eq. (4.32)

$$m^2c^4 = h^2(\nu'^2 + \nu^2 - 2\nu\nu') + m_0{}^2c^4 + 2m_0c^2h(\nu - \nu') \tag{4.33}$$

Subtracting Eq. (4.31) from Eq.(4.33), we get

$$m^2c^2(c^2 - v^2) = -2h^2\nu\nu' + 2h^2\nu\nu'\cos\phi + m_0{}^2c^4 + 2m_0c^2h(\nu - \nu') \tag{4.34}$$

From Einstein's theory of relativity, the variation of mass with velocity can be written as

$$m = \frac{m_0}{\sqrt{1 - \dfrac{v^2}{c^2}}} \tag{4.35}$$

Squaring and rearranging Eq. (4.35), we get

$$m^2(c^2 - v^2) = m_0{}^2c^2 \tag{4.36}$$

Multiplying Eq. (4.36) by c^2, we get

$$m^2c^2(c^2 - v^2) = m_0{}^2c^4 \tag{4.37}$$

Equating the RHS of Eqs. (4.34) and (4.37), we get

$$m_0{}^2c^4 = -2h^2\nu\nu' + 2h^2\nu\nu'\cos\phi + m_0{}^2c^4 + 2m_0c^2h(\nu - \nu') \tag{4.38}$$

Equation (4.38) can be written as

$$2h^2\nu\nu'(1 - \cos\phi) = 2m_0c^2h(\nu - \nu')$$

i.e.

$$h\nu\nu'(1 - \cos\phi) = m_0c^2(\nu - \nu')$$

$$\frac{\nu - \nu'}{\nu\nu'} = \frac{h}{m_0c^2}(1 - \cos\phi)$$

$$\frac{1}{\nu'} - \frac{1}{\nu} = \frac{h}{m_0c^2}(1 - \cos\phi) \tag{4.39}$$

Multiplying both sides by c, we get

$$\frac{c}{\nu'} - \frac{c}{\nu} = \frac{h}{m_0 c}(1 - \cos\phi)$$

$$\lambda' - \lambda = \Delta\lambda = \frac{h}{m_0 c}(1 - \cos\phi) \tag{4.40}$$

Equation (4.40) is known as Compton's shift in wavelength. Using the relation $\cos 2\theta = 1 - 2\sin^2\theta$, one can write $\cos\theta = 1 - 2\sin^2\frac{\theta}{2}$. Substituting the value of $\cos\theta$ in Eq. (4.40), we get

$$\Delta\lambda = \frac{h}{m_0 c}\left(1 - 1 + 2\sin^2\frac{\phi}{2}\right)$$

$$\Delta\lambda = \frac{2h}{m_0 c}\sin^2\frac{\phi}{2} \tag{4.41}$$

Equation (4.41) also gives the value of Compton shift in wavelength. It is independent of the wavelength of the incident photon. It depends only on the scattering angle. The Compton shift in wavelength also increases with the increase of scattering angle.

Case (i) When $\phi = 0°$, $\cos\phi = 1$

$$\Delta\lambda = \frac{h}{m_0 c}(1 - \cos 0°) = \frac{h}{m_0 c}(1 - 1) = 0$$

i.e. when the scattering angle is zero, the Compton shift in wavelength is zero.

Case (ii) When $\phi = 90°$, $\cos\phi = 0$

$$\Delta\lambda = \frac{h}{m_0 c}(1 - \cos 90°) = \frac{h}{m_0 c}(1 - 0) = \frac{h}{m_0 c}$$

Substituting the values of h, m_0, and c, we get

$$\Delta\lambda = \frac{6.626 \times 10^{-34}}{9.1 \times 10^{-31} \times 3 \times 10^{8}} = 0.02426 \text{ Å}$$

i.e. when the scattering angle is 90°, the Compton shift in wavelength is 0.02426 Å.

Case (iii) When $\phi = 180°$, $\cos\phi = -1$

$$\Delta\lambda = \frac{h}{m_0 c}(1 - \cos 180°) = \frac{h}{m_0 c}(1 + 1) = \frac{2h}{m_0 c}$$

$$\Delta\lambda = \frac{2 \times 6.626 \times 10^{-34}}{9.1 \times 10^{-31} \times 3 \times 10^{8}} = 0.0485 \text{ Å}$$

i.e. when the scattering angle is 180°, the Compton shift in wavelength is 0.0485 Å.

The value $h/m_0 c$ is called the **Compton wavelength** of the electron. It is equal to 0.0242 Å. It is not the wavelength of the electron. It represents the Compton shift in wavelength of a photon that scattered the electron through an angle 90°. Compton explained that the unmodified wavelength in Compton scattering is due to the scattering of the photon by the atom as a whole. If m_0 is replaced by the mass of the atom, the change in wavelength is negligible because the mass of the atom is very high.

4.3.3 Direction of the Recoiling Electron

Equations (4.29) and (4.30) are written as

$$mvc \cos\theta = h(\nu - \nu' \cos\phi) \tag{4.42}$$

$$mvc \sin\theta = h\nu' \sin\phi \tag{4.43}$$

Dividing Eq. (4.42) by Eq. (4.43), we get

$$\cot\theta = \frac{(\nu - \nu'\cos\phi)}{\nu'\sin\phi}$$

$$\cot\theta = \frac{1}{\sin\phi}\left(\frac{\nu}{\nu'} - \cos\phi\right) \tag{4.44}$$

Equation (4.39) can be written as

$$\frac{1}{\nu'} - \frac{1}{\nu} = \frac{h}{m_0c^2}(1 - \cos\phi) \tag{4.45}$$

Multiplying Eq. (4.45) by ν, we get

$$\frac{\nu}{\nu'} - 1 = \frac{h\nu}{m_0c^2}(1 - \cos\phi) \tag{4.46}$$

Taking $\alpha = \dfrac{h\nu}{m_0c^2}$, Eq. (4.46) can be written as

$$\frac{\nu}{\nu'} = 1 + \alpha(1 - \cos\phi) \tag{4.47}$$

Substituting the value of $\dfrac{\nu}{\nu'}$ given by Eq. (4.47) in Eq. (4.44), we get

$$\cot\theta = \frac{1}{\sin\phi}(1 + \alpha(1 - \cos\phi) - \cos\phi)$$

$$\cot\theta = \frac{1}{\sin\phi}(1 + \alpha)(1 - \cos\phi) \tag{4.48}$$

Substituting $\sin\phi = 2\sin\dfrac{\phi}{2}\cos\dfrac{\phi}{2}$ and $\cos\phi = 1 - 2\sin^2\dfrac{\phi}{2}$ in Eq. (4.48), we get

$$\cot\theta = \frac{1}{2\sin\dfrac{\phi}{2}\cos\dfrac{\phi}{2}}(1 + \alpha)\left(1 - 1 + 2\sin^2\frac{\phi}{2}\right)$$

$$\cot\theta = \frac{(1 + \alpha) \times 2\sin^2\dfrac{\phi}{2}}{2\sin\dfrac{\phi}{2}\cos\dfrac{\phi}{2}}$$

$$\cot\theta = \frac{(1+\alpha)\times\sin\frac{\phi}{2}}{\cos\frac{\phi}{2}} \tag{4.49}$$

By taking an inverse of Eq. (4.49), we get

$$\tan\theta = \frac{\cot\frac{\phi}{2}}{(1+\alpha)}$$

Substituting the value of α in Eq. (4.49), we get

$$\tan\theta = \frac{\cot\frac{\phi}{2}}{\left(1+\frac{h\nu}{m_0c^2}\right)} \tag{4.50}$$

Equation (4.50) gives the direction of the recoiling electron.

4.3.4 Experimental Verification

The Compton effect is experimentally verified using Bragg's X-ray spectrometer. A monochromatic beam of X-rays produced by the X-ray tube is collimated by two thin slits, S_1 and S_2 as shown in Figure 4.3. The collimated monochromatic X-ray beam is made to incident on a scatterer C such as graphite. The intensity and wavelength of the scattered beam is measured by the spectrometer at different scattering angle. The intensity is plotted against the wavelengths of the scattered beams as shown in Figure 4.4.

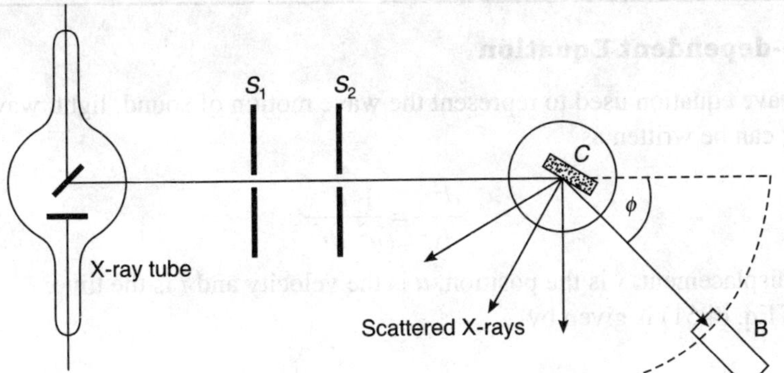

Figure 4.3 Compton effect—experimental verification.

The results obtained by Compton for $\theta = 0°$, 45°, 90° and 135° are shown in Figure 4.4. It is observed that for the scattering angle $\theta = 0°$, only one peak is obtained. For the non-zero scattering angles, two different peaks are obtained, one corresponding to the original wavelength and the other corresponding to the modified wavelength. The difference between the original wavelength and the modified wavelength is known as Compton shift in wavelength. It is found that the Compton shift in wavelength increases with the increase in the scattering angle. The increase of the Compton shift in wavelength with the scattering angle is in accordance with Compton's equation, Eq. (4.40) for Compton shift.

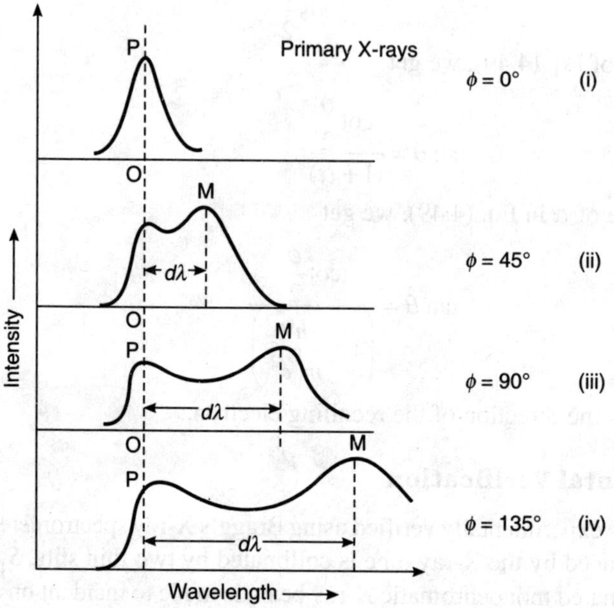

Figure 4.4 Compton effect—a plot of intensity versus wavelength.

4.4 SCHRÖDINGER WAVE EQUATION

4.4.1 Time-dependent Equation

The classical wave equation used to represent the wave motion of sound, light, wave motion in a stretched string can be written as

$$\frac{d^2y}{dx^2} = \frac{1}{u^2}\frac{d^2y}{dt^2} \tag{4.51}$$

where y is the displacement, x is the position, u is the velocity and t is the time.
The solution of Eq. (4.51) is given by

$$y(x, t) = Ae^{-i\omega\left(t-\frac{x}{v}\right)} \tag{4.52}$$

Equation (4.52) can be written as

$$y(x, t) = A\left[\cos\omega\left(t - \frac{x}{v}\right) - i\sin\omega\left(t - \frac{x}{v}\right)\right] \tag{4.53}$$

The imaginary term is irrelevant for the wave motions produced by sound, light and stretched string, etc. So the imaginary term is neglected in classical physics. For the atomic particle, one cannot measure both the position and momentum simultaneously. Therefore, the imaginary term is also taken into account in Eq. (4.53) for the atomic particle. By taking into account of the imaginary terms, the displacement of atomic particle is represented by ψ (psi). Equation (4.51) can be written as

$$\frac{\partial^2\psi}{\partial x^2} = \frac{1}{v^2}\frac{\partial^2\psi}{\partial t^2} \tag{4.54}$$

The solution of Eq. (4.54) can be written as

$$\psi(x,t) = \psi_0 e^{-i\omega\left(t-\frac{x}{v}\right)} \tag{4.55}$$

where ψ_0 is a constant.

Differentiating Eq. (4.55) with respect to t, we get

$$\frac{\partial\psi}{\partial t} = (-i\omega)\psi_0 \exp\left[-i\omega\left(t-\frac{x}{v}\right)\right] \tag{4.56}$$

i.e. $$\frac{\partial\psi}{\partial t} = -i\omega\psi \tag{4.57}$$

where $\psi(x, t) = \psi_0 e^{-i\omega\left(t-\frac{x}{v}\right)}$.

Planck's equation for energy

$$E = h\nu = \frac{h}{2\pi}\times 2\pi\nu = \hbar\omega$$

Substituting the value of ω, given by $E = \hbar\omega$ in Eq. (4.57), we get

$$\frac{\partial\psi}{\partial t} = -i\frac{E}{\hbar}\psi$$

i.e. $$E\psi = i\hbar\frac{\partial\psi}{\partial t} \tag{4.58}$$

In Eq. (4.58), E is the energy operator and it operates on the wave function ψ. The energy operator is given by

$$E = i\hbar\frac{\partial}{\partial t} \tag{4.59}$$

Substituting $\omega = 2\pi\nu$, Eq. (4.55) can be written as

$$\psi(x,t) = \psi_0 e^{-2\pi i\left(\nu t-\frac{\nu x}{v}\right)} \tag{4.60}$$

Substituting $\nu = \frac{E}{h}$ and $\frac{\nu}{v} = \frac{1}{\lambda}$ in Eq. (4.60), we get

$$\psi(x,t) = \psi_0 e^{-2\pi i\left(\frac{E}{h}t-\frac{x}{\lambda}\right)} \tag{4.61}$$

Substituting $\lambda = \frac{h}{p}$ in Eq. (4.61), we get

$$\psi(x,t) = \psi_0 e^{-2\pi i\left(\frac{E}{h}t - \frac{px}{h}\right)} \tag{4.62}$$

$$\psi(x,t) = \psi_0 e^{-i\frac{(Et - px)}{\hbar}} \tag{4.63}$$

where $\hbar = \frac{h}{2\pi} = \frac{6.626 \times 10^{-34}}{2\pi} = 1.054 \times 10^{-34}$ J-s

Differentiating Eq. (4.63) with respect to x, we get

$$\frac{\partial \psi}{\partial x} = i\frac{p}{\hbar}\psi_0 e^{-i\frac{(Et-px)}{\hbar}}$$

$$\frac{\partial \psi}{\partial x} = i\frac{p}{\hbar}\psi \tag{4.64}$$

Differentiating Eq. (4.64) again with respect to x, we get

$$\frac{\partial^2 \psi}{\partial x^2} = i^2\frac{p^2}{\hbar^2}\psi \tag{4.65}$$

From Eq. (4.65), we get

$$p^2\psi = -\hbar^2\frac{\partial^2 \psi}{\partial x^2} \tag{4.66}$$

where p is the momentum operator. Equation (4.66) represents the momentum operator operates on the wave function ψ. From Eq. (4.66), we have

$$p\psi = i\hbar\frac{\partial \psi}{\partial x}$$

The momentum operator p is given by

$$p = i\hbar\frac{\partial}{\partial x} \tag{4.67}$$

The total energy is given by

Total energy = Kinetic energy + Potential energy

$$E = \frac{1}{2}mv^2 + V$$

$$E = \frac{p^2}{2m} + V \tag{4.68}$$

The sum of the kinetic energy and the potential energy is known as Hamiltonian operator. It is represented by H. It is equal to

$$H = \frac{p^2}{2m} + V \tag{4.69}$$

In Eq. (4.68), E is the energy operator and p is the momentum operator. If the operators E and p operate on a wave function ψ, we get

$$E\psi = \frac{p^2\psi}{2m} + V\psi \tag{4.70}$$

Equation (4.70) is written as $E\psi = H\psi$. This equation is known as Eigenvalue equation.

Substituting the value of p given by Eq. (4.67) and the value of E given by Eq. (4.59) in Eq. (4.70), we get

$$-\frac{\hbar^2}{2m}\frac{\partial^2\psi}{\partial x^2} = i\hbar\frac{\partial\psi}{\partial t} - V\psi \tag{4.71}$$

Equation (4.71) is known as Schrödinger's one-dimensional time-dependent equation. In three dimensions, it can be written as

$$\frac{\hbar^2}{2m}\left(\frac{\partial^2\psi}{\partial x^2} + \frac{\partial^2\psi}{\partial y^2} + \frac{\partial^2\psi}{\partial z^2}\right) = -i\hbar\frac{\partial\psi}{\partial t} + V\psi$$

$$-i\hbar\frac{\partial\psi}{\partial t} = \frac{\hbar^2}{2m}\nabla^2\psi - V\psi \tag{4.72}$$

Equation (4.72) is known as Schrödinger's three-dimensional time-dependent equation,

where, $\nabla^2 = \frac{\partial^2}{\partial x^2} + \frac{\partial^2}{\partial y^2} + \frac{\partial^2}{\partial z^2}$. It is known as Laplacian operator.

4.4.2 Time-independent Equation

In stationary state problem, the potential energy of the particle does not depend on time. The force acting on the particle and the potential energy, V also independent of time. Such problems are said to be time-independent problem. From Eq. (4.55), the wave function ψ can be written as

$$\psi(x,t) = \psi_0 e^{-i\omega\left(t - \frac{x}{v}\right)} \tag{4.73}$$

Equation (4.73) can be written as

$$\psi(x,t) = \psi_0 e^{i\omega\frac{x}{v}} e^{-i\omega t} \tag{4.74}$$

By taking $\psi(x) = \psi_0 e^{i\omega\frac{x}{v}}$. Eq. (4.74) can be written as

$$\psi(x, t) = \psi(x)e^{-i\omega t} \tag{4.75}$$

Differentiating Eq. (4.75) with respect to x, we get

$$\frac{\partial\psi(x,t)}{\partial x} = e^{-i\omega t}\frac{\partial\psi(x)}{\partial x} \tag{4.76}$$

Differentiating Eq. (4.76) with respect to x, we get

$$\frac{\partial^2\psi(x,t)}{\partial x^2} = e^{-i\omega t}\frac{\partial^2\psi(x)}{\partial x^2} \tag{4.77}$$

Differentiating Eq. (4.75) with respect to *t*, we get

$$\frac{\partial \psi(x,t)}{\partial t} = (-i\omega)e^{-i\omega t}\psi(x) \tag{4.78}$$

Substituting the values of $\frac{\partial^2 \psi(x,t)}{\partial x^2}, \frac{\partial \psi(x,t)}{\partial t}$ and $\psi(x, t)$ given by Eqs. (4.77), (4.78) and (4.75) respectively in Eq. (4.71), we get

$$-\frac{\hbar^2}{2m} e^{-\omega t} \frac{\partial^2 \psi(x)}{\partial x^2} = i\hbar(-i\omega)e^{-i\omega t}\psi(x) - V\psi(x)e^{-i\omega t} \tag{4.79}$$

$$-\frac{\hbar^2}{2m} \frac{\partial^2 \psi(x)}{\partial x^2} = \hbar\omega\psi(x) - V\psi(x) \tag{4.80}$$

Substituting, $E = \hbar\omega$, we get

$$-\frac{\hbar^2}{2m} \frac{\partial^2 \psi(x)}{\partial x^2} = E\psi(x) - V\psi(x) \tag{4.81}$$

Equation (4.81) is known as one-dimensional time-independent Schrödinger's wave equation. In three dimensions, Eq. (4.81) can be written as

$$-\frac{\hbar^2}{2m}\left(\frac{\partial^2 \psi}{\partial x^2} + \frac{\partial^2 \psi}{\partial y^2} + \frac{\partial^2 \psi}{\partial z^2}\right) = E\psi(x, y, z, t) - V\psi(x, y, z, t) \tag{4.82}$$

Equation (4.82) can be written as

$$-\frac{\hbar^2}{2m}\nabla^2\psi = (E - V)\psi$$

$$\nabla^2\psi + \frac{2m}{\hbar^2}(E - V)\psi = 0 \tag{4.83}$$

Equation (4.83) is known as three-dimensional time-independent Schrödinger's wave equation.

4.4.3 Physical Significance of the Wave Function ψ

1. The wave function ψ is a complex quantity, so one cannot measure it.
2. The wave function relates the particle nature and the wave nature statistically.
3. The square of the wave function is determined by multiplying the wave function by its complex conjugate, i.e. $|\psi|^2 = \psi \times \psi^*$. The square of the wave function is a real quantity.
4. $|\psi|^2$ is the probability of finding the particle in the state, and it is a measure of position probability density. It is represented by the letter, *P*, i.e.
$$P = |\psi|^2 = \psi \times \psi^*$$
5. The probability of finding a particle in a volume $d\tau = dxdydz$ is
$$P = \iiint |\psi|^2 \, dxdydz$$
6. Since $P = \iiint |\psi|^2 \, dxdydz$ is the probability of finding the particle, the value of this integral lies between 0 and 1.

7. If a particle is certainly present, then $P = \iiint |\psi|^2\, dxdydz = 1$. The absence of the particle is given by $P = \iiint |\psi|^2\, dxdydz = 0$. If the probability value is 0.4, it means there is a 40% chance for the presence of the particle.
8. It does not predict the exact location of the particle, but it says where the particle is likely to be.

4.4.4 Application of Schrödinger Wave Equation

One-dimensional potential well problem

Consider a potential well of width L and infinite height. Consider that there is an electron present inside the potential well. Consider the electron is freely moving here and there and it makes collisions with the walls of the container. Consider the elastic collision and hence there is no transfer of energy. The electron cannot escape from the well because it needs infinite energy to escape from the well (Figure 4.5).

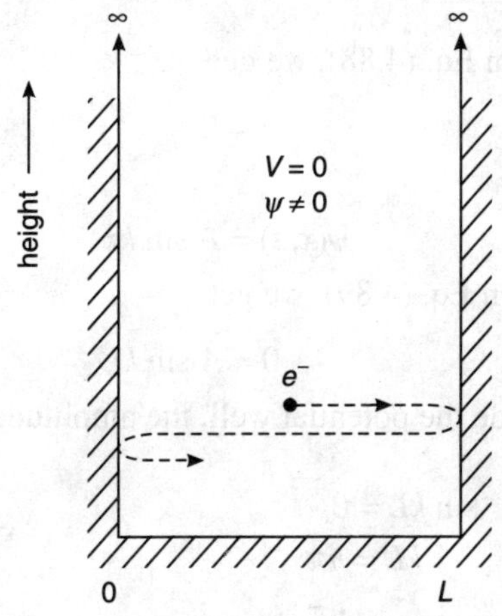

Figure 4.5 Potential well problem.

The potential energy of the freely-moving electron is very small and hence it is negligible, i.e. $V = 0$. The boundary condition for this problem is written as:

(i) Since the electron is present inside the potential well, the wave function of the electron is not equal to zero, i.e.

$$|\psi| \neq 0 \qquad \text{when } 0 < x < L$$

(ii) Outside the potential well and at the wall of the well the electron is not present and hence its wave function is zero, i.e.

$$|\psi| = 0 \qquad \text{when } 0 \geq x \geq L$$

The motion of a freely-moving particle is independent of time and hence the Schrödinger's time-independent wave equation can be written as

$$\frac{\partial^2 \psi}{\partial x^2} + \frac{2m}{\hbar^2}(E - V)\psi = 0 \qquad (4.84)$$

For a freely-moving particle, the potential energy is equal to zero. Substituting, $V = 0$ in Eq. (4.84), we get

$$\frac{\partial^2 \psi}{\partial x^2} + \frac{2m}{\hbar^2} E\psi = 0 \tag{4.85}$$

Taking
$$\frac{2m}{\hbar^2} E = k^2 \tag{4.86}$$

Equation (4.85) can be written as

$$\frac{\partial^2 \psi}{\partial x^2} + k^2 \psi = 0 \tag{4.87}$$

The solution of Eq. (4.87) can be written as

$$\psi(x, t) = A \sin kx + B \cos kx \tag{4.88}$$

where A and B are constants. The values of A and B are evaluated by applying the boundary conditions.

Substituting $\psi = 0$, when $x = 0$, in Eq. (4.88), we get

$$B = 0$$

Equation (4.88) can be written as

$$\psi(x, t) = A \sin kx \tag{4.89}$$

Substituting $\psi = 0$, when $x = L$, in Eq. (4.89), we get

$$0 = A \sin kL$$

Since the electron is present inside the potential well, the amplitude of the wave function A is not equal to zero. Therefore

$$\sin kL = 0$$

i.e.
$$kL = n\pi$$

$$k = \frac{n\pi}{L} \tag{4.90}$$

Substituting Eq. (4.90) in Eq. (4.89), we get

$$\psi(x, t) = A \sin \frac{n\pi}{L} x \tag{4.91}$$

From Eqs. (4.86) and (4.90), we get

$$\frac{n^2\pi^2}{L^2} = \frac{2m}{\hbar^2} E$$

i.e.
$$E = \frac{n^2 h^2}{8mL^2} \tag{4.92}$$

where n is an integer. It takes the values of 1, 2, 3, … etc.

Substituting the values of h, m and $L = 1$ Å, we get

$$E = \frac{n^2 \times (6.626 \times 10^{-34})^2}{8 \times 9.1 \times 10^{-31} \times (1 \times 10^{-10})^2 \times 1.6 \times 10^{-19}}$$

$$= 37.69\, n^2 \text{ eV}$$

Substituting $n = 1, 2, 3, \ldots$ in Eq. (4.92), we get

$$\text{For } n = 1, E_1 = \frac{h^2}{8mL^2} = 37.69 \text{ eV}$$

$$\text{For } n = 2, E_2 = \frac{4h^2}{8mL^2} = 150.76 \text{ eV}$$

$$\text{For } n = 3, E_3 = \frac{9h^2}{8mL^2} = 339.21 \text{ eV}$$

$$\text{For } n = 4, E_4 = \frac{16h^2}{8mL^2} = 603.04 \text{ eV}$$

The values of E_1, E_2, E_3, E_4 shows that the electron is having discrete energy levels and these energy levels are shown in Figure 4.6.

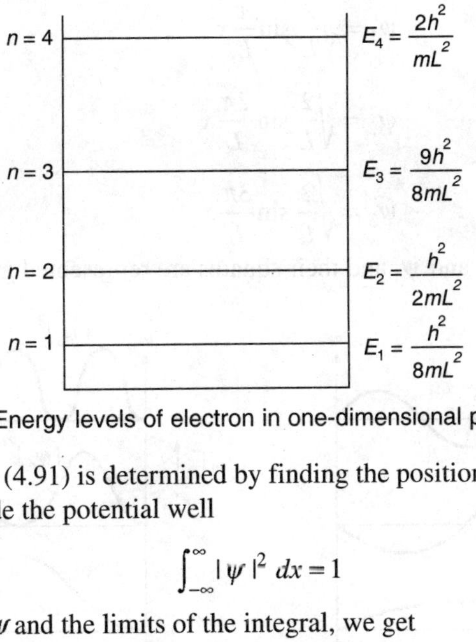

Figure 4.6 Energy levels of electron in one-dimensional potential well.

The value of A in Eq. (4.91) is determined by finding the position probability density. Since the electron is present inside the potential well

$$\int_{-\infty}^{\infty} |\psi|^2\, dx = 1$$

Substituting the values of ψ and the limits of the integral, we get

$$\int_0^L A^2 \sin^2 \frac{n\pi x}{L} dx = 1 \quad (4.93)$$

Using the equation, $\cos 2\theta = 1 - 2\sin^2\theta$, Eq. (4.93) can be written as

$$\frac{1}{2}\int_0^L \left(1-\cos\frac{2n\pi x}{L}\right)dx = \frac{1}{A^2}$$

i.e.

$$\frac{1}{2}\int_0^L dx - \frac{1}{2}\int_0^L \cos\frac{2n\pi x}{L}\,dx = \frac{1}{A^2}$$

Since $\int_0^L \cos\frac{2n\pi x}{L}\,dx = 0$, the above equation can be written as

$$\frac{1}{2}\int_0^L dx = \frac{1}{A^2}$$

i.e.

$$A = \sqrt{\frac{2}{L}} \qquad (4.94)$$

From Eqs. (4.94) and (4.91), we get

$$\psi(x,t) = \sqrt{\frac{2}{L}}\sin\frac{n\pi}{L}x \qquad (4.95)$$

Equation (4.95) represents the wave function of the electron.
Substituting $n = 1, 2, 3, \ldots$ in Eq. (4.95), we get

$$\psi_1 = \sqrt{\frac{2}{L}}\sin\frac{\pi}{L}x$$

$$\psi_2 = \sqrt{\frac{2}{L}}\sin\frac{2\pi}{L}x$$

$$\psi_3 = \sqrt{\frac{2}{L}}\sin\frac{3\pi}{L}x$$

The wavefunctions, ψ_1, ψ_2, and ψ_3 and their squares are represented graphically in Figure 4.7.

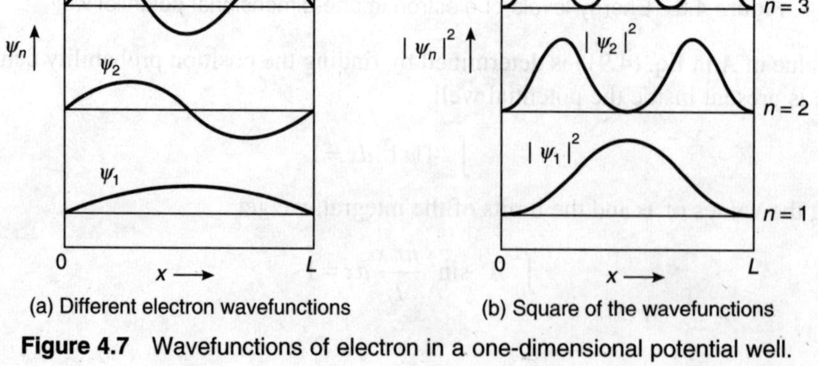

(a) Different electron wavefunctions (b) Square of the wavefunctions

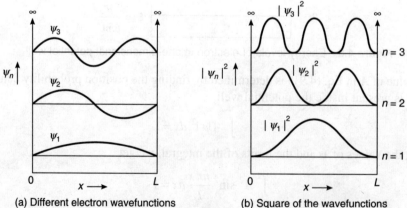

Figure 4.7 Wavefunctions of electron in a one-dimensional potential well.

4.4.5 Eigenvalues and Eigenfunctions

Consider an operator which has a special property such that when it is operating on a function it produces a constant multiplied by that function. Then the function is said to be an **eigenfunction**. The various possible values of the constant value are said to be **eigenvalues**. The equation is said to be an Eigenvalue equation.

Consider the equation

$$\frac{d^2}{dt^2}(\sin \omega t) = -\omega^2(\sin \omega t)$$

where $\frac{d^2}{dt^2}$ is the operator, $\sin \omega t$ is the function and $-\omega^2$ is the constant value. Similarly, the equation $H\psi = E\psi$ has a special property. The Hamiltonian operator H, operating on the wave function produces another wave function ψ multiplied by a constant, E. Such equation is called an **eigenvalue equation**. The wave functions, ψ_1, ψ_2, ψ_3 are called **eigenfunctions** and the energy values, E_1, E_2, E_3 are called **eigenvalues**.

4.5 MATTER WAVES

The radiation possesses dual nature, i.e. particle nature and wave nature as shown in Figure 4.8. Certain properties of the radiations, such as, photoelectric effect, Compton effect, etc. are explained on the basis of the particle nature of radiation. Some other properties of radiation such as interference, reflection, polarization, refraction, etc. are explained on the basis of the properties of the wave nature.

Louis de Broglie in 1924 extended the wave-like property of optics to matter particle also. According to him, the matter particle such as electron, proton and neutron also possesses wave-like characteristics in addition to the particle nature.

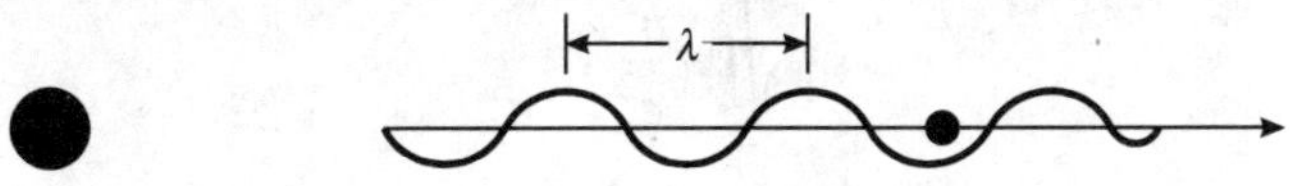

(a) Particle nature (b) Wave nature

Figure 4.8 Particle and wave nature of matter particles.

The following reasons inspired de Broglie to propose the concept of matter waves. These are

1. The universe is symmetrical.
2. The radiation possesses dual nature. If radiation possesses dual nature, the matters will also possess dual nature.

According to de Broglie hypothesis, the wave nature associated with the moving matter particle is known as **matter waves** or **de Broglie waves**. The wavelength of the matter particle is given by

$$\lambda = \frac{h}{p}$$

where p is the momentum and h is the Planck's constant.

4.5.1 Equation for the Wavelength of Matter Waves

Consider an electron which is pictured as a wave in space. The wave function of the electron ψ (pronounced as psi) is written as

$$\psi = \psi_0 \sin 2\pi \nu t \tag{4.96}$$

where t is the time and ν is the frequency. The wave function of the electron at a time $t = t_0$ can be written as

$$\psi = \psi_0 \sin 2\pi \nu_0 t_0 \tag{4.97}$$

According to Einstein's relativity theory, the equation for the transformation of time can be written as

$$t_0 = \frac{t - \frac{vx}{c^2}}{\sqrt{1 - \frac{v^2}{c^2}}} \tag{4.98}$$

Substituting Eq. (4.98) in Eq. (4.97), we get

$$\psi = \psi_0 \sin\left[\frac{2\pi \nu_0}{\sqrt{1 - \frac{v^2}{c^2}}}\left(t - \frac{vx}{c^2}\right)\right] \tag{4.99}$$

The standard equation for the wave motion can be written as

$$y = A \sin\left[\frac{2\pi}{T}\left(t - \frac{x}{u}\right)\right] \tag{4.100}$$

Comparing Eqs. (4.99) and (4.100), we get

$$\frac{2\pi}{T} = \frac{2\pi \nu_0}{\sqrt{1 - \frac{v^2}{c^2}}}$$

i.e.

$$\frac{1}{T} = \nu = \frac{\nu_0}{\sqrt{1 - \frac{v^2}{c^2}}} \tag{4.101}$$

and

$$t - \frac{x}{u} = t - \frac{vx}{c^2}$$

$$\frac{x}{u} = \frac{vx}{c^2}$$

$$u = \frac{c^2}{v} \tag{4.102}$$

From Einstein's equation, the energy can be written as

$$E = mc^2 \tag{4.103}$$

From Planck's equation, the energy can be written as

$$E = h\nu \tag{4.104}$$

Equating Eqs. (4.103) and (4.104), we get

$$h\nu = mc^2 \tag{4.105}$$

At rest, i.e. at $t = t_0$, Eq. (4.105) can be written as

$$h\nu_0 = m_0c^2 \tag{4.106}$$

Equation (4.106) can be written as

$$\nu_0 = \frac{m_0c^2}{h} \tag{4.107}$$

From Eqs. (4.101) and (4.107), we get

$$\nu = \frac{m_0c^2}{h\sqrt{1-\frac{v^2}{c^2}}}$$

i.e.

$$\nu = \frac{mc^2}{h} \tag{4.108}$$

where $m = \frac{m_0}{\sqrt{1-\frac{v^2}{c^2}}}$

The wavelength of the matter wave is determined using the relation

Velocity = wavelength × frequency

From Eqs. (4.102) and (4.108), we get

$$\frac{c^2}{v} = \lambda \times \frac{mc^2}{h}$$

$$\lambda = \frac{h}{mv} = \frac{h}{p} \tag{4.109}$$

Equation (4.109) is known as de Broglie's equation for the wavelength of the matter waves.

4.5.2 Alternate Method

The de Broglie equation for the wavelength of matter particles can also be derived as follows:

From Einstein's equation, the energy is given by

$$E = mc^2$$

From Planck's equation, the energy is given by

$$E = h\nu$$

Equating these two equations, we get

$$h\nu = mc^2$$

i.e. $$h\frac{c}{\lambda} = mc^2$$

i.e. $$\lambda = \frac{h}{mc} = \frac{h}{p} \qquad (4.110)$$

4.5.3 de Broglie Wavelength in terms of Kinetic Energy

The kinetic energy is given by

$$\text{K.E.} = \frac{1}{2}mv^2$$

$$E = \frac{p^2}{2m}$$

$$p = \sqrt{2mE} \qquad (4.111)$$

Substituting the value of p given by Eq. (4.111) in Eq. (4.110), we get

$$\lambda = \frac{h}{\sqrt{2mE}} \qquad (4.112)$$

Equation (4.112) gives the de Broglie's wavelength in terms of kinetic energy.

4.5.4 de Broglie Wavelength of an Electron (de Broglie wavelength in terms of accelerating potential)

Let a specimen is given a potential difference of V volts, the energy acquired by the electron is

$$E = eV \qquad (4.113)$$

The kinetic energy of electron is given by

$$E = \frac{p^2}{2m} \qquad (4.114)$$

Equating Eqs. (4.113) and (4.114), we get

$$\frac{p^2}{2m} = eV$$

$$p = \sqrt{2meV} \qquad (4.115)$$

Substituting Eq. (4.115) in Eq. (4.110), we get

$$\lambda = \frac{h}{\sqrt{2meV}} \qquad (4.116)$$

Equation (4.116) represents the de Broglie's wavelength of an electron. Substituting the values of h, m and e in Eq. (4.116), we get

$$\lambda = \frac{6.626 \times 10^{-34}}{\sqrt{2 \times 9.1 \times 10^{-31} \times 1.602 \times 10^{-19} \times V}} = \frac{12.27}{\sqrt{V}}\,\text{Å}$$

For an applied potential difference of $V = 100$ volts, the wavelength of the electron is given by

$$\lambda = \frac{12.27}{\sqrt{100}} = 1.227 \text{ Å}$$

4.5.5 Wave Number

It is the number of waves present in one unit of length. It is equal to the inverse of the wavelength. It is represented by $\bar{\nu}$ and its unit is m^{-1}. It is given by

$$\bar{\nu} = \frac{1}{\lambda} = \frac{\sqrt{2meV}}{h} \tag{4.117}$$

4.5.6 Properties of Matter Waves

1. The matter waves are produced whenever the matter particle (charged particle or uncharged particle) is in motion, whereas the electromagnetic waves are produced whenever charged particles are in motion, ($\lambda = h/mv$ is independent of charge). This property shows that the matter waves are not electromagnetic radiations. The matter waves are a new kind of waves.
2. The velocity of matter waves is not constant. The velocity of matter waves depends only on the material particle. The velocity of the electromagnetic radiation is a constant.
3. The equation $\lambda = h/mv$ shows that the wavelength of matter waves is large for a lighter particle.
4. The equation $\lambda = h/mv$ shows that the wavelength of matter waves increases with the decrease of velocity of the particle.
5. The wave velocity of the matter wave is greater than the velocity of light. The wave velocity is the speed at which the energy moves through the medium. The wave velocity of a matter wave is obtained as follows:
   Equating the Einstein's equation and Planck's equation for energy, we get
   $$h\nu = mc^2$$
   i.e.
   $$\nu = \frac{mc^2}{h}$$
   Using the relation, velocity = frequency × wavelength
   The wave velocity,
   $$\omega = \nu \times \lambda = \frac{mc^2}{h} \times \lambda$$
   Substituting $\lambda = \dfrac{h}{mv}$, we get
   $$\omega = \frac{mc^2}{h} \times \frac{h}{mv} = \frac{c^2}{v} \tag{4.118}$$
   where v is the particle velocity. It is always less than the velocity of light. This shows that the wave velocity of matter wave is greater than the velocity of light.

6. The particle and the wave properties of a moving particle never appear simultaneously. One can say that the particle has wave-like property and the wave has particle-like property. These two properties linked in such a way that they are inseparable. The matter wave representation is a symbolic representation of what we know about matter.

4.5.7 Experimental Verification of Matter Waves

Louis de Broglie proposed the concept of wave nature of matter particles in 1924. But the experimental evidence for the matter waves was given by two American Scientists Davisson and Germer and independently by G.P. Thomson in 1927.

Davisson and Germer experiment

Davisson and Germer studied about the reflection of electrons from a nickel target. Accidentally the Ni target gets heated so as to produce an anomalous reflection. The reflected intensity has maxima and minima. So Davisson and Germer suspected that the electrons are diffracted and hence the electrons behave like waves in certain conditions.

The experimental set up consists of an electron gun, Ni crystal, a holder to mount the Ni crystal, a circular scale and a Faraday cylinder. The electron gun consists of a filament, F, which is heated by a low-tension battery, LT. The emitted electrons are accelerated by applying a negative potential and collimated by two thin slits, S_1 and S_2. The accelerated fine electron beam emerging from the electron gun is made to incident on a nickel target, N. The nickel target is mounted on a crystal holder, which is capable of rotating with the help of the handle, H, along with the axis of the incident beam. The scattered electrons are detected by an electron detector known as Faraday cylinder, C. The Faraday cylinder can move on the circular graduated scale, S, from 20° to 90° to receive the reflected electrons. The scale is graduated in degrees and the position of the Faraday cylinder is measured from the circular scale. The Faraday cylinder has two walls that are insulated from each other. A retarding potential is applied to the Faraday cylinder, so that it can receive the electrons produced by the electron gun and can avoid the secondary electrons. A galvanometer, G, is attached with the Faraday chamber and it is used to measure the current produced by the reflected electrons collected by the Faraday cylinder (Figure 4.9).

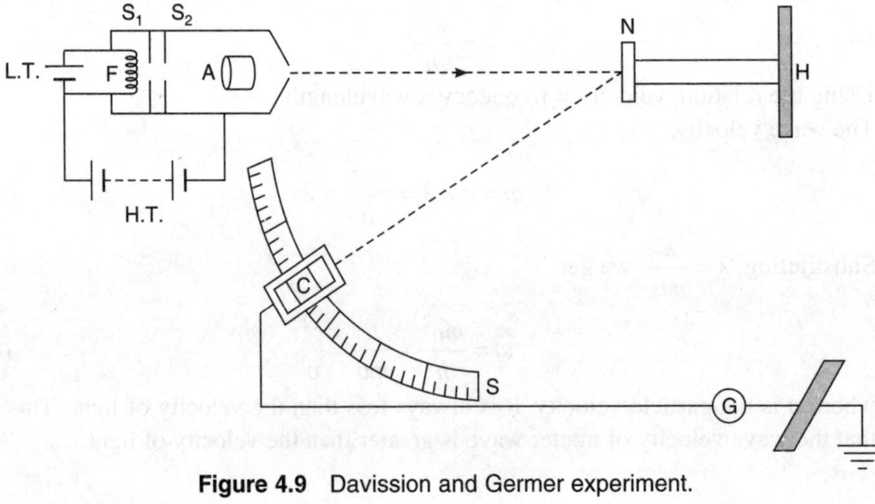

Figure 4.9 Davission and Germer experiment.

The experiment is carried out for different accelerating potentials and hence the scattered electrons are detected with the help of the Faraday cylinder. A graph is plotted between the scattering angle and the number of electrons scattered as shown in Figure 4.10(a). The scattering angle is the angle between the incident beam and the scattered beam and it is measured using the circular scale. The number of electrons scattered is determined by the galvanometer current. It is found that for an accelerating voltage of 54 V, a pronounced peak is obtained at 50°.

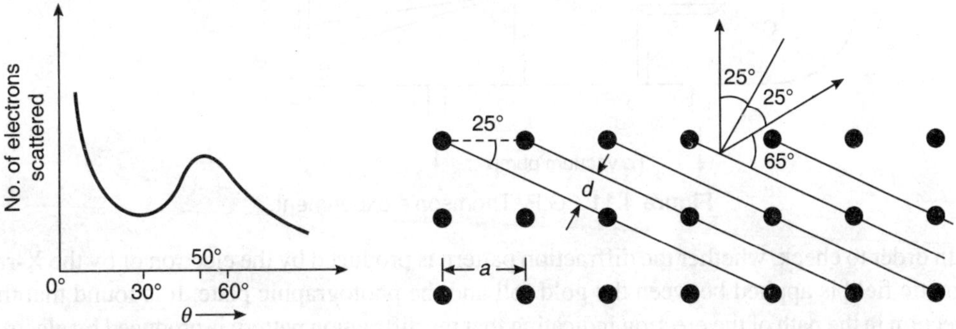

(a) A plot between scattering angle and number of electrons scattered

(b) Calculation of λ using Bragg's equation

Figure 4.10 Scattering of electrons.

The value of the wavelength of the electron is determined using Bragg's equation, $2d \sin \theta = n\lambda$. Since the angle of scattering is 50°, the angle of incidence is 25°. Therefore the glancing angle θ is (90° – 25°) = 65°. The interplanar distance $d = a \sin 25°$. For nickel crystal, $a = 2.15$ Å. Therefore,

$$d = a \sin 25 = 2.15 \times 10^{-10} \times \sin 25° = 0.9086 \times 10^{-10}$$

Substituting in Bragg's equation, we get

$$\lambda = 2 \times 0.9086 \times 10^{-10} \times \sin 65° = 1.647 \text{ Å}$$

The wavelength of the electron is also determined from de Broglie's equation,

$$\lambda = \frac{12.27}{\sqrt{V}} = \frac{12.27}{\sqrt{54}} = 1.669 \text{ Å}$$

The wavelength of electrons obtained from these two methods is in good agreement. This confirms the wave nature of electrons.

G.P. Thomson experiment

G.P. Thomson developed an experimental set up to prove the wave nature of electrons. Thomson observed the diffraction pattern produced by the electron beams accelerated by a potential difference of 10 kV to 50 kV. He obtained a diffraction pattern similar to Debye-Scherrer X-ray diffraction pattern. He was able to calculate the wavelength of electron from the diffraction pattern.

The experimental set up is shown in Figure 4.11. The cathode, C, is heated by a low-tension battery and it emits electrons. The emitted electrons are accelerated by an accelerating potential of nearly 50 kV. The accelerated electrons are passed through a thin slit, S, and the emergent electron beam from the slit is made to fall on a gold or aluminium thin foil, G, of thickness, 10^{-6} m. The diffraction pattern is recorded in a photographic plate, P. The photographic plate is developed and it is found that concentric rings are obtained. The entire apparatus is evacuated to a very high vacuum so as to avoid the loss of energy of the electron due to collision with the gaseous molecules.

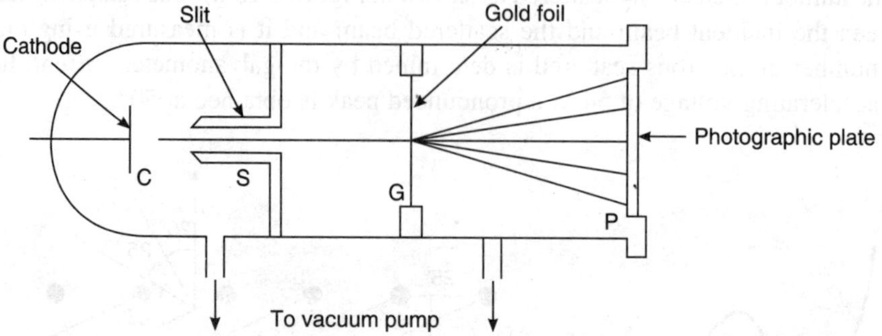

Figure 4.11 G.P. Thomson's experiment.

In order to check, whether the diffraction pattern is produced by the electron or by the X-rays, a magnetic field is applied between the gold foil and the photographic plate. It is found that there is deflection in the path of the electron indicating that the diffraction pattern is produced by electrons. If it is produced by X-rays, there should be no deflection in the path of the X-rays beam is observed, because the X-rays are not affected by electric and magnetic field (Figure 4.12).

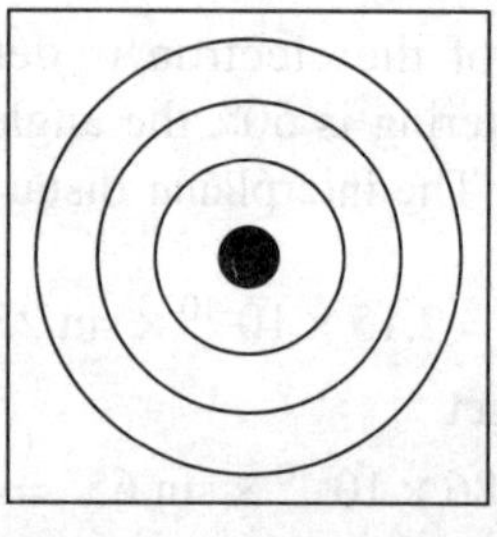

Figure 4.12 Diffraction pattern obtained in G.P. Thomson's experiment.

The circular diffraction pattern confirms the wave nature of electrons since the diffraction is possible only by the wave nature. Thomson was able to calculate the wavelength of the electron from de Broglie equation and the interplanar spacing from Bragg's equation. The interplanar spacing obtained in this method is in agreement with the values obtained from X-ray diffraction.

SOLVED PROBLEMS

4.1 Find the change in wavelength of an X-ray photon when it is scattered through an angle of 135° by a free electron. (Given: $h = 6.63 \times 10^{-34}$ J-s; $m_0 = 9.1 \times 10^{-31}$ kg and $c = 3 \times 10^8$ m s^{-1})

Given data

Planck's constant, $h = 6.63 \times 10^{-34}$ J-s

Mass of the electron, $m_0 = 9.1 \times 10^{-31}$ kg

Velocity of light, $c = 3 \times 10^8$ m s^{-1}
Angle of scattering, $\phi = 135°$

Solution

The change in wavelength, i.e. Compton's wavelength

$$\lambda' - \lambda = \Delta\lambda = \frac{h}{m_0 c}(1 - \cos\phi)$$

$$\Delta\lambda = \frac{6.63 \times 10^{-34}}{9.1 \times 10^{-31} \times 3 \times 10^8}(1 - \cos 135°)$$

$$= 4.1458 \times 10^{-12} \text{ m}$$

The change in wavelength is 4.1458×10^{-12} m

4.2 A photon of wavelength 2Å is made to incident on an electron at rest at an angle of 90°. Calculate (i) the Compton shift, (ii) the wavelength of the scattered photon, (iii) the energy of the recoiling electron and (iv) the angle at which the recoil electron appears.

Given data
Wavelength of the photon, $\lambda = 2$Å
Angle of scattering, $\varphi = 90°$

Solution

(i) Compton shift, $$\Delta\lambda = \frac{h}{m_0 c}(1 - \cos\phi)$$

$$= \frac{6.626 \times 10^{-34}}{9.1 \times 10^{-31} \times 3 \times 10^8}(1 - \cos 90°)$$

$$= 0.02427 \text{ Å}$$

(ii) Wavelength of the scattered photon,

$$\lambda' = \Delta\lambda + \lambda$$

$$= 0.02426 \text{ Å} + 2 \text{ Å}$$

$$= 2.02427 \text{ Å}$$

(iii) Energy of the recoiling electron $= h\nu - h\nu'$

$$= \frac{hc}{\lambda} - \frac{hc}{\lambda'} = hc\left(\frac{1}{\lambda} - \frac{1}{\lambda'}\right)$$

$$= 6.626 \times 10^{-34} \times 3 \times 10^8 \times \left(\frac{1}{2 \times 10^{-10}} - \frac{1}{2.02427 \times 10^{-10}}\right)$$

$$= 1.191 \times 10^{-17} \text{ J} = 74.477 \text{ eV}$$

(iv) The angle at which the recoiling electron appears,

$$\tan 0 = \frac{h\nu \sin\phi}{(h\nu - h\nu' \cos\phi)}$$

Substituting $\varphi = 90°$, we get

$$\tan\theta = \frac{h\nu \sin 90°}{(h\nu - h\nu' \cos 90°)}$$

$$\theta = 45°$$

(i) The Compton's shift is 0.02427 Å.
(ii) The wavelength of the scattered photon = 2.02427 Å.
(iii) The energy of the recoiling electron = 74.477 eV.
(iv) The angle at which the recoiling electron appears = 45°.

4.3 The energy of the incident photon is 1 MeV. If it strikes an electron at an angle of 60°, calculate (i) the Compton shift and (ii) the wavelength of the scattered photon.

Solution

(i) Compton shift, $\Delta\lambda = \frac{h}{m_0 c}(1 - \cos\phi)$

$$= \frac{6.626 \times 10^{-34}}{9.1 \times 10^{-31} \times 3 \times 10^{8}}(1 - \cos 60°)$$

$$= 1.213 \times 10^{-12} \text{ m}$$

The energy of the incident photon,

$$E = h\nu = \frac{hc}{\lambda}$$

$$\lambda = \frac{6.626 \times 10^{-34} \times 3 \times 10^{8}}{1 \times 1.6 \times 10^{-19} \times 10^{6}}$$

$$= 1.242 \times 10^{-12} \text{ m}$$

(ii) The wavelength of the scattered photon,

$$\lambda' = \Delta\lambda + \lambda$$

$$= 1.213 \times 10^{-12} + 1.242 \times 10^{-12}$$

$$= 2.455 \times 10^{-12} \text{ m}$$

(i) The Compton's shift = 0.01213 Å
(ii) The wavelength of the scattered photon = 2.455×10^{-12} m

4.4 Calculate the number of photons emitted in one second by a 60 W sodium vapour lamp emitting wavelength of 5893 Å.

Solution

Energy of the photon = $h\nu$

$$= \frac{hc}{\lambda} = \frac{6.626 \times 10^{-34} \times 3 \times 10^{8}}{5893 \times 10^{-10}}$$

$$= 3.373 \times 10^{-19} \text{ J} = 2.108 \text{ eV}$$

The number of photons emitted per second

$$= \frac{\text{output power}}{\text{energy of one photon}}$$

$$= \frac{60}{3.373 \times 10^{-19}} = 1.778 \times 10^{20} \text{ photons per second}$$

The number of photons emitted = 1.778×10^{20} photons per second.

4.5 Calculate the momentum, energy and mass of a photon of wavelength 10 Å.

Solution

(i) Energy = $h\upsilon = \dfrac{hc}{\lambda} = \dfrac{6.626 \times 10^{-34} \times 3 \times 10^{8}}{10 \times 10^{-10}}$ J = 1.9878×10^{-16} J

= 1.242 keV

(ii) Momentum, $p = \dfrac{h}{\lambda} = \dfrac{6.626 \times 10^{-34}}{10 \times 10^{-10}} = 6.626 \times 10^{-25}$ kg m s^{-1}

(iii) Mass = $\dfrac{h}{\lambda c} = \dfrac{6.626 \times 10^{-34}}{10 \times 10^{-10} \times 3 \times 10^{8}} = 2.2086 \times 10^{-33}$ kg

Energy = 1.9878×10^{-16} J = 1.242 keV

Momentum = 6.626×10^{-25} kg m s^{-1}

Mass = 2.2086×10^{-33} kg

4.6 Calculate the wavelength associated with an electron subjected to a potential difference of 1.25 kV.

Given data

Potential difference applied = 1.25 kV

Solution

The de Broglie wavelength of the electron,

$$\lambda = \frac{h}{\sqrt{2meV}}$$

$$\lambda = \frac{12.27}{\sqrt{V}} \text{Å}$$

$$\lambda = \frac{12.27}{\sqrt{1.25 \times 10^{3}}} \text{Å}$$

$$= 0.347 \text{Å}$$

The de Broglie wavelength of the electron = 0.347 Å

4.7 Calculate the de Broglie wavelength for a beam of electrons whose energy is 45 eV.

Given data

Energy of the electron = 45 eV

Solution

The de Broglie wavelength of the electron,

$$\lambda = \frac{h}{\sqrt{2mE}}$$

$$\lambda = \frac{6.626 \times 10^{-34}}{\sqrt{2 \times 9.1 \times 10^{-31} \times 45 \times 1.6 \times 10^{-19}}}$$

$$= 1.830 \text{ Å}$$

The de Broglie wavelength of the electron = 1.830 Å

4.8 Calculate the de Broglie wavelength of an electron moving with a velocity of 10^7 m s^{-1}.

Given data

Velocity of the electron = 10^7 m s^{-1}

Solution

The de Broglie wavelength of the electron,

$$\lambda = \frac{h}{mv}$$

$$\lambda = \frac{6.626 \times 10^{-34}}{9.1 \times 10^{-31} \times 10^{7}}$$

$$= 0.728 \text{ Å}$$

The de Broglie wavelength of the electron = 0.728 Å

4.9 Calculate the de Broglie wavelength of a proton accelerated through a potential difference of 1000 V. Given Planck's constant = 6.626 × 10^{-34} J-s, mass of the proton, m_p = 1.67 × 10^{-27} kg and charge of the electron, e = 1.6 × 10^{-19} J.

Given data

Potential difference applied = 1000 V

Planck's constant, h = 6.626 × 10^{-34} J-s

Mass of the proton, m_p = 1.67 × 10^{-27} kg

Charge of the electron, e = 1.6 × 10^{-19} J

Solution

The de Broglie wavelength of the proton,

$$\lambda = \frac{h}{\sqrt{2meV}}$$

$$\lambda = \frac{6.626 \times 10^{-34}}{\sqrt{2 \times 1.67 \times 10^{-27} \times 1000 \times 1.6 \times 10^{-19}}}$$

$$= 9.064 \times 10^{-13} \text{ m}$$

The de Broglie wavelength of proton = 9.064 × 10^{-13} m

4.10 A particle is moving in one-dimensional infinite potential well of width 25 Å. Calculate the probability of finding the particle within a small interval of 0.05 Å at the centre of the box when it is in its state of least energy.

Given data

Width of the potential well = 25 Å

Interval, $\Delta x = 0.05$ Å

Solution

Probability of finding the particle

$$P = \int_{-\infty}^{\infty} |\psi|^2 \, dx$$

Substituting the value of ψ, we get

$$P = \int_{-\infty}^{\infty} A^2 \sin^2 \frac{n\pi x}{L} dx$$

Substituting the value of A, where $A = \sqrt{\frac{2}{L}}$

$$P = \int_0^L \frac{2}{L} \sin^2 \frac{n\pi x}{L} dx$$

Since the particle is in the least energy, $n = 1$ and it is at the centre, $x = L/2$, the width of the well, $L = 25$ Å. The particle is in a small interval, $dx = 0.05$ Å. Substituting the values, we get

$$P = \int_0^{25\times10^{-10}} \frac{2}{25\times10^{-10}} \left(\sin^2 \frac{\pi}{2}\right) \times 0.05 \times 10^{-10} = 0.004$$

The probability of finding the particle is 0.004.

4.11 An electron is confined to a one-dimensional potential well of width 1 Å. Calculate the lowest energy of the electron. Express the result in electron volt.

Given data

Width of the potential well = 1Å

Solution

The energy of the electron in a one-dimensional potential well

$$E = \frac{n^2 h^2}{8mL^2}$$

For the lowest energy value, $n = 1$. Therefore,

$$E = \frac{h^2}{8mL^2}$$

Substituting the values of h, m and L, we get

$$E = \frac{(6.626 \times 10^{-34})^2}{8 \times 9.1 \times 10^{-31} \times (1 \times 10^{-10})^2} = 6.03 \times 10^{-18} \text{ J}$$

$$E = \frac{6\times10^{-18}}{1.6\times10^{-19}}\ \text{eV} = 37.69\ \text{eV}$$

The lowest energy of the electron = 37.69 eV

4.12 Consider two electrons which are confined to a potential well of width 1 Å . Calculate the lowest energy of the system.

Given data

Width of the potential well = 1 Å

Solution

According to the Pauli's exclusion principle, the first two electrons occupy the ground state. Therefore, $n = 1$.

The total energy of the system

$$E = 2\frac{n^2h^2}{8mL^2}$$

For the lowest energy value, $n = 1$. Therefore,

$$E = \frac{2h^2}{8mL^2}$$

Substituting the values of h, m and L, we get

$$E = \frac{2\times(6.626\times10^{-34})^2}{8\times9.1\times10^{-31}\times(1\times10^{-10})^2} = 1.206\times10^{-17}\ \text{J}$$

$$E = \frac{1.206\times0^{-17}}{1.6\times10^{-19}}\ \text{eV} = 75.38\ \text{eV}$$

The lowest energy of the system = 75.38 eV

4.13 A one-dimensional box of length 1 Å has three electrons. Calculate the lowest energy of the system and state the quantum numbers of the electrons.

Given data

Width of the potential well = 1 Å

Solution

According to the Pauli's exclusion principle, the first two electrons occupy the ground state ($n = 1$) and the second electron occupy the next state ($n = 2$).

The total energy of the system

$$E = 2\frac{h^2}{8mL^2} + \frac{2^2h^2}{8mL^2}$$

i.e.

$$E = \frac{6h^2}{8mL^2}$$

Substituting the values of h, m and L, we get

$$E = \frac{6 \times (6.626 \times 10^{-34})^2}{8 \times 9.1 \times 10^{-31} \times (1 \times 10^{-10})^2} = 3.618 \times 10^{-17} \text{ J}$$

$$E = \frac{3.616 \times 0^{-17}}{1.6 \times 10^{-19}} \text{eV} = 226.15 \text{ eV}$$

Quantum Numbers are,

(i) $n = 1, l = 0, m_L = 0, m_S = +\frac{1}{2}$

(ii) $n = 1, l = 0, m_L = 0, m_S = -\frac{1}{2}$

(iii) $n = 2, l = 0, m_L = 0, m_S = +\frac{1}{2}$

The lowest energy of the systems = 226.15 eV

4.14 The lowest energy of a particle in an infinite potential well with a width of 100 Å is 0.025 eV. What is the mass of the particle?

Given data

Width of the well = 100 Å

Lowest energy = 0.025 eV

Solution

The energy of the potential well

$$E = \frac{n^2 h^2}{8mL^2}$$

For the lowest energy, $n = 1$. Substituting $n = 1$, we get

$$E = \frac{h^2}{8mL^2}$$

The mass of the particle

$$m = \frac{h^2}{8EL^2}$$

Substituting the values of h, E, and L, we get

$$m = \frac{(6.626 \times 10^{-34})^2}{8 \times 0.025 \times 1.6 \times 10^{-19} \times (100 \times 10^{-10})^2}$$

$$= 1.37 \times 10^{-31} \text{ kg}$$

The mass of the particle is 1.37×10^{-31} kg

4.15 Find the energy density of a black body radiation at $T = 6000$ K in the range of 450 nm to 460 nm.

Given data

Wavelength region = 450 nm to 460 nm

Temperature = 6000 K

Solution

(i) For $\lambda = 450$ nm, $\nu = \dfrac{c}{\lambda} = \dfrac{3 \times 10^8}{450 \times 10^{-9}} = 6.67 \times 10^{14}$ Hz

(ii) For $\lambda = 460$ nm, $\nu = \dfrac{c}{\lambda} = \dfrac{3 \times 10^8}{460 \times 10^{-9}} = 6.52 \times 10^{14}$ Hz

The average value of ν is 6.595×10^{14} Hz.

The energy density

$$\rho_\nu = \frac{8\pi h\nu^3}{c^3} \frac{1}{\exp\left(\dfrac{h\nu}{kT}\right) - 1}$$

Substituting the values of h, c, k, T and ν, we get

$$\rho_\nu = \frac{8\pi \times 6.626 \times 10^{-34} \times (6.595 \times 10^{14})^3}{(3 \times 10^8)^3} \frac{1}{\exp\left(\dfrac{6.626 \times 10^{-34} \times 6.595 \times 10^{14}}{1.38 \times 10^{-23} \times 6000}\right) - 1}$$

$$= 9.077 \times 10^{-16} \text{ J m}^{-3}$$

The energy density of the black body = 9.077×10^{-16} J m^{-3}.

SHORT QUESTIONS

1. What is meant by a black body?
2. Mention the characteristics of a black body spectrum.
3. State Kirchhoff's law.
4. State Stefan–Boltzmann law.
5. State Wien's displacement law.
6. State Rayleigh–Jeans' law.
7. Write the postulates of Planck's quantum-free electron theory of radiation.
8. Write the Planck's formula for black body radiation and explain the terms.
9. Deduce Wien's displacement formula from Planck's equation.
10. Deduce Rayleigh and Jeans' law from Planck's equation.

11. Write the Planck's quantum theory of radiation.
12. Mention the advantages of Planck's quantum theory.
13. Mention the properties of photon.
14. State Compton effect.
15. Write the Compton equation and explain the terms.
16. Write the equation for the direction of the recoiling electron and hence explain the terms.
17. Write the Schrödinger time-dependent equation and explain the terms.
18. Write the Schrödinger time-independent equation and explain the terms.
19. Write the physical significance of the wave function, ψ.
20. What are eigenfunctions?
21. What are eigenvalues?
22. What are matter waves?
23. What are the reasons that prompted de Broglie to propose the concept of matter waves?
24. Deduce the wavelength of the matter waves.
25. Deduce the de Broglie wavelenth in terms of kinetic energy.
26. Derive the de Broglie wavelength of an electron.
27. Deduce the de Broglie wavelength of electron in terms of voltage.
28. List out the properties of de Broglie wavelength.

DESCRIPTIVE TYPE QUESTIONS

1. (i) Describe the Planck's quantum theory of radiation and hence deduce the Planck's equation for the black body radiation.
   (ii) Deduce Rayleigh and Jeans' law from Planck's equation.
   (iii) Deduce Wien's displacement law from Planck's equation.
2. (i) What is Compton effect?
   (ii) Derive the mathematical derivation for the Compton shift in wavelength.
   (iii) Explain the experimental verification for the Compton's effect.
3. Derive the Schrodinger's time-dependent wave equation and hence derive the time-independent wave equation.
4. Solve the one-dimensional potential well problem and hence obtain expressions for the wavefunction of the electrons and the energy of the electrons.
5. What are the matter waves? Derive an expression for the wavelength of the matter waves. Enumerate the properties of the matter waves.
6. Explain in detail with neat sketch the Davisson and Germer experiments.
7. Explain in detail the G.P. Thomson's experiment for the matter waves.

PROBLEMS

1. Calculate the de Broglie wavelength of the electrons in the following cases:
   (i) an electron accelerated by a potential difference of 1000 V
   (ii) an electron of energy 100 eV
   (iii) an electron having velocity 10^6 m s^{-1}
2. The first order maximum of electron diffraction in a metal crystal occurred ($d = 0.9086$ Å) at a glancing angle of 65°. Determine the de Broglie wavelength of electron and their velocity.
3. Calculate the wavelength associated with a thermal neutron of energy 0.025 eV.
4. An α particle has energy equal to 10^7 eV. Calculate the wavelength to which this corresponds.
5. Neutrons are in equilibrium with matter at 300 K. Calculate the energy in eV of a neutron and its de Broglie wavelength.
6. The width of an infinite potential well is 10 Å. Calculate the first three energy levels in terms of eV for an electron.
7. A photon of energy 1 MeV is made to collide with an electron at rest at an angle of 60°. What is Compton shift in wavelength? What is the wavelength of the scattered photon?
8. Calculate the Compton shift, when a photon is made to collide on an electron at the following angles, namely (i) 45°, (ii) 60°, (iii) 90°, (iv) 135°, (v) 180°.
9. A particle of energy 10 keV makes collision with an electron at rest at the angle of 60°. Determine the (i) Compton shift, (ii) Wavelength of the scattered photon, (iii) energy gained by the electron, and (iv) direction of the recoiling electron.
10. In a Compton scattering experiment, the X-ray photon is scattered by a electron at 60° and the energy of the recoiling electron is 2 keV. Calculate the wavelength of the incident photon.
11. Determine the energy and momentum of a photon of wavelength 1 Å.
12. The energy of a photon is 1 MeV. Calculate its wavelength.

CHAPTER 5

ELECTRON OPTICS

5.1 INTRODUCTION

An instrument that uses the principle of optics is known as an optical instrument. The instruments such as photographic camera, telescope, microscope, spectrometer, etc. are called optical instruments. A microscope is used to produce an enlarged image of an object, using the principle of optics. An electron microscope uses the electron beam. The electron beam can be focused and converged in a way that is similar to ordinary light rays. In addition the electrons behave as a wave. The wave nature of electron is utilized in the electron microscope. The branch of science that deals with the focusing, magnification and transmission of electron is known as electron optics. In this chapter, the principle, construction and working of a metallurgical microscope, optical microscope, scanning electron microscope and transmission electron microscope are discussed.

5.2 METALLURGICAL MICROSCOPE

5.2.1 Principle

A metallurgical microscope works on the principle of the reflection of light. The light reflected from a metal surface is passed through an objective lens and an eyepiece lens and these two lenses magnify the image.

5.2.2 Construction

A metallurgical microscope is shown in Figure 5.1. It consists of an incandescent xenon or halogen lamp, *S*. It has an objective lens and an eyepiece lens. In between the objective lens and the eyepiece lens there is a plane glass reflector inclined at an angle of 45°. The specimen holder lies just below the objective lens.

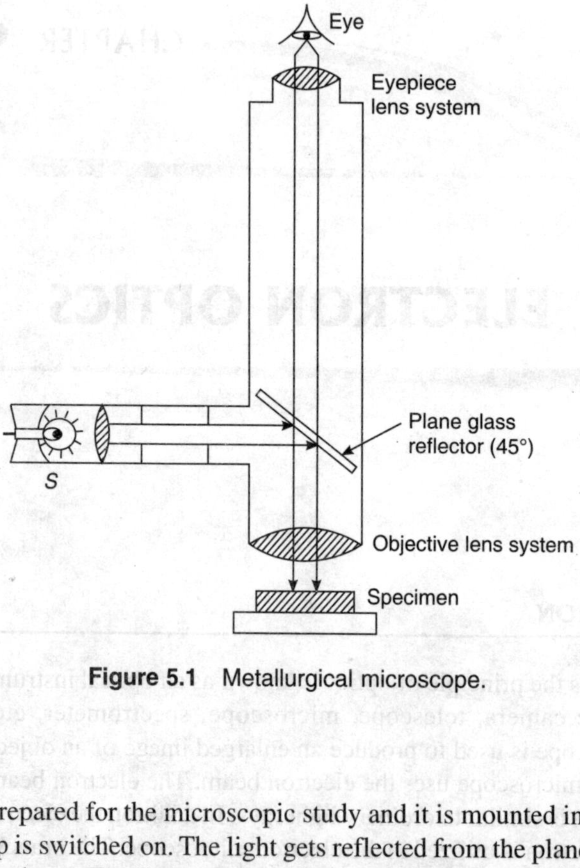

Figure 5.1 Metallurgical microscope.

5.2.3 Working

The metal sample is prepared for the microscopic study and it is mounted in the specimen holder. The incandescent lamp is switched on. The light gets reflected from the plane glass reflector and it travels towards the specimen. The light reflected from the specimen is viewed from the eyepiece lens. The objective lens has a turret mount, which contains lenses with different magnifications such as 5x, 10x, 20x, 40x and 100x. Similarly, the eyepiece lens also has lenses with different magnifications 5x, 10x, 15x ... etc. Lenses with suitable magnification are selected from the eyepiece lens and the turret of the objective lens, the magnification is varied from 20x to 2000x. The total magnification is the product of the magnification of eyepiece lens and the magnification of the objective lens.

Generally, in a metallurgical microscope for the magnification above 1000x, an oil-immersed objective lens with magnification 100x is used. A drop of the oil having high refractive index (the refractive index of the cedar oil is 1.5) is applied between the objective lens and the specimen and then viewed.

5.2.4 Uses

1. Using a metallurgical microscope, the surface analysis such as the grain size of steel and carbon steel is studied. From the surface morphology, the percentage of the carbon content in carbon steel is determined.
2. It is used for the study of the changes in the grain size due to the heat treatment of carbon steel.
3. It is used to find the composition of certain metals based on the grain distribution in the surface of the material.

5.3 ELECTROSTATIC ELECTRON AND ELECTROMAGNETIC ELECTRON LENSES

The ordinary light is magnified or focused by an ordinary lens made of glasses. But the electron beams are magnified and focused with the help of electrostatic and electromagnetic lenses.

5.3.1 Electrostatic Electron Lens

The simplest form of the electrostatic lens consists of two co-axial cylinders, A and B as shown in Figure 5.2. These two cylinders are kept at different potentials. The cylinder B is kept at some higher potential than the cylinder A. The electron beam passing through these cylinders travels normal to the equipotential surfaces. The electron beams are focused at a focal point, F. The focal length can be varied by varying the potential difference applied to these cylinders. In order to get good focusing of electron beams more than two lenses are used.

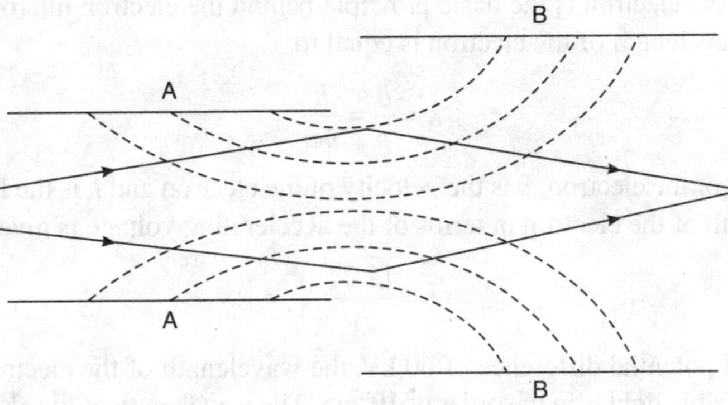

Figure 5.2 Electrostatic electron lens.

5.3.2 Electromagnetic Electron Lens

In the electromagnetic electron lens, the magnetic field is used to focus the electron beams. Several types of electromagnetic lenses are available. One of them is shown in Figure 5.3. In Figure 5.3, AB represents the section of the electromagnet, p and q represent the gaps and the dots represent the section of the wire. A soft iron shield surrounds the magnet except the gaps p and q. When electron beams are passed through these magnetic fields, the electron beams get focused. The magnetic lenses are mostly used in microscopes whenever an intense and very fast electron beams are required.

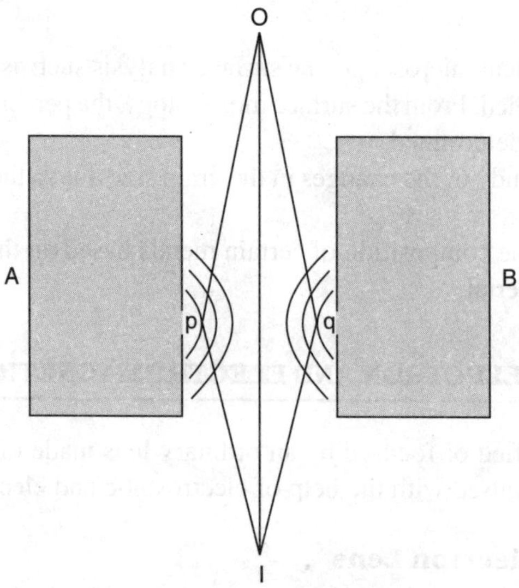

Figure 5.3 Electromagnetic electron lens.

5.4 ELECTRON MICROSCOPE

5.4.1 Principle

The wave nature of the electron is the basic principle behind the electron microscope. According to de Broglie, the wavelength of the electron is equal to

$$\lambda = \frac{h}{p} = \frac{h}{mv} \tag{5.1}$$

where m is the mass of the electron, v is the velocity of the electron and h is the Planck's constant.

The wavelength of the electron in terms of the accelerating voltage is given by

$$\lambda = \frac{12.27}{\sqrt{V}} \text{ Å} \tag{5.2}$$

For an applied potential difference of 60 kV, the wavelength of the electron is 0.05 Å. The wavelength of the visible light is in the order of 10^{-7} m. The wavelength of the electron is 10^5 times shorter than the wavelength of the visible light. The resolving power of any instrument is inversely proportional to the wavelength of the source used. Therefore, a high resolving power can be achieved by using electrons as a source.

5.4.2 Construction

The construction of an electron microscope is similar to that of an optical microscope. An optical microscope and an electron microscope are shown in Figure 5.4 for comparison. An optical microscope consists of an incandescent lamp, S, a condensing lens L_1, magnifying lenses L_2 and L_3, an intermediate screen S_1 and another screen S_2. Similarly an electron microscope, also consists of an electron gun, E, the condensing lens L_1, magnifying lenses, L_2 and L_3, intermediate screen S_1 and a phosphor screen S_2. It has an optical window W to view the image on the phosphor screen. The window, W has an ordinary lens having a magnification of 10x. The electron microscope works at very high vacuum. Therefore it is enclosed in a rigid metal frame, and a high degree of vacuum is maintained with a high speed diffusion pump. In electron microscope, the electrostatic and electromagnetic lenses are used. It has the provision to introduce the object slide and the photographic plate from outside.

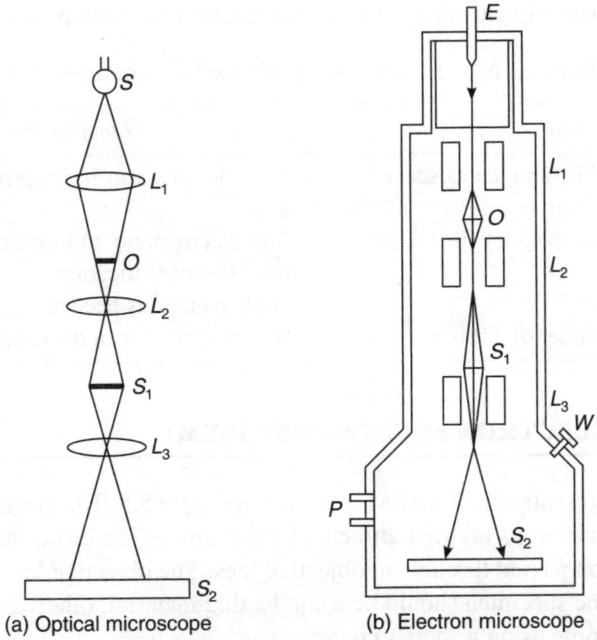

Figure 5.4 Comparison between an optical microscope and an electron microscope.

5.4.3 Working

In an electron microscope, monochromatic beam of electron is produced by the electron gun. This electron beam is collimated and passed through a condensing lens L_1. The lens L_1 condenses the electron beam. The condensed electron beam is passed through the object. The object is mounted on a thin film of cellulose-supported wire gauge. The electron beam emerging from the object is passed through the magnifying lens L_2. The lens L_2 magnifies the image 100 times. Then the image is made to fall on an intermediate screen S_1. Then the electron beam is passed through the magnifying lens L_3. It also magnifies the image again by 100 times. Then the image is made to fall on a phosphor screen S_2. The image formed on the screen S_2 is viewed through a window W. The window W has an optical lens. It magnifies the image 10 times and hence the total magnification becomes 10^5 times.

5.4.4 Advantages

1. The magnification of the electron microscope is in the order of 10^5 and its resolving power is in the order of 10Å to 100 Å.

2. The structural details of virus, proteins, etc. are studied with electron microscopes.
3. In metallurgy, it is used for the study of surface analysis of metals.
4. It is used for the surface study of nanostructured thin films.

5.5 DIFFERENCE BETWEEN AN OPTICAL MICROSCOPE AND AN ELECTRON MICROSCOPE

The difference between an optical microscope and an electron microscope is displayed in Table 5.1.

Table 5.1 Difference between an optical microscope and an electron microscope

Optical microscope	*Electron microscope*
Uses the light produced by an incandescent lamp as a source.	Uses the electron beam produced by an electron gun.
Ordinary lens are used to magnify the image.	The electrostatic and electromagnetic lenses are used for magnification.
No need of vacuum.	High vacuum is needed.
Magnification is in the order of 1500.	Magnification is in the order of 10^5 times.

5.6 SCANNING ELECTRON MICROSCOPE (SEM)

The diagrammatic representation of a SEM is shown in Figure 5.5. The electrons are produced by the electron gun. The electrons are first attracted by the anode. The electrons are condensed by a condenser lens and then passed through an objective lens. The objective lens focuses the electron beam on the object. The specimen should be a conducting material; otherwise a thin gold layer is deposited over the sample using a sputter coater.

A set of small coils, called scan coils, are located inside the objective lens. A scan generator is used to operate the scan coils. The scan generator produces a varying voltage, and this varying electric field induces a magnetic field. The magnetic fields produced by the scan coil deflect the electron beam back and forth in a controlled pattern called raster. The raster is similar to that of the raster in television receiver.

The varying voltage produced by the scan generator is also applied to the set of deflection coils in CRO. These coils produce magnetic field and hence they deflect the electron beam that incident on the CRT screen. The deflection of the electron beam in the CRT screen is the same as the deflection of the electron beam incident on the sample.

When the electrons are made to strike on the sample, it produces secondary electrons. These secondary electrons are detected by a detector known as Everhart-Thornly detector. The detector has Faraday cage or collector screen at the front. Faraday cage is a wire mesh or a metal ring with a positive 300 V applied to it so as to attract the secondary electrons.

The secondary electrons are accelerated by 12 kV in the Faraday cage and made to incident on a scintillator. The scintillator is a metal-coated disk that acts as a collector. When the electrons strike the scintialltors, they are converted into photons of light. The photons are transmitted through the light pipe and then the photons are made to fall on a photomultiplier tube. The photomultiplier tube converts the photons into highly amplified electrical signal. The output voltage is then amplified by a preamplifier. The preamplifier output voltage is then applied to the grid of the CRT and modulates/changes the intensity of the spot on the CRT screen. Finally an image is produced on the CRT screen that corresponds to the topography of the sample.

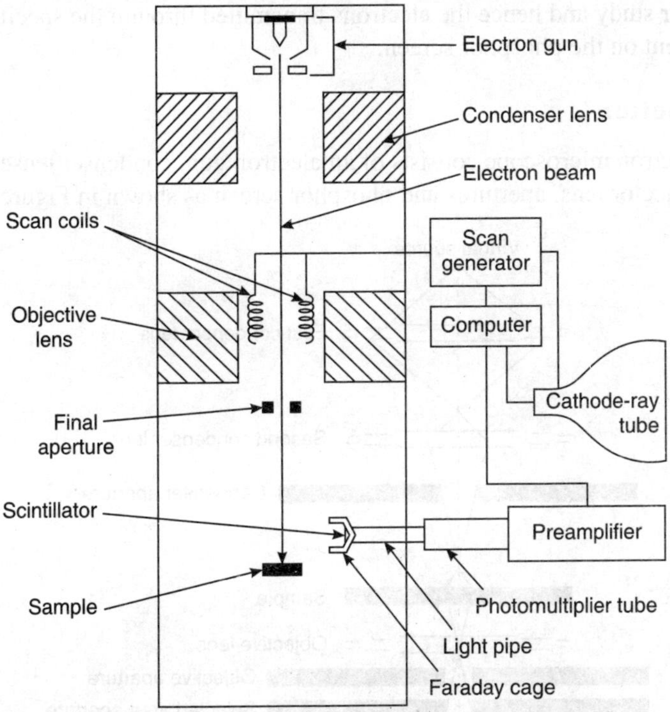

Figure 5.5 Scanning electron microscope (SEM).

5.6.1 Advantages

Elemental analysis

SEM is used for elemental analysis in conjunction with EDS (Energy Dispersive Spectrum). It is used to detect the low atomic number elements such as carbon and oxygen.

Topography

The surface analysis of the specimen and its texture can be studied using SEM.

Morphology

It is used to identify the size, shape and arrangement of the particles which make up the specimen in atomic scale.

Other applications

1. In plastics, SEM can be used to examine surface irregularities or fracture area in the specimen.
2. It is used to measure the thickness (in cross section) of thin coating.

5.7 TRANSMISSION ELECTRON MICROSCOPE (TEM)

5.7.1 Principle

The transmission electron microscope works on the principle of passing a beam of electrons through the specimen under study and hence the electrons transmitted through the specimen are collected and made to incident on the phosphor screen.

5.7.2 Construction

A transmission electron microscope consists of an electron gun, condenser lenses, sample holder, objective lens, projector lens, apertures and phosphor screen as shown in Figure 5.6.

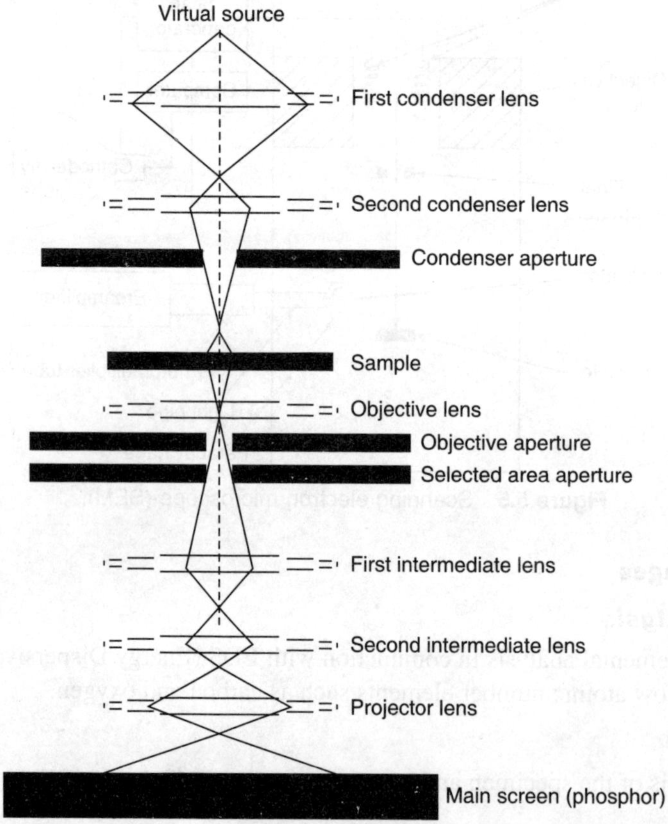

Figure 5.6 Transmission Electron Microscope (TEM).

5.7.3 Working

The TEM column is shown in Figure 5.7 and the corresponding lenses in TEM are shown in Figure 5.6. The electron gun produces a stream of monochromatic electrons. The electron is accelerated by an accelerating voltage of 20 kV to 100 kV. The resolution of TEM depends on the accelerating voltage; the higher the accelerating voltage, the greater is the resolution. The stream of electrons emerging from the filament has a spot size of 50 μm and it is focused into the first condenser lens. The first condenser lens reduces the spot size of electron beam into 1 μm.

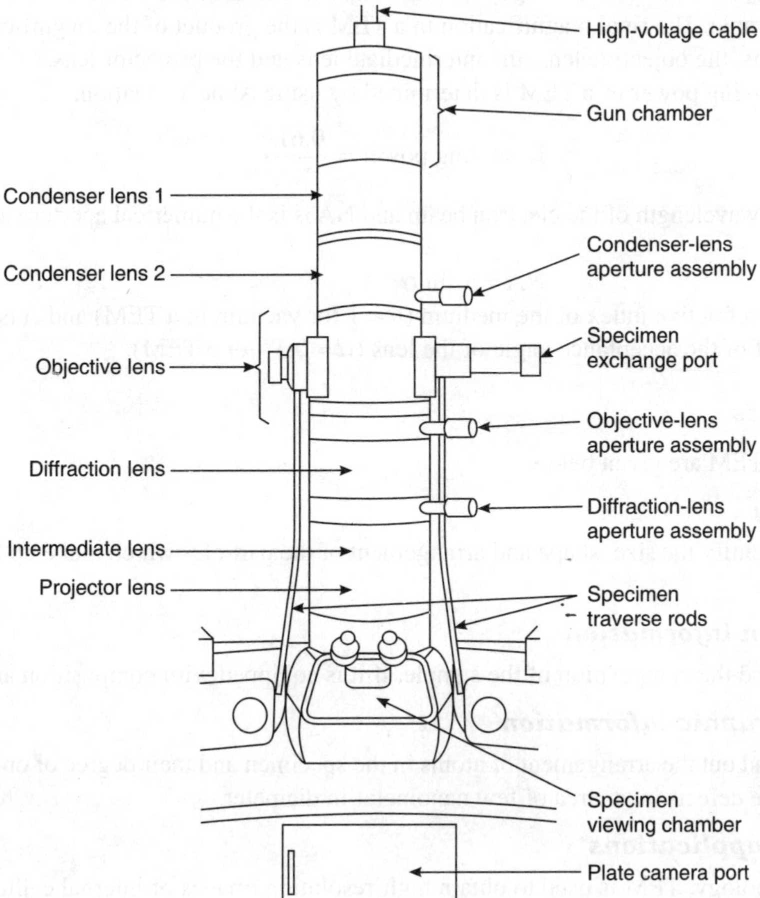

Figure 5.7 TEM column.

The second condenser lens is usually controlled by intensity or brightness knob. It controls the size of the electron beam from 1 to 10 μm. The electron beam is restricted by the condenser aperture. The condenser lens aperture protects the specimen from too many stray electrons, which can contribute excessive heat and X-ray production.

The specimen is inserted into the objective lens. The specimen can be moved both in *X*- and *Y*-directions. Nowadays, most modern TEM has *Z*-direction movement and tilting and rotating capabilities. The electron beam is made to incident on the sample.

A part of the incident electron beam is transmitted through the sample. The objective lens focuses the transmitted electron beam into an image. The objective aperture is used to enhance the contrast by blocking out high-angle diffracted electrons. The selected area aperture enables the user to examine the periodic diffraction of electrons by ordered arrangement of atoms in the sample. The image gets enlarged when it is passing through the intermediate and projector lenses. The image is made to strike the phosphor screen and the user can be allowed to see the image. The darker areas of the image represent that only a few electrons have transmitted through the sample, whereas the lighter areas of the image represent that more number of electrons have transmitted through the sample. The final magnification in a TEM is the product of the magnification of each magnifying lens, the objective lens, the intermediate lens and the projector lens.

The resolving power of a TEM is determined by using Abbe's equation,

$$\text{Resolving power} = \frac{0.61\lambda}{\text{NA}} \tag{5.3}$$

where λ is the wavelength of the electron beam and NA is is the numerical aperture and it is equal to

$$\text{NA} = n \sin \alpha$$

where n is the refractive index of the medium ($n \approx 1$ for vacuum in a TEM) and α is the aperture angle, one-half of the acceptance angle of the lens ($\alpha \approx 0.3°$ for a TEM).

5.7.4 Merits

The merits of TEM are given below.

Morphology

It is used to identify the size, shape and arrangement of the particles which make up the specimen in atomic scale.

Composition information

It is used to find the composition of the sample, if it is equipped with composition analyzer.

Crystallographic information

It is used to find out the arrangement of atoms in the specimen and their degree of order, detection of atomic scale defects in an area of few nanometer in diameter.

Biological applications

1. In biology, TEM is used to obtain high-resolution images of internal cellular structures of animal and plant tissues, as well as microorganisms.
2. TEM is used to obtain the structures of small particulate specimens such as viruses, phages and DNA.
3. The special preparation of samples allows us to use TEM for the localization of elements, enzymes and proteins. It is also used for the study of membrane interfaces and the structure of micromolecules.

SHORT QUESTIONS

1. What is an optical instrument?
2. What is an optical microscope?
3. What is a metallurgical microscope?

4. What is an electromagnetic electron lens?
5. What is an electrostatic electron lens?
6. What is an electron microscope?
7. What are the differences between an electron microscope and an optical microscope?
8. What is the principle of a scanning electron microscope?
9. What is the principle of transmission electron microscope?
10. What are the uses of an electron microscope?
11. What are the uses of a scanning electron microscope?
12. What are the uses of a transmission electron microscope?

DESCRIPTIVE TYPE QUESTIONS

1. Explain the principle, construction and working of an electron microscope with a neat sketch.
2. Explain with a neat sketch the principle, construction and working of a scanning electron microscope.
3. Explain with a neat sketch the principle, construction and working of a transmission electron microscope.
4. Explain the principle, construction and working of a metallurgical microscope with a neat sketch.

CHAPTER 6

ACOUSTICS

6.1 INTRODUCTION

Acoustics is the branch of science that deals with the study of sound. Acoustics is important for creating musical instruments or concert halls or surround sound stereo or hearing aids. Different types of acoustics are (i) general acoustics, (ii) architectural acoustics, (iii) medical acoustics, (iv) musical acoustics and (v) underwater acoustics. *General acoustics* is the study of the science of sound waves. It is the study of the production, propagation and receiving of sound. *Architectural acoustics* is the study of design buildings and other spaces that have pleasing sound quality and safe sound levels. Architectural acoustics includes the design of concert halls, classrooms and even heating systems. *Medical acoustics* is the study and use of acoustics to diagnose and treat different types of ailments by the physician. The study of medical acoustics includes the use of ultrasound and other acoustical techniques to learn how different types of sound interact with cells, tissues, organs and entire organisms. *Musical acoustics* is the study of the science of music. Musical acoustics deals with the study of the production, travel and receiving of musical sound. *Underwater acoustics* is the study of natural and man-made underwater sounds. It is the study of the production and propagation of sound underwater.

This chapter deals about the basic concepts about acoustics of buildings. The topics like reverberation, reverberation time, Sabine's formula, absorption coefficient, and its measurement are discussed in this chapter.

6.2 CLASSIFICATION OF SOUND

Sound waves are classified into three categories based on their frequencies, namely, (i) infrasonic, (ii) audible sound and (iii) ultrasound. The infrasound and ultrasound are not audible by us. The sound waves having frequency less than 20 Hz are said to be **infrasound**. The sound waves, whose frequency is lying between 20 Hz and 20kHz are said to be **audible sound**. The sound waves those having frequency greater than 20 kHz are said to be **ultrasonic waves**.

The audible sound is classified into two categories, namely, (i) musical sound and (ii) noise. The sound waves that produce pleasing effect on our ear is said to be musical sound. The musical sound has the following properties.

1. The waveform of the musical sound is regular in shape.
2. The waveform has periodic changes in its amplitude.
3. There are no sudden changes in its amplitude.

The sound produced by the piano, flute, guitar, violin, etc. are the examples for the musical sound. The waveforms of different musical sound are represented in Figure 6.1. A musical sound is produced by successive similar pulses having uniform shape followed by one another.

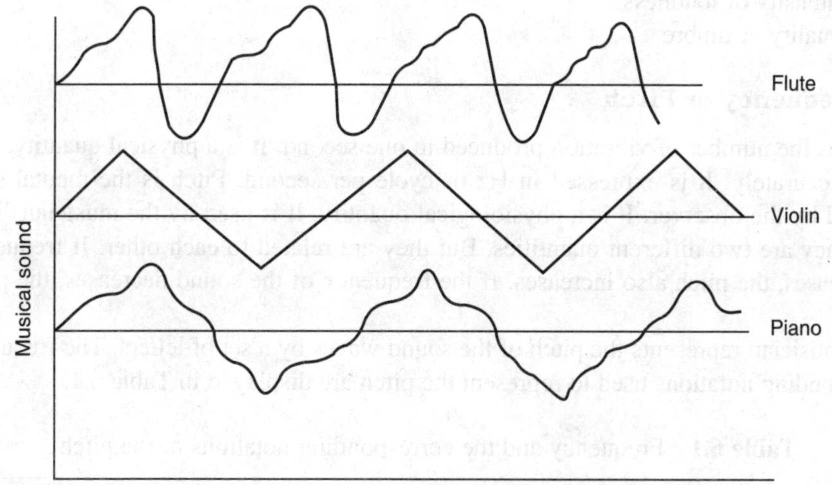

Figure 6.1 Waveforms of musical sounds

An unpleasant sound produced in our ear is said to be a noise. A noise has the following properties:

1. The waveform of a noise is irregular in shape.
2. There is no periodicity.
3. There are some sudden changes in their amplitude.

The waveform produced by a noise is shown in Figure 6.2.

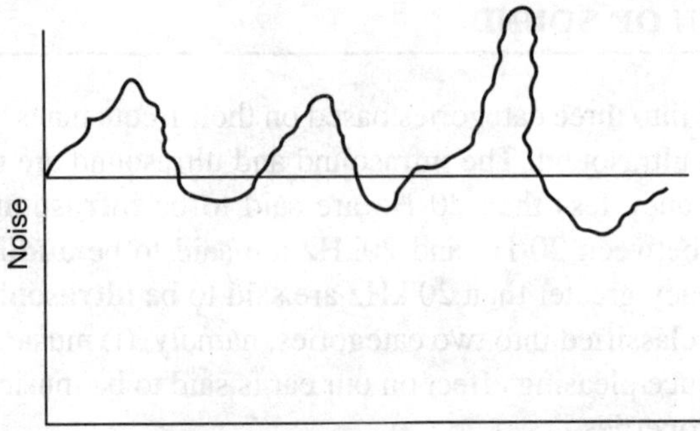

Figure 6.2 Waveform of a noise.

6.3 CHARACTERIZATION OF MUSICAL SOUND

The characteristics of the musical sound are as follows:

1. Frequency or pitch
2. Intensity or loudness
3. Quality or timbre

6.3.1 Frequency or Pitch

Frequency is the number of vibration produced in one second. It is a physical quantity. It can be measured accurately. It is expressed in Hz or cycle per second. Pitch is the mental sensation experienced by the observer. It is a physiological quantity. It is used by the musician. The pitch and frequency are two different quantities. But they are related to each other. If frequency of a sound increases, the pitch also increases. If the frequency of the sound decreases, the pitch also decreases.

The musician represents the pitch of the sound waves by a set of letters. The frequency and the corresponding notations used to represent the pitch are displayed in Table 6.1.

Table 6.1 Frequency and the corresponding notations of the pitch

Frequency (Hz)	256	288	320	342	384	426	480	512
Symbol	Middle C	D	E	F	G	A	B	Top C
Indian names	sa	ri	ga	ma	pa	dha	ni	sa

6.3.2 Intensity and Loudness

The quantity of sound energy flowing through unit area of cross-section in one second is called the intensity of the sound. It is given by the expression:

$$I = \frac{Q}{tA} \tag{6.1}$$

where Q is the quantity of sound energy, A is the area of cross-section, and t is the time. Equation (6.1) can be written as:

$$I = \frac{P}{A} \tag{6.2}$$

where P is the power and it is equal to Q/t. The unit for the intensity of the sound is W m^{-2}. It is a physical quantity. It can also be measured accurately.

Loudness is the degree of sensation conveyed to our ear. It varies from one observer to another observer. It is a physiological quantity. The loudness of a sound is not directly measured. The relative value of the loudness is measured rather than the absolute measurement.

The intensity and the loudness are two different quantities. But they are related to each other. If intensity of the sound increases, loudness also increases and vice versa.

Weber–Fechner law

Loudness is the degree of sensation conveyed to our ear. The Weber–Fechner law states that the loudness of a sound L is directly proportional to the logarithm of the intensity I of the sound. The loudness of the sound is written as:

$$L \propto \log I$$

$$L = K \log I \tag{6.3}$$

where K is a constant. Equation (6.3) is known as Weber–Fechner law.

Consider the intensity of the sound is doubled, i.e., $I = 2I_0$. Then the loudness is given by

$$L = K \log(2I)$$

$$L = K(\log 2 + \log I) = K \log 2 + K \log I$$

$$L = K \log 2 + L_1$$

$$L - L_1 = K \log 2 \tag{6.4}$$

From Eq. (6.4), we infer that if the intensity of the sound is doubled, irrespective of its initial value, there is a constant increase in the loudness of the sound.

Phon

Phon is the unit for loudness. Generally, the absolute value of the loudness is not measured. The loudness of a sound is measured by comparing with the loudness of a standard source of frequency 1000 Hz. To measure the loudness of a given source, its loudness is compared with the loudness produced by the standard source of frequency 1000 Hz. The two sources are heard alternatively, and the intensity of the standard source is adjusted so that its loudness matches with the loudness of the given source. Then the intensity level of the standard source is determined. If it is N dB, the equivalent loudness of the sound is N phons.

It should be noted that the interval of loudness corresponding to one decibel on the intensity scale is called one phon. The loudness is expressed by phon, whereas the intensity level is expressed by decibel.

The loudness level (LL) in phon for a frequency of 1000 Hz is:

$$\text{LL} = 10\log\left(\frac{I}{I_0}\right) \tag{6.5}$$

where I_0 is the intensity of the standard source, and it is equal to 10^{-12} W m^{-2}.

Intensity level of a sound wave

Generally, the relative intensity of the sound has more practical importance than the absolute intensity. Therefore, the relative intensity of the sound is measured, rather than the absolute intensity. For measuring the relative intensity of sound, the intensity of the sound wave produced by 1000 Hz is chosen as the standard sound intensity. The sound with frequency 1000 Hz produces an intensity of 10^{-12} W m^{-2}. This value is chosen as the standard sound intensity because it is the minimum audible sound intensity.

The relative intensity or intensity level of a sound is defined as the logarithmic value of the ratio between the intensity of the given sound and the intensity of the standard sound intensity. It is represented by the letter β.

$$\beta = \log_{10}\left(\frac{I}{I_0}\right) \tag{6.6}$$

where I is the intensity of the given sound and I_0 is the standard sound intensity. The unit for the intensity level is bel, named after Graham Bell, the inventor of telephone.

Consider, $I = I_0$, then $\beta = \log_{10}\left(\frac{I_0}{I_0}\right) = 0$ bel

$I = 10I_0$, then $\beta = \log_{10}\left(\frac{10I_0}{I_0}\right) = \log_{10} 10 = 1$ bel

$I = 100I_0$, then $\beta = \log_{10}\left(\frac{100I_0}{I_0}\right) = \log_{10} 100 = 2$ bel

Since the unit bel is one of a larger unit, another unit known as decibel is used. One decibel is one tenth of a bel (i.e., 1 decibel = 1/10 bel). It is represented by 1 dB. In terms of decibel, the relative intensity can be written as:

$$\beta = 10\log_{10}\left(\frac{I}{I_0}\right) \tag{6.7}$$

where I is the intensity of the given sound and I_0 is the standard sound intensity.

The intensity level of the audible sound varies from 0 dB to 120 dB. The intensity level of the minimum audible sound is 0 dB. It is called the threshold of audibility. The intensity level of the maximum audible sound is 120 dB. It is called threshold of feeling or pain threshold, since it produces pain in our ear. Table 6.2 gives the approximate relative intensity levels of different sounds in decibel.

Table 6.2 Approximate relative intensity levels of different sounds in decibels

Typical sound levels (in decibels)	
Threshold of audibility	0
3 m from human breathing	10
Average whisper 4 feet away	20
Soft radio in apartment; average residence	35
Average office	45
Moderate restaurant background	50
Noisy office or department store	55
Average busy street / two people talking	60
Stenographic room	70
City subway	80
Loud automobile horn 23 feet away	100
Express train passing at high speed	105
Threshold of feeling (varies with frequency)	120
Threshold of painful sounds; limit of ear's endurance	130
3 m from a jet engine	160

Meaning of 0 dB: Consider the intensity of sound I is equal to the standard intensity I_0 ($I_0 = 10^{-12}$ W m^{-2}). Then the relative intensity is

$$\beta = 10\log_{10}\left(\frac{I}{I_0}\right) = 10\log_{10}\left(\frac{I_0}{I_0}\right) = 10\log_{10}1 = 0 \text{ dB}$$

The relative intensity of the standard intensity is equal to 0 dB. It is called as threshold of audibility.

Physical significance of 1 dB: Let us calculate what percentage change of intensity is represented as 1 dB. For this, take $\beta = 1$ dB. The relative intensity is given by

$$\beta = 10\log_{10}\left(\frac{I}{I_0}\right) \tag{6.8}$$

Substituting, $\beta = 1$ dB, we get

$$1\,\text{dB} = 10\log_{10}\left(\frac{I}{I_0}\right)$$

Rearranging the above equation, we get

$$\log_{10}\left(\frac{I}{I_0}\right) = 0.1$$

$$\left(\frac{I}{I_0}\right) = 10^{0.1} = 1.26.$$

i.e.,

$$I = 1.26\,I_0 \tag{6.9}$$

From Eq. (6.9), we infer that an increase of 1 dB corresponds to an increase of 26% of the intensity of the sound. Similarly, an increase of 2 dB, corresponds to an increase of $(1.26)^2 I_0 = 1.5876 I_0$, i.e., 58% of the sound intensity.

Sound pressure level (SPL)

The intensity level of a sound is also expressed in terms of pressure known as sound pressure level (SPL). The intensity level of a sound is given by

$$\beta = 10 \log \left(\frac{I}{I_0} \right) \tag{6.10}$$

Since $I \propto P^2$, Eq. (6.10) can be written as:

$$\text{SPL} = 10 \log \left(\frac{P^2}{P_0^{\,2}} \right)$$

$$\text{SPL} = 20 \log \left(\frac{P}{P_0} \right) \tag{6.11}$$

The sound pressure level is defined as 20 times the logarithmic value of the ratio between the pressure of the sound P and the standard pressure P_0. The standard pressure chosen for this purpose is $P_0 = 2 \times 10^{-5}$ N m^{-2}. The sound pressure level of a quiet office is 30 dB, ordinary conversation is 40 dB, street noise is 60 dB, a truck noise is 100 dB and near the jet engine, the noise is 160 dB.

Sone

The loudness is also expressed in another unit called sone. The unit phon is not a true scale of loudness, i.e., the loudness of 60 phon is not twice as that of 30 phon, hence sone is used to represent the loudness. One sone is the loudness of 1000 Hz sound of 40 dB sound pressure level. Above 40 dB SPL, for every increase of 10 dB SPL, the sone scale gets doubled. Table 6.3 gives the SPL in dB and the corresponding values of loudness in sone.

Table 6.3 Sound pressure level in dB and the corresponding values of loudness in sone

SPL in dB	Loudness in sone
120	256
110	128
100	64
90	32
80	16
70	8
60	4
50	2
40	1
30	0.5
20	0.2
10	0.085
0	

Phon and sone are related by the following relation:

$$\log S = 0.03\,P - 12 \qquad (6.12)$$

where S is the loudness in sone and P is the loudness in phon.

Quality or timbre

Quality is the characteristics of a musical sound. The property that enables us to identify the sound produced by a particular musical instrument is called quality or timbre. A musical sound is produced by a combination of different frequencies such as fundamental frequency and overtones. For example, the musical sound produced by one instrument is a combination of frequencies such as $2n$, $4n$, $6n$, $8n$... and so on, whereas the sound produced by another musical instrument may be a combination of n, $3n$, $5n$, $7n$... and so on, where n is the frequency of the fundamental note. Due to the different combinations of frequencies of a musical sound, one can easily identify the sound produced by a particular musical instrument. It should be noted that a closed organ pipe produces overtones whose frequencies are odd multiples of the fundamental frequency. The overtones produced by the open organ pipe consist of integral multiples of fundamental frequency. Hence the sounds produced by the open organ pipe are of better quality than that of the closed organ pipe.

If a musical sound is consist of only one frequency, say, the fundamental frequency, then it will produce dull effect to our ear. The pleasing effect of the musical sound is produced only due to the combination of frequencies. Even though more than one musical instrument is played at a time in a hall, one can easily identify the sound produced by a particular musical instrument because of the timbre or quality of sound.

6.4 ACOUSTICS OF BUILDINGS

The study of the acoustical characteristics of a building, lecture hall, cinema hall, and auditorium is called as the acoustics of building. The acoustics of building is of more interest to musician, architect, speakers, listeners and civil engineers. Until the nineteenth century, no importance was given to the acoustical characteristics of the building. Hence many halls and auditorium constructed before nineteenth century has acoustical problem. An auditorium (Fogg Art Museum Hall) constructed during the year 1900 in the Harvard University also had acoustical problem. The audiences were not able to hear the sound clearly. Wallace C. Sabine, Professor in Physics of that university was asked to study about the problem of the auditorium and rectify the defect. Professor Sabine and his team made a systematic and scientific study about the auditorium. The study of acoustic of building was given birth by Professor Sabine and his team.

According to Sabine, a hall is said to be acoustically fit, then it should satisfy the following conditions:

1. The sound produced by the speaker should bc sufficiently loud.
2. The tonal quality of the sound should not be changed.
3. There should be no overlapping between any two successive sound produced by the speaker.

In order to satisfy these conditions, one should consider the following factors before constructing an auditorium:

(i) Reverberation
(ii) Loudness
(iii) Focusing due to wall and ceiling
(iv) Echo
(v) Echelon effect
(vi) Extraneous noise
(vii) Resonance within the building

In this section, we shall discuss about the reverberation and Sabine's formula for reverberation time and absorption coefficient. The remaining factors are discussed in Section 6.8.

6.4.1 Reverberation

The existence of sound or prolongation of sound even after the source is cut-off is known as reverberation. Generally, the sound produced by the speaker in the auditorium undergoes nearly three to three hundred times reflection from wall, floor, ceiling, etc. before it becomes inaudible. Due to this multiple reflection, the sound exists for some time even after the source is cut-off. This property of existing of the sound due to reflection even after the source is cut-off is called reverberation.

The time taken by the sound to become inaudible is called reverberation time. Sabine did an experiment using an open organ pipe. He excited a tuning fork of 512 Hz, and using the open organ pipe experiment, he found that the sound becomes inaudible when its intensity gets reduced to one-millionth of its initial value. From this, one can define the term reverberation time as the time taken by the sound wave to reduce its intensity to one-millionth of its initial value.

A lecture hall requires the reverberation time of 0.5 s, a music hall or a concert hall requires a reverberation time of 1.1 s. The reverberation time of a cinema hall depends upon its volume. A cinema hall with smaller volume requires a reverberation time of 1.1 s to 1.5 s, whereas the cinema hall with larger volume requires the reverberation time up to 2.3 s.

Factors affecting the reverberation time: The following factors affect the reverberation time of a hall. They are,

(i) *Reflection of sound*: The sound produced in a hall gets reflected by ceiling, wall, floor and so on. If the reflection coefficient of these materials is high, more amount of sound gets reflected. Hence the sound exists for a longer period of time. The existence of sound for longer time increases the reverberation time.

(ii) *Absorption of sound:* The sound produced in a hall gets absorbed by the absorbing materials such as curtain, cushion, etc. present in the hall. The absorption of sound by the absorbing materials reduces the existence of the sound, and it reduces the reverberation time.

(iii) *Intensity of sound:* The intensity of sound is related to the loudness of the sound. If the intensity of the sound increases, the loudness of the sound also increases. Generally, the sound with high loudness exists for a longer period of time. Therefore, the increase of the intensity of sound increases the reverberation time.

(iv) *Frequency of sound:* The reverberation time also depends upon the frequency of sound. Generally, the absorption coefficient increases with the increase in the frequency of sound. The sound with higher frequency is easily absorbed by the absorbing materials. Therefore, the increase in the frequency of the sound reduces its reverberation time.

(v) *Volume of the hall:* The reverberation time of hall is directly proportional to the volume of the hall. A hall with larger volume requires a larger value of the reverberation time. According to Sabine's formula,

$$T = \frac{0.162V}{\Sigma aS} \tag{6.13}$$

Therefore, the reverberation time of the hall is directly proportional to the volume.

6.5 SABINE'S FORMULA FOR REVERBERATION TIME

Sabine's formula for the reverberation time is derived by assuming that the sound is uniformly distributed inside that hall. The derivation has following three stages:

1. The first step is to calculate the average sound energy received by the volume dV in one second.
2. The second step is to calculate the rate of absorption, rate of growth and the rate of emission of sound.
3. The third step is to formulate a mathematical equation using the rate of emission, rate of growth and rate of absorption, and then to solve the differential equation. By solving the differential equation, mathematical equations for growth and decay of sound are obtained. From the equation for the decay of sound, the equation for the reverberation time is obtained.

Step 1: Consider a wall AB and a small elemental area dS in the wall. Draw two semicircles with radii r and $r + dr$ by taking O as centre. Shade a small portion that makes an angle $d\theta$ with the centre.

The arc length of the shaded portion is dr, and its radial length of the shaded portion is $rd\theta$. The area of the shaded portion is $rd\theta \cdot dr$. Let the shaded portion is rotated through an angle $d\phi$ by taking O as centre. The distance travelled by the shaded portion is $r \sin \theta d\phi$.

The volume traced by the shaded portion is:

$$dV = r^2 \sin \theta d\theta \, d\phi \, dr \tag{6.14}$$

Assume that there is a sound source at centre O, and it is producing sound energy. Let E be the sound energy density of the room. The term sound energy density is defined as the sound energy available per unit volume of the room.

$$\text{The sound energy available in the volume } dV = Er^2 \sin \theta d\theta \, d\phi \, dr \tag{6.15}$$

$$\text{The sound energy available per unit solid angle} = \frac{Er^2 \sin \theta d\theta d\phi dr}{4\pi} \tag{6.16}$$

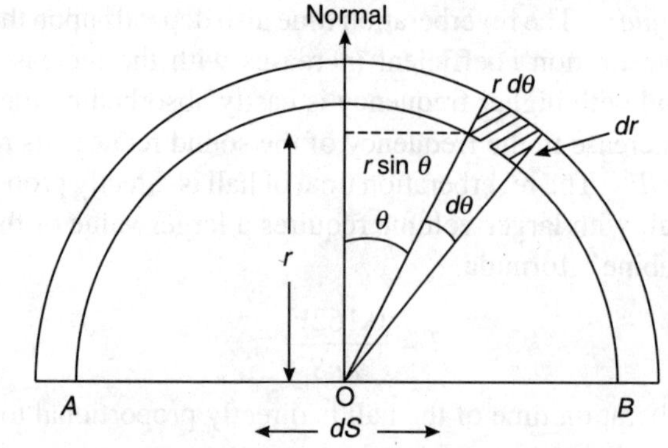

Figure 6.3 Sabine's formula derivation.

where 4π is the maximum value of the solid angle.

$$\text{The sound energy available in the volume } dV = \frac{Er^2 \sin\theta d\theta d\phi dr}{4\pi} \times \frac{dS\cos\theta}{r^2} \quad (6.17)$$

where $\frac{dS\cos\theta}{r^2}$ is the solid angle subtended by the surface dS.

$$\text{Total sound energy received by } dV \text{ in one second} = \frac{EdS}{4\pi}\int_0^C dr \int_0^{2\pi} d\phi \int_0^{\frac{\pi}{2}} \sin\theta\cos\theta d\theta \quad (6.18)$$

where C is the velocity of sound. It is the distance travelled by the sound in one second. The sound energy received by dV in one second

$$= \frac{EdS}{4\pi}\int_0^C dr \int_0^{2\pi} d\phi \times \frac{1}{2}\int_0^{\frac{\pi}{2}} 2\sin\theta\cos\theta d\theta \quad (6.19)$$

Substituting the value of the integrals $\int_0^C dr = C$, $\int_0^{2\pi} d\phi = 2\pi$ and $\int_0^{\frac{\pi}{2}} 2\sin\theta\cos\theta d\theta = \int_0^{\frac{\pi}{2}} \sin 2\theta d\theta = \left(-\frac{\cos 2\theta}{2}\right)_0^{\pi/2} = 1$, we get

$$\text{Sound energy received by } dV \text{ in one second} = \frac{EdS}{4}C \quad (6.20)$$

Step 2: Let a be the absorption coefficient of the wall. The sound energy absorbed by dV in one second $= \frac{EC}{4} adS$ (6.21)

A hall or an auditorium has a number of absorbing surfaces such as wall, window, ceiling, curtain, etc. By taking this into account, the rate of absorption of sound by dV from Eq. (6.21) can be written as:

$$= \frac{EC}{4}(a_1 dS_1 + a_2 dS_2 + a_3 dS_3 + \cdots) \tag{6.22}$$

where a_1, a_2, a_3,... are the absorption coefficients of first, second, third surfaces respectively. Similarly, dS_1, dS_2, dS_3, represents the surface areas of the first, second, third surfaces respectively. By taking $A = a_1 dS_1 + a_2 dS_2 + \cdots + a_n dS_n = \Sigma a dS$, Eq. (6.22) can be written as:

$$\text{Rate of absorption of sound by } dV = \frac{EC}{4}A \tag{6.23}$$

where A is the total absorption. It is represented in a unit m^2 sabine and total surface area dS is expressed in m^2.

Consider a source is emitting sound energy. Let E_m be the maximum sound energy density emitted by the source.

The rate of emission is given by

$$P = \frac{E_m C}{4}A \tag{6.24}$$

The maximum energy density E_m is given by

$$E_m = \frac{4P}{CA} \tag{6.25}$$

Let E be the sound energy density, and V be the volume of the hall. The total sound energy available in the hall is EV.

$$\text{The rate of growth of sound} = \frac{d}{dt}(EV) = V\frac{dE}{dt} \tag{6.26}$$

Step 3: Let us formulate a mathematical relation using rate of emission of sound, rate of absorption of sound and rate of growth of sound. i.e.,

Rate of emission of sound = Rate of absorption of sound
+ Rate of growth of sound

$$P = \frac{EC}{4}A + V\frac{dE}{dt} \tag{6.27}$$

Dividing both sides of Eq. (6.27) by V, we get

$$\frac{dE}{dt} + \frac{CA}{4V}E = \frac{P}{V} \tag{6.28}$$

Taking $\frac{CA}{4V} = \alpha$, Eq. (6.28) can be written as:

$$\frac{dE}{dt} + \alpha E = \frac{4P}{CA}\alpha \tag{6.29}$$

where $\frac{1}{V} = \frac{4\alpha}{CA}$. Multiplying Eq. (6.29) by $e^{\alpha t}$, we get

$$\left(\frac{dE}{dt} + \alpha E\right)e^{\alpha t} = \frac{4P}{CA}\alpha e^{\alpha t} \tag{6.30}$$

Equation (6.30) can be written as:

$$\frac{d}{dt}(Ee^{\alpha t}) = \frac{4P}{CA}\alpha e^{\alpha t} \tag{6.31}$$

Integrating Eq. (6.31), we get

$$Ee^{\alpha t} = \frac{4P}{CA}e^{\alpha t} + K \tag{6.32}$$

where K is an integration constant.

Case I: Growth of sound

Consider the sound source is switched off. Before switching on the source, $t = 0$, and $E = 0$. Substituting, these boundary conditions in Eq. (6.32), we get

$$K = -\frac{4P}{CA} \tag{6.33}$$

From Eq. (6.32) and Eq: (6.33), we get

$$Ee^{\alpha t} = \frac{4P}{CA}e^{\alpha t} - \frac{4P}{CA}$$

$$Ee^{\alpha t} = E_m(e^{\alpha t} - 1) \tag{6.34}$$

where $E_m = \frac{4P}{CA}$. Dividing both sides of Eq. (6.34) by $e^{\alpha t}$, we get

$$E = E_m(1 - e^{-\alpha t}) \tag{6.35}$$

Equation (6.35) represents the growth of sound energy. If a graph is plotted between E and t, using Eq. (6.35), an exponentially increasing curve, as shown in Figure 6.4, is obtained. This shows that the sound grows exponentially.

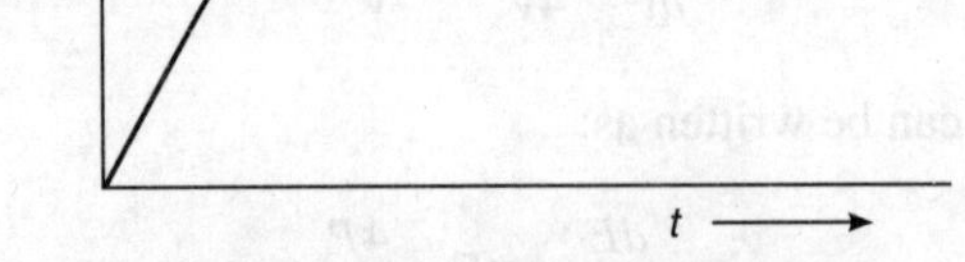

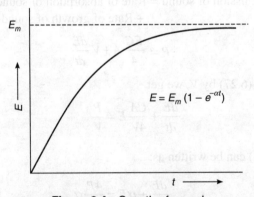

Figure 6.4 Growth of sound.

Case II: Decay of sound

Consider the source is producing sound and then it is switched off. The rate of emission of sound becomes zero, i.e., $P = 0$. At the time of switching off the source $t = 0$ s, and $E = E_m$. Substituting these boundary conditions in Eq. (6.32), we get, $K = E_m$

Substituting $P = 0$ and $K = E_m$ in Eq. (6.32), we get

$$Ee^{\alpha t} = E_m$$

i.e.,
$$E = E_m e^{-\alpha t} \tag{6.36}$$

If a graph is plotted between E and t, then a decaying curve will be obtained. From Figure 6.5, we infer that the sound decays exponentially.

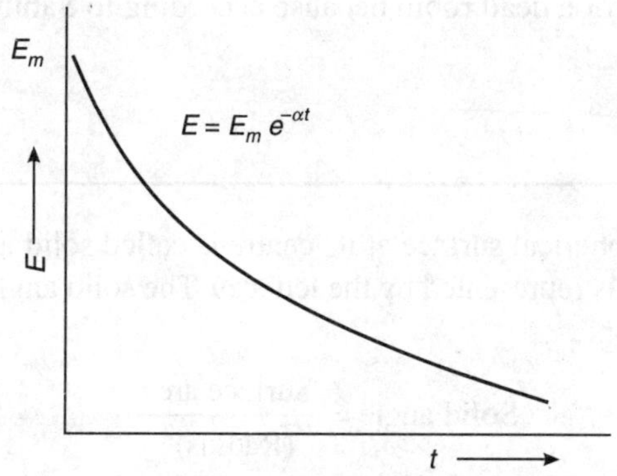

Figure 6.5 Decay of sound.

Reverberation time

The term reverberation time is defined as the time taken by a sound to reduce its intensity to one-millionth of its initial value. That is, when $E = \dfrac{E_m}{10^6}$, then $t = T$. Substituting these conditions in Eq. (6.36), we get

$$\frac{E_m}{10^6} = E_m e^{-\alpha T}$$

$$10^{-6} = e^{-\alpha T}$$

i.e.,
$$10^6 = e^{\alpha T}$$

Taking log on both sides, we get

$$\alpha T = \ln 10^6$$

Substituting $\alpha = \dfrac{CA}{4V}$ in the above equation, we get

$$\frac{CA}{4V} T = 2.302 \times 6$$

Substituting, $C = 340$ m s^{-1} (the velocity of sound at 20°C is 340 ms^{-1}), we get,

$$T = \frac{2.302 \times 6 \times 4 \times V}{340 \times A} = \frac{0.1625V}{A} \tag{6.37}$$

Equation (6.37) gives the reverberation time of a hall. In Eq. (6.37), A represents the total absorption ($A = \Sigma aS$) and V is the volume of the hall. This equation is called Sabine's mathematical expression for the reverberation time.

Drawback of Sabine's relation

1. A dead room is a room that absorbs the entire sounds produced inside it. For a dead room $A = 1$, and $T = 0$ s. Sabine's mathematical expression for the reverberation time is not applicable for a dead room because according to Sabine's relation, $T \neq 0$, when $A = 0$.

6.6 SOLID ANGLE

The angle subtended by a spherical surface at its centre is called solid angle. Its unit is steradian. Its maximum value is 4π. It is represented by the letter ω. The solid angle of a spherical surface is determined using the relation:

$$\text{Solid angle} = \frac{\text{Surface area}}{(\text{Radius})^2} \tag{6.38}$$

For a spherical surface, the surface area is $4\pi r^2$. Therefore,

$$\text{Solid angle} = \frac{\text{Surface area}}{(\text{Radius})^2} = \frac{4\pi r^2}{r^2} = 4\pi \tag{6.39}$$

Consider a spherical surface AB as shown in Figure 6.6. Let O be its centre and r be the radius of the spherical surface. Then the solid angle of the surface AB is:

$$\omega = \frac{\text{Area of the surface AB}}{(\text{Radius})^2} = \frac{\text{Area AB}}{r^2}$$

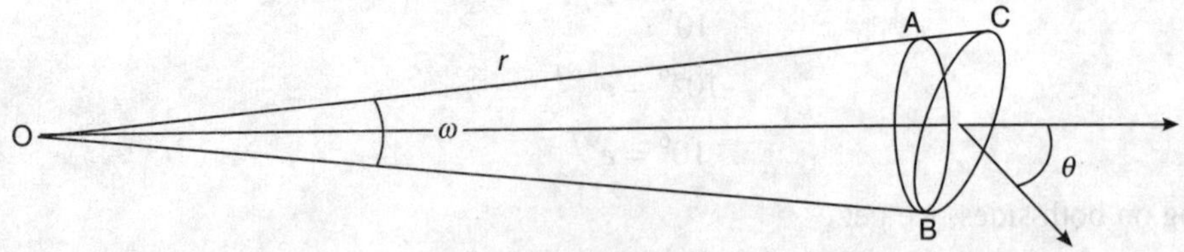

Figure 6.6 Solid angle.

The solid angle of surface BC inclined at an angle θ is:

$$\omega = \frac{\text{Area of the surface BC}}{(\text{Radius})^2} = \frac{\text{Area AB} \times \cos\theta}{r^2} = \frac{ds \cos\theta}{r^2} \tag{6.40}$$

where θ is the angle of inclination. In general, the solid angle of a surface is given by $\omega = \frac{ds \cos\theta}{r^2}$, where ds is the surface area, θ is the angle of inclination and r is the radius.

6.7 ABSORPTION COEFFICIENT AND ITS MEASUREMENT

The absorption of sound varies from one material to another material (Table 6.4). The absorption of sound by different materials is expressed by the term absorption coefficient. An open window is considered as a perfect absorber of sound because it does not reflect any sound, and it transmits the entire sound incident on it.

Table 6.4 Absorption coefficients of some materials

Material	*Absorption coefficients in O.W.U.*
Marble	0.01
Brick wall (30-cm thick)	0.03
Brick wall painted	0.016
Carpets	0.15–0.30
Wooden floor	0.06
Glass	0.02
Ordinary chair	0.17
Human body	0.43–0.47
Plastered ceiling	0.04

The absorption coefficient is defined as the reciprocal of the area of the sound absorbing surface that absorbs the same quantity of sound as that of 1 m^2 of an open window. For example, if 5 m^2 of a material absorbs the same quantity of sound absorbed by 1 m^2 of an open window, then the absorption coefficient of the material is 1/5 = 0.2. Sound absorption coefficient describes the efficiency of the material or the surface to absorb the sound. It is also defined as the ratio between the sound intensity absorbed I_a and sound intensity incident I_i. The unit for the absorption coefficient is sabine or open window unit (O.W.U). The absorption coefficient is frequency dependent and the value of the absorption coefficient of some sound absorbing materials is listed in Table 6.4. The total absorption is given by $A = \Sigma aS$ and its unit is m^2 sabine.

6.7.1 Measurement of Absorption Coefficient

The following methods are used to measure the absorption coefficient of a material:

Method I

The absorbing materials available in the hall are removed. The reverberation time of an empty hall is determined. Let it be T_1 second. The reverberation time is given by

$$T_1 = \frac{0.16V}{\Sigma aS} \tag{6.41}$$

or
$$\frac{1}{T_1} = \frac{\Sigma aS}{0.16V} \tag{6.42}$$

The given material, whose absorption coefficient is to be determined, is placed inside the hall. The reverberation time is again noted. Let it be T_2. The reverberation time is given by

$$\frac{1}{T_2} = \frac{\Sigma aS + a_1 S_1}{0.16V} \tag{6.43}$$

where a_1 is the absorption coefficient of the material and S_1 is its surface area.
Subtracting Eq. (6.42) from Eq. (6.43), we get

$$\frac{1}{T_2} - \frac{1}{T_1} = \frac{a_1 S_1}{0.16V} \tag{6.44}$$

By rearranging Eq. (6.44), we get

$$a_1 = \frac{0.16V}{S_1}\left[\frac{1}{T_2} - \frac{1}{T_1}\right] \tag{6.45}$$

Equation (6.45) gives the absorption coefficient of the material. By knowing the values of T_1, T_2, S_1 and V, the absorption coefficient a_1 is determined.

Method II

This method is used to determine the absorption coefficient of a material, whenever the ratio of the powers of two different sources and their reverberation times are known. Let P_1/P_2 be the ratio of the powers of two different sources, T_1 and T_2 are the reverberation times of theses sources. The maximum energy density of these two sources can be written as:

$$E_{m1} = \frac{4P_1}{CA} \tag{6.46}$$

$$E_{m2} = \frac{4P_2}{CA} \tag{6.47}$$

where C is the velocity of sound and A is the total absorption. The reverberation time of these two sounds are T_1 and T_2 seconds respectively. It means these two sources take T_1 and T_2 seconds respectively to become inaudible. When they become inaudible, the energy density of these two sources is equal, and let it be E_0. The value of the energy density E can be written as:

$$E_0 = E_{m1} e^{-\alpha T_1} \tag{6.48}$$

$$E_0 = E_{m2} e^{-\alpha T_2} \tag{6.49}$$

From Eq. (6.48) and Eq. (6.49), we get

$$E_{m1} e^{-\alpha T_1} = E_{m2} e^{-\alpha T_2} \tag{6.50}$$

i.e.,
$$\frac{E_{m1}}{E_{m2}} = e^{\alpha(T_1 - T_2)} \tag{6.51}$$

From Eq. (6.46) and Eq. (6.47), we get

$$\frac{E_{m1}}{E_{m2}} = \frac{P_1}{P_2} \tag{6.52}$$

From Eq. (6.51) and Eq. (6.52), we get

$$\frac{P_1}{P_2} = e^{\alpha(T_1 - T_2)} \tag{6.53}$$

Taking log on both sides, we get

$$\ln\left(\frac{P_1}{P_2}\right) = \alpha(T_1 - T_2) \tag{6.54}$$

i.e.,

$$\alpha = \frac{\ln P_1 - \ln P_2}{T_1 - T_2} \tag{6.55}$$

Substituting the value of α as, $\alpha = \dfrac{CA}{4V}$ in Eq. (6.55), we get,

$$\frac{CA}{4V} = \frac{\ln(P_1/P_2)}{T_1 - T_2}$$

Substituting $A = \Sigma aS = a\Sigma S$, we get

$$a = \frac{4V}{C\Sigma S} \times \frac{\ln(P_1/P_2)}{T_1 - T_2} \tag{6.56}$$

The absorption coefficient of the material can be determined using Eq.(6.56).

6.8 FACTORS AFFECTING THE ACOUSTICS OF BUILDING AND THEIR REMEDIES

The acoustics of a building is affected by (i) reverberation time, (ii) loudness, (iii) focussing due to walls and ceiling, (iv) echo, (v) echelon effect, (vi) extraneous noises and (vii) resonance within the building. We shall discuss about these factors.

Reverberation time and its importance

The time taken by a sound wave to reduce its intensity to one-millionth of its initial value is known as reverberation time. The reverberation time of a hall should not be too large or it should not be too low. If it is too high, a speaker may deliver more number of words within that time. So, there may be a overlapping between the successive words produced by the speaker, and the audiences are not able to hear the sound clearly. If the reverberation time is too low, the sound will become inaudible within a short time. Therefore, the reverberation time of the hall should be an optimum value.

Remedy

1. If the reverberation time of a hall is too high, then it is brought into an optimum value by placing sound absorbing materials in the required places.
2. If the reverberation time is too low, sound reflecting materials should be placed inside the auditorium so as to reflect more amount of sound, and to increase the reverberation time.
3. The concert hall or theatre is constructed so that the reverberation time of the hall is made as uniform whether it is empty for rehearsal or filled with audience. For this, the seats can be backed with plush, which has the same coefficient of absorption as that of audience.

Loudness

The loudness of a sound also affects the acoustics of the building. If a sound is louder, it exists for a longer time, and the reverberation time of the hall is very high. The sound with low loudness exists for a short period of time, and hence its reverberation time is low. The loudness of the sound is directly related to the reverberation time. The increase or decrease of the reverberation time due to loudness affects the acoustics of the building. The loudness can be made as uniform so that audience can enjoy the performance of the stage.

Remedy

1. Nowadays public address system is used to increase or decrease the volume of the sound.
2. Plane surfaces are useful in increasing the loudness by reflecting the sound.
3. Sound reflecting boards should be placed in suitable places so as to increase the reflection of sound.

Focussing due to wall and ceiling

A plane surface distributes the sound energy uniformly. Curved surfaces, do not distribute the sound uniformly, focuses the sound at a particular place and it will produces harmful effect to the audience. Care should be taken while constructing curved surfaces. If the radius of curvature of the curved surface is equal to the height as shown in Figure 6.7(a), then there is an undesirable concentration of sound. In order to avoid bad focussing due to the curved surface, the radius of curvature should be made as twice as that of the height [Figure 6.7(b)] and it will distribute the sound uniformly.

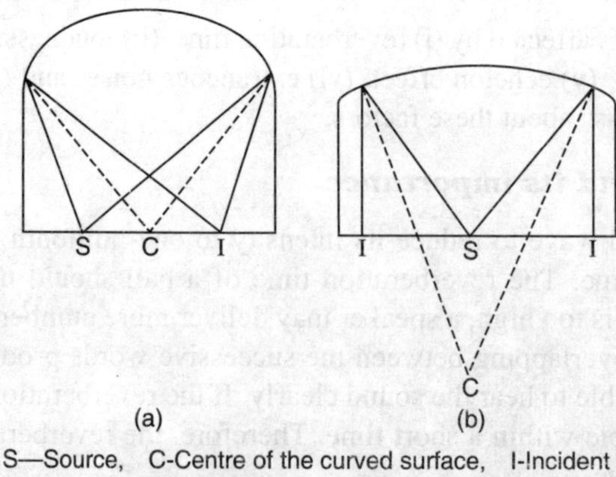

S—Source, C-Centre of the curved surface, I-Incident beam

Figure 6.7 Reflection of sound from curved surfaces for which the radius of curvature is (a) equal to the height and (b) twice as that of the height.

Remedy

1. The bad focussing of sound is avoided using plane surfaces.
2. The radius of the curved surface is made as twice as that of the height so as to avoid bad focussing.
3. A model of the building should be constructed before constructing a new building. The model is floated in a ripple tank and the reflection of water waves in the model is studied. The sound is also propagating as a wave, the construction of a model and the study of the reflection of the water wave is also useful to study about the reflection of sound.

Echo

The repetition of sound produced due to the reflection from a large surface like wall, hill or mountain is called echo. Consider an observer is producing a sound, and it gets reflected by an obstacle. The sound travels towards the observer and the observer, hears the sound again. Let d be the distance between the observer and the obstacle, v is the sound velocity, and t is the time taken by the sound for the to and fro motion, then the velocity of the sound is given by

$$v = \frac{2d}{t} \tag{6.57}$$

where t is the time taken by the sound for the to and fro motion. Substituting $t = 1/10$ s, it is the minimum time required to distinguish between any two sounds and $v = 340$ m s^{-1}, it is the velocity of sound at 20°C, then $d = vt/2 = [340 \times (1/10)]/2 = 17$ m. This shows that if the distance between the observer and the obstacle is less than 17 m, then echo is produced. The echo affects the acoustics of the building. Due to the production of echo, the audiences are not able to hear the sound produced by the speaker clearly.

Remedy

1. The production of echo is avoided by keeping the distance between the speaker and the reflecting surface greater than 17 m.

Echelon effect

Whenever a person is walking on a hard paved path besides equally spaced paling (paling is a regularly spaced design in ceiling), the sound produced by the footstep is followed by an echo. This effect is called echelon effect. The echelon effect is noticeable when a person is walking along a staircase with a steel tipped shoe. Consider the tread of the staircase is 0.3 m as shown in Figure 6.8. The distance travelled by the sound for the to and fro motion is 0.6 m. The velocity of sound is 350 m s^{-1} at 30°C. The frequency of the sound produced is 350/0.6 = 583 Hz.

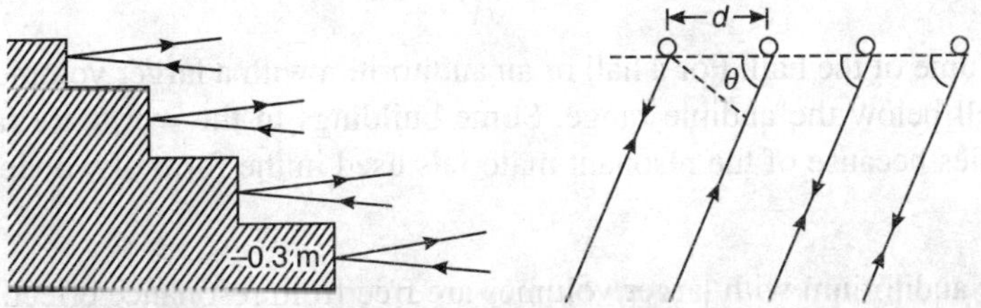

Figure 6.8 Echelon effect.

Remedy

1. The echelon effect is minimized by covering the staircase using carpet.
2. The echelon effect is minimized by constructing the spacing of the paling irregular.

Extraneous noises

The sound produced outside the hall that disturbs the audience in the hall or an auditorium is called extraneous noises. There are two types of extraneous noises:

1. Air-borne sound
2. Structure-borne sound

Air-borne sound: The sound produced outside the hall or an auditorium that propagates through air is called air-borne sound. A vehicle moving in the adjacent road of the auditorium or hall produces an unwanted sound. It will propagate through air and reaches the auditorium. This sound produces disturbances to the audience present in the hall or auditorium.

Remedy

1. The air-borne sound is minimized by using a double or treble doors and windows and avoiding the passage of sound.
2. By air conditioning the room and closing all the doors and windows, the air-borne sound is minimized.

Structure-borne sound: The sound produced in the adjacent room of the hall and auditorium that propagates through the structure of the building is called structure-borne sound. Moving desks in the adjacent room and flowing of water through tap are the examples for the structure-borne sound. These sounds are propagated through the structure or wall of the building.

Remedy

1. The structure-borne sound is minimized by constructing rigid structures.
2. By introducing discontinuity in the wall, the propagation of sound from one hall to another is minimized.
3. The sound produced in a water pipe is minimized by using rubber gaskets.

Resonance within the building

Resonance is the process of producing an amplified sound at certain frequencies. Due to the resonance effect, the sound gets amplified and a loud sound is produced. The resonance of sound is also one of an unwanted phenomenon. The resonant frequency of a hall is given by

$$\nu = \frac{1}{\sqrt{V}} \tag{6.58}$$

where V is the volume of the hall. For a hall or an auditorium with a larger volume, the resonance frequency lies well below the audible range. Some buildings in the world are famous for their acoustical properties because of the resonant materials used in the form of wooden paneling.

Remedy

1. Halls or auditorium with larger volumes are free from resonance effect.
2. The resonance effect is produced only at certain frequency. Sound absorbing materials, which absorb that particular frequency, may be used to reduce the resonance effect.

SOLVED PROBLEMS

6.1 It is found, in an experiment, that 5 m^2 area of a given material absorbs the same amount of sound energy as that absorbed by 1 m^2 area of open window. Calculate the absorption coefficient of the given material

Given data

Area of the material = 5 m^2

Area of the open window = 1 m^2

Solution

$$\text{Absorption coefficient} = \frac{1}{\text{Area of the given material}} = \frac{1}{5} = 0.2 \text{ O.W.U.}$$

Absorption coefficient of the material = 0.2 O.W.U.

6.2 If the intensity of sound produced by thunder is 0.2 W m^{-2}, calculate the intensity level in decibels.

Given data

Intensity of the sound = 0.2 W m^{-2}

Solution

$$\text{Intensity level of sound } \beta = 10\log\left(\frac{I}{I_0}\right) = 10\log\left(\frac{0.2}{10^{-12}}\right) = 113.01 \text{ dB}$$

The intensity level of the sound produced by thunder is 113.01 dB.

6.3 The intensity of a sound wave is doubled. By how many decibels will the intensity level increases?

Given data

Initial intensity = I

Final intensity = $2I$

Solution

$$\text{Intensity level of sound } \beta_1 = 10\log\left(\frac{I}{I_0}\right).$$

$$\text{Intensity level of sound after doubling its intensity } \beta_2 = 10\log\left(\frac{2I}{I_0}\right).$$

$$\text{Increase in intensity level} = \beta_2 - \beta_1 = 10\log\left(\frac{2I}{I_0}\right) - 10\log\left(\frac{I}{I_0}\right)$$

$$= 10\log\left(\frac{2I}{I}\right) = 3.0103 \text{ dB}$$

The increase in intensity level when the sound intensity is doubled = 3.0103 dB.

6.4 The average reverberation time of a hall is 1.5 second and the area of the interior surface is 3340 m^2. If the volume of the hall is 12000 m^2, find the absorption coefficient.

Given data

Volume of the hall = 12000 m^2

Area of the interior surface = 3340 m^2

Reverberation time of the hall = 1.5 s

Solution

$$\text{Reverberation time } T = \frac{0.162V}{\sum aS} = \frac{0.162V}{a\sum S}$$

$$\text{Average absorption coefficient } a = \frac{0.162V}{T\sum S} = \frac{0.162 \times 12000}{1.5 \times 3340} = 0.388 \text{ O.W.U.}$$

Average absorption coefficient = 0.388 O.W.U.

6.5 For an assembly hall of volume 3000 m^3 and inner surface area 1300 m^2, the average absorption coefficient of sound is 0.1 sabine. Find the reverberation time. What area of the wall should be covered by curtains so as to reduce the reverberation time to 2.5 s. Given that absorption coefficient of curtain cloth is 0.5 sabine.

Given data

Volume of the hall = 3000 m^2

Inner surface area = 1300 m^2

Average absorption coefficient of sound = 0.1 O.W.U.

Absorption coefficient of curtain cloth = 0.5 O.W.U.

Reverberation time required T_2 = 2.5 s

Solution

$$\text{Reverberation time of the hall } T_1 = \frac{0.162V}{a\sum S}$$

$$= \frac{0.162 \times 3000}{0.1 \times 1300} = 3.738 \text{ s}$$

$$\text{Reverberation time required after using curtain } T_2 = \frac{0.162V}{a\sum S + a_1 S_1}$$

where a_1 and S_1 are the absorption coefficient and surface area of curtain cloth

$$a\sum S + a_1 S_1 = \frac{0.162V}{T_2}$$

$$S_1 = \frac{1}{a_1}\left(\frac{0.162V}{T_2} - a\sum S\right)$$

$$S_1 = \frac{1}{0.5}\left(\frac{0.162 \times 3000}{2.5} - 0.1 \times 1300\right) = 128.8 \text{ m}^2$$

Area of the curtain required = 128.8 m^2

6.6 A cinema hall has a volume of 8000 m^3, and the total absorption of 500 O.W.U. raises the absorption by 150 O.W.U. due to audience. Calculate the change in reverberation.

Given data

Volume of the cinema hall = 8000 m^3

Total absorption in the hall = 500 O.W.U.

Increase in absorption due to audience = 150 O.W.U.

Solution

$$\text{Reverberation of the hall } T_1 = \frac{0.162V}{A}$$

$$= \frac{0.162 \times 8000}{500} = 2.592 \text{ s}$$

Total absorption of the hall with audience = Total absorption of the hall without audience + Absorption by audience

$$A = 500 + 150$$

$$= 650 \text{ O.W.U.}$$

$$\text{Reverberation of the hall } T_2 = \frac{0.162V}{A}$$

$$= \frac{0.162 \times 8000}{650} = 1.9938 \text{ s}$$

Change in reverberation time = $T_1 - T_2 = 2.592 - 1.9938 = 0.5982$ s

Change in reverberation = 0.5982 s

6.7 *A music hall has a volume of 8500 m^3. If the reverberation time required is 1.05 s, what should be the total absorption in the hall?*

Given data

Volume of the music hall = 8500 m^3

Reverberation time required = 1.05 s

Solution

$$\text{Reverberation of the hall } T = \frac{0.162V}{A}$$

$$\text{Total absorption in the hall } A = \frac{0.162V}{T}$$

$$= \frac{0.162 \times 8500}{1.05}$$

$$= 1311.428 \text{ m}^2 \text{ sabine}$$

Total absorption in the hall A = 1311.428 m^2 sabine

6.8 An auditorium having dimension of 30 m length, 20 m breadth and 10 m height has an average sound absorption coefficient of 0.4. What is its reverberation time?

Given data

Length of the auditorium = 30 m

Breadth of the auditorium = 20 m

Height of the auditorium = 10 m

Average absorption coefficient = 0.4

Solution

Volume of the hall = Length × Breadth × Thickness = 30 × 20 × 10 = 6000 m^3

Surface area = Area of floor + Area of ceiling + Area of four walls

Surface area = (30 × 20) + (30 × 20) + (30 × 10 + 20 × 10 + 30 × 10 + 20 × 10) = 2200 m^2

$$\text{Reverberation time } T = \frac{0.162V}{a\sum S} = \frac{0.162 \times 6000}{0.4 \times 2200} = 1.104 \text{ s.}$$

Reverberation time of the hall = 1.104 s.

6.9 A hall of volume 4000 m^3 has a reverberation time of 4.2 s. How will you reduce the reverberation time of the hall to 2.1 s, using a material whose sound absorption coefficient is 0.48 O.W.U.?

Given data

Volume of the hall = 4000 m^3

Reverberation time of the hall T_1 = 4.2 s

Reverberation time required T_2 = 2.1 s

Sound absorption coefficient = 0.48 O.W.U.

Solution

$$\text{Reverberation time } T_1 = \frac{0.162V}{a\sum S}$$

or

$$\frac{1}{T_1} = \frac{\sum aS}{0.162V} \quad \text{(i)}$$

The required reverberation time is given by,

$$T_2 = \frac{0.162V}{\sum aS + a_1 S_1}$$

or

$$\frac{1}{T_2} = \frac{\sum aS + a_1 S_1}{0.162V} \quad \text{(ii)}$$

where $a_1 S_1$ is the absorption of the absorbing material. Subtracting Eq. (i) from Eq. (i), we get

$$\frac{1}{T_2} - \frac{1}{T_1} = \frac{a_1 S_1}{0.162V}$$

Rearranging the above equation, we get

$$S_1 = \frac{0.162V}{a_1}\left(\frac{1}{T_2} - \frac{1}{T_1}\right)$$

Substituting the values of T_1, T_2, a_1, and V, we get

$$S_1 = \frac{0.162 \times 4000}{0.48}\left(\frac{1}{2.1} - \frac{1}{4.2}\right)$$

$$S_1 = 321.428 \text{ m}^2$$

By introducing 321.428 m^2 of the absorbing material, the reverberation time of the hall is reduced from 4.2 s to 2.1 s.

6.10 A room of volume 600 m^3 has the seating capacity for 80 persons. The wall area of the room is 220 m^2, the floor area is 120 m^2 and the ceiling area is 120 m^2. The absorption coefficients for the walls, the ceiling and the floor area are 0.03, 0.08 and 0.06 O.W.Us respectively. The absorption coefficient for audience is 0.4367 per person. Calculate the reverberation time, (i) when the hall is empty and (ii) when the hall is filled with full capacity of audience.

Given data

Volume of the hall = 600 m^3

Wall area = 220 m^2

Floor area = 120 m^2

Ceiling area = 120 m^2

Absorption coefficient of wall = 0.03 O.W.U.

Absorption coefficient of floor = 0.08 O.W.U.

Absorption coefficient of ceiling = 0.06 O.W.U.

Absorption coefficient of audience = 0.4367 O.W.U.

Solution

(i) *When the hall is empty*

$$\text{Reverberation time } T = \frac{0.162V}{\Sigma aS}$$

$$\text{Total absorption} = \Sigma aS = 220 \times 0.03 + 120 \times 0.08 + 120 \times 0.06$$

$$= 23.4 \text{ m}^2 \text{ sabine}$$

$$T = \frac{0.162 \times 600}{23.4} = 4.1538 \text{ s.}$$

(ii) *When the hall is filled with audience*

$$\text{Reverberation time } T = \frac{0.162V}{\Sigma aS + a_1S_1}$$

$$\text{Total absorption} = \Sigma aS + a_1S_1 = 23.4 + 0.4367 \times 80 = 58.336 \text{ m}^2 \text{ sabine}$$

$$T = \frac{0.162 \times 600}{58.336} = 1.666 \text{ s.}$$

Reverberation time of the hall, when it is empty = 4.1538 s

Reverberation time of the hall, when it is filled with audience = 1.666 s

SHORT QUESTIONS

1. What is meant by acoustics?
2. Write about different types of acoustics.
3. How are sounds classified?
4. How is audible sound classified?
5. What is a musical sound? What are the properties of musical sound?
6. What is a noise? What are the properties of noise?
7. What are the characteristics of musical sound?
8. What is frequency? What is its unit?
9. What is pitch?
10. Define intensity of sound? What is its unit?
11. What is meant by loudness of sound?
12. State Weber–Fechner law.
13. Define the term phon.
14. What is meant by intensity level of sound?
15. What is the meaning of 0 dB?
16. Write the physical significance of 1 dB.
17. What is sone? Write the relation between sone and phon.
18. What is meant by acoustics of buildings?
19. What are the conditions needed for acoustically fit hall?
20. What is reverberation?
21. What is meant by reverberation time?
22. What are the factors affecting the reverberation time?
23. What is meant by total absorption? What is its unit?
24. Write the drawback of Sabine's relation.
25. What is solid angle? What is its unit?
26. Define the term absorption coefficient. What is its unit?
27. What are the factors affecting the acoustics of building?
28. How reverberation time affects the acoustics of buildings? How will you minimize it?
29. How loudness affects the acoustics of buildings? How will you minimize it?
30. Explain the focussing due to walls and ceiling. What are the remedies?
31. What is echelon effect? What are the remedies to minimize the echelon effect?
32. What are extraneous noises?
33. What is air-borne sound? How is it minimized?
34. What is structure-borne sound? How will you minimize it?

35. What is an echo? How it affects the acoustics of building?
36. What is meant by resonance? How will you minimize resonance effect?

DESCRIPTIVE TYPE QUESTIONS

1. Derive Sabine's mathematical expression for the reverberation time.
2. What is meant by absorption coefficient? Explain the experimental methods used to measure the absorption coefficients.
3. Describe in brief the factors affecting the acoustics of buildings.
4. Write a short note on the following:
   (a) Sone
   (b) Phon,
   (c) Intensity level
   (d) Sound pressure level

PROBLEMS

1. Calculate the intensity level, in decibel, produced by a source of sound of intensity 1.26×10^{-12} W m^{-2}. (**Ans** $\beta = 1.0037$ dB)
2. A hall has a volume of 500 m^3. It is required to have a reverberation time of 1.5 s. What should be the total absorption in the hall? (**Ans** $A = 54$ m^2 sabine)
3. The amplitude of a sound wave is doubled. By how many decibels will the intensity level increases? (**Ans** $\beta = 6.02$ dB)
4. A hall has a volume of 5000 m^3. It is required to have reverberation time of 1.5 s. What should be the total sound absorption in the hall? (**Ans** $A = 540$ m^2 sabine)
5. A hall has a volume 1000 m^3 and a sound absorbing surface of area 400 m^2. If the average absorption coefficient of the hall is 0.2, what is the reverberation time of the hall? (**Ans** $T = 2.025$ s)
6. Calculate the intensity level of a turbine whose sound intensity is 100 W m^{-2}, when it is in operation. Given that the standard intensity level is 10^{-12} W m^{-2}. (**Ans** $\beta = 140$ dB)
7. Calculate the intensity level of a supersonic plane just leaving a run way having sound intensity of about 10000 Wm^{-2}. (**Ans** $\beta = 160$ dB)
8. The time of reverberation of an empty hall without and with 500 persons in the audience is 1.5 s and 1.4 s respectively. Find the time of reverberation with 1000 persons in the hall. (**Ans** $T = 1.3125$ s)
9. How many times more intense a sound with an intensity level 20 dB than a sound with intensity level of 0 dB? (**Ans** 100 times)

10. Calculate the reverberation time of a hall of volume 1500 m^3 having a seating capacity for 120 persons, (i) When the hall is empty and (ii) With full capacity of audience. The area of different surfaces and their absorption coefficients are given below:

Surface	*Area*	*Coefficient of absorption in O.W.U.*
Plastered wall	112 m^2	0.03
Wooden floor	130 m^2	0.06
Plastered ceiling	170 m^2	0.04
Wooden doors	20 m^2	0.06
Cushioned chairs	120 Nos.	0.50 per chair
Audience	120	0.4367 per person

(**Ans** $T_1 = 3.0697$ s, $T_2 = 1.847$ s)

CHAPTER 7

ULTRASONICS

7.1 INTRODUCTION

The sound waves are classified into three types on the basis of their frequencies. They are, namely (i) infrasonic waves, (ii) audible sound waves and (iii) ultrasonic waves. If the frequency of the sound waves is less than 20 Hz, then they are said to be *infrasonic waves*. If the frequency of the sound waves lies between 20 Hz and 20 kHz, then they are said to be *audible sound waves*. The sound waves having frequency greater than 20 kHz are called *ultrasonic waves*.

The ultrasonic waves are produced by the following methods, namely (i) mechanical generator or Galton's whistle, (ii) magnetostriction method and (iii) piezoelectric method. The magnetostriction method is based on the principle of magnetostriction effect. The piezoelectric method is based on the principle of inverse piezoelectric effect.

7.2 TYPES OF ULTRASONIC WAVES

Ultrasonic waves are classified into four types based on their propagation through the medium. They are, namely

1. Transverse ultrasonic waves or shear waves
2. Longitudinal or compressional ultrasonic waves
3. Surface or Rayleigh waves
4. Lamb waves or plate waves

7.2.1 Transverse Ultrasonic Waves

In the transverse ultrasonic waves, the particle of the medium vibrates perpendicular to the direction of the propagation of the wave as shown in Figure 7.1.

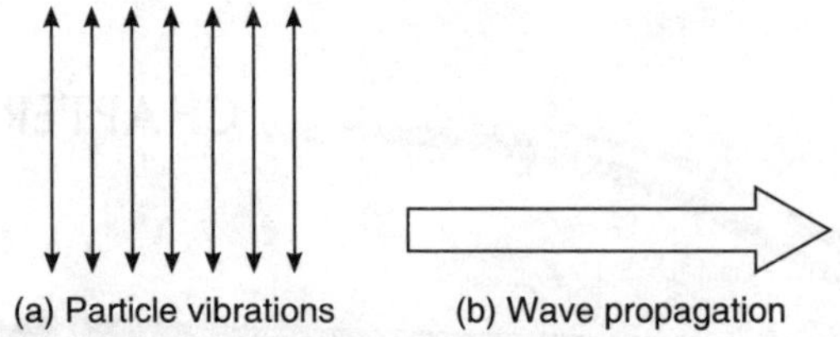

Figure 7.1 Transverse ultrasonic waves.

The transverse wave is also called shear wave. The velocity of the transverse ultrasonic wave is nearly 50% of the velocity of the longitudinal ultrasonic wave in the same medium. The velocity of a shear wave is given by

$$v_T = \sqrt{\frac{\eta}{\rho}} \tag{7.1}$$

where η is the rigidity modulus of the medium and ρ is the density of the medium.

The transverse wave can propagate only in solids, because in liquids and gases, this type of wave is greatly attenuated (Figure 7.2).

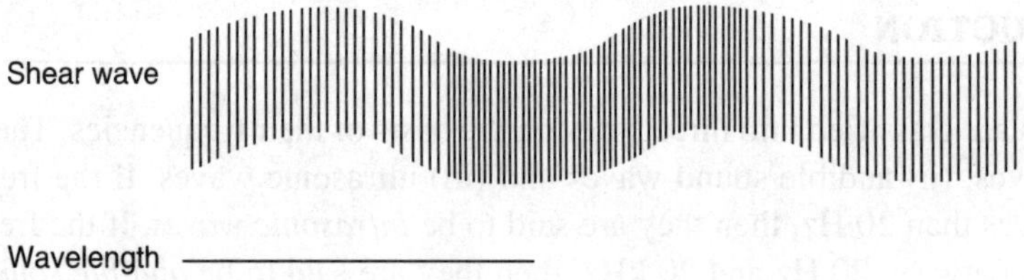

Figure 7.2 Wave motion in transverse wave.

7.2.2 Longitudinal Ultrasonic Waves

In the longitudinal ultrasonic waves, the particle of the medium vibrates parallel to the direction of the propagation of the wave as shown in Figure 7.3.

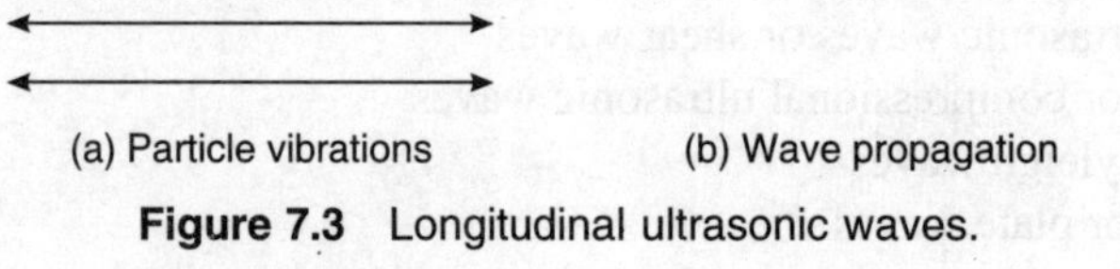

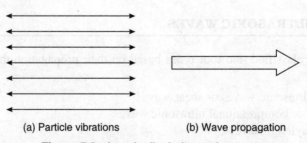

Figure 7.3 Longitudinal ultrasonic waves.

The velocity of the longitudinal wave propagating through a solid material is given by

$$v_L = \sqrt{\frac{Y}{\rho}} \tag{7.2}$$

where Y is the Young's modulus and ρ is the density of the solid.

The ultrasonic wave produces alternate compression and rarefaction while passing through the liquid or gaseous medium. Longitudinal motion is the most common form of sound transmission. The longitudinal waves can propagate in solids, liquids and gases (Figure 7.4).

Figure 7.4 Wave propagation in longitudinal ultrasonic waves.

7.2.3 Surface or Rayleigh Waves

In certain elastic medium, the ultrasonic wave passes only along the surface of the medium and hence they are called surface waves or Rayleigh waves. The surface waves were first discovered by Rayleigh and hence they are also known as Rayleigh waves. The surface wave does not propagate through the entire thickness of the material. The surface waves are used to study about the surface defects in the materials. The surface waves are analogous to water waves created on the surface of the water. The particles in water wave vibrate both in longitudinal and transverse waves, whereas in Rayleigh waves, the particles are having elliptical vibration. The particle vibration and the direction of motion of the wave are illustrated in Figure 7.5. The velocity of the surface wave is, $v = 0.9\ v_T$, where v_T is the velocity of the transverse wave.

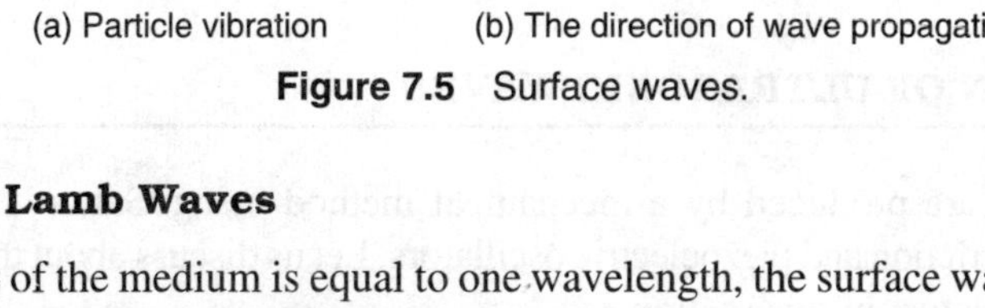

(a) Particle vibration (b) The direction of wave propagation

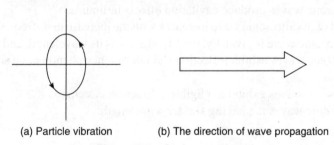

Figure 7.5 Surface waves.

7.2.4 Plate or Lamb Waves

When the thickness of the medium is equal to one wavelength, the surface wave does not exist and the medium vibrates whenever the ultrasonic waves are passed through it. This type of wave is said to be plate wave or Lamb wave. The Lamb wave occurs in two different modes, namely (i) symmetrical mode and (ii) asymmetrical mode.

If the displacement of the particles is elliptical on the surface and longitudinal in nature, then it is said to be symmetrical Lamb wave. If the displacement is elliptical on the surface and transverse in nature then it is said to be asymmetrical Lamb wave. The symmetrical and asymmetrical Lamb waves are illustrated in Figure 7.6 and Figure 7.7 respectively. The velocity of Lamb wave depends on the frequency, incident angle, thickness and the material property, whereas the velocity of the longitudinal, transverse and surface waves depends on the material property alone.

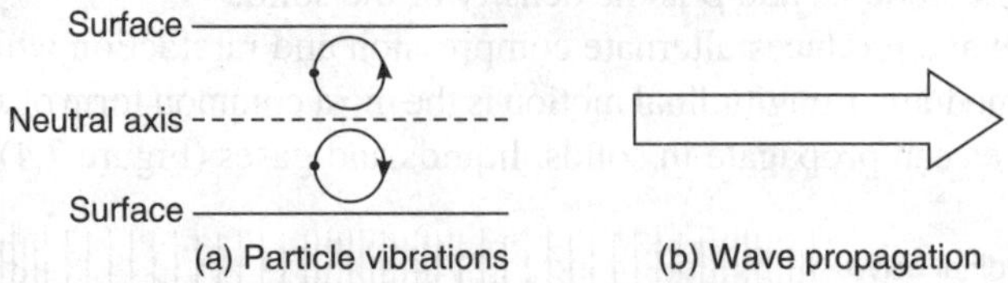

Figure 7.6 Symmetrical Lamb wave.

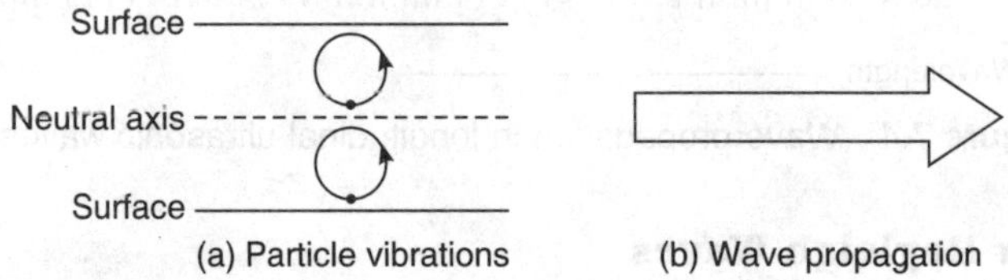

Figure 7.7 Asymmetrical Lamb wave.

7.3 PROPERTIES OF ULTRASONIC WAVES

1. The frequency of the ultrasonic waves is greater than 20 kHz.
2. The ultrasonic waves are highly energetic. Because of their high frequency, they have intensity up to 10 kW m^{-2}.
3. Like ordinary sound waves, ultrasonic waves produce alternate compression and rarefaction, when they are propagating through liquid and gaseous medium.
4. The ultrasonic waves produce cavitation effects in liquids.
5. The speed of the ultrasonic wave increases with the increase in its frequency. The velocity of the ultrasonic wave is given by, $v = f\lambda$, where λ is its wavelength and f is its frequency.
6. The ultrasonic waves exhibit reflection and interference phenomenon similar to ordinary light waves.
7. The ultrasonic waves exhibit negligible diffraction effect.
8. The ultrasonic waves are having shorter wavelength.

7.4 PRODUCTION OF ULTRASONIC WAVES

The ultrasonic waves are produced by a mechanical method using Galton's whistle. It is also produced by magnetostriction and piezoelectric oscillators. Let us discuss about the magnetostriction and piezoelectric oscillators in this section.

7.4.1 Magnetostriction Effect

Whenever a ferromagnetic material like nickel is subjected to a static magnetic field, there is a change of the dimension of the ferromagnetic material. Consider that the length of the ferromagnetic material increases due to the application of the magnetic field. If the polarity of the magnetic field is reversed, the length of the ferromagnetic material decreases. Instead of applying a static magnetic field, if a variable magnetic field is applied to a ferromagnetic material, the length of the material continuously changes. During one half cycle of the magnetic field, the length of the ferromagnetic material increases and during the other half cycle of the magnetic field, the length decreases.

Therefore, when an alternating magnetic field is applied to a ferromagnetic material, there is a continuous change of the dimension of the material and hence the material is set into vibration. This phenomenon is known as **magnetostriction effect.** The magnetostriction effect is demonstrated in Figure 7.8.

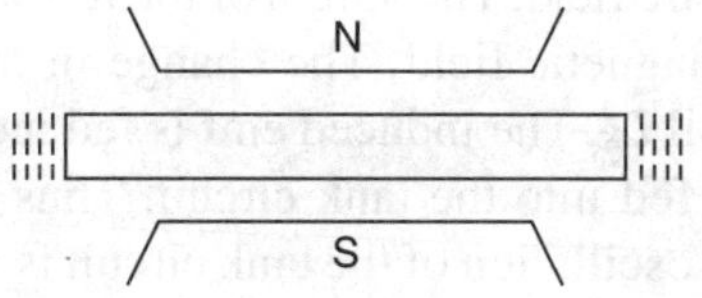

Figure 7.8 Magnetostriction effect in a ferromagnetic material.

Magnetostriction method

(a) Principle

The magnetostriction oscillator is based on the principle of magnetostriction effect. When a ferromagnetic material is subjected to a varying magnetic field, the length of the ferromagnetic rod changes and the ferromagnetic rod is set into resonant vibration, whenever the frequency of the tank circuit coincides with the frequency of the vibration of the ferromagnetic rod.

(b) Construction

A ferromagnetic rod AB made up of Ni is clamped in the middle, C as shown in Figure 7.9. Coil of wires L_1 and L_2 are winded at the ends of A and B. One end of the coil L_2 is connected to the base of an NPN transistor and the other end of the coil L_2 is connected to the emitter and the negative terminal of a battery. A variable capacitor C_1 is connected across the coil L_1. One end of the variable capacitor is connected to the collector circuit, whereas the other end of the variable capacitor is connected to the positive end of the battery through a milliammeter and a key *K*.

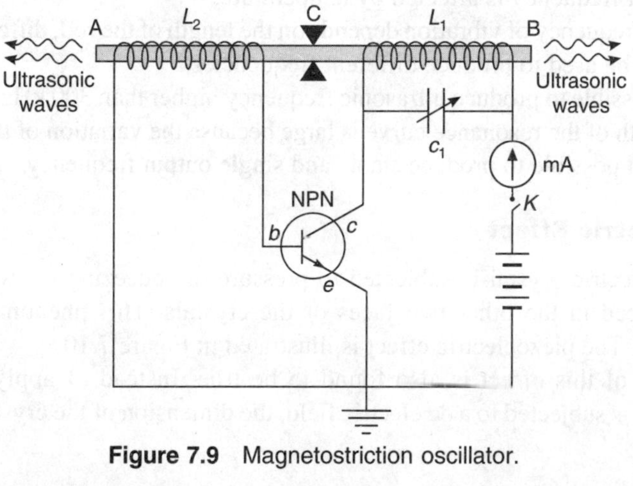

Figure 7.9 Magnetostriction oscillator.

(c) Working

The magnetostriction oscillator is shown in Figure 7.9. When the key is closed, the tank circuit (LC circuit) is set into oscillation with a frequency of vibration given by

$$f = \frac{1}{2\pi\sqrt{L_1 C_1}} \tag{7.3}$$

where L_1 is the value of the inductance of the coil L_1 and C_1 is the value of the capacitance of the capacitor, C_1.

An alternating emf is produced due to the vibration of the tank circuit. This alternating emf induces an alternating magnetic field. The length of the ferromagnetic material gets changed due to the induced alternating magnetic field. The change in the length of the ferromagnetic material induces an emf in the coil L_2. The induced emf is fed into the base of the transistor and hence it gets amplified and it is fed into the tank circuit. Thus the oscillation is continuously maintained. The frequency of the oscillation of the tank circuit is varied by adjusting the variable capacitance value. If the frequency of the tank circuit matches with the frequency of vibration of the ferromagnetic rod, ultrasonic waves are produced and the ultrasonic waves are emitted from the ends of the rod as shown in Figure 7.9. The frequency of the ultrasonic waves produced is given by

$$f = \frac{1}{2l}\sqrt{\frac{Y}{\rho}} \tag{7.4}$$

where l is the length of the ferromagnetic material, Y is the Young's modulus and ρ is the density of the ferromagnetic material.

(d) Advantages

1. Using this method, the ultrasonic waves having frequency up to 300 kHz can be produced.
2. The construction of the oscillator circuit is very simple.
3. The cost of construction is low.

(e) Drawbacks

1. The output frequency is affected by temperature.
2. Since the frequency of vibration depends on the length of the rod, different ferromagnetic rod has to be used to produce different frequencies.
3. It is not possible to produce ultrasonic frequency higher than 300 kHz using this method.
4. The breadth of the resonance curve is large because the variation of the elastic constant. So it is not possible to produce stable and single output frequency.

7.4.2 Piezoelectric Effect

Whenever a piezoelectric crystal is subjected to pressure or squeezing or twisting, the electric charges are developed in the other two faces of the crystals. This phenomenon is known as **piezoelectric effect**. The piezoelectric effect is illustrated in Figure 7.10.

The converse of this effect is also found to be true. Instead of applying pressure, if a piezoelectric crystal is subjected to a dc electric field, the dimension of the crystal changes, i.e. the dimension either increases or decreases. If the polarity of the electric field is reversed, the changes in the dimension are also reversed. Instead of applying a dc electric field, if an ac field is applied, the dimension changes continuously. During one half cycle of the electric field, the dimension of the crystal increases and during the other half cycle of the electric field the dimension of the crystal decreases. Therefore, whenever an alternating electric field is applied to a piezoelectric crystal, the dimension of the crystal changes continuously and hence the crystal is set into vibration. This phenomenon is known as **inverse piezoelectric effect**. The inverse piezoelectric effect is shown in Figure 7.11.

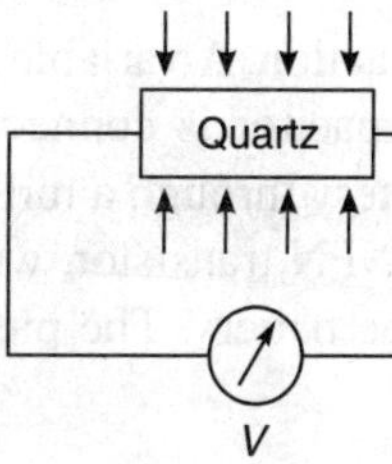

Figure 7.10 Piezoelectric effect.

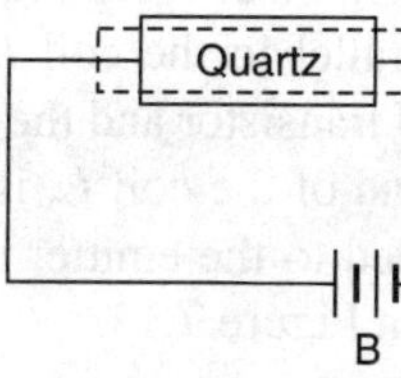

Figure 7.11 Inverse piezoelectric effect.

X-cut and Y-cut crystals

The piezoelectric crystals crystallize in a rhombohedron crystal structure. The rhombohedron-shaped crystal is shown in Figure 7.12(a). The cross-sectional view of the crystal consists of six faces and looks like a hexagon and it is shown in Figrue 7.12(b). The line joining any two opposite corners of the crystal is known as **electric axis or *X*-axis**. The line joining the midpoint of the opposite faces of the crystal is known as **mechanical axis or *Y*-axis**. If a piece of crystal is cut perpendicular to the *X*-axis, then it is said to be an ***X*-cut crystal**. If the crystal is cut perpendicular to the *Y*-axis then it is said to be ***Y*-cut crystal**. The *X*-cut crystals are used to produce the longitudinal ultrasonic waves, whereas the transverse ultrasonic waves are produced by the *Y*-cut crystals.

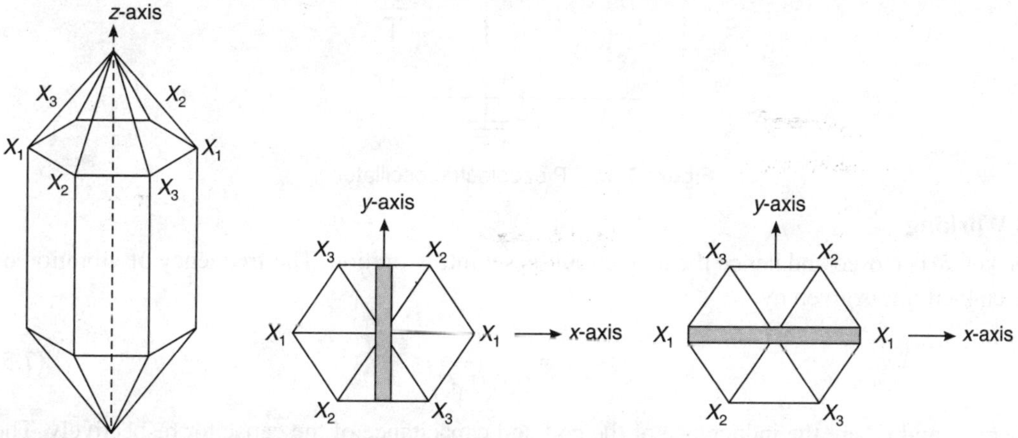

Figure 7.12 Piezoelectric effect.

Piezoelectric oscillator

(a) Principle

The piezoelectric oscillator is based on the principle of inverse piezoelectric effect. Whenever a piezoelectric crystal is subjected to an alternating voltage, the piezoelectric crystal is set into vibration. The resonant vibration of the piezoelectric crystal takes place whenever the frequency of vibration of the piezoelectric crystal matches with the frequency of vibration of the tank circuit and hence it produces ultrasonic waves.

(b) Construction

The coils L_1, L_2 and L_3 are coupled together by means of transformer action. A variable capacitor, C_1, is connected parallel to the coil L_1. One end of the variable capacitor is connected to the collector of the NPN transistor and the other end is connected to a battery through a milliammeter and a key, K. One end of the coil L_2 is connected to the base of the NPN transistor, whereas the other end is connected to the emitter and the negative terminal of the battery. The piezoelectric oscillator is shown in Figure 7.13.

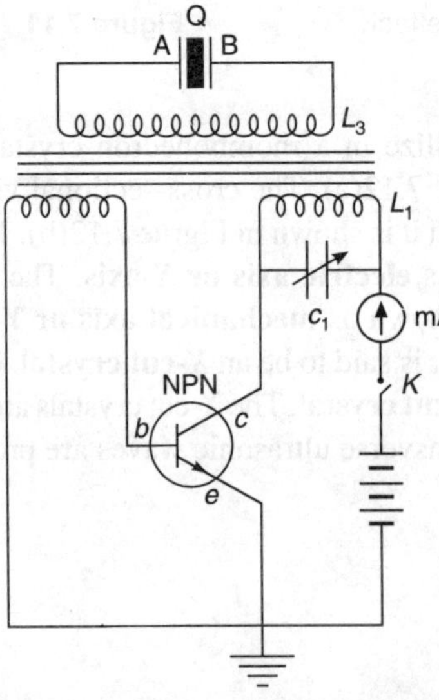

Figure 7.13 Piezoelectric oscillator.

(c) Working

The key K is closed and hence the tank circuit is set into vibration. The frequency of vibration of the tank circuit is given by

$$f = \frac{1}{2\pi\sqrt{L_1 C_1}} \tag{7.5}$$

where L_1 and C_1 are the inductance of the coil and capacitance of the capacitor respectively. The oscillation of the tank circuit produces an alternating emf in the tank circuit. An emf is induced by mutual induction in the coil L_2. This emf is fed into the transistor as an input into the base–emitter terminal. The transistor is used to ensure the correct phase of the feedback. The input is amplified and fed into the collector circuit. Due to the transformer action an emf is induced in the coil L_3.

This induced emf is applied to the piezoelectric crystal. Due to this alternating emf, the piezoelectric crystal is set into continuous vibration. The resonant vibration takes place, only when the frequency of the tank circuit coincides with the frequency of the piezoelectric crystal. During the resonant vibration, the piezoelectric crystal produces ultrasonic waves.

The frequency of the ultrasonic waves produced when the crystal is placed along its thickness between the plates A and B is given by

$$f = \frac{1}{2t}\sqrt{\frac{Y}{\rho}} \tag{7.6}$$

The frequency of the ultrasonic waves produced when the crystal is placed along its length between the plates A and B is given by

$$f = \frac{1}{2l}\sqrt{\frac{Y}{\rho}} \tag{7.7}$$

where t is the thickness of the crystal, l is the length of the crystal, Y is the Young's modulus of the piezoelectric crystal and ρ is the density of the piezoelectric crystal.

The velocity of the ultrasonic wave produced is given by

$$v = \sqrt{\frac{Y}{\rho}} \tag{7.8}$$

where Y is the Young's modulus of the piezoelectric crystal and ρ is the density of the piezoelectric crystal.

(d) Advantages

1. The maximum frequency of the ultrasonic wave produced using this method is 500 MHz.
2. The output frequency is independent of temperature and humidity.
3. A stable and constant output frequency is produced because the breadth of the resonance curve is small.
4. It is more efficient than magnetostriction oscillator.

(e) Drawbacks

1. The piezoelectric oscillator is costly.
2. The cutting and shaping of the piezoelectric crystal is difficult.

7.5 DETECTION OF ULTRASONIC WAVES

The ultrasonic waves are detected by the following methods, namely

1. Acoustic grating method
2. Kundt's tube method
3. Thermal detection
4. Sensitive flame
5. Piezoelectric detection

7.5.1 Acoustic Grating Method

Consider a container consisting of a liquid. An ultrasonic transducer is placed in the lower end of the container. If ultrasonic waves are passed through the liquid, alternate compression and rarefactions are produced. The compressed portion acts as an opaque medium for the ordinary light and the rarefied portion acts as a transparent medium for the ordinary light. If the container is illuminated by a source of light, it undergoes diffraction and the diffraction pattern is obtained on a screen. So, the container acts as a grating. Therefore, it is said to be an acoustic grating (Figure 7.14).

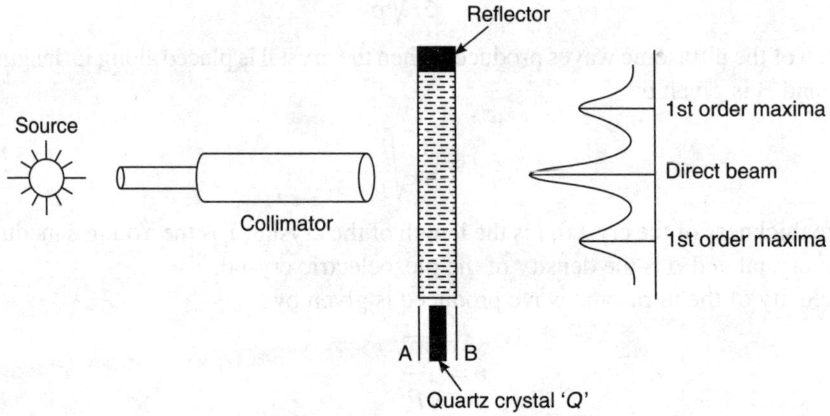

Figure 7.14 Acoustic grating.

The condition for the diffraction pattern is

$$2d \sin \theta = n\lambda \tag{7.9}$$

where n is the order of diffraction, λ is the wavelength of the light used, θ is the angle of diffraction and d is the distance between any two alternate compressions or rarefactions. The distance between any two compressions or any two rarefactions is equal to half of the wavelength.

i.e. $$\frac{\lambda_0}{2} = d \quad \text{or} \quad 2d = \lambda_0 \tag{7.10}$$

where λ_0 is the wavelength of the ultrasonic waves. Substituting Eq. (7.10) in Eq. (7.9), we get

$$\lambda_0 \sin \theta = n\lambda \tag{7.11}$$

By measuring the diffraction angle, θ, the wavelength of the ultrasonic wave is calculated from Eq. (7.11). By knowing the frequency of the ultrasonic wave, the velocity of the ultrasonic wave is calculated using the relation,

$$v = \lambda_0 f \tag{7.12}$$

Equation (7.12) is used to measure the velocity of ultrasonic wave in the liquid medium using acoustic grating method.

7.5.2 Kundt's Tube Method

Kundt's tube apparatus consists of a 1 m long, 3 to 4 cm diameter glass tube as shown in Figure 7.15. Kundt's tube is used to measure the velocity of sound in laboratory. It is also used to detect the ultrasonic waves having longer wavelength. One end of the Kundt's tube is fitted with a cork and a movable piston. On the other end an ultrasonic transducer is placed. The lycopodium powder is sprinkled between the movable piston and the ultrasonic transducer. When ultrasonic waves are produced, it passes through the Kundt's tube apparatus and gets reflected back from the movable piston. The movable piston is adjusted so that a stationary wave is produced. When the stationary wave is formed, the lycopodium powder collects in the form of heaps at nodal points and blown off at the antinodal points. The distance between any two nodal or antinodal points is measured and hence the wavelength of the ultrasonic wave (λ_0) is determined using the relation,

$$\lambda_0 = 2d \tag{7.13}$$

where d is the distance between any two adjacent nodals or antinodal points. By measuring the wavelength of the ultrasonic wave, its velocity is determined using the relation,

$$v = \lambda f \tag{7.14}$$

where λ is the wavelength and f is the frequency of the ultrasonic waves.

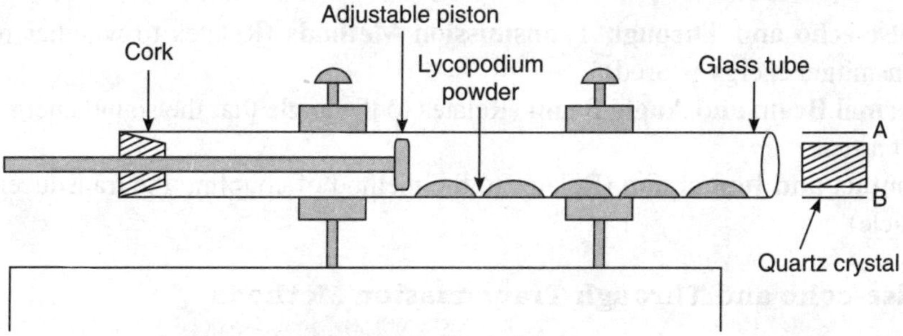

Figure 7.15 Kundt's tube.

7.5.3 Thermal Detection

Ultrasonic waves produce alternate compression and rarefaction while passing through a gaseous or a liquid medium. The compression of particles produces heat energy and the rarefaction produces the cooling effect. Consider a platinum resistance thermometer placed in the path of the ultrasonic wave. Platinum is sensitive even to a small change in temperature. If the platinum resistance thermometer is moved in the path of the ultrasonic wave, at the node there is a change of temperature and at the antinodes temperature remains constant. By measuring the temperature at the node using the platinum resistance thermometer, the ultrasonic waves are detected.

7.5.4 Sensitive Flame

When a sensitive flame is moved in the path of the ultrasonic wave, the flame will flicker when it crosses the node and there is no change in the flame occurs when it crosses the antinodes. The ultrasonic wave is detected using a sensitive flame by observing this change.

7.5.5 Piezoelectric Detection

Consider that an ultrasonic wave is made to fall on a pair of opposite faces of a piezoelectric crystal. Since the ultrasonic waves are highly energetic waves, positive and negative charges are developed in the other pair of opposite faces, which are perpendicular to the first pair. The charges are amplified and detected (Figure 7.16).

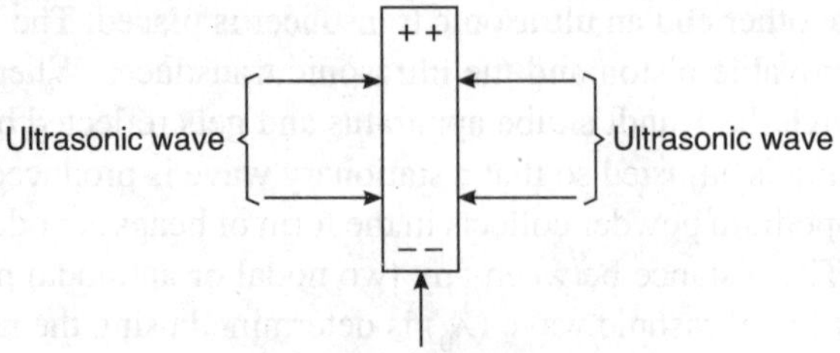

Figure 7.16 Piezoelectric detection.

7.6 ULTRASONIC INSPECTION TECHNIQUES

The ultrasonic inspection techniques are commonly divided into three primary types as follows.

(i) **Pulse-echo and Through Transmission Methods** (Relates to whether reflected or transmitted energy is used)

(ii) **Normal Beam and Angle Beam** (Relates to the angle that the sound energy enters the test article)

(iii) **Contact and Immersion** (Relates to the method of coupling the transducer to the test article)

7.6.1 Pulse-echo and Through Transmission Methods

Pulse-echo method

This method is based on the principle of reflection of the ultrasonic waves. The ultrasonic wave in the form of a pulse is produced and it is passed through the specimen under study as shown in Figure 7.17(a). The pulse gets reflected from the specimen and it is also received by the same transducer. The time difference between the transmitted and the received signal is determined from a CRO. The ultrasonic velocity through the specimen is determined using the relation

$$v = \frac{2d}{t} \tag{7.15}$$

where d is the thickness of the specimen. The factor 2 is introduced because the ultrasonic wave gets propagated through the specimen two times.

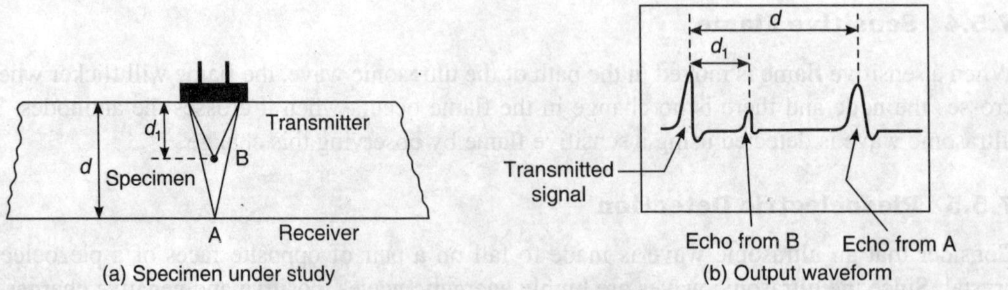

Figure 7.17 Pulse-echo method.

Through transmission method

This method is based on the principle of the transmission of the ultrasonic wave through the specimen. In this method, two different transducers are arranged in such a way that they are fixed in the pair of opposite surfaces. These two surfaces should be highly polished, flat and made perfectly parallel to each other. One transducer is used to transmit the ultrasonic waves, whereas the other is used to receive the ultrasonic waves as shown in Figure 7.18(a). By measuring the time taken by the ultrasonic waves to pass through the specimen from one transducer to the other transducer, the ultrasonic velocity is determined using the relation

$$v = \frac{d}{t} \tag{7.16}$$

where d is the thickness of the specimen.

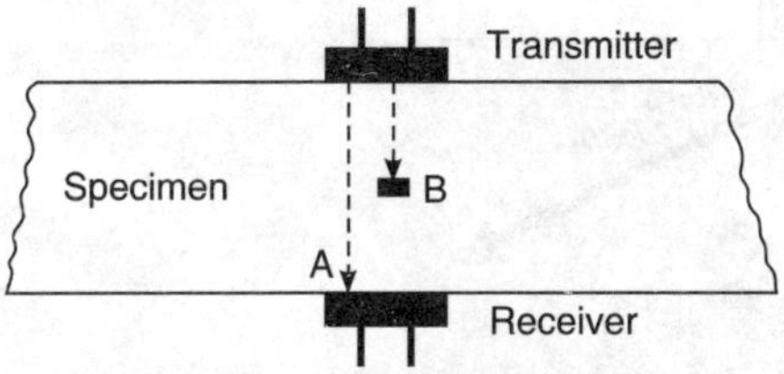

(a) Specimen under study

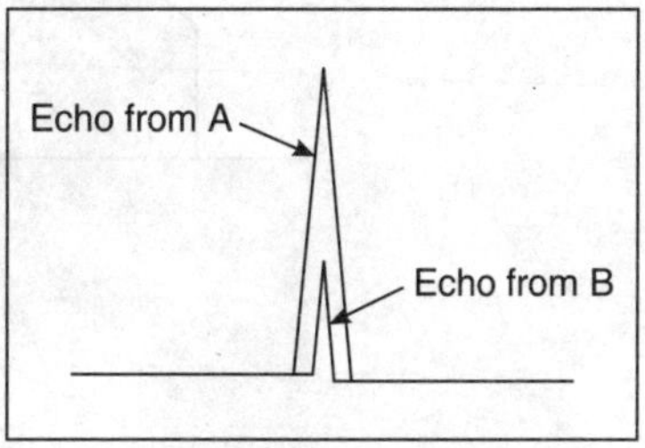

(b) Output waveform

Figure 7.18 Transmission method.

7.6.2 Normal Beam and Angle Beam Methods

Normal beam method

In normal beam method, the probe is kept directly over the specimen and the ultrasonic wave is introduced at an angle of 90° through the test specimen. The normal beam testing method is shown in Figure 7.19. The pulse-echo method is one of the normal beam methods.

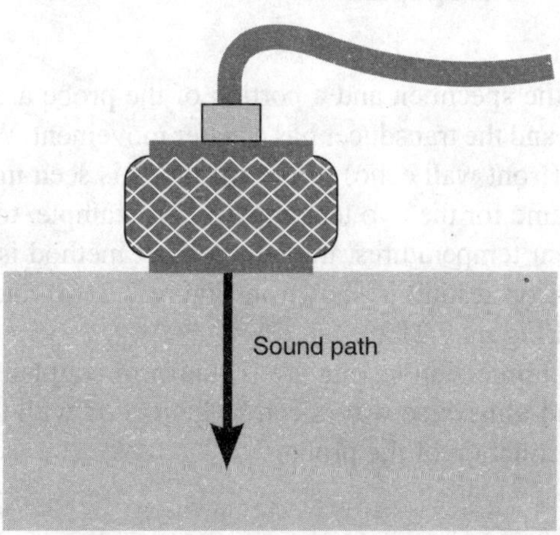

Figure 7.19 Normal beam method.

Angle beam method

In angle beam testing, the sound beam is introduced into the test specimen at some angle other than 90° as shown in Figure 7.20. It is used to access the areas that are not accessible by normal beam method. The shear wave probe is used for the angle beam inspection techniques.

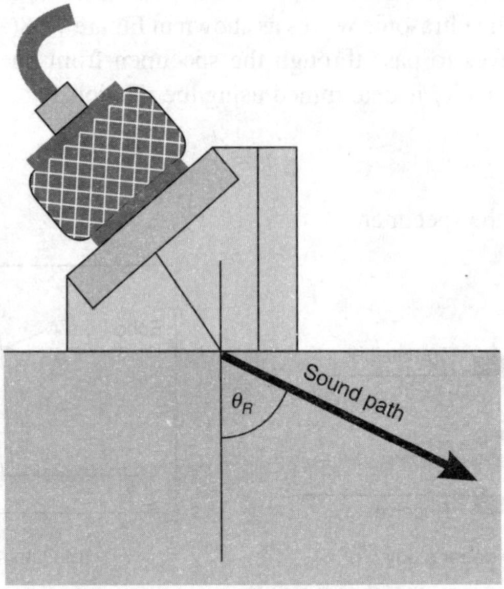

Figure 7.20 Angle beam method.

7.6.3 Contact and Immersion Methods

Contact method

In the contact method, the transducer is kept in contact with the specimen. Figure 7.17(a) is an example of the contact method. In contact testing a couplant such as water, oil or a gel is applied between the transducer and the testing specimen.

Non-contact method

In the non-contact method, the specimen and a portion of the probe are immersed in water. The water bath acts as a couplant and the transducer has a better movement. With immersion testing, an echo from the front surface (front wall echo) of the specimen is seen in the signal, but otherwise signal interpretation is the same for the two techniques. For example, to determine the ultrasonic velocity of liquids at different temperatures, the non-contact method is used. The diagrammatic representation of the immersion testing is shown in Figure 7.21(a) and the results obtained are shown in Figure 7.21(b) and Figure 7.21(c).

The advantages of the immersion testing are (i) uniform couplant conditions are obtained, and (ii) both longitudinal and transverse waves can be generated with the same probe simply by changing the angle of the inclination of the probe.

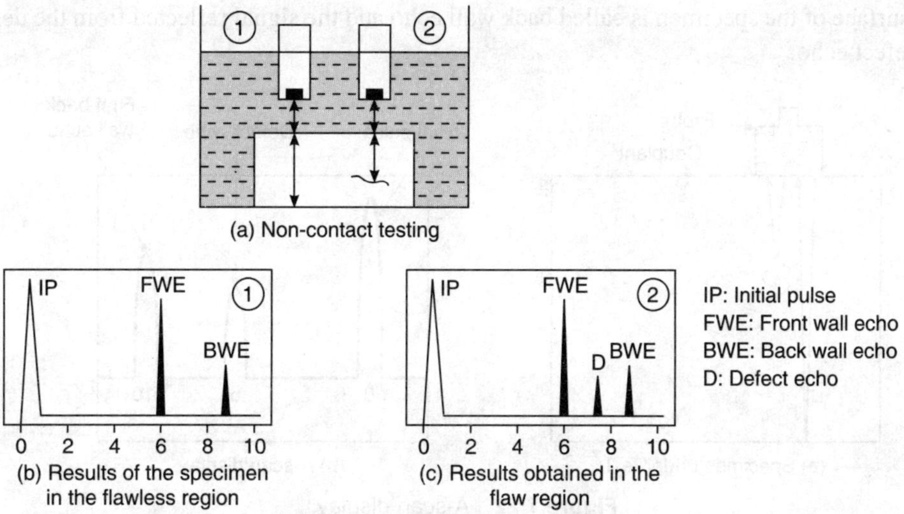

Figure 7.21 Immersion testing.

7.6.4 Application of Ultrasonic Testing

The ultrasonic testing is used for the following applications.

1. It is used for the detection of flaws in the materials such as metals, etc.
2. It is used for the measurement of corrosion thickness in a material.
3. The grain size in metals can be estimated using ultrasonic testing methods.
4. The depth of case hardening is measured using ultrasonic testing.
5. It is used for the estimation of void content in composites and plastics.

7.7 SCAN DISPLAYS

The ultrasonic data collected can be displayed in different formats. They are

(i) A-scan (or, A-scope)
(ii) B-scan (or B-scope)
(iii) C-scan (or, C-scope).

7.7.1 A-scan

A-scan is one of the simplest methods of displaying the results of ultrasonic testing. The echoes can be displayed just as in the ordinary oscilloscope. It is a plot of the time in the *X*-axis, and the amplitude in the *Y*-axis. In A-scan the probe is placed stationary in one position. The time-measuring instrument such as CRO is synchronized with the transmitter/receiver system for the A-scan display. This is mostly used display in the ultrasonic testing of the specimen. From the time difference between the transmitted and the reflected signals, the depth of the flaw is determined using mathematical calculations. Figure 7.22 illustrates the A scan testing of specimen and display. In Figure 7.22(b), the transmitted signal is represented as initial pulse, the pulse reflected from the bottom surface of the specimen is called back wall echo and the signal reflected from the defect is called defect echo.

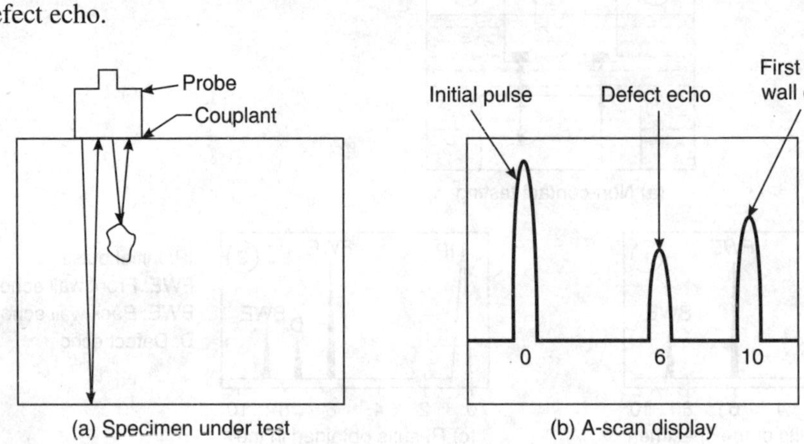

Figure 7.22 A-scan display.

7.7.2 B-scan

In B-scan, the echo signal is used to control the brightness of the spot on the screen instead of applying it to the Y-plate of the CRO. The resultant data produces a cross sectional view of the specimen. The position is plotted against amplitude or time of flight. It provides the reflector depth in the cross section and from the plot, the linear dimension can be determined. Figure 7.23 illustrates B-scan testing of specimen and B-scan display. The brightness of the echo is proportional to the echo amplitude. In B-scan the probe is moved along a line, i.e. in one direction. A number of dots will appear on the screen at any one time. The entire picture will be created by storage oscilloscope or by a long persistent screen.

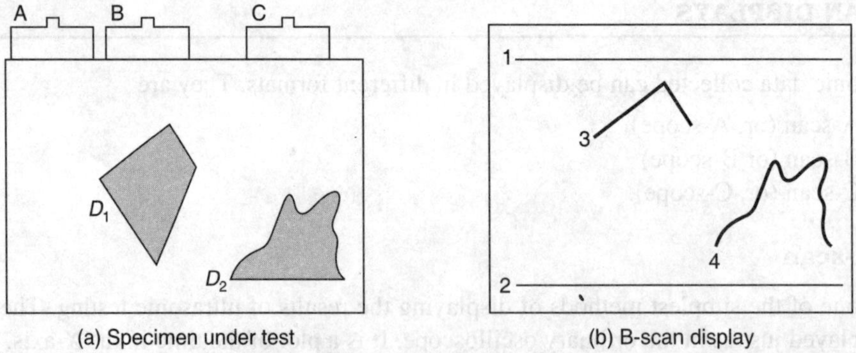

Figure 7.23 B-scan display.

7.7.3 C-scan

In C-scan, the probe is moved from one end to the other end of the specimen along the horizontal position. It records the echoes from the internal portions of the test piece as a function of the position. The C-scan display presents the plan type view of the test specimen and discontinuities.

Flaws are seen in C-scan display. C-scan is used in conjunction with A-scan for depth determination. The C-scan display is produced by an automated data acquisition system, such as in immersion scanning.

In C-scan any pulse that appears between the transmitted pulse and the back wall echo is gated out and produces a plane projection of all the echoes within the gate length. The C-scan display is useful in corrosion detection in pipes and pressure vessels. The intensity of the reflected signal is recorded as a variation in time shading. The absence of shading refers to the presence of flaw. The C-scan testing and display are shown in Figure 7.24.

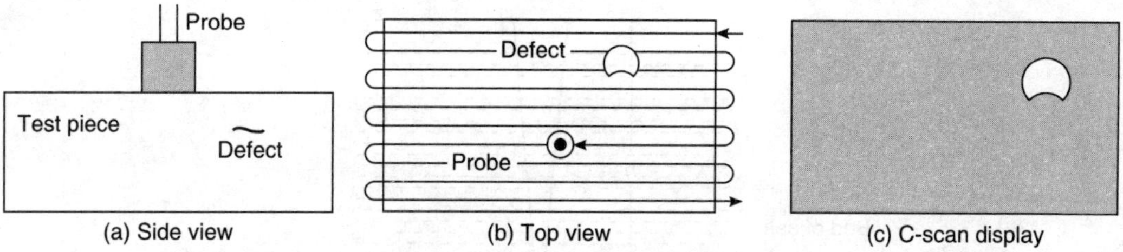

Figure 7.24 C-scan display.

7.8 CAVITATIONS

Whenever highly energetic ultrasonic waves are propagated through liquid medium, gas bubbles are formed. This process is said to be *cavitations*. The cavitation effect is also produced by lowering the pressure in a liquid. Whenever the pressure in a liquid is reduced, the boiling point also gets reduced. If the pressure of the liquid is reduced sufficiently, the liquid gets boiled without being heated. When the pressure of the liquid is lowered, gas bubbles are formed. The formation of gas bubbles in a liquid when its pressure is reduced is known as cavitations. The bubbles, when they explode, produce high temperatures (nearly 10,000°C and high pressures 50,000 lbs per inch) and high pressures are produced in the surroundings. These bubbles when explode remove the dirt and do cleaning due to the production of high temperatures.

7.9 APPLICATIONS OF ULTRASONIC WAVES

7.9.1 Industrial Applications

Depth of sea

The ultrasonic waves exhibit very feeble diffraction effect and they are not absorbed by the sea water. On the basis of this property ultrasonic waves can be transmitted through water for a longer distance. Consider a piezoelectric transducer is used to produce the ultrasonic waves and it is transmitted towards the bed of the sea water as shown in Figure 7.25. The ultrasonic wave reflected from the bed of the sea water is collected and the time taken for the to and fro motion is noted. Then by knowing the velocity of ultrasonic wave in sea water, the depth of sea is determined using the relation given by

$$\text{Depth of the sea, } h = \frac{v \times t}{2} \tag{7.17}$$

where v is the velocity of the ultrasonic wave in sea water, and t is the time taken for the to and fro motion.

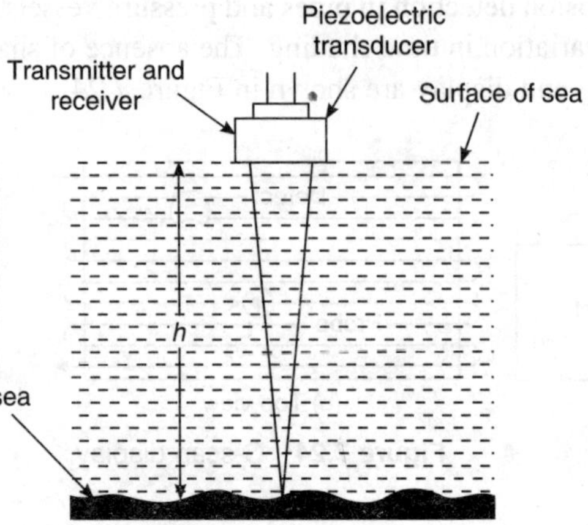

Figure 7.25 Depth of sea.

Sonar

Sonar stands for **SO**und **N**avigation **A**nd **R**anging. The ultrasonic waves are used to find out the submerged submarines, icebergs and so on. The ultrasonic waves undergo negligible diffraction effect and they are not absorbed by the sea water. On the basis of this property, the ultrasonic waves are passed through the sea water. Consider the ultrasonic waves which are passed through the sea water using a transducer (Figure 7.26). If there is any submerged submarine under the sea, the reflected signal from the submerged submarine is collected by the receiving transducer. The time taken for the to and-fro motion is noted. By knowing the velocity of the ultrasonic wave in the sea water, the position and the depth of the submerged submarine is located. Sonar is also used to find out shoal of fishes, icebergs and so on. If any submarine is moving under the sea, using the Doppler shift principle, the velocity of the moving submarine and its depth is determined.

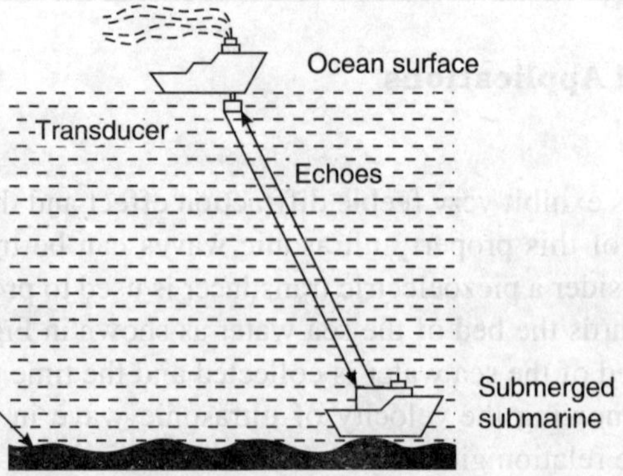

Figure 7.26 Sonar.

Ultrasonic welding

When high energetic ultrasonic waves are made to incident on two pieces of materials kept in contact under pressure produces the materials melts and hence create a joint. This process is called **ultrasonic welding**. The ultrasonic welding is called **cold welding** because during welding stress is induced and no heat energy is evolved. Two portions are joined together at a very low temperature. Ultrasonic welding is commonly used to weld plastics and for joining dissimilar materials. The frequency of the ultrasonic waves used for this purpose is in the range of 20 kHz to 70 kHz (usually 20 kHz, 30 kHz, 35 kHz, 40 kHz and 70 kHz are used). The ultrasonic welding is used for the following fields:

1. The ultrasonic welding is used for cell phones, consumer electronics, disposable medical tools, toys, etc.
2. It is used to carry out small welds in malleable metals such as Al, Cu, Ni.
3. It is used to join any two dissimilar materials.

(a) Aerospace industry

1. In aerospace industry, it is used to join thin sheet gauge metals and other lightweight materials. It is used to weld aluminium (Al) in aerospace industry. Welding on Al is not easily possible using other techniques, because of its high thermal conductivity.
2. It is used to weld composite materials and carbon fibre.

(b) Automobile industry

1. Ultrasonic welding is used for welding of plastic and electric components such as door panels, lamps, air ducts, steering wheels, upholstery and engine components.
2. Ultrasonic welding has low cycle times, automation, low capital cost and flexibility.
3. It does not damage the surface finish, which is a crucial consideration for car-makers.

(c) Medical industry

The items such as arterial filters, anesthesia filters, blood filters, dialysis tubes, pipettes, cardiometry reservoirs, blood/gas filters and face makers can be made using ultrasonic welding.

(d) Textile industry

Hospital gowns, sterile garments, masks, and textiles for clean rooms can be sealed and sewn by ultrasonic welding.

(e) Packaging industry

1. Sealing of containers, tubes and blister packs are some common applications of ultrasonic welding in packaging industry.
2. Packaging of dangerous materials such as explosives, fireworks and other reactive chemicals are made using ultrasonic welding.

(f) Food industry

Milk and juice containers are often sealed using ultrasonic welding.

Ultrasonic cutting and drilling

The ultrasonic machining tool is made of a ferromagnetic rod. Coil of wires are wounded on the ferromagnetic rod. An alternating voltage of frequency f is passed through the coil of wire. It induces alternating magnetic field and hence the rod vibrates at ultrasonic frequency. If the length of the ferromagnetic rod is chosen such that it can produce resonant frequency ($2f$) for an applied frequency of f, then the rod makes resonant vibration. The machining tool is placed over the material, which is to be cut or a drill is to be made and the rod is made to vibrate at resonant frequency. A solid abrasive material in the aqueous form is transferred into the place in which the material is to be cut or a drill is to be made, then vibration of the machining tool transfers very high energy to the abrasive powder. The abrasive powder makes collision with the work piece and hence it removes the minute particles. Due to this process, the material is cut or a precise hole is made on the material. Using this process the drill is made in brittle materials like glass and hard materials like diamond, alloys and the semiconducting silicon wafer are machined using ultrasonic method.

Ultrasonic cleaning

The ultrasonic waves are used for the purpose of cleaning in industry. Generally for cleaning, acetone or the soap solution is used. In ultrasonic cleaning, the materials to be cleaned are taken in an ultrasonic cleaning bath that contains a cleaning solution such as soap solution and an ultrasonic generator. The ultrasonic waves produced by the generator are passed through the cleaning solution. The ultrasonic waves produce bubbles due to the cavitation effect whenever they are passing through the liquid. When these gas bubbles explode, a very high amount of energy is released. This energy torn offs the dirt or the contaminants present in the material. This method is very useful to clean any delicate materials. The jewellers use this method to clean jewellery.

Ultrasonic soldering

The ultrasonic wave is used to solder on aluminium. It is not possible to solder on aluminium using an ordinary soldering iron due to the presence of oxide material over aluminium. Soldering is made on aluminium using a special soldering iron, which vibrates at the frequency of the ultrasonic waves. The passage of ultrasonic wave on aluminium metal removes the oxide layer and then soldering is made on aluminium without flux.

Non-destructive testing (NDT)

One of the advantages of the ultrasonic testing is the non-destructive testing of the specimen. The ultrasonic NDT is used for the flaw detection. The diagrammatic representation of an ultrasonic flaw detector is shown in Figure 7.27. The master timer produces the triggering signals at a regular interval of time. Whenever a triggering signal is received, the pulse generator produces an alternating pulse of electrical signal. This electrical signal is applied to a transducer, which in turn produces ultrasonic waves. The ultrasonic wave passes through the specimen and gets reflected from the specimen. The reflected ultrasonic wave is received by the same transducer and it is received by a signal amplifier. The signal amplifier amplifies the reflected signal and it is fed into the Y plate of the CRO. The master timer also sends the triggering signal to the time base circuit of the CRO and hence a pattern is formed between the amplitude of the reflected signal and time as shown in Figure 7.27.

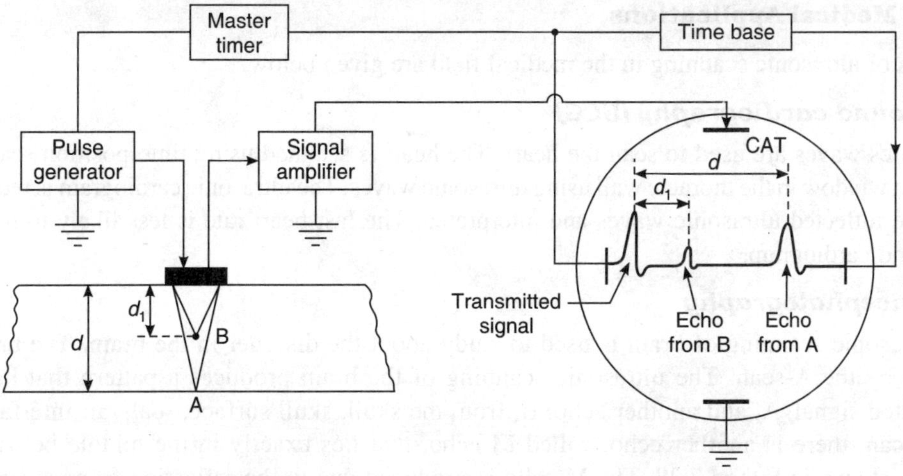

Figure 7.27 Non-destructive testing of specimen.

The output wave in the CRO consists of the transmitted signal and the pulse reflected from the bottom surface of the specimen A. The time difference between the two peaks is used to measure the velocity of the ultrasonic wave which can be given as

$$v = \frac{2d}{t} \tag{7.18}$$

where t is the time difference between the transmitted and reflected signal and d is the thickness of the specimen. If there is any discontinuity like a flaw in the specimen, then another echo (reflected signal) is seen between the transmitted signal and the signal reflected from the bottom surface. The time difference between the transmitted signal and the reflected signal from the flaw is used to find the depth of the flaw using the relation,

$$d_1 = \frac{vt}{2} \tag{7.19}$$

where d_1 is the depth of the flaw, v is the velocity of the ultrasonic wave in the specimen and t is the time taken.

(a) Advantages of ultrasonic NDT

1. The ultrasonic testing can be made from any accessible surface.
2. The ultrasonic test can be made in a small specimen.
3. Ultrasonic equipment is compact and portable.
4. The testing provides result immediately. It should be interpreted carefully and decision should be taken accordingly.
5. It is one of the cheapest methods.
6. Permanent records can be maintained.

(b) Drawbacks of ultrasonic NDT

1. The skilled operators are needed to interpret the results.
2. Irregular shape and rough specimen cannot be studied.
3. Flaw estimation requires some standard specimen.

7.9.2 Medical Applications

The uses of ultrasonic scanning in the medical field are given below.

Ultrasound cardiography (UCG)

Ultrasonics waves are used to scan the heart. The heart is scanned using time-position scanning through a window in the thoracic wall using ultrasonic waves. The ultrasonic cardiogram is recorded using the reflected ultrasonic waves and interpreted. The fast heart rate is less likely to register ultrasound cardiogram.

Echoencephalography

The ultrasonic scanning of brain is used to study about the disorder in the brain. The brain is scanned by the A-scan. The ultrasonic scanning of the brain produces a pattern that has the transmitted signal, A, and another echo, B, from the skull, skull surface, scalp/air interface. In this A-scan, there is another echo, called M echo, that lies exactly in the middle between A and B as shown in Figure 7.28. The M echo is produced due to the reflection from several co-planar interfaces at about the same distance from the probe. If there is any departure of M echo, it indicates that there is an abnormality, like bleeding in one side of the brain due to injury or stroke, intracranial tumors, hydrocephalus.

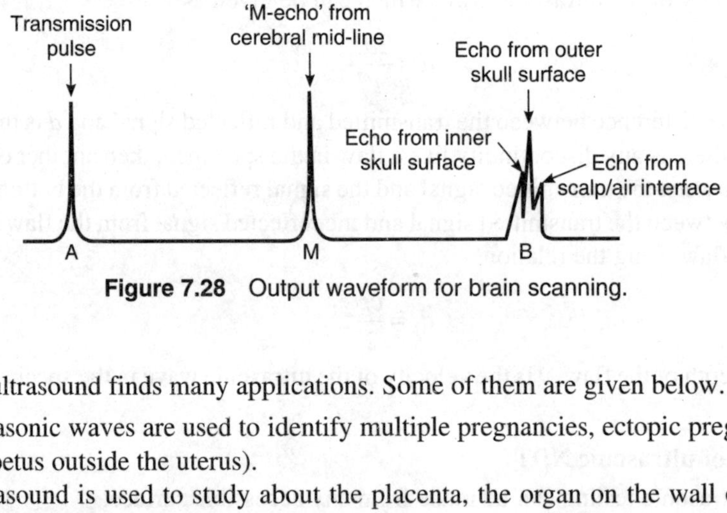

Figure 7.28 Output waveform for brain scanning.

Obstetrics

In obstetrics, ultrasound finds many applications. Some of them are given below.

1. Ultrasonic waves are used to identify multiple pregnancies, ectopic pregnancy (growth of foetus outside the uterus).
2. Ultrasound is used to study about the placenta, the organ on the wall of the uterus in which the foetus is nourished.

Other applications

1. The dentist uses the ultrasonic drill to produce a small hole in the teeth. The ultrasonic drill is used to reduce the pain and to improve the definition of the drilled area.
2. Another medical use of ultrasonics is cutting through tissue with a knife edge. An advantage of this technique for the patient is reduced bleeding from coagulation of blood vessels and the ease at which the knife blade cuts from reduced friction and increased sharpness.

Doppler effect

When a source, the medium and the observer are in relative motion, the frequency of the sound received by the observer is different from the frequency emitted by the source. This is known as Doppler effect.

Derivation

Consider that a source is producing a sound. Let f be the frequency and c be the velocity of the sound. Consider that an observer is at rest. The wavelength of the sound produced by the source when it is stationary is given by

$$\lambda = \frac{c}{f} \tag{7.20}$$

Consider that the source is moving with a velocity v towards the observer as shown in Figure 7.29. The wavelength of the sound when the source is moving is given by

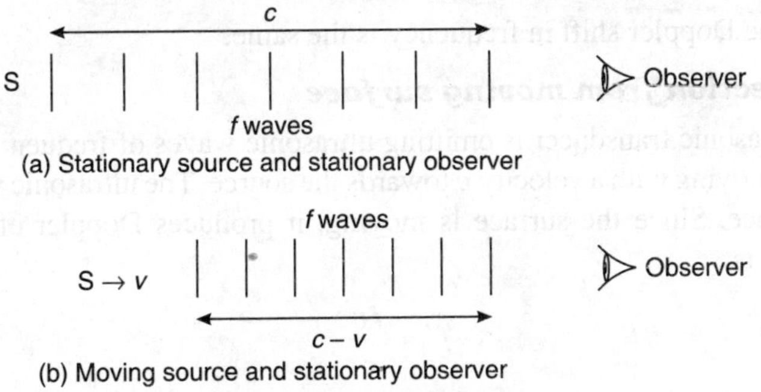

Figure 7.29 Doppler effect.

$$\lambda' = \frac{c - v}{f} \tag{7.21}$$

The apparent frequency is

$$f' = \frac{c}{\lambda'} = \frac{cf}{(c - v)} \tag{7.22}$$

The change in frequency is

$$f' - f = \frac{cf}{(c - v)} - f$$

$$\Delta f_1 = \frac{fv}{c - v}$$

Since $c >> v$, the above equation can be written as,

$$\Delta f_1 = \frac{fv}{c} \tag{7.23}$$

Now consider that the source is at rest and the observer is moving towards the source. The wavelength of the sound emitted by the source is

$$\lambda = \frac{c}{f}$$

i.e.

$$f = \frac{c}{\lambda} \tag{7.24}$$

The apparent frequency of the sound observed by the observer is

$$f' = \frac{c+v}{\lambda}$$

Substituting $\lambda = \dfrac{c}{f}$, we get

$$f' = \frac{(c+v)f}{c} \tag{7.25}$$

The change in frequency is

$$f' - f = \frac{(c+v)f}{c} - f$$

$$\Delta f_2 = \frac{fv}{c} \tag{7.26}$$

In these two cases the Doppler shift in frequency is the same.

Ultrasound reflection from moving surface

Consider that an ultrasonic transducer is emitting ultrasonic waves of frequency f. Assume that a reflecting surface is moving with a velocity v towards the source. The ultrasonic waves get reflected by the moving surface. Since the surface is moving, it produces Doppler effect. The Doppler shifted frequency, f' is

$$f' = \frac{fv}{c} \tag{7.27}$$

The frequency of the reflected sound is f'. It is similar to a moving virtual source emitting a sound of frequency f' (Figure 7.30). The receiver is the stationary object and hence the Doppler shifted frequency is

$$f'' = \frac{fv}{c} \tag{7.28}$$

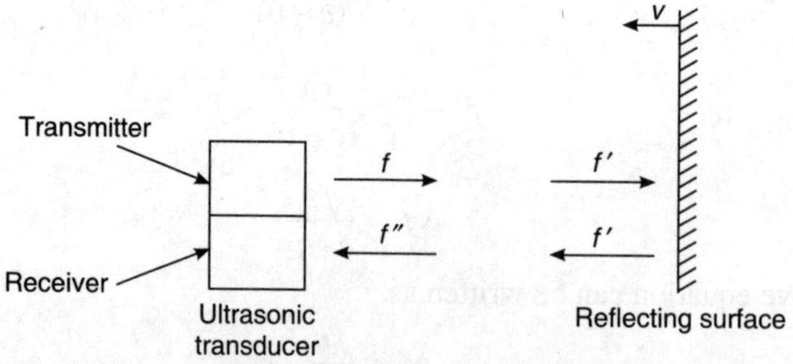

Figure 7.30 Reflection of ultrasound.

Therefore, the total Doppler shifted frequency is

$$f' + f'' = \frac{fv}{c} + \frac{fv}{c}$$

$$\Delta f = 2\frac{fv}{c} \tag{7.29}$$

The sound reflected by a moving surface and it is received by a stationary transducer which has a Doppler shifted frequency of $\Delta f = 2\dfrac{fv}{c}$.

Measurement of the Doppler shifted frequency

The continuous Doppler shift produced by the moving object in our body is measured by using the block diagram as shown in Figure 7.31. The functions of various blocks are given below.

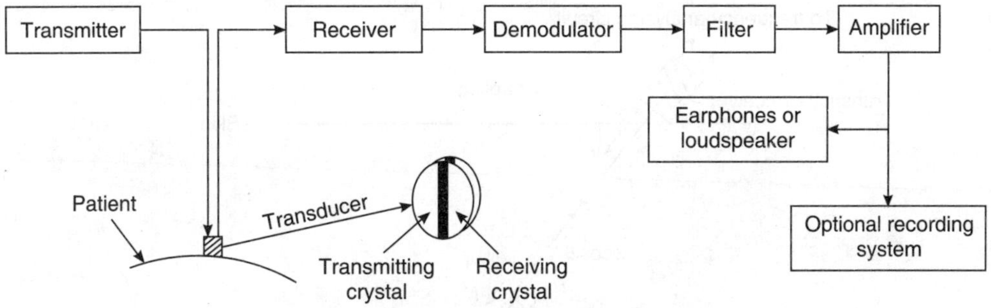

Figure 7.31 Doppler shift measurement.

The transducer is kept over the patient body. In between the skin and the transducer, a couplant in the form of a gel is applied so as to avoid the air-gap between the transducer and the skin. The transmitter emits a continuous voltage. When the transducer is excited by a continuous voltage, it produces ultrasonic waves. The ultrasonic wave is gone through the patient body and it gets reflected from the body.

The reflected signal is received by the receiver circuit. The reflected signal has two components. One component is due to the reflection of the ultrasonic wave from the stationary object and the other component is due to the reflection from a moving object. The demodulator mixes the frequencies and hence produces beat frequency. The beat frequency is equal to the difference between these two frequencies. It is actually equal to the Doppler shifted frequency. The filter circuit is used to remove the unwanted frequencies and passes the Doppler shifted frequency through it. The amplifier amplifies the Doppler shifted signal. The output of the signal is obtained from a CRO monitor, or a plotter or a headphone or a photographic camera and so on.

A well-experienced person identifies the Doppler shifted frequency by hearing the sound from a headphone. The foetal heart movement produces a sound of galloping horse and the placenta produces a sound similar to rushing of winds through trees.

7.9.3 Application of Ultrasonic Waves Using Doppler Shift Principle

Blood flow measurement

The continuous Doppler shift is used to find out the speed of the blood in the veins and arteries. Consider that an ultrasonic transducer is positioned on the skin at an angle of θ to the blood vessel. If the ultrasonic wave is passed through the skin, it gets reflected back by the moving object like blood and the stationary object like blood vessel. The blood is moving with a velocity of v and the transducer is inclined at an angle of θ. Therefore, the blood has a component $v \cos \theta$ along the beam direction (Figure 7.32). So, the Doppler shift produced by the movement of the blood is

$$\Delta f = \frac{2 f v \cos\theta}{c} \tag{7.30}$$

where v is the speed of the blood, c is the speed of the ultrasonic wave through the human body, f is the frequency of the transducer, θ is the angle of inclination of the probe with blood vessel and Δf is the Doppler shifted frequency.

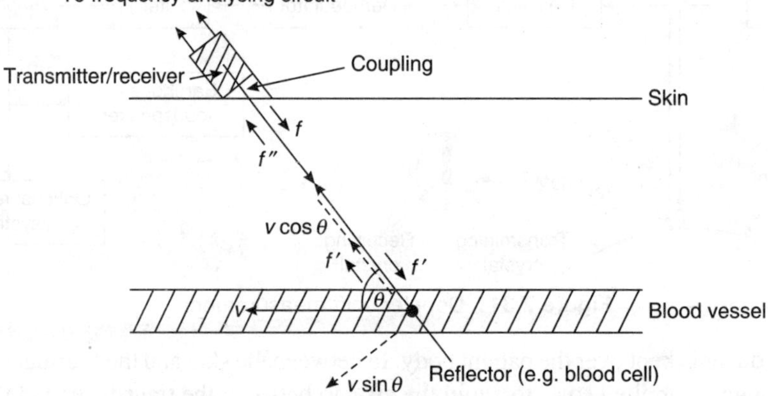

Figure 7.32 Measurement of the speed of the blood.

This method has some advantages. They are (i) there is no need of surgical penetration of the vessel and (ii) continuous readings can be made without any risk or discomfort. Using this method, the direction of the blood flow is not identified. But by using a colour Doppler flowmeter, one can identify the direction of the flow.

Foetus heart movement

The continuous Doppler shift is used to study about the foetus heart movement and hence the growth of the child. The block diagram for the measurement of the foetus heart movement using Doppler shift is shown in Figure 7.33. The frequency of the transducer used for this purpose is 2–3 MHz. The transducer is placed in the mother's abdomen. A couplant is also applied between the abdomen and the transducer.

The transmitter emits a continuous electrical signal and it is applied to an ultrasonic transducer. The ultrasonic transducer emits ultrasonic waves. The ultrasonic wave passes through the body and it gets reflected from the body. The reflected signal is received by the receiver. The received signal has two components—the first component reflected by the stationary object and the second one reflected by a moving object such as the pulsating heart of the baby and so on. The demodulator mixes the frequency and produces a beat frequency. The beat frequency is the Doppler shifted frequency. The filter removes all the unwanted frequency except the Doppler shifted frequency. This Doppler shifted frequency is amplified by the amplifier. The output devices such as CRT monitor or a computer screen, or a plotter or a photographic camera is used to see the image.

From this one can study the heart movement and the blood flow of the unborn child. It is also used to study the blood flow in the mother's body.

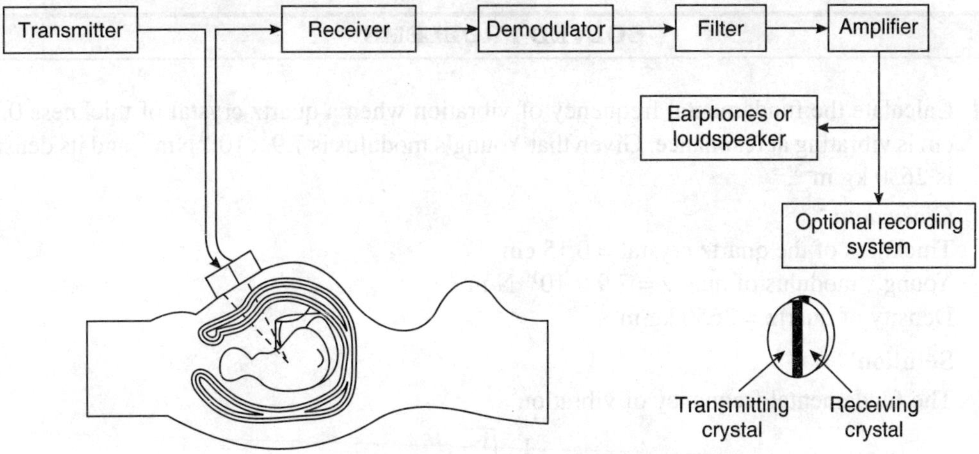

Figure 7.33 Foetus heart movement.

Sonogram

Sonogram is a computerized picture taken by passing ultrasonic waves through the interior parts of our body. It is safer and it offers more details than X-rays. It is painless and the person receiving the sonogram will not be inconvenient or made to feel uncomfortable. The advantages of taking sonogram are given below:-

1. It is used to study about the growth of the unborn child.
2. It is used to detect multiple pregnancies.
3. It is used to know about the birth defects at a very early stage.
4. The parents can even learn the sex of their unborn child.
5. It is used to identify the pelvic bleeding and discomfort.
6. It is used to find the source of menstrual problem.
7. It is used to identify cysts and locate cancer cells.
8. It is used to treat prostrate and cancers in men.

7.9.4 Other Applications of Ultrasonic Waves

Biological effect of ultrasonics

1. Ultrasonics has a very destructive effect upon small living organisms. Small fish in very high numbers have been killed due to high-power echo ranging and sounding devices.
2. The bacteria can be destroyed by ultrasonics. The bacteria in milk have been reduced by the application of ultrasonics. This indicates that milk can be sterilized by ultrasonics.

Emulsification due to ultrasonics

An emulsion is a suspension of fine particles of a liquid. The emulsions are produced by violent agitations. Ultrasonic waves can be used to produce emulsions.

1. If two immiscible liquids, such as water and gasoline, are placed in a container and subjected to intense sound vibrations, an emulsion will be produced.
2. Ultrasonics can be used to produce alloys of iron and lead, aluminium and lead, aluminium and cadmium which are immiscible in the liquid state.

SOLVED PROBLEMS

7.1 Calculate the fundamental frequency of vibration when a quartz crystal of thickness 0.15 cm is vibrating at resonance. Given that Young's modulus is 7.9×10^{10} Nm^{-2} and its density is 2650 kg m^{-3}.

Given data

Thickness of the quartz crystal = 0.15 cm

Young's modulus of quartz = 7.9×10^{10} N m^{-2}

Density of quartz = 2650 kg m^{-3}

Solution

The fundamental frequency of vibration

$$f = \frac{1}{2t}\sqrt{\frac{Y}{\rho}}$$

$$= \frac{1}{2 \times 0.15 \times 10^{-2}}\sqrt{\frac{7.9 \times 10^{10}}{2650}}$$

$$= 1.8199 \text{ MHz}$$

The fundamental frequency of vibration of the crystal is 1.8199 MHz.

7.2 Calculate the fundamental frequency and the first overtone emitted by a quartz crystal of thickness 1 mm in a piezoelectric oscillator. Given, Young's modulus for quartz = 7.9×10^{10} N m^{-2} and $\rho = 2650$ kg m^{-3}.

Given data

Thickness of the quartz crystal = 1 mm = 1×10^{-3} m

Young's modulus of quartz = 7.9×10^{10} N m^{-2}

Density of quartz = 2650 kg m^{-3}

Solution

The frequency of vibration

$$f_1 = \frac{p}{2t}\sqrt{\frac{Y}{\rho}}$$

For the fundamental frequency, $p = 1$ and for the first overtone, $p = 2$.

$$= \frac{1}{2 \times 1 \times 10^{-3}}\sqrt{\frac{7.9 \times 10^{10}}{2650}}$$

$$= 2.7299 \times 10^{6} \text{ Hz}$$

The frequency of the first overtone

$$f_2 = \frac{p}{2t}\sqrt{\frac{Y}{\rho}}$$

$$= \frac{2}{2 \times 1 \times 10^{-3}} \sqrt{\frac{7.9 \times 10^{10}}{2650}}$$

$$= 5.459 \times 10^6 \text{ Hz}$$

The fundamental frequency of vibration of the crystal is 2.7299 MHz.
The frequency of the first overtone of the crystal is 5.459 MHz.

7.3 Calculate the velocity of the ultrasonic wave passing through an acoustic grating experiment using the following data: Wavelength of the light used is 589.3 nm, frequency of the ultrasonic transducer is 100 MHz, and the angle of diffraction in the first order is 2°15′.

Given data

Wavelength of the light used, $\lambda = 589.3 \text{ nm} = 5.893 \times 10^{-7} \text{ m}$
Frequency of the ultrasonic transducer = 100 MHz = 1×10^8 Hz
Order of diffraction = 1
Angle of diffraction = 2°15′

Solution

The Bragg's equation is

$$2d \sin \theta = n\lambda$$

Substituting the values, we get

$$2d \sin 2°15' = 5.893 \times 10^{-7}$$

$$d = 7.505 \text{ μm}$$

The wavelength of the ultrasonic wave

$$\lambda = 2d = 2 \times 7.505 \times 10^{-6} = 1.501 \times 10^{-5} \text{ m}$$

The velocity of the ultrasonic wave

$$v = f\lambda = 1 \times 10^8 \times 1.501 \times 10^{-5}$$

$$= 1501 \text{ m s}^{-1}$$

The velocity of the ultrasonic wave = 1501 m s^{-1}.

7.4 An ultrasonic transducer of 2 MHz is used to measure the blood speed. The probe is inclined at an angle of 30° and the blood is moving with a speed of 3 m s^{-1}. If the velocity of the ultrasonic wave through the human body is 800 m s^{-1}, calculate the Doppler shifted frequency.

Given data

Frequency of the transducer = 2 MHz
Speed of blood = 3 m s^{-1}
Velocity of Ultrasonic wave = 800 m s^{-1}
Angle of inclination of the probe = 30°

Solution

Doppler shifted frequency

$$\Delta f = \frac{2 f v \cos \theta}{c}$$

Since the probe is inclined at an angle of 30°, the ultrasonic waves make an angle of 30° with the blood. Substituting the values, we get

$$\Delta f = \frac{2 \times 2 \times 10^6 \times 3 \times \cos 30°}{800} = 1.299 \times 10^4 \text{ Hz}$$

The Doppler shifted frequency is 0.01299 MHz.

7.5 Determine the velocity of the ultrasonic wave produced by a piezoelectric oscillator. The density of quartz crystal is 2650 kg m^{-3} and the Young's modulus of quartz is 7.9×10^{10} N m^{-2}.

Given data

Young's modulus of quartz = 7.9×10^{10} N m^{-2}

Density of quartz = 2650 kg m^{-3}

Solution

The velocity of the ultrasonic waves

$$v = \sqrt{\frac{Y}{\rho}}$$

$$= \sqrt{\frac{7.9 \times 10^{10}}{2650}}$$

$$= 5459.97 \text{ m s}^{-1}$$

The velocity of the ultrasonic waves = 5459.97 m s^{-1}.

SHORT QUESTIONS

1. What are ultrasonic waves?
2. Mention different types of ultrasonic wave propagation.
3. What is a transverse ultrasonic wave?
4. What is a longitudinal ultrasonic wave?
5. What are surface waves?
6. What are Lamb waves?
7. Mention the properties of ultrasonic waves.
8. What is magnetostriction effect?
9. What is the principle of magnetostriction oscillator?
10. What are the advantages of a magnetostriction method?
11. What are the drawbacks of magnetostriction method?
12. What is piezoelectric effect?
13. What is inverse piezoelectric effect?
14. What are *X*-cut and *Y*-cut crystals?

15. Mention the principle of a piezoelectric oscillator.
16. What are the advantages of the piezoelectric method?
17. What are the drawbacks of the piezoelectric method?
18. Mention the methods used to detect ultrasonic waves.
19. Write about the thermal detection of ultrasonic waves.
20. Write about the piezoelectric detection of ultrasonic waves.
21. Write about the principle behind the flaw detection of ultrasonic waves.
22. Mention the names of ultrasonic inspection techniques.
23. What is a pulse echo method?
24. What is normal beam method?
25. What is through transmission method?
26. What is angle beam method?
27. What are contact method and non-contact method of ultrasonic inspection technique?
28. What are the applications of ultrasonic testing?
29. What are the different formats for displaying the ultrasonic testing results?
30. What is A-scan?
31. What is B-scan?
32. What is C-scan?
33. Describe how the depth of sea is determined using ultrasonic waves.
34. What is sonar?
35. Write about the ultrasonic welding.
36. Write about ultrasonic cutting and drilling.
37. How are ultrasonic waves used for cleaning?
38. Write about ultrasonic soldering.
39. What is non-destructive testing?
40. Write about ultrasonic cardiography.
41. Mention the uses of ultrasonics in obstetrics.
42. What is echoencephalography?
43. Mention the dental use of ultrasonic wave.
44. Define Doppler effect.
45. Write about the Doppler shifted frequency due to the reflection of ultrasonic wave from a moving surface.
46. What is sonogram?
47. Mention the biological applications of ultrasonic waves.
48. Mention the emulsification due to ultrasonic wave.
49. What are cavitations? What are its uses?

DESCRIPTIVE TYPE QUESTIONS

1. (a) Define magnetostriction effect.
   (b) Describe the principle, construction and working of a magnetostriction oscillator.
   (c) What are the advantages and drawbacks of a magnetostriction oscillator?
2. (a) Define piezoelectric effect.
   (b) Describe with neat sketch the principle, construction and working of a piezoelectric oscillator.
   (c) Mention the advantages and drawbacks of a piezoelectric oscillator.
3. (a) Explain how the velocity of ultrasonic wave is determined using acoustic grating.
   (b) Discuss the different methods used for ultrasonic wave detection.
4. (a) Discuss the different types of ultrasonic inspection methods.
   (b) Discuss the different methods of displaying the results of ultrasonic inspection.
5. (a) Discuss the industrial/engineering applications of ultrasonic waves.
   (b) Describe the medical applications of ultrasonic waves.
6. What is NDT? Describe the ultrasonic non-destructive method used for flaw detection.

PROBLEMS

1. Calculate the thickness of a quartz crystal to produce ultrasonic waves of frequency 20×10^6 Hz. The density and Young's modulus of quartz are 2650 kg m^{-3} and 80×10^9 Pa respectively.
2. The frequency produced by a magnetostriction oscillator using nickel is 0.024 MHz. The density of Nickel is 8908 kg m^{-3}. If the length of the ferromagnetic rod is 10 cm in length, calculate its Young's modulus.
3. An acoustic grating is illuminated by a light of wavelength 600 nm and it is used to detect the ultrasonic waves. The frequency of the ultrasonic transducer used is 100 MHz and the first order diffraction occurs at $2°6'$, calculate the wavelength and the velocity of the ultrasonic waves.
4. Calculate the depth of the sea using the following data: Velocity of the ultrasonic wave in sea water is 1498 m s^{-1}. The time taken for the to and fro motion of the ultrasonic wave is 2.86 s.
5. An ultrasonic flaw detector receives an echo from a flaw. The time difference between the transmitted pulse and the reflected pulse from the flaw is 5 μs. The thickness of the specimen is 10 cm and the velocity of ultrasonic wave through the specimen is 5000 m s^{-1}. Calculate the depth of flaw from the transducer.

CHAPTER 8

LASER (PHOTONICS)

8.1 INTRODUCTION

Einstein proposed the existence of two different types of emission, namely *stimulated emission* and *spontaneous emission* in 1917 on the basis of the quantum theory. Stimulated emission was used by Townes in 1954 for the construction of MASER, an acronym for Microwave Amplification by Stimulated Emission of Radiation. In 1958, Schawlow and Townes extended the principle of MASER to the optical region, which resulted in the development of a light-amplifying device known as Laser.

Laser is an acronym for Light Amplification by Stimulated Emission of Radiation. The first laser was invented by Maiman in 1960. He first demonstrated the working of a Ruby laser. After this invention, laser action was produced in liquid, gases, semiconductors, dyes, etc. In this chapter, let us discuss about the production of laser beam, its properties and applications.

8.2 PRINCIPLES OF LASERS

The basic principle behind the laser operation is the absorption and emission of light. Let us discuss about the absorption and emission of light in the following section.

8.2.1 Absorption of Light

Consider a system consisting of two energy levels, E_1 and E_2 (Figure 8.1). Let E_1 be the lower energy level and E_2 be the upper energy level. The lower energy state is called ground state, whereas the upper energy state is called excited state. Let N_1 and N_2 be the populations of lower and upper energy levels. Let the energy difference between these two energy levels be $h\nu$ (i.e., $E_2 - E_1 = h\nu$). Let a photon of energy $h\nu$ is incident on this system. The atoms present in the lower energy states absorb the incident photon and get excited to the higher energy level. This process is said to be absorption of light (Figure 8.1).

The process of bringing an atom from a lower energy level to a higher energy level by the absorption of light is also called excitation of atoms or optical pumping or absorption of light.

Let ρ_ν be the energy density (energy of the incident photon per unit volume) of the incident photon, then the rate of absorption, R_{12}, is given by

$$R_{12} \propto \rho_\nu$$
$$\propto N_1$$

By introducing the proportionality constant B_{12}, the rate of absorption can be written as

$$R_{12} = B_{12}\,\rho_\nu N_1 \tag{8.1}$$

where B_{12} is known as the coefficient for absorption or Einstein's coefficient.

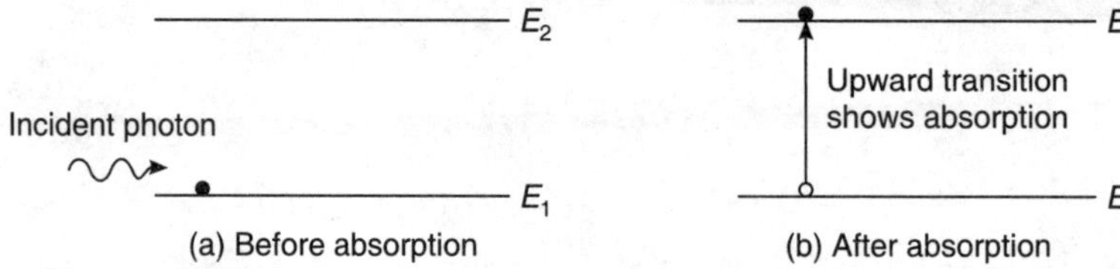

Figure 8.1 Absorption of light.

8.2.2 Emission of Light

The atom in the excited state exists in that state only for a short period of time. The average time spent by an atom in the excited state is known as lifetime of an atom. The lifetime of an atom in the excited state is in the order of 10^{-8} s. In some energy levels the atoms exist for a longer time, even up to 10^{-3} s. These energy levels are called metastable state. After spending a short period of time in the excited state, it automatically returns to the lower energy state by emitting the excess of energy possessed by it. This process is said to be the emission of light. There are two types of emissions. They are

1. Spontaneous emission and
2. Stimulated emission

Let us discuss about these two types of emissions.

Spontaneous emission

Consider an atom in the excited state. It will exist in that state only for a short period of time. After that time, it returns to the lower energy state. If the atom lying in the upper energy level returns to the lower energy level without any external stimulation or inducement, then the emission is said to be spontaneous emission (Figure 8.2).

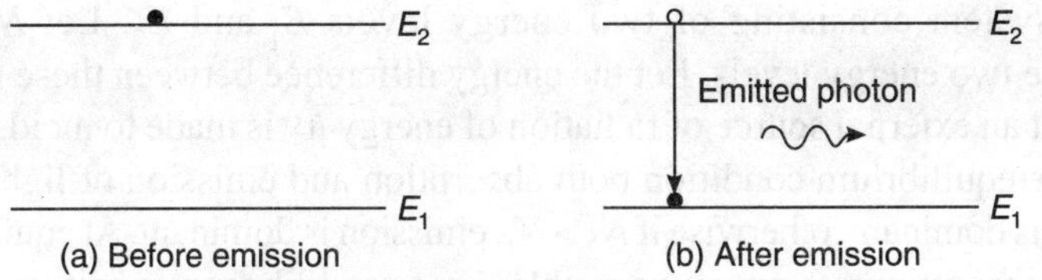

Figure 8.2 Spontaneous emission.

The rate of spontaneous emission, R_{21}(spont), can be written as

$$R_{21}(\text{spont}) \propto N_2$$

By introducing the proportionality constant A_{21}, the rate of spontaneous emission can be written as

$$R_{21}(\text{spont}) = A_{21} N_2 \tag{8.2}$$

where A_{21} is the spontaneous emission coefficient or Einstein's coefficient.

Stimulated emission of light

Consider an atom in the metastable state. The lifetime of a metastable state is in the order of 10^{-3} s. So, an atom exists for a longer time in the metastable state. At that time, if an external source of radiation having energy $h\nu$ (where $E_2 - E_1 = h\nu$) is incident on the system, then the atom is stimulated to emit a radiation of energy $h\nu$ and hence it returns to the ground state (Figure 8.3). This phenomenon is called stimulated emission of light.

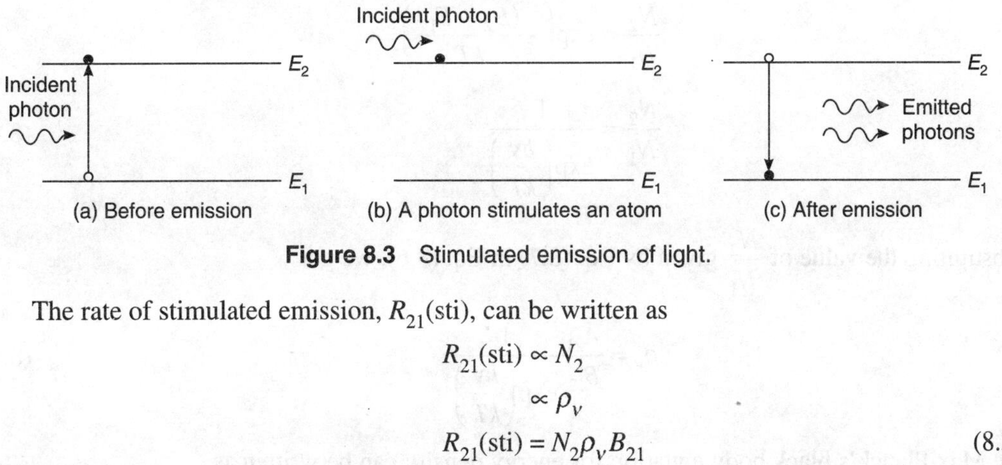

Figure 8.3 Stimulated emission of light.

The rate of stimulated emission, R_{21}(sti), can be written as

$$R_{21}(\text{sti}) \propto N_2$$

$$\propto \rho_\nu$$

$$R_{21}(\text{sti}) = N_2 \rho_\nu B_{21} \tag{8.3}$$

where ρ_ν is the energy density of the incident radiation and B_{21} is the Einstein's coefficient or the coefficient for stimulated emission of light.

8.3 EINSTEIN'S EXPLANATION FOR STIMULATED EMISSION

Einstein, in 1917, derived a mathematical expression to show the existence of stimulated emission of radiation. Let us derive the mathematical expression given by him.

Consider a system consisting of two energy levels E_1 and E_2. Let N_1 and N_2 be the populations of these two energy levels. Let the energy difference between these two energy levels is, $E_2 - E_1 = h\nu$. Let an external source of radiation of energy $h\nu$ is made to incident on the system. Before reaching the equilibrium condition both absorption and emission of lights are possible. If $N_1 > N_2$, absorption is dominant, otherwise if $N_2 > N_1$, emission is dominant. At equilibrium condition, both emission and absorption are equally possible, i.e. at equilibrium condition

$$\text{rate of absorption} = \text{rate of emission} \tag{8.4}$$

In those days, only the spontaneous emission was known to the scientists. Taking only the spontaneous emission, the above equation can be written as

$$B_{12}\,\rho_\nu N_1 = A_{21}\,N_2 \tag{8.5}$$

$$\rho_\nu = \frac{A_{21}}{B_{12}}\frac{N_2}{N_1} \tag{8.6}$$

The value of $\frac{N_2}{N_1}$ is obtained from the Maxwell–Boltzmann distribution function and it is given by

$$\frac{N_2}{N_1} = \frac{\exp\left(-\frac{E_2}{kT}\right)}{\exp\left(-\frac{E_1}{kT}\right)}$$

$$\frac{N_2}{N_1} = \exp\left(-\frac{(E_2 - E_1)}{kT}\right)$$

$$\frac{N_2}{N_1} = \frac{1}{\exp\left(\frac{h\nu}{kT}\right)} \tag{8.7}$$

Substituting the value of $\frac{N_2}{N_1}$ given by Eq. (8.7) in Eq. (8.6), we get

$$\rho_\nu = \frac{A_{21}}{B_{12}}\frac{1}{\exp\left(\frac{h\nu}{kT}\right)} \tag{8.8}$$

The Max Planck's black body equation for energy density can be written as

$$\rho_\nu = \frac{8\pi h\nu^3}{c^3}\frac{1}{\exp\left(\frac{h\nu}{kT}\right) - 1} \tag{8.9}$$

Comparing Eq. (8.8) and Eq. (8.9), one can infer that these two equations are not equal. So, Eq. (8.5) can be rearranged by introducing the stimulated emission of light as follows:

$$\text{rate of absorption} = \text{rate of emission}$$

$$B_{12}\,\rho_\nu N_1 = A_{21}\,N_2 + B_{21}\,\rho_\nu N_2 \tag{8.10}$$

Rearranging Eq. (8.10), we get

$$B_{12}\,\rho_\nu N_1 - B_{21}\,\rho_\nu N_2 = A_{21}\,N_2$$

i.e.

$$\rho_\nu (B_{12}\,N_1 - B_{21}N_2) = A_{21}\,N_2$$

$$\rho_\nu = \frac{A_{21}N_2}{B_{12}N_1 - B_{21}N_2} \tag{8.11}$$

Dividing both the numerator and the denominator by N_2, we get

$$\rho_\nu = \frac{A_{21}}{B_{12}\dfrac{N_1}{N_2} - B_{21}} \tag{8.12}$$

Substituting the value of $\dfrac{N_1}{N_2}$ given by Eq. (8.7) in Eq. (8.12), we get

$$\rho_\nu = \frac{A_{21}}{B_{12}\exp\left(\dfrac{h\nu}{kT}\right) - B_{21}} \tag{8.13}$$

Comparing Eq. (8.9) and Eq. (8.13), we get

$$B_{12} = B_{21}$$

$$\frac{A_{21}}{B_{12}} = \frac{8\pi h\nu^3}{c^3} \tag{8.14}$$

The agreement of Eq. (8.13) with Eq. (8.9) shows that there exists another kind of radiation known as stimulated emission of light.

The ratio of stimulated emission to spontaneous emission can be written as

$$\frac{\text{stimulated emission}}{\text{spontaneous emission}} = \frac{B_{21}N_2\rho_\nu}{A_{21}N_2} = \frac{B_{21}}{A_{21}}\rho_\nu \tag{8.15}$$

Substituting

$$\frac{B_{21}}{A_{21}} = \frac{B_{12}}{A_{21}} = \frac{c^3}{8\pi h\nu^3}$$

and

$$\rho_\nu = \frac{8\pi h\nu^3}{c^3}\frac{1}{\exp\left(\dfrac{h\nu}{kT}\right) - 1}$$

we get

$$\frac{\text{stimulated emission}}{\text{spontaneous emission}} = \frac{1}{\exp\left(\dfrac{h\nu}{kT}\right) - 1} \tag{8.16}$$

Equation (8.16) gives the ratio of stimulated emission to spontaneous emission.

8.4 DIFFERENCES BETWEEN STIMULATED AND SPONTANEOUS EMISSIONS

The differences between the spontaneous and stimulated emissions are displayed in Table 8.1.

Table 8.1 Differences between spontaneous and stimulated emissions

Stimulated emission	*Spontaneous emission*
It is induced by an external radiation.	It is a spontaneous process. It is automatically produced.
It requires inverted population.	No need of population inversion.
It produces a coherent source of light.	It produces incoherent source of light.
The photons emitted by the stimulated emission travels in the direction of the incident photon.	The photon emitted by the spontaneous emission travels in any direction.
The rate of emission depends upon the concentration of the upper level and the energy density of the incident beam. For example, laser beam.	The rate of emission depends only upon the concentration of the upper level. For example, light emitted by sodium vapour lamp.

8.5 POPULATION INVERSION

Consider a system consisting of two energy levels. At normal condition, the lower energy level contains more number of atoms per unit volume, whereas the higher energy level has less number of atoms per unit volume. If a graph is plotted between energy, E, and the concentration, N, using the Maxwell–Boltzmann's equation, a curve as shown in Figure 8.4(a) is obtained. From this curve, one can observe that the concentration of the lower energy level is higher than the concentration of the upper energy level. In order to produce emission from the higher energy level, the population of the higher energy level should be made as high as possible. If the population of the higher energy level is more than the population of the lower energy level, then it is said to be **inverted population** or **population inversion**.

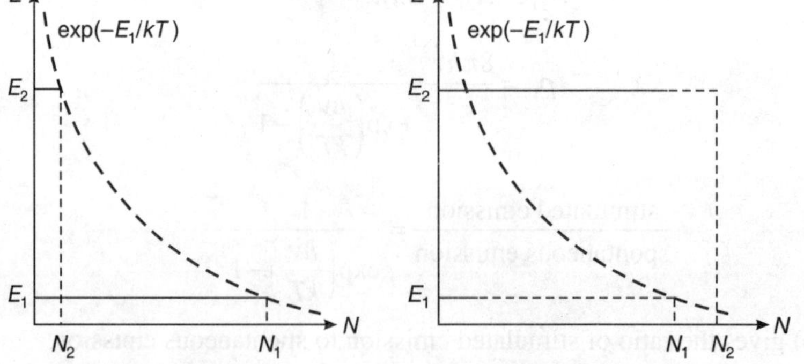

(a) At normal condition (b) After the population inversion

Figure 8.4 A plot between concentration and energy using Maxwell–Boltzmann distribution function.

The emission of the laser beam takes place only when the rate of emission is greater than the rate of absorption, i.e.

$$\text{rate of emission} > \text{rate of absorption}$$

$$A_{21}N_2 + B_{21}\,\rho_\nu N_2 > B_{12}\,\rho_\nu N_1 \tag{8.17}$$

For laser oscillation to be dominant, only the stimulated emission is taken into account. Therefore, by omitting the spontaneous emission, we get

$$B_{21}\,\rho_\nu N_2 > B_{12}\,\rho_\nu N_1 \tag{8.18}$$

Cancelling the common terms, we get

$$N_2 > N_1$$

This is the condition required for laser oscillation. This condition is known as population inversion. The population inversion can be produced by the following methods. They are

(i) Optical pumping or photon excitation
(ii) Electron excitation
(iii) Inelastic atom–atom collision and
(iv) Chemical reaction

8.5.1 Photon Excitation

Photon excitation is the process of exciting an atom from a lower energy level to a higher energy level by photon of energy $h\nu$. It is also called optical pumping. The photon excitation is shown in Figure 8.5(a). The photon excitation is used in solid state lasers such as ruby laser and Nd: YAG laser.

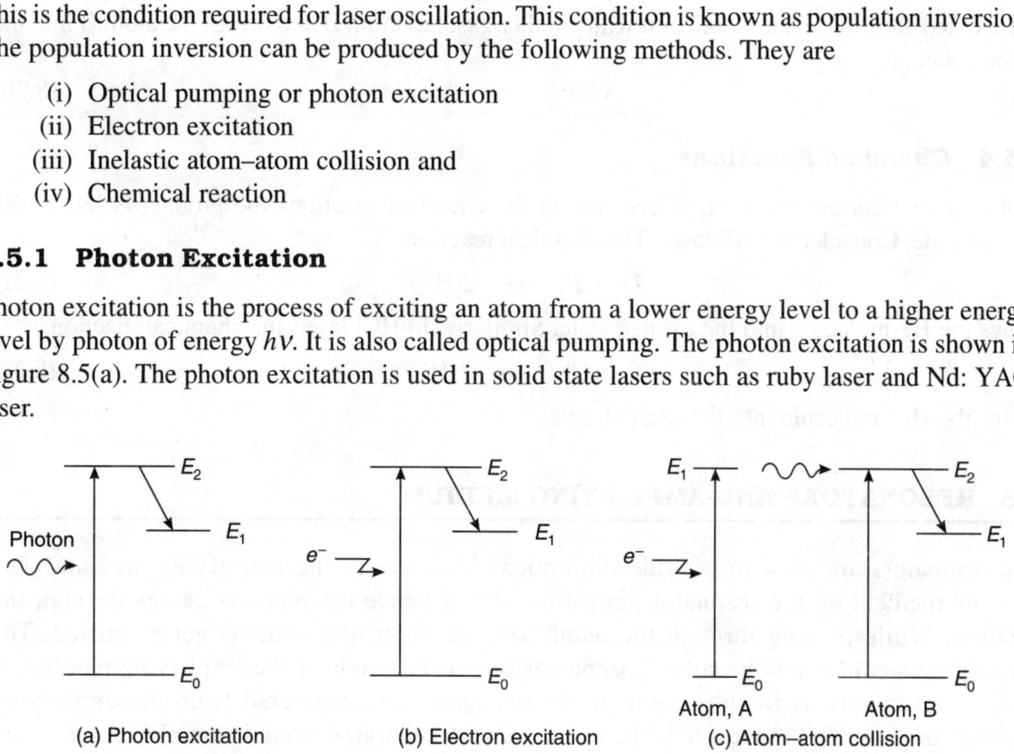

Figure 8.5 Different methods of producing population inversion.

8.5.2 Electron Excitation

In the case of gaseous lasers, the basic principle behind the laser operation is gaseous discharge. To produce gaseous discharge, a high amount of electric field is applied. The electron present in the electric field makes collisions with the gaseous atoms and gets excited to the higher energy level. This process of excitation is known as electron excitation. In argon laser, the electron excitation is used. The electron excitation is demonstrated in Figure 8.5(b).

8.5.3 Inelastic Atom–Atom Collision

In some gaseous laser, a mixture of gases is used. In these lasers, due to the electron bombardment one gaseous atom gets excited into the higher energy level. The excited gaseous atom makes collision with another gaseous atom and brings it into the excited state. This process of excitation is known as inelastic atom–atom collision and it is shown in Figure 8.5(c). The inelastic atom–atom collision is used in He–Ne and CO_2 lasers.

The inelastic atom–atom collision can be written mathematically as follows. Consider there is a mixture of two gaseous atoms, say A and B. The gaseous atom A can be raised to the excited state by the collision of electrons, i.e.

$$A + e^- \rightarrow A^* + e^- \tag{8.19}$$

The excited atom A*, makes collision with another gaseous atom B and brings the atom B into the excited state, i.e.

$$A^* + B \rightarrow A + B^* \tag{8.20}$$

8.5.4 Chemical Reactions

In the case of certain chemical lasers, due to the chemical reactions the product is left in the excited state. Consider the HF laser. The chemical reaction

$$H_2 + F_2 \rightarrow 2(HF)^* \tag{8.21}$$

brings the HF molecule into the excited state. Similarly, in HCl laser, the chemical reaction

$$H_2 + Cl_2 \rightarrow 2(HCl)^* \tag{8.22}$$

brings the HCl molecule into the excited state.

8.6 RESONATORS AND AMPLIFYING MEDIUM

The resonators are used to provide the optical feedback to the amplifying medium. The photons incident on the resonator get reflected and hence the photons passes through the medium. While passing through the amplifying medium, the photons get amplified. The amplifying medium may be solid, gaseous or liquid. The ends of the amplifying medium is used as resonators as in ruby laser or the resonators are separated from the amplifying medium as shown in Figure 8.6. If the resonators are separated from the amplifying medium several possible configurations may be used. Some commonly used configurations are shown in Figure 8.7. They are (i) plane–parallel ($r_1 = r_2 = \infty$), (ii) concave–concave with large radius (r_1, $r_2 >> L$, radius typically ~ 20 m), (iii) concave–concave ($r_1 = r_2 = L$, it is called confocal resonator), (iv) concave–plane ($r_1 = L$, $r_2 = \infty$, it is called hemispherical). These resonators have some advantages and drawbacks. Aligning the plane–parallel resonator is very difficult. The mirrors should be flat within $\lambda/100$. If the mirrors are strictly parallel, the radiation beam makes the maximum usage of the laser radiation. The mirrors in confocal resonators are easy to align. In gas laser, if large output is required, the large radius resonators are used.

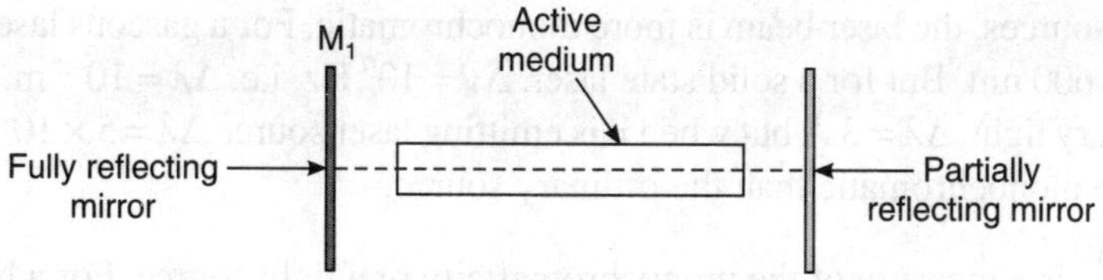

Figure 8.6 Amplifying medium.

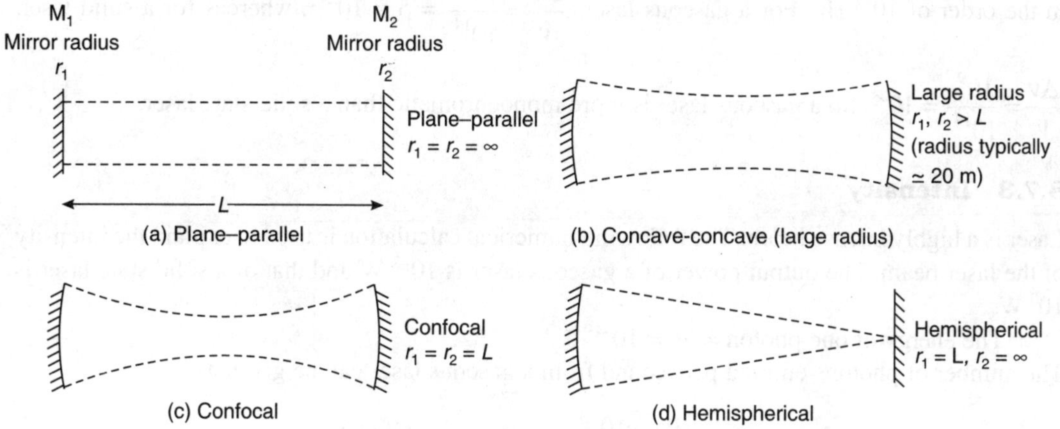

Figure 8.7 Resonators.

8.7 DISTINCT PROPERTIES OF LASER

Laser differs from the ordinary light with respect to some properties. They are

(i) Directionality (ii) Monochromaticity
(iii) Intensity and (iv) Coherence

8.7.1 Directionality

Generally a light source travels in a straight line up to a distance of d^2/λ, where d is the diameter of the aperture, and λ is the wavelength of the light used, then the light will spread. The angular spread of a light is given by $\Delta\theta = \lambda/d$. For an ordinary light, the angular spread is 1 m when it is travelling a distance of 1 m. But for a laser beam, the angular spread is 1 mm by 1 m. Due to this reason, one can send the laser beam from earth to moon with a few km angular spread. The small divergence of the laser beam, allows us to measure the distance between moon and earth. The radius of the spread is $r = \Delta\theta \cdot D$ and the area of the spread of the laser beam is $\pi(D \cdot \Delta\theta)^2$, where D is the distance between the source and screen and $\Delta\theta$ is the angular spread.

8.7.2 Monochromaticity

Generally, a light source is said to be monochromatic source, when the bandwidth $\Delta\nu = 0$. Strictly speaking, including the laser source, no source is found to be perfect monochromatic. Comparing with the available sources, the laser beam is more monochromatic. For a gaseous laser, $\Delta\nu = 500$ Hz, i.e. $\Delta\lambda = 10^{-7}$ m at 600 nm. But for a solid state laser, $\Delta\nu = 10^9$ Hz, i.e. $\Delta\lambda = 10^{-2}$ m. For ruby, when it is emitting ordinary light, $\Delta\lambda = 3$ Å but when it is emitting laser source $\Delta\lambda = 5 \times 10^{-4}$ Å. This shows that a laser is more monochromatic than the ordinary source.

The ratio $\frac{\Delta\nu}{\nu}$ is a measure of the monochromaticity of a light source. For a light source, ν is in the order of 10^{14} Hz. For a gaseous laser, $\frac{\Delta\nu}{\nu} = \frac{500}{10^{14}} = 5 \times 10^{-12}$, whereas for a solid laser, $\frac{\Delta\nu}{\nu} = \frac{10^9}{10^{14}} = 10^{-5}$. So a gaseous laser is more monochromatic than a solid state laser.

8.7.3 Intensity

Laser is a highly intense beam. The following numerical calculation is used to explain the intensity of the laser beam. The output power of a gaseous laser is 10^{-3} W and that of a solid state laser is 10^9 W.

The energy of one photon = $h\nu = 10^{-19}$ J

The number of photons emitted per second from a gaseous laser can be given as

$$\frac{P}{h\nu} = \frac{10^{-3}}{10^{-19}} = 10^{16} \text{ photons per second}$$

The number of photons emitted per second from a solid state laser is given by

$$\frac{P}{h\nu} = \frac{10^9}{10^{-19}} = 10^{28} \text{ photons per second}$$

The number of photons emitted per second in a thermal radiation is given by

$$\frac{P}{h\nu} = \frac{2\pi}{\lambda^2} \frac{1}{\exp\left(\frac{h\nu}{kT}\right) - 1} \Delta\nu\Delta A \tag{8.23}$$

Substituting, $\lambda = 600$ nm, $\Delta\lambda = 100$ nm, $\Delta A = 1$ cm^2 and $T = 1000$°C, the number of photons emitted $= 10^{12}$ photons per second. This numerical calculation shows that the laser beam emits more number of photons per second and hence it is found to be more intense.

8.7.4 Coherence

Two sources are said to be coherent sources, if they have same phase or constant phase difference, same frequency, same amplitude, same wavelength, and same direction of emission. There are two types of coherent sources, namely (i) temporal coherence and (ii) spatial coherence.

If two sources have coherent property only for a short period of time, then it is said to be temporal coherence. If two sources are having coherent property for the entire period of their travel, it is said to be spatial coherence. The coherence property is used to produce interference pattern. The term coherence length, L_c, is used to find out how long the two sources are maintaining the coherence property. The coherence length is given by

$$L_c = cT_c \qquad (8.24)$$

where c is the velocity of light and T_c is the coherence time.

Laser beams are highly coherent compared to the ordinary source of light because the stimulated emission produces photons with definite phase relationship with each other. Therefore, the laser beams are in same phase and of the same frequency. The construction of hologram using laser also indicates that the laser is coherent.

8.8 TYPES OF LASERS

Lasers are classified into different types based on the nature of the amplifying medium used. They are

(i) Solid state lasers
(ii) Gaseous lasers
(iii) Liquid lasers
(iv) Semiconductor lasers
(v) Dye lasers and
(vi) Chemical lasers, etc.

Let us discuss in detail about the construction and working of some lasers.

8.8.1 Ruby Laser

The ruby laser is a solid state laser and it was first demonstrated by T.H. Maiman in 1960.

Construction

The ruby laser uses a ruby rod as an amplifying medium. The ruby rod is an Al_2O_3 crystal doped with Cr_2O_3. Only 0.05 wt% of Cr^{3+} is doped with Al_2O_3. The colour of the ruby rod for 0.05 wt% of Cr^{3+} doping is pink.

A ruby laser uses a ruby rod of length 4 cm and diameter 5 mm. The two ends of the ruby rod are made as optically plane and parallel. They are made as highly polished surfaces so that one end is fully reflecting surface, whereas the other end is partially reflecting surface. A helical xenon lamp is made to surrounds the quartz tube. The ruby rod is kept inside a quartz container. Air is circulated through the apparatus as a coolant. The diagrammatic representation of a ruby laser apparatus is shown in Figure 8.8.

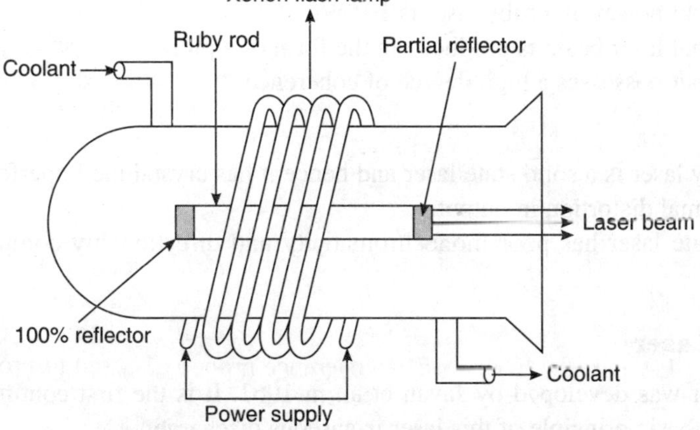

Figure 8.8 Ruby laser.

Working

When an electric field is applied to the flash lamp it produces a flash of light. The flash of light is absorbed by Cr^{3+} ions, and hence it gets excited to higher energy levels. The flash of light produced by the xenon lamp is a white light and it has a broad spectrum of wavelength. If Cr^{3+} absorbs 0.42 μm wavelengths it gets excited into the second excited state labeled as $^4F_1$ and if it absorbs a wavelength of 0.55 μm, the Cr^{3+} gets excited to the first excited state labeled as $^4F_2$. The excited Cr^{3+} atom exists in that excited state only for a short period of time. After spending a short time in the excited state, it returns to the metastable state. When it is making a transition from the metastable state to the ground state, it emits laser radiations of wavelengths 0.6943 μm and 0.6927 μm respectively. The emitted lights move within the amplifying medium and get reflected by resonators and finally emerge out as a red beam from partially reflecting surface (Figure 8.8).

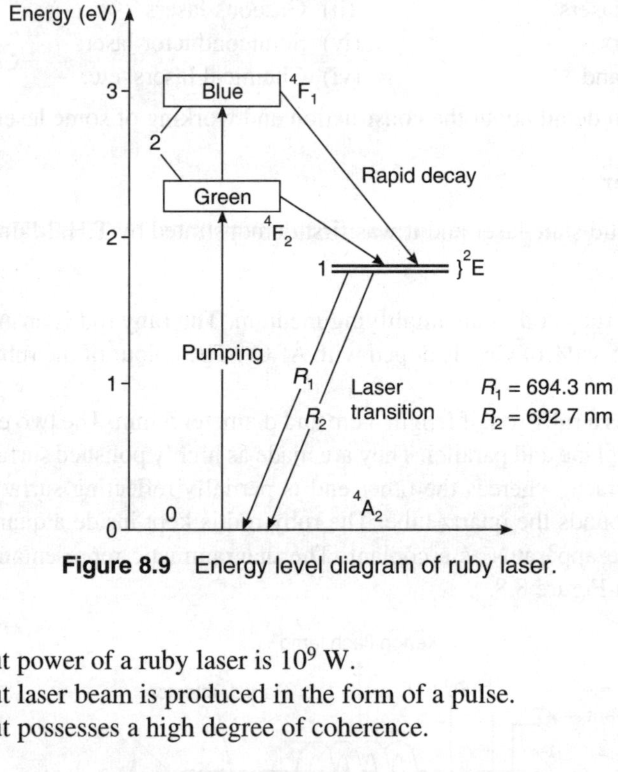

Figure 8.9 Energy level diagram of ruby laser.

Advantages

(a) The output power of a ruby laser is 10^9 W.
(b) The output laser beam is produced in the form of a pulse.
(c) The output possesses a high degree of coherence.

Drawbacks

(a) The ruby laser is a solid state laser and hence it has crystalline imperfection, scattering and thermal distortion in output.
(b) Solid state laser has poor monochromaticity and directionality compared to gaseous lasers.

8.8.2 He–Ne Laser

Helium–neon laser was developed by Javan et al. in 1961. It is the first continuously-operated gaseous laser. The basic principle of this laser is gaseous discharge.

Construction

Helium and neon gases are used as an amplifying medium. The helium and neon gases are mixed in the ratio of 10:1 in a quartz container of length 80 cm and inner diameter 1.5 cm. Helium gas is taken at a pressure of 1 Torr (1 Torr = 1 mm of Hg) and neon gas is taken at a pressure of 0.1 Torr. In a helium–neon laser either a Fabry–Perot resonator or a confocal resonator is used. The apparatus has the provision to apply high electric fields. The diagrammatic representation of a helium–neon laser apparatus is shown in Figure 8.10 and it consists of Fabry–Perot resonator with suitable provision for adjustment.

Working

When electric field is applied, the electrons present in the electric field make collision with the atoms and hence helium atoms get excited to the higher energy levels, 2^1S and 2^3S as shown in Figure 8.11. The excited helium atoms make collision with the neon atoms and bring the Ne atoms into the excited states 2s and 3s. The energy levels 2s and 3s for Ne are the metastable states and the Ne atoms are directly pumped into these energy levels. Since the neon atoms are excited directly into the levels 2s and 3s, these energy levels are more populated than the lower energy levels 2p and 3p. Therefore, the population inversion is achieved between 3s and 3p, 3s and 2p, and 2s and 2p. The transition between these levels produces wavelengths of 3.39 μm, 0.6328 μm and 1.15 μm respectively.

The energy levels of helium are represented by L-S representation, whereas neon energy levels are represented by Paschen's notations.

Advantages

(a) The output power of He–Ne laser is in the order of a few mW.
(b) Continuous output laser beam can be produced.
(c) The monochromaticity and directionality of gaseous laser are high.

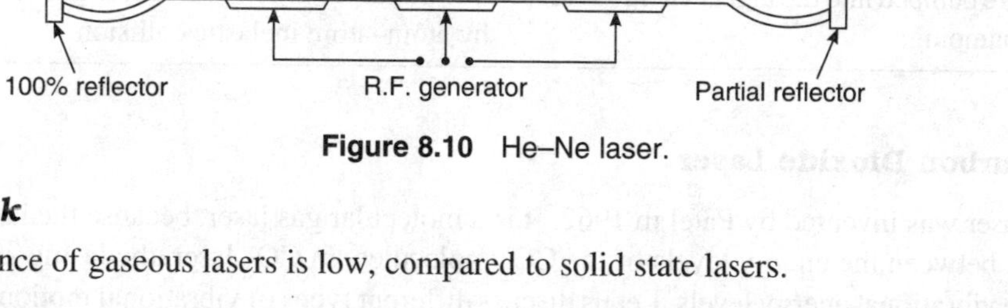

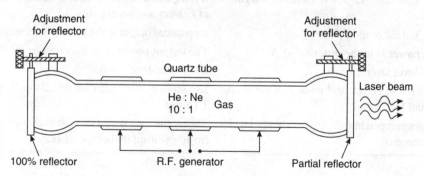

Figure 8.10 He–Ne laser.

Drawback

The coherence of gaseous lasers is low, compared to solid state lasers.

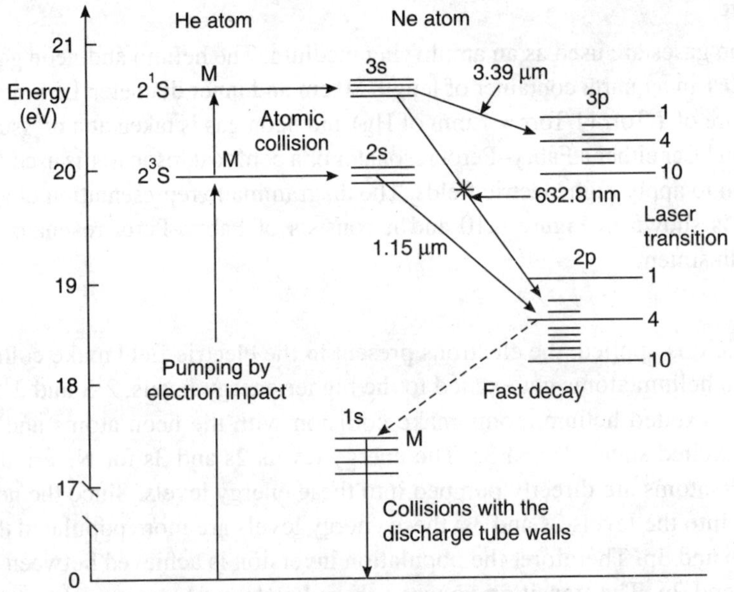

Figure 8.11 Energy level diagram for He–Ne laser.

Differences between ruby and He–Ne lasers

The differences between the ruby laser and the He–Ne laser are listed in Table 8.2.

Table 8.2 Differences between ruby and He–Ne lasers

Ruby laser	*He–Ne laser*
It is a solid state laser. Its active medium is a ruby rod.	It is a gaseous laser. Its active medium is a mixture of helium and neon gases.
It produces pulsed output.	It produces output in the form of a continuous wave.
The output power is in the order of 10^9 W.	The output power is in the order of 10^{-3} W.
It is a three-level laser.	It is a four-level laser.
Its output is red in colour and its wavelengths are 694.3 nm and 692.7 nm.	Its output wavelengths are 632.8 nm, 3.39 µm and 1.15 µm.
The atoms are pumped into the higher energy levels by optical pumping.	The atoms are excited into the higher energy levels by atom–atom inelastic collision.

8.8.3 Carbon Dioxide Laser

The CO_2 laser was invented by Patel in 1962. It is a molecular gas laser, because the lasing action takes place between the energy levels of the CO_2 molecules. In CO_2 laser, the lasing action takes place in the vibrational energy levels. Let us discuss different types of vibrational motion possessed by a CO_2 molecule.

Different types of vibrations of CO_2 molecules

The CO_2 molecule is represented as O=C=O. The CO_2 molecule possesses vibrational motion and rotational motion. There are three types of vibrational motion, namely

(a) Symmetric mode of vibration
(b) Bending mode of vibration and
(c) Asymmetric mode of vibration

(a) Symmetric mode of vibration

In symmetric mode of vibration both the oxygen molecules vibrate either towards or away from the carbon molecule, which is fixed and hence the C-O bond length is kept as constant during vibration. The frequency of vibration is said to be symmetric stretching frequency. The symmetric mode of vibration is shown in Figure 8.12(a).

(b) Bending mode of vibration

In this type of vibration both carbon and oxygen molecules vibrate along vertical direction and another direction which is perpendicular to the plane of this paper. The frequency of this vibration is known as bending frequency. The bending mode of vibration is shown in Figure 8.12(b).

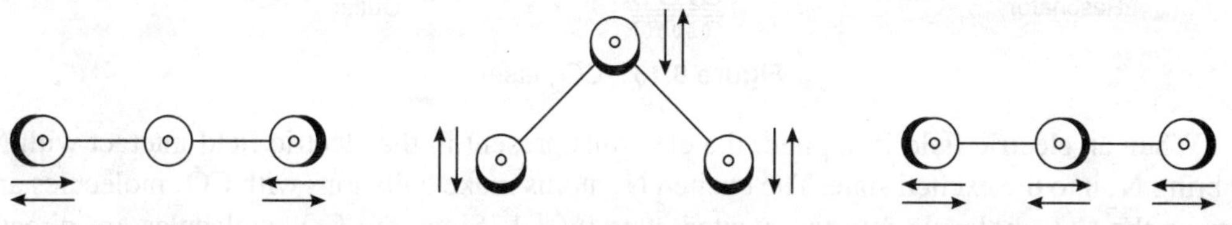

Figure 8.12 Different modes of vibration of CO_2 molecule.

(c) Asymmetric mode of vibration

In asymmetric mode of vibration both the oxygen and the carbon molecules are vibrating along the horizontal direction. Therefore, the C—O bond lengths are not equal. The frequency of vibration is said to be asymmetric stretching frequency. The asymmetric mode of vibration is shown in Figure 8.12(c).

The energy levels produced by these vibrations are called vibrational energy levels. The vibrational energy levels are represented by $n\,m^l\,p$, where n, m and p are the quantum numbers. They take the values of 0, 1, 2, 3, ... and so on. The superscript l is the total angular momentum for the vertical and perpendicular directions of bending mode of vibrations. The quantum numbers $1\,0^0\,0$; $2\,0^0\,0$ respectively represent the first and second energy levels produced by symmetric mode of vibration. Similarly, the quantum numbers $0\,1^0\,0$, $0\,2^0\,0$ represent the first and second energy levels of bending mode of vibration. The superscript 0 (zero) represents the total angular momentum value. The first and second energy levels of asymmetric mode of vibration are represented by $0\,0^0\,1$ and $0\,0^0\,2$ respectively.

CO_2 laser—construction

The diagrammatic representation of a CO_2 laser apparatus is shown in Figure 8.13. It is made of a quartz tube of length 5 m and diameter 2.5 cm. It contains inlets for allowing CO_2, N_2, and He gases. The end faces of CO_2 laser apparatus are fitted with NaCl Brewster window. The confocal resonator made of Si crystal and coated with Al is used. The apparatus has provision to provide high electric field.

Working

The basic principle of CO_2 laser is gaseous discharge. So, it is energized by a high ac or dc voltage. The CO_2 laser uses three gases, namely CO_2, N_2 and He as shown in Figure 8.13. These three gases are passed through the container in the ratio of 0.33 Torr of CO_2 (1 Torr = 1mm of Hg), 1.2 Torr of N_2 and 7 Torr of He. A continuous flow of these three gases are maintained and the used gases are passed through the outlet, because during the laser operation some unwanted product such as CO is produced.

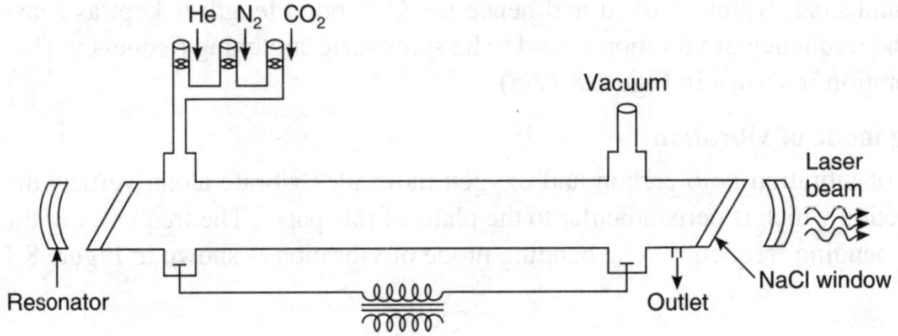

Figure 8.13 CO_2 laser.

When an electric field is applied, the electrons present in the electric field interact with N_2 and bring N_2 into the excited state. The excited N_2 atoms make collisions with CO_2 molecules and it brings the CO_2 molecule into the excited state 0 0^0 1. Since the CO_2 molecules are directly pumped into the 0 0^0 1 energy level, the energy levels 0 2^0 0 and 1 0^0 0 are less populated compared to the energy level 0 0^0 1. Therefore, the population inversion is achieved between 0 0^0 1 and 1 0^0 0, 0 0^0 1 and 0 2^0 0 and hence the laser transition takes place as shown in Figure 8.14. The laser transition between 0 0^0 1 and 1 0^0 0 energy levels produces a wavelength of 10.6 μm and the transition between 0 0^0 1 and 0 2^0 0 produces a wavelength of 9.6 μm. The wavelength 10.6 μm is more dominant over the wavelength 9.6 μm.

In CO_2 laser, helium has two roles, namely (i) to depopulate the energy levels 0 2^0 0 and 1 0^0 0 and (ii) a coolant. The energy levels 1 0^0 0, 0 0^0 1, 0 2^0 0 and ground levels are closely spaced. Therefore, the population of the energy levels 0 2^0 0 and 1 0^0 0 should be decreased, otherwise it will increase the population of these levels and hence the population inversion is lost. The helium gas is used to depopulate the energy levels 0 2^0 0 and 1 0^0 0. During the laser operation, a high amount of heat is produced and hence the CO_2 molecule gets heated. In order to produce the laser oscillation continuously, the CO_2 molecule should be kept as cool as possible. Since helium is one of the best heat conductors, it is used as a coolant in order to keep the CO_2 molecule cool.

Advantages

(a) The output power even up to 60 kW is achieved using CO_2 laser.
(b) The efficiency of CO_2 laser is about 30%.
(c) The CO_2 laser produces a continuous output.
(d) Since a high amount of energy is used, CO_2 laser is used in industry for welding and cutting of metals.

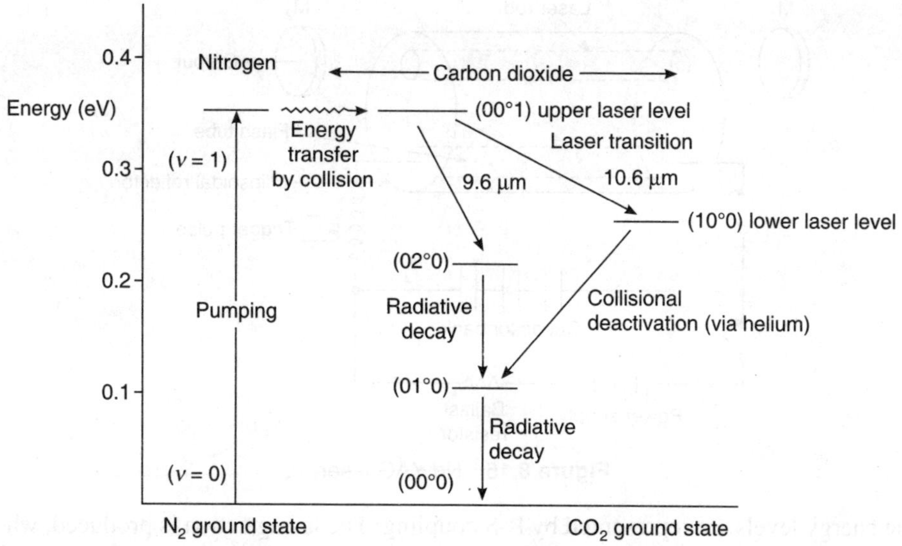

Figure 8.14 Energy level diagram of a CO_2 laser.

Drawback

It produces harmful effect to our eye and causes other damages. Therefore, stringent precautions are necessary in working even with a low-powered CO_2 laser.

8.8.4 Nd:YAG Laser

The Nd:YAG laser is a solid state laser. It is a four-level laser. Neodymium (Nd^{3+}) is a rare earth material and it is an active medium. It is doped with the host material Yttrium Aluminium Garnet ($Y_3Al_5O_{12}$). YAG is an acronym for Yttrium Aluminium Garnet.

Construction

The YAG (Yttrium Aluminium Garnet) is used as an amplifying medium. Nd^{3+} ions are doped with Yttrium Aluminium Garnet. The two end surfaces of the Nd:YAG rod are made as highly reflecting, perfectly parallel and silvered. One end is made as fully reflecting surface, whereas the other end is used as partially reflecting surface. In some lasers separate external mirrors are used as resonators. Xenon or krypton flash lamp is used as an optical source.

The lasing medium and a linear xenon lamp are kept inside a reflecting elliptical cavity. The experimental arrangement of Nd:YAG laser is shown in Figure 8.15. The xenon flash lamp is connected to a capacitor bank, a trigger pulse generator and a power supply.

Working

When the capacitor is discharging, the trigger pulse generator produces a triggering pulse. The xenon lamp produces a flash of light, when it receives the triggering pulse. The xenon flash lamp emits a white light that has a broad spectrum. Nd^{3+} ions absorb this radiation and get excited to the higher energy levels.

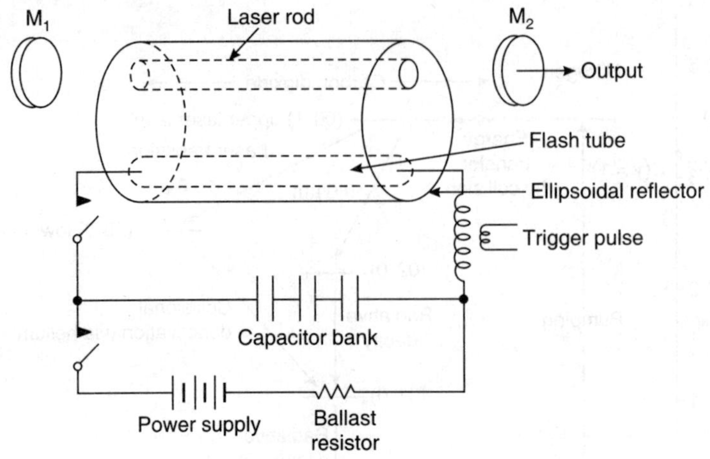

Figure 8.15 Nd:YAG laser.

The energy levels are represented by L-S coupling. The lasing action is produced, whenever the Nd^{3+} atoms are returning from the energy levels $4F_{3/2}$ to $4I_{11/2}$ as shown in Figure 8.16. The laser transition produces a wavelength of 1.064 μm. During the operation of this laser, a high amount of heat is produced by the flash tube and hence the laser rod gets heated. To avoid the damage resulting from this heating, air is forced to fall on the laser rod. For high output power laser, it is necessary to cool the laser rod using water cooling arrangement.

Advantages

(a) The solid state lasers are easy to maintain and capable of generating high peak powers.
(b) The output is a pulse when xenon flash lamp is used. A continuous output may be produced by using quartz-halogen lamp.
(c) The output power is nearly 2×10^4 W.

Drawbacks

(a) The solid state lasers are rugged.
(b) The actual power efficiency is 0.1%.

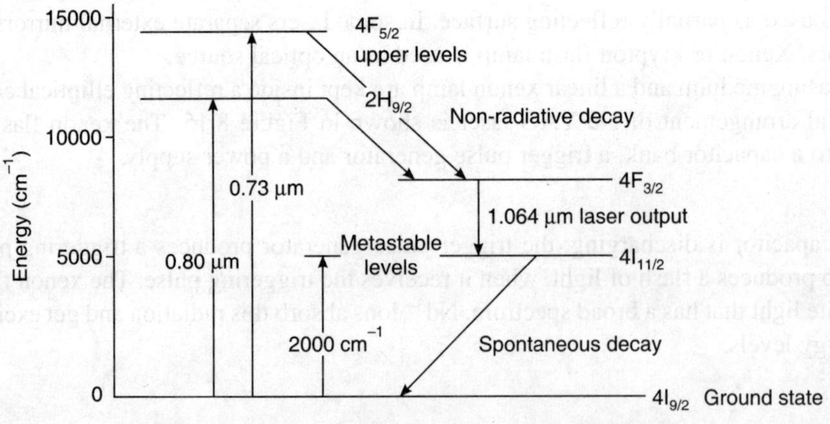

Figure 8.16 Energy level diagram of a Nd:YAG laser.

8.8.5 Semiconductor Laser

The semiconductor laser was first constructed by Hall et al. in 1962. The active medium in a semiconductor laser is a *p-n* junction diode and hence the semiconductor laser is also called diode laser.

Principle

The semiconductor laser is based on the principle of the emission of the recombination energy in the form of light in certain semiconducting materials such as GaAs, GaAlAs, etc. By heavily doping the *p-n* junction diode and by applying a high current density, the stimulated emission of light is produced and hence the laser operation is made possible in the semiconductor laser.

Construction

The *p-n* junction of a semiconductor laser is prepared using *n*-GaAs and *p*-GaAs. The height, width and breadth of this laser are 1 mm × 1 mm × 1 mm respectively. The upper and bottom surface of the *p-n* junction are used for ohmic contact, the front and back surfaces are used as resonators. For resonators, the end faces are made perfectly parallel to each other and highly polished. One end is made fully reflecting, whereas the other end is made partially reflecting. The remaining two surfaces are made rough, so as to prevent the leakage of light through these faces. The diode is forward biased, with a voltage nearly equal to the band gap voltage of the material. The *p*-type and *n*-type layers are heavily doped.

Working

The heavy doping of the *p*-type and *n*-type regions produces population inversion in the depletion region of the diode. The diode is forward biased with a voltage equal to the energy gap voltage, $V = E_g/e$ and hence the recombination of the holes and electrons takes place. In order to produce and maintain the stimulated emission continuously, a high current density nearly in the order of 20 kA/cm^2 is applied. Due to the high current density, more and more number of electrons are injected into the *n*-region and more number of holes are injected into the *p*-region and hence the continuous emission of light is made possible. Since, more number of charge carriers are injected, this laser is also called injection laser. The output laser beam emerges from the *p-n* junction of the partially reflecting surface (Figure 8.17). The output wavelength is given by

$$\lambda = \frac{hc}{E_g}$$

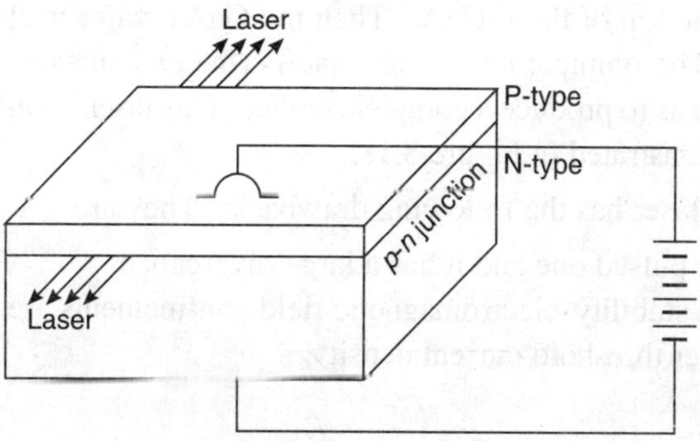

Figure 8.17 Semiconductor laser.

For GaAs, $E_g = 1.44$ eV and the output wavelength emitted by a diode laser made up of GaAs is

$$\lambda = \frac{hc}{E_g} = \frac{6.626 \times 10^{-34} \times 3 \times 10^8}{1.44 \times 1.6 \times 10^{-19}} = 862.7 \text{ nm}$$

Advantages

(a) The diode laser is small in size, low cost and very compact.
(b) The efficiency of diode laser is 10%. The efficiency can be increased even up to 100% by lowering the temperature into liquid nitrogen temperature (77 K).
(c) The output can be tuned to any desired wavelength by varying the band gap of the material.

Drawbacks

(a) Since the size is very small, preparing the cavity resonator is very difficult.
(b) The *p-n* junction width is nearly 1 μm. The laser beam emitted from this junction is having a width of nearly 25 to 40 μm. Due to the increase in the width of the output beam, it has poor coherence and monochromaticity.

8.9 HOMOJUNCTION AND HETEROJUNCTION LASER

In a *p-n* junction, if *p* and *n* type regions are made from the same single crystalline semiconducting material then it is said to be homojunction. For example, consider a GaAs single crystal which is used to prepare both *n* and *p* regions, and thus known as **homojunction**. If two or more than two different materials are used for the preparation of the *n*-type and *p*-type regions, for example, *n*-GaAs and *p*-GaAlAs, then it is said to be **heterojunction**. Semiconducting diode laser can be prepared either as a homojunction device or as a heterojunction device.

8.9.1 Homojunction Laser

Consider, a homojunction laser is prepared by taking an *n* type GaAs single crystal. The upper surface is made *p*-type by suitable doping. Zinc is one of the Group II elements in the periodic table. If Zn is diffused from the upper surface of *n*-GaAs and as Ga is a Group III element in the periodic table, then Zn forms bonding with Ga atoms and Zn becomes an acceptor impurity. This produces a *p*-GaAs on the top of the *n*-GaAs. Then the GaAs wafer is cleaved and sliced into a number of *p-n* devices. The ohmic contacts, resonators (the end surfaces are made as polished) and heat sink are made so as to produce a complete device. This laser is called homojunction laser. These processes are demonstrated in Figure 8.18.

The homojunction laser has the following drawbacks. They are

(a) The output is a pulsed one and it has a large divergence.
(b) Coherence and stability, electromagnetic field confinements are poor.
(c) It needs a higher threshold current density.

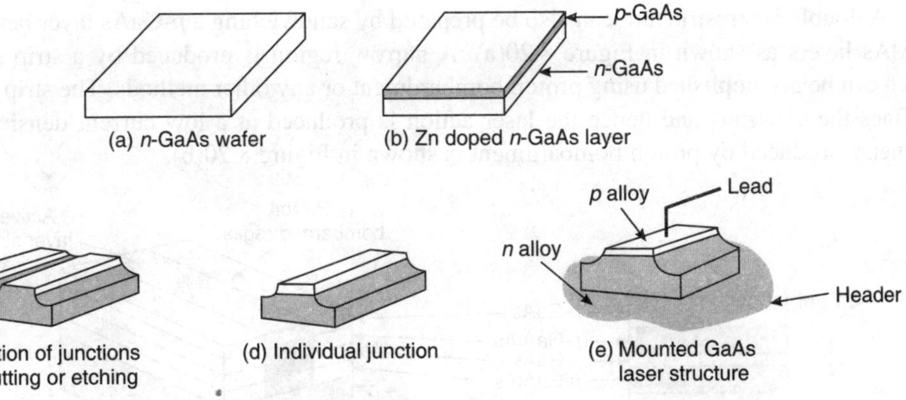

Figure 8.18 Fabrication of homojunction laser.

8.9.2 Heterojunction Laser

A heterojunction laser is prepared by depositing a *p*-GaAs layer of thickness 1 μm over an *n*-GaAs substrate and then *p*-GaAlAs is coated over the *p*-GaAs as shown in Figure 8.19(a). The energy band diagram of the device at equilibrium condition without applying any bias and with forward bias is shown in Figure 8.19(b) and Figure 8.19(c) respectively.

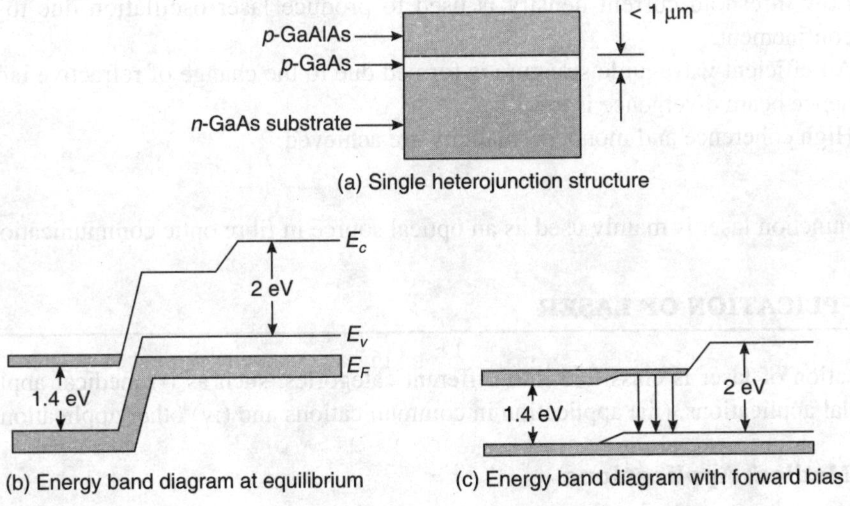

Figure 8.19 Heterojunction laser.

The injection of charge carriers into the *p-n* junction (electrons into the *n* region and holes into the *p* regions), confines the electrons to a narrow region. The population inversion is achieved at a lower current density because of the confinement of electrons. The GaAs–GaAlAs heterostructure has a higher efficiency for injection of electrons than the hole injection. The electrons cannot overcome the barrier potential at the GaAs–GaAlAs junction. In GaAs, the laser action takes place mainly in the *p* region of the device due to higher efficiency for electron injection. The change of refractive indices due to coating of two different layers such as GaAlAs and GaAs provide a waveguide effect for the photon confinement.

A double heterostructure can also be prepared by sandwiching a *p*-GaAs layer between two GaAlAs layers as shown in Figure 8.20(a). A narrow region is produced by a strip geometry, which can be accomplished using proton bombardment or any other methods. The strip geometry confines the electrons and hence the laser action is produced at a low current density. A strip geometry produced by proton bombardment is shown in Figure 8.20(b).

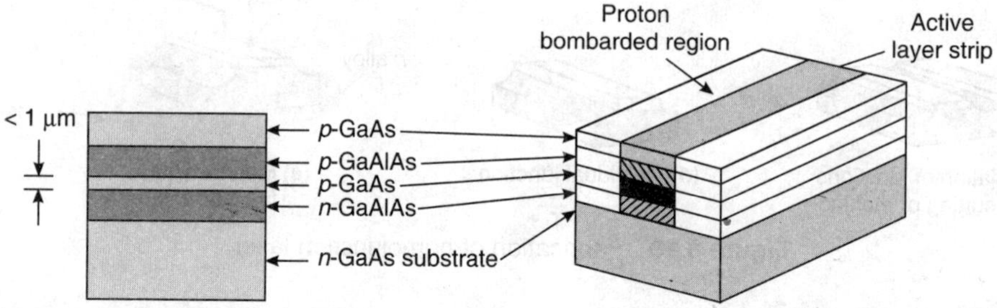

(a) Double heterojunction structure

(b) A strip geometry of double heterojunction structure

Figure 8.20 Heterojunction laser.

Advantages

(a) Continuous wave operation at the room temperature is possible.
(b) Low threshold current density is used to produce laser oscillation due to electron confinement.
(c) An efficient waveguide structure is formed due to the change of refractive indices and hence beam divergence is low.
(d) High coherence and monochromaticity are achieved.

Use

The heterojunction laser is mainly used as an optical source in fibre optic communication.

8.10 APPLICATION OF LASER

The application of laser is classified into different categories, such as (i) medical applications, (ii) industrial applications, (iii) application in communications and (iv) other applications.

8.10.1 Medical Applications

Treatment of tumours

The malignant tumours absorb the laser beam strongly. This property is used to remove tumours successfully.

Dental studies

The decayed area of the teeth has rough surfaces and appears dark and dull in colour. The decayed portion of our teeth absorbs more amount of laser beam. This property is used to remove the caries (decayed portion) of the teeth. Ruby laser is generally used for this purpose.

Surgical applications

The laser beam is used to cut the blood vessel and tissue without producing bleeding. This makes surgery easier. It is used for the operation like the removal of a portion of liver lobe, which was previously impossible due to inevitable haemorrhage.

Treatment of retina

If tear develops, it passes from vitreous body through the hole pushing the retinal cell away from choroids and it produces partial blindness. This is said to be retinal photocoagulation. The retina is spot welded using a laser beam and brought back into the original position.

Removal of urinary stone

Laser in conjunction with fibre endoscope used to vapourize blood clots and to remove urinary stones.

Other applications

(i) The helium–neon laser is used to cure tropic ulcers, poorly healing wounds.
(ii) In X-ray radiography, laser is used as a line-up device and it is used to vapourize samples from teeth, bone, internal organs and so on. These vapourized compounds are used for spectrographic analysis.

8.10.2 Industrial Applications

In industry laser is used for cutting, welding, drilling and surface hardening.

Cutting

In order to cut a metal, the laser beam is made to fall on the metal. The light energy made to fall on the metal gets converted into heat energy. Due to this heat energy the metal may gets heated. If the heating is sufficient, the metal melts and then vapourised. Then the metal is cut into two portions.

Welding

Two different materials are brought together and then a laser beam is made to fall on them. The two metals are heated together to a compatible condition and fuse them together.

Drilling

In order to drill a hole in a metal, a laser beam is made to fall on the required portion of the metal. Due to the intensed heating of laser on the metal, a fine hole is produced.

Surface hardening

The laser beam is used to increase the hardness of the outer surface of steel. The laser beam is made to fall on the outer surface of a material and hence it melts the outer portion of the materials and then it is cooled. Due to this the hardness of the outer surface is increased. This process is said to be laser surface hardening.

8.10.3 Communication

The laser is used in optical fibre communication. The higher bandwidth is one of the advantages of the laser beam. The bandwidth is directly proportional to the rate at which the information can be sent. Due to the higher bandwidth, large information can be sent simultaneously. No crosstalk, no leakage, safer communication, small in size, light weight are some of the other advantages of communication using laser beam.

8.10.4 Other Applications

Nuclear fusion

Fusion is the process of producing a heavier nucleus by combining any two lighter nuclei. To produce the fusion reaction, a very high temperature in the order of 10^8 K is required. When a laser beam with high output power is made to incident on the pellets of the reactants, the incident laser beam produces sufficient heat energy required for the fusion reaction. The fusion reaction with laser beam is possible because the laser is highly energetic.

Separation of isotopes

The separation of isotopes is possible using the laser beam. Consider the uranium ore. It has two isotopes, namely U^{235} and U^{238}. These two isotopes are having different excited energy levels, i.e. the excited states of U^{235} and U^{238} are different. In order to separate U^{235} from U^{238}, a laser beam having a wavelength of $\lambda = \dfrac{hc}{(E_2 - E_1)}$, is made to incident on the uranium ore, where E_1 and E_2 represent the first and the second energy levels of U^{235} respectively. The U^{235} isotope gets excited to the energy level, E_2. Then by exciting the U^{235} isotope from E_2 to E_3, it is brought into the energy level E_3. Similarly by using suitable laser beams, the U^{235} isotope is brought into a detector and hence it is separated.

Velocity of light

By doing the older experiments, used to measure the velocity of light, using a laser beam the velocity of light can be determined accurately.

Defense

1. Laser beams are used to guide the missiles. The laser guided missiles hit the target with a very high accuracy.
2. Laser weapons are used to disable the enemy weapon and to destroy it.
3. LIDAR is an acronym for **LI**ght **D**etection **A**nd **R**anging. It uses the laser beam in RADAR to detect the direction, velocity, and position of an object precisely. The light produced by a pulse laser is made to return after reflection from the moving object. The time delay of the laser beam for the round-trip is used to find the range of the target. Using the Doppler shift principle, the velocity of the moving object is determined.

Chemistry

Laser beam is used to accelerate a chemical reaction. It is also used to produce a new compound by simply breaking an atomic bond.

Photography

Laser beams are used to produce hologram, a lensless photography. The image is recorded by means of interference of lights.

Optical data storage

Since laser beam can be focused into a small spot using a convex lens, it is used to store more information in the optical data storage medium. It is used to write and read in the compact discs (CDs).

SOLVED PROBLEMS

8.1 A He–Ne laser produces an output power of 5 mW. If it emits light of wavelength 632.8 nm, calculate the number of photons emitted by the laser in one second.

Given data

Output power, $P = 5$ mW

Wavelength, $\lambda = 632.8$ nm

Solution

$$\text{Energy of one photon, } h\nu = \frac{hc}{\lambda} = \frac{6.626 \times 10^{-34} \times 3 \times 10^{8}}{632.8 \times 10^{-9}}$$

$$= 3.141 \times 10^{-19} \text{ J} = 1.96 \text{ eV}$$

$$\text{Number of photons emitted} = \frac{\text{output power}}{\text{energy of one photon}}$$

$$= \frac{5 \times 10^{-3}}{3.141 \times 10^{-19}}$$

$$= 1.591 \times 10^{16} \text{ photons per second}$$

The number of photons emitted per second by a He–Ne laser is 1.591×10^{16} photons per second.

8.2 A transition between the energy levels E_2 and E_1 produces a light of wavelength 632.8 nm, calculate the energy of the emitted photons.

Given data

Wavelength, $\lambda = 632.8$ nm

Solution

$$\text{Energy of the emitted photon, } E = h\nu = \frac{hc}{\lambda} = \frac{6.626 \times 10^{-34} \times 3 \times 10^{8}}{632.8 \times 10^{-9}}$$

$$= 3.141 \times 10^{-19} \text{ J} = 1.96 \text{ eV}$$

The energy of the photon = 1.96 eV.

8.3 A system has three energy levels E_1, E_2 and E_3. The energy levels E_1 and E_2 are at 0 eV and 1.4 eV respectively. If the lasing action takes place from the energy level E_3 to E_2, and emits a light of wavelength 1.15 μm, find the value of E_3.

Given data

The value of first energy level, $E_1 = 0$ eV

Value of second energy level, $E_2 = 1.4$ eV

Wavelength, $\lambda = 1.15$ μm

Solution

$$\text{Energy of one photon, } h\nu = \frac{hc}{\lambda} = \frac{6.626 \times 10^{-34} \times 3 \times 10^8}{1.15 \times 10^{-6}}$$

$$= 1.728 \times 10^{-19}\ \text{J} = 1.079\ \text{eV}$$

$$\text{The energy value of } E_3 = E_2 + h\nu$$

$$= 1.4\ \text{eV} + 1.079\ \text{eV}$$

$$= 2.479\ \text{eV}$$

The energy of $E_3 = 2.479$ eV.

8.4 A laser transition takes place from an energy level at 3.2 eV to another level at 1.6 eV. Calculate the wavelength of the laser beam emitted.

Given data

The value of higher energy level $E_1 = 3.2$ eV

The value of lower energy level $E_2 = 1.6$ eV

Solution

Energy difference, $E_2 - E_1 = 3.2 - 1.6 = 1.6$ eV

$$\text{Wavelength, } \lambda = \frac{hc}{E} = \frac{6.626 \times 10^{-34} \times 3 \times 10^8}{1.6 \times 1.6 \times 10^{-19}}$$

$$= 7.7648 \times 10^{-7}\ \text{m}$$

The wavelength of the photon, $\lambda = 7.7648 \times 10^{-7}$ m.

8.5 The band gap of GaAs is 1.42 eV. What is the wavelength of the laser beam emitted by a GaAs diode laser.

Given data

Band gap of GaAs = 1.42 eV

Solution

Wavelength of laser emitted by GaAs,

$$\lambda = \frac{hc}{E} = \frac{6.626 \times 10^{-34} \times 3 \times 10^8}{1.42 \times 1.6 \times 10^{-19}}$$

$$= 8.749 \times 10^{-7}\ \text{m}$$

The wavelength of the laser emitted by GaAs, l = 8.749×10^{-7} m

8.6 Calculate the relative population of the energy levels N_1 and N_2 at 300 K, $\lambda = 500$ nm.

Given data

Temperature, $T = 300$ K

Wavelength = 500 nm

Solution

From Maxwell and Boltzmann law, the relative population is given by

$$\frac{N_1}{N_2} = \frac{\exp\left(-\frac{E_1}{kT}\right)}{\exp\left(-\frac{E_2}{kT}\right)} = \exp\left(-\frac{E_1 - E_2}{kT}\right) = \exp\left(\frac{h\nu}{kT}\right)$$

Substituting the values of T and λ, we get

$$\frac{N_1}{N_2} = \exp\left(\frac{hv}{kT}\right) = \exp\left(\frac{hc}{\lambda kT}\right) = \exp\left(\frac{6.626 \times 10^{-34} \times 3 \times 10^8}{500 \times 10^{-9} \times 1.38 \times 10^{-23} \times 300}\right)$$

$$= \exp (96.029)$$

$$= 5.068 \times 10^{41}$$

The relative population between N_1 and N_2 is 5.068×10^{41}.

8.7 Examine the possibility of stimulated emission at 300 K, and $\lambda = 600$ nm.

Given data

Temperature, $T = 300$ K

Wavelength = 600 nm

Solution

The ratio between the stimulated emissions to spontaneous emission is given by

$$\frac{\text{stimulated emission}}{\text{spontaneous emission}} = \frac{1}{\exp\left(\frac{hv}{kT}\right) - 1}$$

Substituting the values of T and λ, we get

$$\frac{\text{stimulated emission}}{\text{spontaneous emission}} = \frac{1}{\exp\left(\frac{6.626 \times 10^{-34} \times 3 \times 10^8}{600 \times 10^{-9} \times 1.38 \times 10^{-23} \times 300}\right) - 1}$$

$$= \frac{1}{\exp(80.024) - 1} = 1.762 \times 10^{-35}$$

The ratio between stimulated emission and spontaneous emission is 1.762×10^{-35}. Therefore, the stimulated emission is not possible in this condition.

8.8 Calculate the efficiency of a He–Ne laser, if it produces an output power of 5 mW and if it is operated with a current of 10 mA at 3 kV.

Given data

Output power, $P = 5$ mW $= 5 \times 10^{-3}$ W

Current, $I = 10$ mA $= 10 \times 10^{-3}$ A

Voltage, $V = 3$ kV $= 3 \times 10^3$ V

Solution

$$\text{Efficiency} = \frac{\text{output power}}{\text{input power}} \times 100\%$$

$$= \frac{5 \times 10^{-3}}{10 \times 10^{-3} \times 3 \times 10^3} \times 100\%$$

$$= 0.016667\%$$

The efficiency of the laser = 0.016667%.

8.9 A laser beam emits an output power of 1 mW. If it is focused as a spot having a diameter of 1 μm, calculate the intensity of the laser beam.

Given data

Output power, $P = 1\ \text{mW} = 1 \times 10^{-3}\ \text{W}$

Diameter = 1 μm

Radius, $r = 0.5\ \mu\text{m} = 0.5 \times 10^{-6}\ \text{m}$

Solution

$$\text{Intensity of laser} = \frac{\text{power}}{\text{area of cross section}}$$

$$= \frac{1 \times 10^{-3}}{\pi(0.5 \times 10^{-6})^2}$$

$$= 1.273 \times 10^9\ \text{W m}^{-2}$$

The intensity of the laser = $1.273 \times 10^9\ \text{W m}^{-2}$

8.10 A laser beam of wavelength 632.8 nm is made fall on a wall that lies at a distance of 5 m and if it produces a spot having a diameter of 1 mm, calculate the angular spread and the area of the angular spread.

Given data

Wavelength, $\lambda = 632.8\ \text{nm} = 632.8 \times 10^{-9}\ \text{m}$

Distance, $D = 5\ \text{m}$

Diameter, $d = 1\ \text{mm}$

Solution

$$\text{Angular spread, } \Delta\theta = \frac{\lambda}{d} = \frac{632.8 \times 10^{-9}}{1 \times 10^{-3}} = 6.328 \times 10^{-4}\ \text{radian}$$

$$\text{Radius of the spread, } r = (D \times \Delta\theta) = \left(D\frac{\lambda}{d}\right) = 5 \times 6.328 \times 10^{-4} = 3.164\ \text{mm}$$

Area of the spread

$$= \pi (D \times \Delta\theta)^2 = \pi\left(D\frac{\lambda}{d}\right)^2 = \pi (5 \times 6.328 \times 10^{-4})^2$$

$$= 3.145 \times 10^{-5}\ \text{m}^2$$

Angular spread = 6.328×10^{-4} radian

Area of the spread = $3.145 \times 10^{-5}\ \text{m}^2$

SHORT QUESTIONS

1. What is absorption of light?
2. What is meant by emission of light?
3. What is stimulated emission?
4. What is spontaneous emission?

5. Write the mathematical expression for the stimulated and spontaneous emissions and explain the terms.
6. Distinguish between spontaneous and stimulated emission.
7. What is meant by the population inversion?
8. What is optical pumping?
9. What is electron excitation?
10. What is inelastic atom–atom collision?
11. What is a resonator?
12. Mention about the different types of resonators used for laser production.
13. Mention the distinct properties of laser.
14. What is monochromaticity?
15. What is intensity of laser?
16. What is meant by directionality of laser?
17. What is coherence?
18. Mention the different types of lasers.
19. What are the advantages and drawbacks of ruby laser?
20. What is a solid state laser? Give an example.
21. What are the advantages and drawbacks of He–Ne laser?
22. Distinguish between a ruby laser and a He–Ne laser.
23. Mention the different types of vibration in CO_2 laser.
24. What are the advantages and drawbacks of CO_2 laser.
25. What are the advantages and drawbacks of a Nd:YAG laser.
26. What is the principle of a diode laser?
27. What are the advantages and drawbacks of a diode laser?
28. State some of the applications of lasers in engineering and industry.
29. What is holography?
30. What is the role of nitrogen in CO_2 laser?
31. What are the roles of helium in CO_2 laser?
32. Give some applications of laser in medical field.
33. What are the differences between a homojunction and a heterojunction laser?
34. What are the applications of laser in communications?
35. What is meant by LIDAR? Give its use.
36. Give the applications of LASER in material processing.
37. What are the differences between holography and photography?

DESCRIPTIVE TYPE QUESTIONS

1. Derive Einstein's relation for the stimulated emission of radiation and hence find the relation between the stimulated emissions and the spontaneous emission.
2. Describe the principle, construction and working of a ruby laser. What are its merits and demerits?
3. Explain with neat sketches, the principle, construction and working of a He–Ne laser. What are its merits and demerits?
4. What is a molecular gas laser? Describe the different types of vibration of a CO_2 molecule. Describe the principle, construction and working of a CO_2 laser.
5. Explain with neat sketches the principle, construction, working and energy level diagram of Nd:YAG laser.
6. Explain the principle, construction and working of a semiconductor laser with neat sketches. What are its merits and demerits?
7. What is a hologram? Explain the process of construction and reconstruction of a hologram. What are the advantages of holography?

PROBLEMS

1. Determine the wavelength of the laser output from a GaAs *p-n* junction laser. Assume that $E_G = 1.4$ eV for GaAs, Planck's constant $h = 6.626 \times 10^{-34}$ Js and velocity of light $c = 3 \times 10^8$ m s^{-1}.
2. The output wavelength of a CO_2 laser is 10.6 μm. If it produces an output of 1 kW, how many photons are emitted in one minute?
3. A laser beam is emitted by a transition from an energy level E_2 to the energy level E_1. If the energy level E_2 is 2.4×10^{-19} J, and E_1 is 0 J, calculate the wavelength of the laser beam emitted.
4. A gaseous laser is operated by 2 A current and 230 V. If it produces an output of 10 mW, what is the efficiency of the laser?
5. Examine the possibility of stimulated emission in the microwave region at $\lambda = 20$ cm and $T = 300$ K.
6. A tungsten filament is heated to 2000 K and it emits an average frequency of 5×10^{14} Hz, calculate the relation between the stimulated emission and the spontaneous emission.
7. Calculate the intensity of the laser beam, if it is focused in an area of 1 mm^2 and its output power is 5 mW.
8. For InP laser diode, the wavelength of light emission is 1.55 μm. What is its band gap in eV?
9. A laser source emits light of wavelength 0.621 μm and has an output of 35 mW. Calculate how many photons are emitted per minute by this laser source.
10. A ruby laser of output wavelength 694.3 nm and spot diameter of 1 mm is focused on the moon from the earth. The distance between the earth and the moon is 384,400 km. Find the angular spread, radius of the spread and area of the spread.

CHAPTER 9

FIBRE OPTICS

9.1 INTRODUCTION

Fibre optics is a branch of science that deals with the propagation of light through optical fibre. In optical fibres, the light is launched in one end of the fibre and it is passed through the other end without any loss of signals. The light passes through the optical fibre due to the total internal reflection of light. The light undergoes total internal reflection more than one lakh times within one metre length of the fibre. The light propagating through the fibre is explained by means of the ray optics.

9.2 BASIC PRINCIPLE OF FIBRE OPTICS

The basic principle behind the propagation of light through the optical fibre is total internal reflection. The refractive index is one of the most important optical parameters of a medium. It is defined by

$$\text{Refractive index of the medium} = \frac{\text{velocity of light in vacuum}}{\text{velocity of light in a medium}} \quad (9.1)$$

i.e.

$$n = \frac{c}{v} \quad (9.2)$$

The refractive index of air is 1.0002, water is 1.333, vacuum is 1, fused silica is 1.452, and grown glass is 1.517.

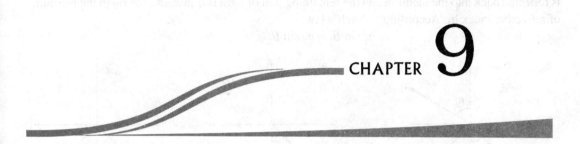

Consider that a ray of light is passing through a medium of refractive index n_1 to another medium of refractive index n_2 as shown in Figure 9.1. Let $n_1 > n_2$. Some part of the incident beam is reflected back into the medium and the remaining part of light is transmitted through the medium of refractive index n_2. According to Snell's law,

$$n_1 \sin \theta_1 = n_2 \sin \theta_2 \tag{9.3}$$

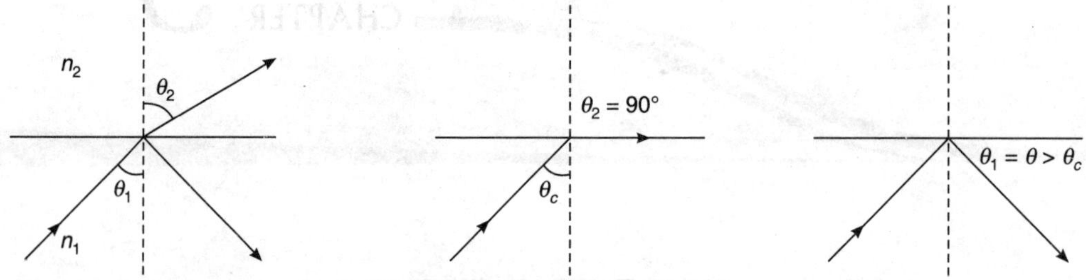

(a) Some part of the incident beam is transmitted through n_2, when $\theta_1 < \theta_c$ (b) Refracted ray passes through the core-cladding interface, when $\theta_1 = \theta_c$ (c) Total internal reflection occurs when $\theta_1 > \theta_c$

Figure 9.1 Reflection of light.

If the angle of incidence θ_1 is slightly increased, the angle of refraction, θ_2 also increases. The angle of refraction becomes 90° (i.e. $\theta_2 = 90°$) at some particular value of the angle of incidence. If the angle of incidence is further increased, the entire incident light gets reflected back into the medium and there is no light to be refracted into the medium of refractive index n_2. This type of reflection of light within the medium is known as **total internal reflection**. The minimum angle of incidence required to produce the total internal reflection is known as critical angle. It is represented by θ_c.

At critical angle

$$\theta_1 = \theta_c \text{ and } \theta_2 = 90°$$

Substituting in Eq. (9.3), we get

$$n_1 \sin \theta_c = n_2 \sin 90°$$

$$\sin \theta_c = \frac{n_2}{n_1}$$

$$\theta_c = \sin^{-1}\left(\frac{n_2}{n_1}\right) \tag{9.4}$$

Equation (9.4) gives the value of the critical angle (θ_c) for the total internal reflection.

9.3 CONSTRUCTION OF OPTICAL FIBRE

An optical fibre consists of a core at the centre and it is cladded by another layer known as cladding. The cladding is surrounded by a protective layer known as sheath. The core is generally constructed using germanium doped silica glass, cladding is prepared using nearly pure silica glass and sheath is prepared using ultraviolet cured plastic material. The refractive index of the core (n_1) is made as slightly greater than the refractive index of the cladding (n_2) as in Figure 9.2. The sheath is generally used to provide the protection against abrasion and external forces. The sheath is generally a coloured so as to enable the user to distinguish it from fibre.

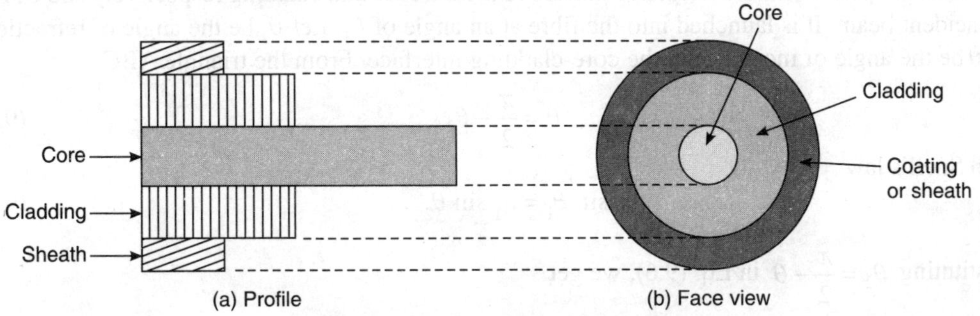

Figure 9.2 Construction of optical fibre.

9.4 ACCEPTANCE ANGLE AND NUMERICAL APERTURE

Consider that OA is the light ray launched into the fibre as shown in Figure 9.3. The light launched into the fibres gets refracted and passed through the core and it is incident on the core-cladding interface. The angle of incidence at the core-cladding interface should be greater than the critical angle. Let θ_1 be the angle of incidence in the air-core interface, θ_2 be the angle of refraction and θ be the angle of incidence at the core-cladding interface. From the triangle ABC,

$$\theta = \frac{\pi}{2} - \theta_2$$

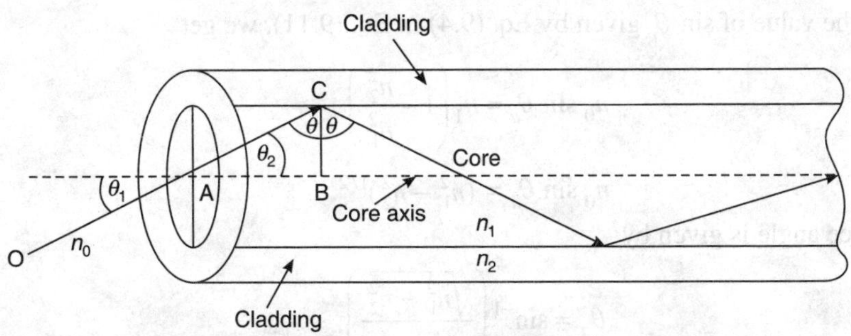

Figure 9.3 Transmission of light through optical fibres.

In order to make the angle of incidence (θ) at the core-cladding interface greater than the critical angle (θ_C), θ_2 should be low and hence the angle of incidence at the air-core interface (θ_1) should be minimum. The angle of incidence at the air-core interface that makes the light transmitted into the core to be incident in the core-cladding interface at an angle greater than the critical angle is known as the **acceptance angle**.

The value of the acceptance angle is limited by the angle of incidence at the core-cladding interface. The cone of angle at which the light is launched into the fibre that makes the light to be incident at an angle greater than the critical angle at the core-cladding interface is said to be the **acceptance angle.** It is represented by the letter, θ_a. Let us derive a mathematical relation for it.

Let n_0, n_1 and n_2 be the refractive indices of the air, core and cladding respectively and OA be the incident beam. It is launched into the fibre at an angle of θ_1. Let θ_2 be the angle of refraction. Let θ be the angle of incidence at the core-cladding interface. From the triangle ABC

$$\theta = \frac{\pi}{2} - \theta_2 \tag{9.5}$$

From Snell's law

$$n_0 \sin \theta_1 = n_1 \sin \theta_2 \tag{9.6}$$

Substituting $\theta_2 = \frac{\pi}{2} - \theta$ in Eq. (9.6), we get

$$n_0 \sin \theta_1 = n_1 \sin\left(\frac{\pi}{2} - \theta\right) \tag{9.7}$$

i.e.

$$n_0 \sin \theta_1 = n_1 \cos \theta \tag{9.8}$$

Using the relation,

$$\sin^2\theta + \cos^2 \theta = 1$$

we get,

$$\cos \theta = (1 - \sin^2\theta)^{1/2} \tag{9.9}$$

Substituting the value of cos θ in Eq.(9.8), we get

$$n_0 \sin \theta_1 = n_1(1 - \sin^2\theta)^{1/2} \tag{9.10}$$

Consider the light is launched in such a way that the angle of incidence at C is θ_c, then $\theta_1 = \theta_a$, and $\theta = \theta_c$, Eq. (9.10) can be written as

$$n_0 \sin \theta_a = n_1(1 - \sin^2\theta_c)^{1/2} \tag{9.11}$$

Substituting the value of sin θ_c given by Eq. (9.4) in Eq. (9.11), we get

$$n_0 \sin \theta_a = n_1\left(1 - \frac{n_2^2}{n_1^2}\right)^{1/2} \tag{9.12}$$

$$n_0 \sin \theta_a = (n_1^2 - n_2^2)^{1/2} \tag{9.13}$$

The acceptance angle is given by

$$\theta_a = \sin^{-1}\left(\frac{\sqrt{n_1^2 - n_2^2}}{n_0}\right) \tag{9.14}$$

Equation (9.14) gives the value of the acceptance angle. The rays incident within a cone of half angle θ_a will be collected and propagated by the fibre. This cone of half angle θ_a is known as acceptance angle.

The light gathering capacity of the fibre or how much light is accepted by the fibre is represented by numerical aperture (NA) of the fibre. The numerical aperture of the fibre is defined as

$$\text{Numerical aperture (NA)} = n_0 \sin \theta_a = (n_1^2 - n_2^2)^{1/2} \tag{9.15}$$

The numerical aperture is also represented by relative refractive index differences, Δ as

$$\mathrm{NA} = (n_1^2 - n_2^2)^{1/2} = n_1\sqrt{2\Delta} \tag{9.16}$$

where $\Delta = \dfrac{n_1^2 - n_2^2}{2n_1^2} \approx \dfrac{n_1 - n_2}{n_1}$ (i.e. when $n_1 = n_2$), Δ is known as refractive index difference.

9.5 LIGHT PROPAGATION IN OPTICAL FIBRE

Consider that a ray of light is launched into the fibre. Let θ_1 be the angle of incidence of the light ray at the air-core interface. Let θ_2 be the angle of refraction and θ be the angle of incidence at the core-cladding interface. If the angle of incidence θ at the core-cladding interface is greater than the critical angle, then the light ray passes through the core by means of total internal reflection and it reaches the other end of the fibre. The light will be guided through the core, provided the semi-angle of the core should be less than θ_a.

The light ray within the core is transmitted either by means of reflection or by means of refraction of light. The light propagates through the single mode step index fibre and the multimode step index fibre by means of the reflection of light as shown in Figure 9.4. The light travels in a zigzag path due to the reflection of light.

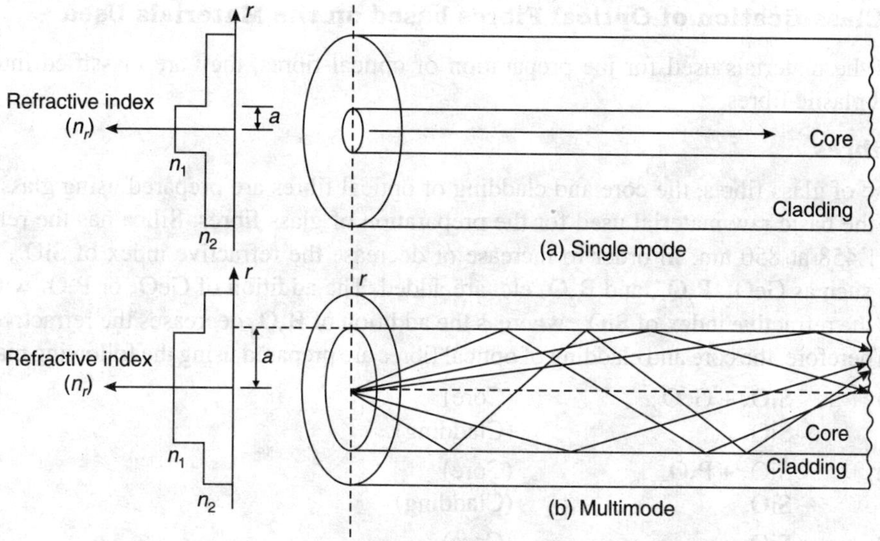

Figure 9.4 Light propagation in step index fibre.

In the case of the multimode graded index fibre, the light propagates through the core by means of refraction of light as shown in Figure 9.5. In a graded index fibre, the refractive indices of the core vary radially, i.e. the refractive index is maximum at the centre of the core and minimum at the core-cladding interface. In between these two, the refractive index varies radially. The light takes a curved path because of refraction of light through the core having variable refractive indices.

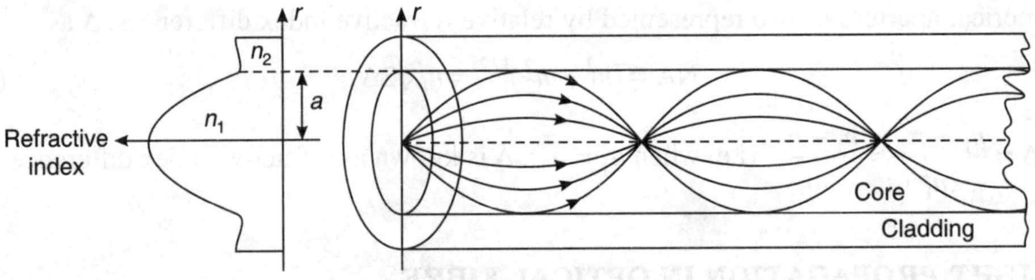

Figure 9.5 Propagation of light in multimode graded index fibre.

9.6 CLASSIFICATION OF OPTICAL FIBRES

The materials used to prepare the optical fibres should satisfy the following conditions:

(i) It should guide the light effectively.

(ii) The losses due to scattering of light, absorption, attenuation and dispersion should be low.

Only two materials, namely glass and plastics are satisfying these conditions. Therefore, the optical fibres are prepared using glass and plastics. The optical fibres are classified based on the materials used, number of modes propagating through optical fibres and the refractive index profile.

9.6.1 Classification of Optical Fibres based on the Materials Used

Based on the materials used for the preparation of optical fibres, they are classified into glass fibres and plastic fibres.

Glass fibres

In the case of glass fibres, the core and cladding of optical fibres are prepared using glass. Silica (SiO_2) is the basic raw material used for the preparation of glass fibres. Silica has the refractive index of 1.458 at 850 nm. In order to increase or decrease the refractive index of SiO_2, doping materials such as GeO_2, P_2O_5, and B_2O_3 etc. are added. The addition of GeO_2 or P_2O_5 with SiO_2 increases the refractive index of SiO_2, whereas the addition of B_2O_3 decreases the refractive index of SiO_2. Therefore, the core and cladding of optical fibres are prepared using the following materials.

(i)	$SiO_2 + GeO_2$	(Core)
	SiO_2	(Cladding)
(ii)	$SiO_2 + P_2O_5$	(Core)
	SiO_2	(Cladding)
(iii)	SiO_2	(Core)
	$SiO_2 + B_2O_3$	(Cladding)

In a glass fibre the difference between the refractive indices of core and cladding is low and hence the numerical aperture and the acceptance angles are low.

Plastic fibres

In the case of plastic fibres, the core and cladding are prepared using plastic materials. For example, in a plastic fibre, the core is made of polysterene (n = 1.60) and the cladding is made of polymethylmethacrylate (n = 1.49). In a plastic fibre, if the core is prepared using polymethylmethacrylate (n = 1.49) and the cladding is made up of its co-polymer (n = 1.4).

In a plastic fibre since the difference between the refractive indices of core and cladding is high (nearly, $n_1 - n_2 \approx 0.9$) and hence the numerical aperture (NA ≈ 0.5) and the acceptance angle ($\phi_a \approx 30°$) are high.

9.6.2 Classification of Optical Fibres based on the Refractive Index Profile

Based on the refractive index profile, the optical fibres are classified into two types. They are (i) step index fibre and (ii) graded index fibre.

Step index fibre

The refractive indices of the core and cladding of a step index fibre is given below:

$$\begin{aligned} &\text{Core} && n(r) = n_1 \quad \text{when } r < a \\ &\text{Cladding} && n(r) = n_2 \quad \text{when } r \geq a \end{aligned} \tag{9.17}$$

where a is the core radius and r is the radial distance.

From Eq. (9.17), one can infer that the refractive indices of the core and cladding are constant, then the fibre is said to be a step index fibre. In a step index fibre, the light travels in a zigzag path as shown in Figure 9.6.

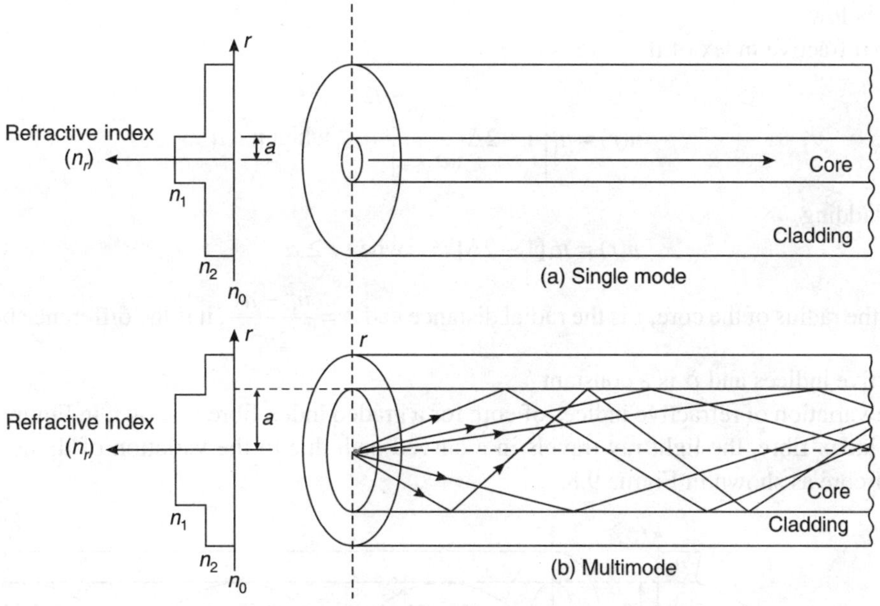

Figure 9.6 Step index fibre.

A step index fibre can be prepared in single mode and multimode. The single mode step index fibre and multimode step index fibre are shown in Figure 9.6. The variation of refractive indices of air, cladding and core for single mode and multimode step index fibres are shown in Figure 9.6. For step index fibre, the refractive indices of air, cladding and core vary in step by step.

Only one signal is passed through a single mode step index fibre, whereas more than one signal is passed through multimode step index fibres. The cross-sectional view of the single mode step index fibres is shown in Figure 9.7. The inter-modal dispersion is high in the case of a step index fibre.

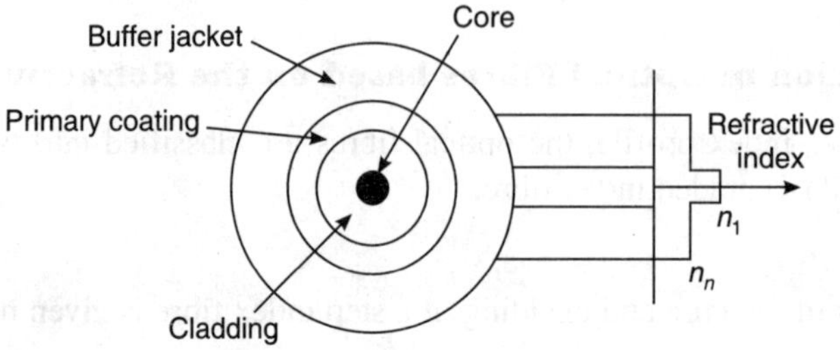

Figure 9.7 Cross-sectional view of single mode step index fibre.

Graded index fibre

In a graded index fibre the refractive index of the core varies with radial distance. The refractive index is maximum (n_1) at the centre of the core and it is minimum (n_2) at the core-cladding interface. In between the centre of the core and the core-cladding interface, the refractive index varies with radial distance, and hence it lies between n_1 and n_2. The refractive indices of the core and cladding are given below.

The refractive index of the core is

$$n(r) = n_1\left[1 - 2\Delta\left(\frac{r}{a}\right)^{\alpha}\right]^{1/2} \quad \text{when } r < a \tag{9.18}$$

and for cladding

$$n(r) = n_1[1 - 2\Delta]^{1/2} \quad \text{when } r \geq a \tag{9.19}$$

Here a is the radius of the core, r is the radial distance and $\Delta = \dfrac{n_1^2 - n_2^2}{2n_1^2}$, it is the difference between the refractive indices and α is a constant.

The variation of refractive indices of core for a graded index fibre is shown in Figure 9.8. In a graded index fibre, the light ray travels in a curved path due to the variation of the refractive indices of core as shown in Figrue 9.8.

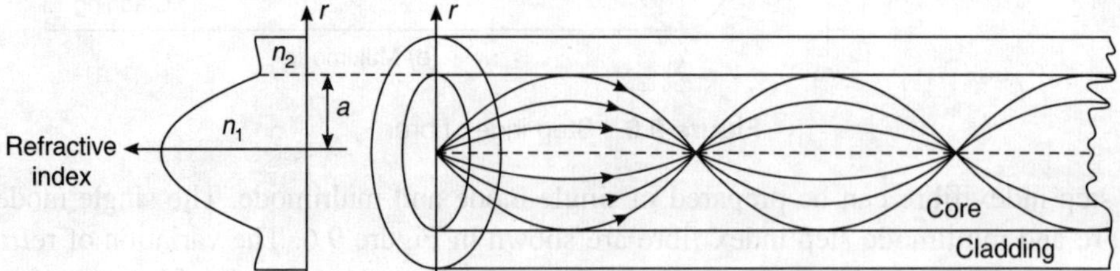

Figure 9.8 Multimode graded index fibre.

The variations of the refractive indices of the core with different α values are shown in Figure 9.9. For $\alpha = \infty$, the refractive index of the core is constant, for $\alpha = 1$, the variation of the refractive index is in the form of triangular shape and for $\alpha = 2$, the parabolic variation of the refractive index for core is obtained. In the case of graded index fibre, α is chosen as 2.

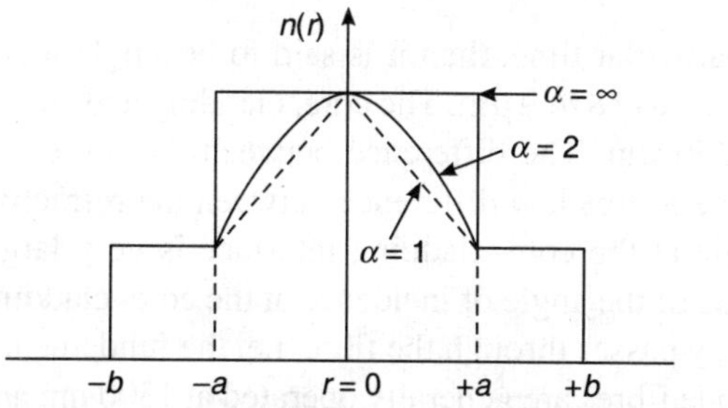

Figure 9.9 Variations of refractive indices of the core with different values of α.

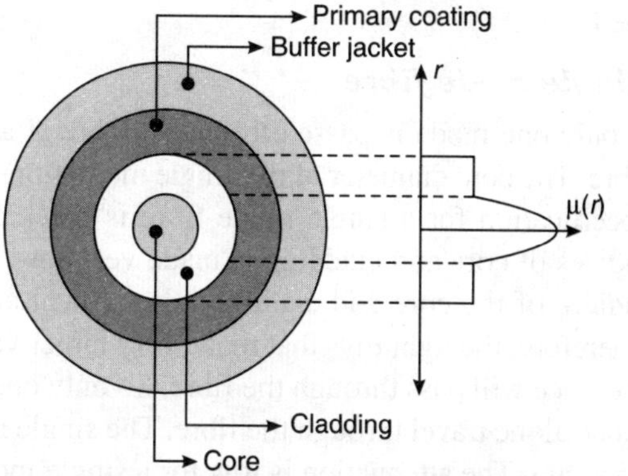

Figure 9.10 Cross-sectional view of multimode graded index fibres.

The cross-sectional view of the multimode graded index fibres is shown in Figure 9.10. The intermodal dispersion in graded index fibre is low.

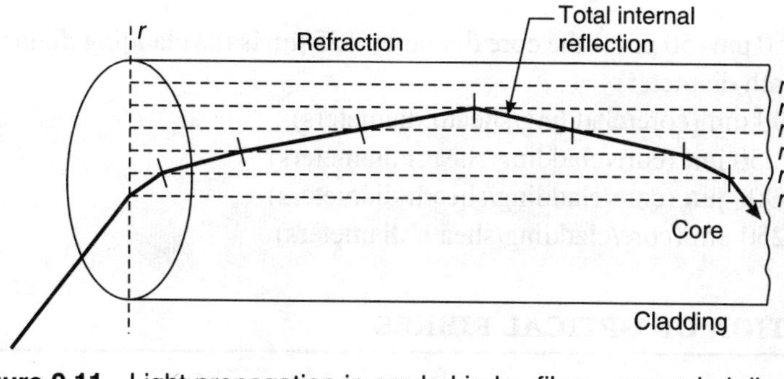

Figure 9.11 Light propagation in graded index fibre—expanded diagram.

A curved path of light in graded index fibre is explained by considering an expanded diagram as shown in Figure 9.11. The core consists of a number of layers having gradual variation in the refractive index. Whenever the light is launched into the fibre, it undergoes refraction from one layer to the other layer and there is a change in the path of the light ray. This will produce a curved path whenever the light is passing through a number of layers (more than 100 layers).

9.6.3 Classification of Optical Fibres based on the Number of Modes Propagating

The optical fibres are classified into two types on the basis of the number of modes propagating through the fibres. They are (i) single mode fibre and (ii) multimode fibre.

The number of modes propagating through an optical fibre is

$$N = \frac{0.5 \times d \times (\text{NA})^2}{\lambda} \tag{9.20}$$

where d is the diameter of the fibre core, NA is the numerical aperture and λ is the wavelength of the light.

Single mode fibre

If only one mode is passed through a fibre at a particular time, then it is said to be single mode fibre. The core diameter of the single mode fibre is nearly 8 to 9 μm. The core, cladding and sheath specification for a single mode fibre is 8.5/125/250 μm. The difference between the refractive indices of core and cladding is made very low. Due to this low difference between the refractive indices of the core and cladding, the critical angle at the core–cladding interface is very large. Therefore, the light rays that make very larger value of the angle of incidence at the core–cladding interface will pass through the fibre. So only one ray passes through the fibre, i.e. the fundamental mode alone travel through the fibre. The single mode fibres are generally operated at 1300 nm and 1550 nm. The attenuation is low for a single mode fibre at 1550 nm wavelength operation.

Multimode fibre

If more than one mode is passed through the fibre, then it is said to be multimode fibre. A multimode fibre has a core diameter of 50 μm or greater. A large number of signals are passed through the multimode fibre, because of its large diameter. The multimode fibres are available in three different sizes.

(i) 50/125/250 μm (50 μm is the core diameter, 125 μm is the cladding diameter and 250 μm is the sheath diameter)
50/125/900 μm (core/cladding/sheath diameters)

(ii) 62.5/125/250 μm (core/cladding/sheath diameters)
62.5/125/900 μm (core/cladding/sheath diameters)

(iii) 100/140/250 μm (core/cladding/sheath diameters)

9.7 PREPARATION OF OPTICAL FIBRES

The preparation of optical fibre is a two-stage process. The first stage is to prepare a rod called preform that contains both core and cladding materials and the second stage is drawing of the fibre. The preparation methods are generally classified into two types, namely (i) liquid phase (melting) method and (ii) vapour phase oxidation method. Let us discuss the double crucible method and the modified chemical vapour deposition method in this section.

9.7.1 Double Crucible Method

The method uses a double crucibles, one crucible is arranged inside the other one. The double crucible apparatus is shown in Figure 9.12. These two crucibles have separate nozzles at the base. The core and the cladding materials are taken in pure form either in the form of rod or in the form of powder. The core material is taken in the inner crucible and the cladding material is taken in the outer crucible. The double crucible apparatus is kept inside a vertical furnace and it is heated to around 800° to 1200°C. Due to the high temperature the core and cladding materials will melt and they are allowed to flow through the nozzles simultaneously. The core and cladding materials are then passed through a bath containing a plastic material. A protective layer or sheath is coated over core and cladding. Then it is passed through a curing furnace so as to cure the fibre and then the fibre is winded in a take-up drum. The single mode fibre prepared using sodium borosilicate glass by this technique has an attenuation of 3 dB/km.

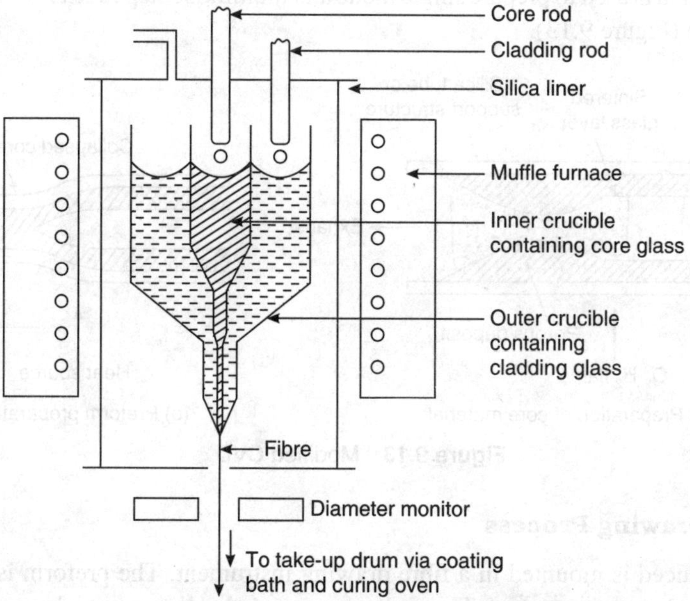

Figure 9.12 Double crucible apparatus.

9.7.2 Modified Chemical Vapour Deposition Technique (MCVD)

This method is a vapour phase oxidation process. This method uses a silica tube having a length of 1 m and inner diameter of 15 mm. The oxyhydrogen flame that produces a temperature of 1500°C is moved from the one end of the silica tube to the other end. The reactants are the vapours of halides of the required compound and O_2 gas. The deposition of the material is made using the following reactions.

$$\underset{\text{(vapour)}}{SiCl_4} + \underset{\text{(gas)}}{O_2} \xrightarrow{\text{Heat}} \underset{\text{(solid)}}{SiO_2} + \underset{\text{(gas)}}{2Cl_2}$$

$$\underset{\text{(vapour)}}{GeCl_4} + \underset{\text{(gas)}}{O_2} \xrightarrow{\text{Heat}} \underset{\text{(solid)}}{GeO_2} + \underset{\text{(gas)}}{2Cl_2}$$

The core material is deposited in the inner portion of the tube. The required compounds in the form of vapours of halides are passed through one end of the silica tube. The exhaust gases are passed through the other end. The oxyhydrogen (O_2-H_2) flame is passed from the one end of the tube to the other end. Due to the chemical reaction a soot of $GeO_2 + SiO_2$ is formed and coated in the inner portion of the tube. Similarly, the O_2-H_2 flame is moved from one end to the other end until the required thickness of the core material is obtained. By varying the concentration of the reactants ($SiCl_4$, $GeCl_4$), the variation in the refractive index of each layer of the coating is achieved.

After getting the required thickness the silica material is collapsed into a rod using a flame at a temperature of around 1700°C to 1900°C. This collapsed rod is said to be a preform. The center portion of the preform contains the core material and the outer silica tube is used as cladding material. This method is used to prepare single mode and multimode step index fibres and multimode graded index fibre (Figure 9.13).

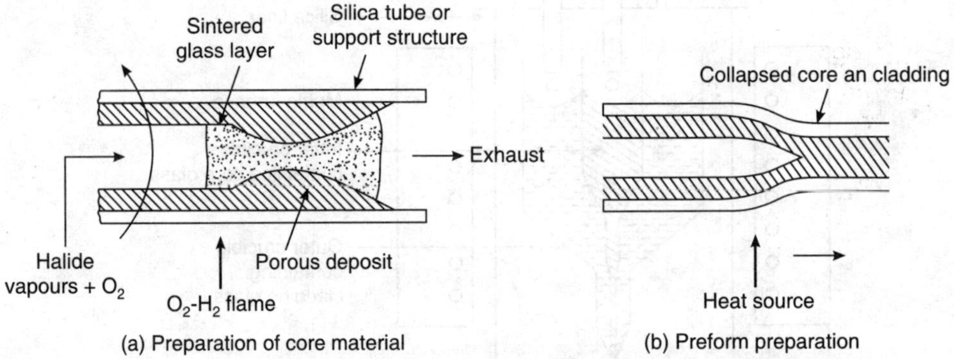

Figure 9.13 Modified CVD.

9.7.3 Fibre Drawing Process

The preform produced is mounted in a fibre drawing instrument. The preform is heated using a muffle furnace, so that it gets melted. The molten core and cladding materials are allowed to flow

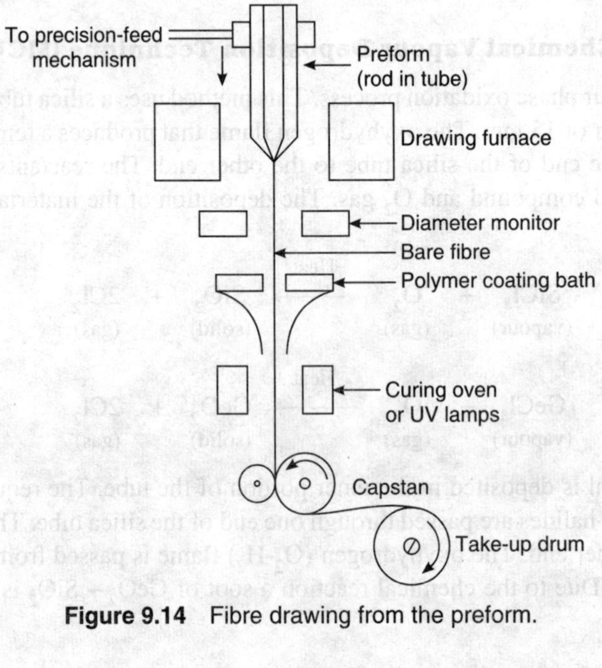

Figure 9.14 Fibre drawing from the preform.

through a thickness monitor and then through a bath containing a plastic material. A protective layer is coated to the core–cladding material while it is flowing through the plastic bath. Then the fibre is cured using UV lamp or thermally and then it is winded in a take-up drum as shown in Figure 9.14. Nearly 20–30 km long fibre can be prepared using a preform of length of 1 m in 2–3 hours.

9.8 LOSSES IN OPTICAL FIBRES

The loss of light, when it is propagating through the fibre, takes place due to the following factors, namely

(i) Attenuation
(ii) Dispersion
(iii) Bending losses
(iv) Fresnel's reflection losses and
(v) Mismatch in numerical aperture and core radius

9.8.1 Attenuation

Attenuation is the decrease of the magnitude of the power of the light beam. It is represented in dB/unit length. The attenuation of the optical signal with wavelength is shown in Figure 9.15.

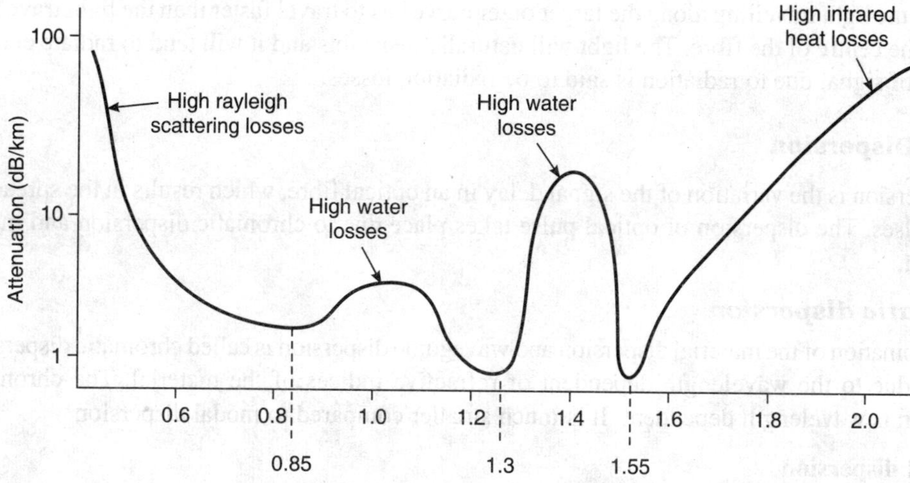

Figure 9.15 Typical attenuation versus wavelength curve of glass fibre.

Figure 9.15 shows that the attenuation is low at 850 nm, 1300 nm and 1550 nm. These three wavelengths are said to be optical windows. The loss of signal at 850 nm is higher (around 2 to 3 dB km^{-1}) than losses at 1300 nm and 1550 nm. Therefore, devices have been developed to operate and exploit the low attenuation of 1300 nm and 1550 nm wavelengths. The attenuation in glass fibre may be due to absorption losses, scattering losses, and radiation losses.

Absorption losses

The absorption of light due to the impurities and the hydroxyl ion (OH^-) is said to be absorption losses. Preparing the optical fibres with 100% pure glass is very difficult. Even though steps were taken to prepare optical fibres with 100% pure glass, the optical fibres have some impurities such as copper, iron and nickel. These impurities and the hydroxyl ion (OH^-) absorb the light.

Scattering losses

Scattering of light takes place in the optical fibres for two reasons, viz. (i) Rayleigh scattering and (ii) imperfection scattering.

The scattering of light due to the variation of density is called Rayleigh scattering and is proportional to λ^{-4}. Glass is an amorphous material. In amorphous materials, there is irregular arrangement of atoms. The irregularity in the atomic arrangement at the core–cladding interface acts as an imperfection. The scattering of light from this imperfection is said to be imperfection scattering. Due to the imperfection scattering of lights, there is a loss of optical signal. This loss is said to be scattering loss.

Radiation losses

The radiation losses in optical fibres are caused by bending and microbending in fibres. Consider that a fibre has microbending. The light passing through the centre of the fibre in the microbending region travels a shorter distance than the light ray traveling along the larger outer curve of the fibre. So, the light travelling along the larger outer curve has to travel faster than the light travelling through the centre of the fibre. The light will naturally resist this and it will tend to radiate energy. The loss of signal due to radiation is said to be radiation losses.

9.8.2 Dispersion

The dispersion is the variation of the signal delay in an optical fibre, which results in the spreading of the pulses. The dispersion of optical pulse takes place due to chromatic dispersion and modal dispersion.

Chromatic dispersion

The combination of the material dispersion and waveguide dispersion is called chromatic dispersion. It arises due to the wavelength dependent of refractive indices of the material. The chromatic dispersion is wavelength dependent. It is much smaller compared to modal dispersion.

Material dispersion

The material dispersion is due to the wavelength dependence of the refractive indices of glass and consequently the group velocity of the optical signal. A LED emits a broad spectrum of light. Whenever a LED is used as a source, the broad spectrum contains a number of wavelengths. These different wavelengths of light are travelling through optical fibre at different speeds because the refractive indices of the glass vary with wavelengths. Because of the difference in speeds, the light reaches the other end of the fibre at different times. The difference in the group velocity produces pulse spreading. The spreading of light due to the wavelength dependence of refractive indices is known as material dispersion.

Waveguide dispersion

The waveguide dispersion is also wavelength dependent and it arises due to the small size of the core diameter. If a ray of light is launched into a single mode fibre, then some part of the light travels through the cladding, as the core thickness of a single mode fibre is low. The light passing through the cladding travels with a high speed because of the low refractive index of cladding, whereas the light travels through core with a less speed. This type of loss of signal is said to be waveguide dispersion. The waveguide dispersion is negligible in multimode fibres.

The variation of chromatic dispersion with wavelength is shown in Figure 9.16; the material dispersion has a positive slope of change whereas the waveguide dispersion has a negative slope of change. The resultant chromatic dispersion is represented in Figure 9.16 in dotted lines. The zero value of the resultant chromatic dispersion is known as zero-dispersion wavelength. At this wavelength, there is no dispersion loss. Since the chromatic dispersion is the time delay per unit length of the signal due to the change in refractive indices with wavelength ($dn/d\lambda$), it is represented in ps/nm/km.

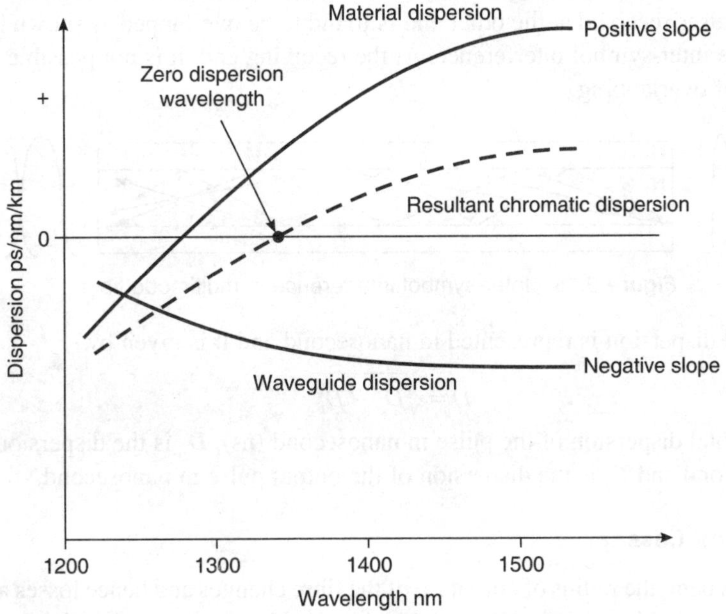

Figure 9.16 Wavelength dependence of chromatic dispersion.

Modal dispersion

The modal dispersion is generally produced in optical fibres those having high numerical apertures and a large core diameter. Generally, the multimode fibre has a large diameter. So, the modal dispersion is produced in multimode fibres. In multimode operation, the lowest mode travel with slowest group velocity and the highest mode travel with higher group velocity. It is due to the most of the higher mode signals are passed through cladding and the lower mode signals are passed through the core. These modes therefore take different times to travel through the fibre and reach the other end of the fibre at different times. This phenomenon is called as modal *dispersion* or *intermodal dispersion*.

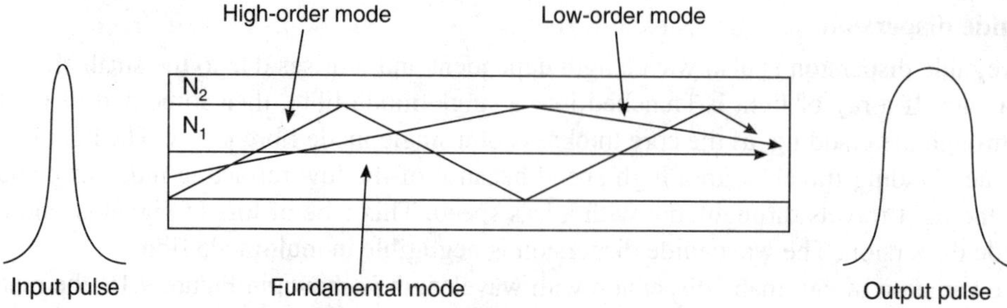

Figure 9.17 Modal dispersion in optical fibres.

The light ray that travels through the centre of the fibre is known as a fundamental mode. The light rays that travel shorter distance down the length of the fibre are called as lower order mode and the light rays that travels longer distance down the length of the fibre is called as higher order mode.

If two or more input pulses that are very close together are launched at one end of a multimode fibre, then the pulses received at the other end is found to be overlapped as shown in Figure 9.18. This is said to be inter-symbol interference. At the receiving end, it is not possible to identify the signal because of overlapping.

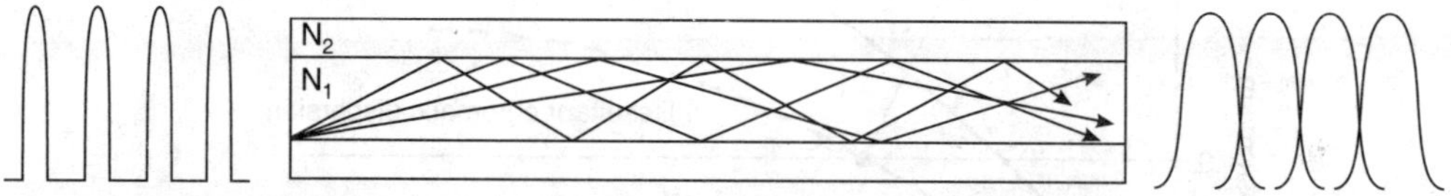

Figure 9.18 Inter-symbol interference in multimode fibre.

The modal dispersion is represented in nanosecond and it is given by

$$D = \sqrt{D_o^2 - D_i^2} \tag{9.21}$$

where D is the total dispersion of the pulse in nanosecond (ns), D_o is the dispersion of the output pulse in nanosecond and D_i is the dispersion of the output pulse in nanosecond.

9.8.3 Bending Losses

When the fibre is bent, the radius of curvature of the fibre changes and hence losses are introduced. These losses are said to be 'bending losses'. The bending losses are produced due to two types of bending. They are (i) microbending and (ii) macrobending.

Microbending

Microbending is a small scale bending or distortion in the fibre. The losses due to microbending in a fibre are known as microbending losses. The microbending is produced by imperfections in the cladding, ripples in the cladding/core interface, tiny cracks in the fibres and external force. Microbending is produced when a sharp object falls on the fibre cable. Due to the microbending the light ray may incident at an angle less than the critical angle and it will be refracted from cladding. The microbending is shown in Figure 9.19. The actual microbending may not be visible apparently and the losses may be temperature related, tensile related or crush related. The microbending loss can be reversible, once the cause is removed.

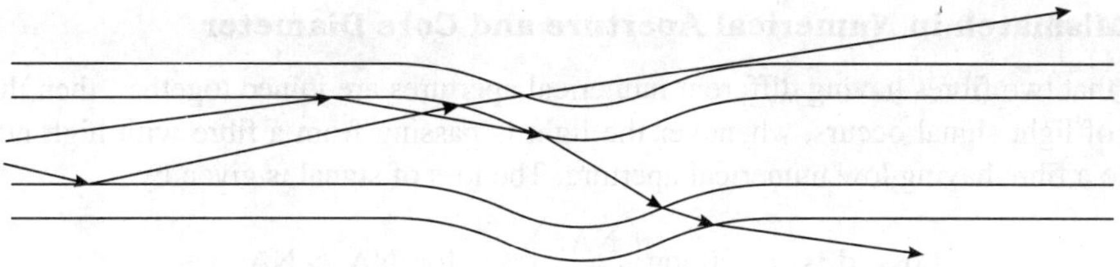

Figure 9.19 Microbending in optical fibres.

Macrobending

Macrobending is the large scale bending produced in the fibre cable. This loss is produced, if the cable is installed with a bend in it that has a radius less than the minimum bending radius. The light ray travelling through the core/cladding interface strike at an angle less than the critical angle and will be lost into the cladding. These losses are reversible once the bends are straightened. The macrobending is shown in Figure 9.20. This loss is wavelength dependent.

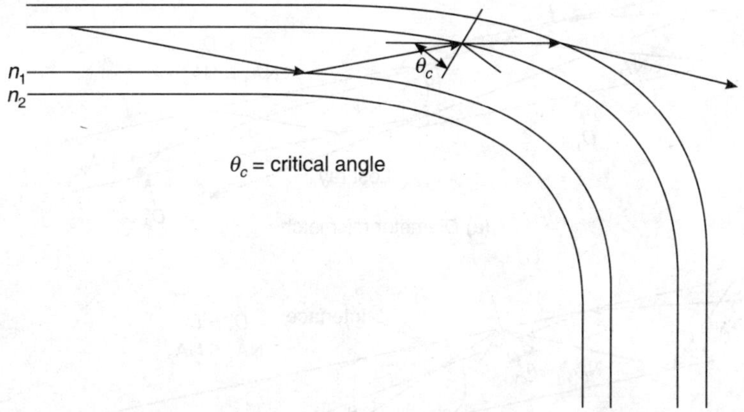

Figure 9.20 Macrobending in optical fibres.

9.8.4 Fresnel Reflection Loss

Consider that light is travelling from one medium to the other medium, if there is a change of refractive indices of the medium, some parts of the light gets reflected and the remaining parts of the light gets transmitted through the fibres. This type of reflection is called Fresnel reflection and the loss of light due to Fresnel reflection is called Fresnel reflection loss. Nearly 4% of the incident light gets reflected due to Fresnel reflection. The Fresnel's reflection occurs as shown in Figure 9.21, whenever the light is passing through the fibres where there is a joint in the fibre and if there is a mismatch in the refractive indices of the joined fibres.

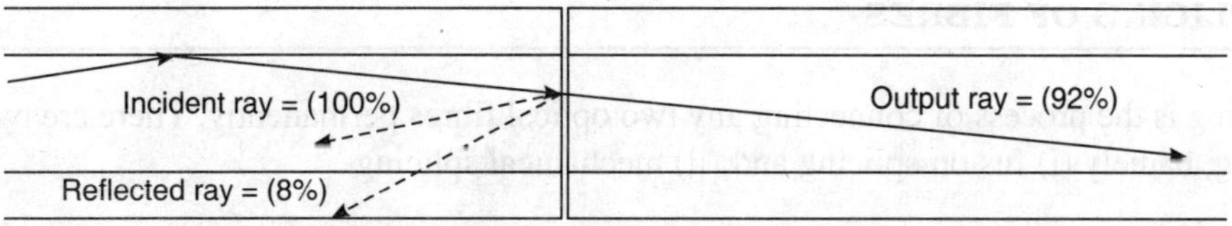

Figure 9.21 Fresnel reflection loss.

9.8.5 Mismatch in Numerical Aperture and Core Diameter

Consider that two fibres having different numerical apertures are joined together, then there will be a loss of light signal occurs, whenever the light is passing from a fibre with high numerical aperture to a fibre having low numerical aperture. The loss of signal is given by

$$\text{Loss (dBs)} = -20 \log \left(\frac{NA_1}{NA_2} \right) \qquad \text{for } NA_1 > NA_2 \qquad (9.22)$$

Similarly, if any two fibres having different diameters are joined together, the light travelling from a fibre with a larger diameter to a fibre with a smaller diameter, there is a loss of signal. The loss of signal is given by

$$\text{Loss (dBs)} = -20 \log \left(\frac{D_1}{D_2} \right) \qquad \text{for } D_1 > D_2 \qquad (9.23)$$

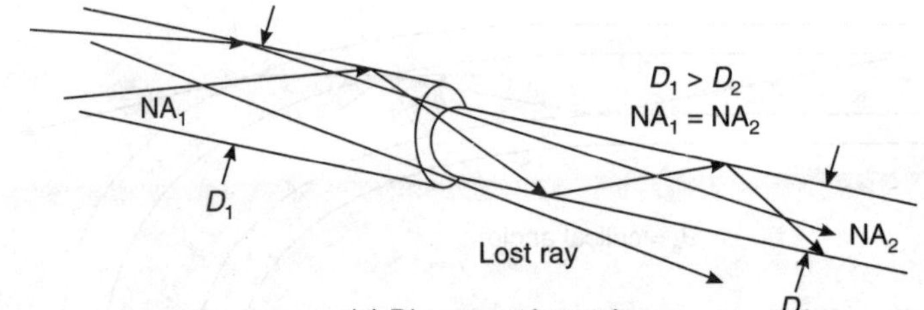

(a) Diameter mismatch

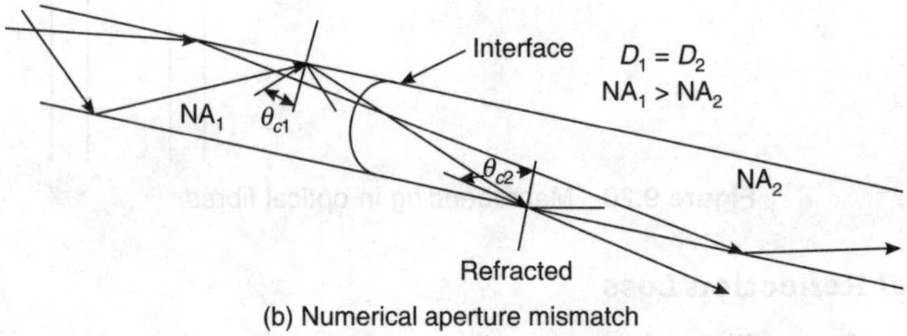

(b) Numerical aperture mismatch

Figure 9.22 Loss of optical signal due to mismatch in NA and diameters.

If light travels from a fibre having a low numerical aperture to the other fibre having high numerical aperture or from a smaller diameter fibre to a larger diameter fibre, there is no loss of signals. Since the numerical aperture is depending upon the diameter of the fibre and if there is a numerical aperture mismatch, then there is a diameter mismatch and vice versa (Figure 9.22).

9.9 SPLICING OF FIBRES

The splicing is the process of connecting any two optical fibres permanently. There are two types of splicing, namely (i) fusion splicing and (ii) mechanical splicing.

9.9.1 Fusion Splicing

Fusion splicing is a process of fusing the two ends of the optical fibres together using a fusion arc. The splicing process typically involves the following six steps.

(i) Stripping the fibre ends with a stripping tool
(ii) Preparing the end faces with a fibre optic cleaver
(iii) Inserting the fibre into the fusion splicer and aligning the fibres by fusion splicer
(iv) Fusing the fibres using an electric arc ignited between two electrodes
(v) Analyzing the finished splice
(vi) Protecting and storing the splice

The diagrammatic representation of a fusion splice is shown in Figure 9.23. It has (i) a fusion welder, (ii) fibre holders and positioners, (iii) alignment device and (iv) optical performance checking.

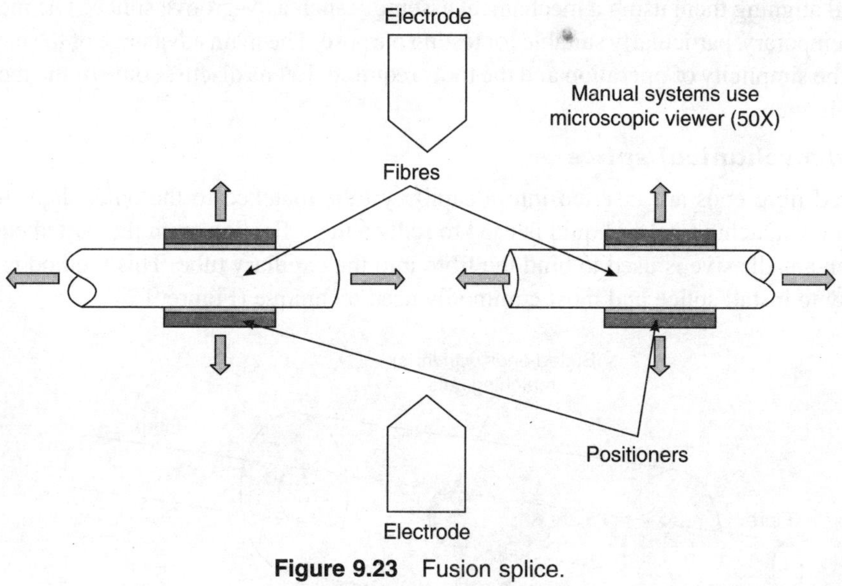

Figure 9.23 Fusion splice.

The ends of the fibres to be fused together are initially prepared for fusion process. The preparation of fibre involves the process such as precision cleaving, removal of a small portion of the cable sheath or buffer tube and thoroughly cleaning the exposed fibre. After preparing the fibres, they are separated at an appropriate distance and a moderate arc is used for about one second to clean and round the edges. The two ends of the fibres are kept in the fibre holders and they are positioned properly using a computer-controlled positioning system. Manual alignment system is also used, which involves a microscope or a video camera to magnify the fibre ends that enables the operator to align them properly. Then the two ends are brought close together and fusing arc is used to melt the ends. The surface tension of the molten glass tends to align the fibres. A protective plastic coating such as epoxy is coated for atmospheric protection and then a mechanical protection such as a heat-shrink sleeve or mechanical clip is fitted.

Advantages

(i) It produces low-loss bond between the fibres.
(ii) The fibre joints are free from air gaps and inclusion.
(iii) It is one of a permanent method of joining the fibres.
(iv) The fusion splicing is more reliable.
(v) The joined fibres have a tensile strength comparable to that of the original fibres.

Drawback

Fusion splicing equipment is costly.

9.9.2 Mechanical Splicing

It is a process of joining the ends of two different fibres by a mechanical method. Before joining the two pieces of fibres, they are cleaned and stripped. The cleaned fibre ends are carefully butt together and aligning them using a mechanical assembly such as V-groove splice. The mechanical splicing is temporary, particularly suitable for testing purpose. The main advantage of the mechanical splicing is the simplicity of operation and the tools required. Let us discuss some of the mechanical splicing technique.

Capillary mechanical splice

The prepared fibre ends are inserted into a capillary tube matched to the outer diameter of the fibre. An index matching gel or liquid is used to reduce the reflection from the butted ends of the fibres. Often an adhesive is used to bind the fibre into the capillary tube. This method is a cheap, simple, easy to install splice and most commonly used technique (Figure 9.24).

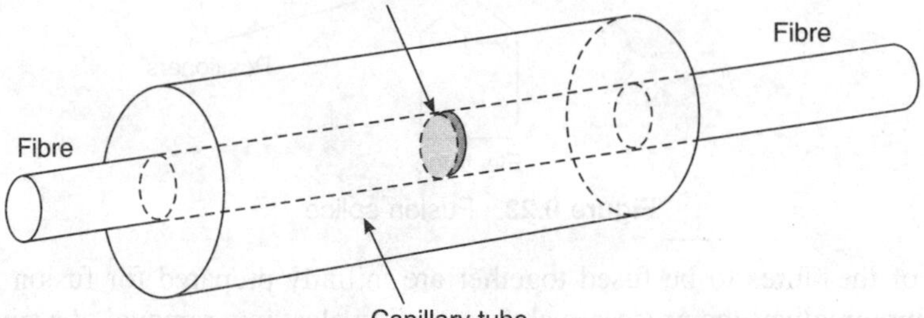

Figure 9.24 Capillary mechanical splice.

V-groove splice

A V-groove splice is shown in Figure 9.25. Some V-groove splices use self-aligning springs and lever-activated rollers for dual axis control. The two ends of the fibres are confined between two self-aligning V-groove plates. A rapid curing index matching glue is applied to the ends of the fibres for a permanent connection. The splice connector is then crimped on to the two buffer tubes. The splice loss of less than 0.1 dB is possible, if correct adhesive is used.

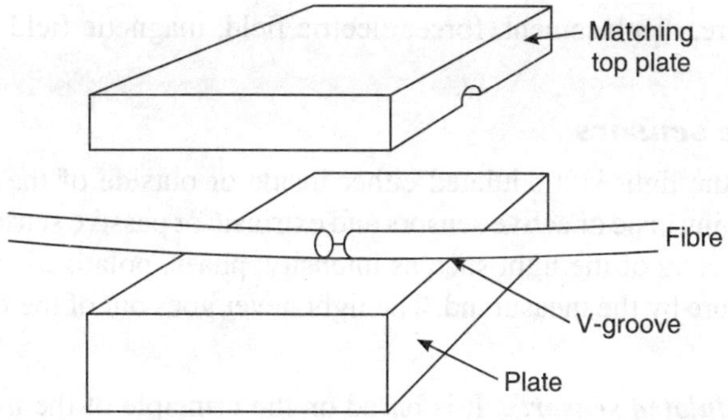

Figure 9.25 V-groove splices.

Ultraviolet splice

The ultraviolet splice is one of the mechanical splices used for low-loss splice. The ultraviolet elastomeric splicing process is shown in Figure 9.26. The ends of the fibres to be joined are prepared initially for splicing. The preparation involves precision cleaving, removal of a small portion of the cable sheath or buffer tube and thoroughly cleaning the exposed fibre. One or two drops of the ultraviolet cured adhesive are applied into the ends of the fibre before introducing into the splice. The two ends of the fibres are inserted into splice until they butt against each other and then cured by the application of ultraviolet light source. The splice loss less than 0.2 dB can be produced using this method.

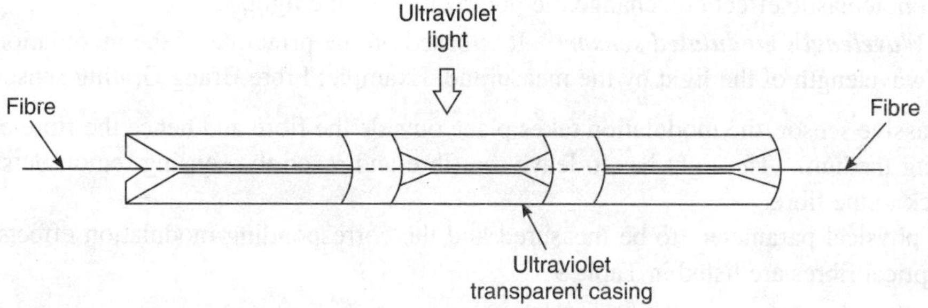

Figure 9.26 Mechanical field splicing kit.

Advantage

Since these joints are purely mechanical and flames are not used, they are useful in hazardous areas.

Drawbacks

(i) A high value of splice loss may be introduced due to an imperfect butt joint.
(ii) These joints are suitable for short-term, short-distance or temporary purposes.

9.10 FIBRE OPTIC SENSOR

A sensor is a device that converts one form of a physical quantity into another different measurable parameter so that it can be measured more accurately and conveniently. The fibre optic sensors are used to measure pressure, displacement, force, electric field, magnetic field, acceleration, liquid level, flow rate etc.

Active and passive sensors

In fibre optic sensors, the light is modulated either inside or outside of the fibre and hence the fibres are classified into intrinsic or active sensors and extrinsic or passive sensors. In active sensors any one physical parameters of the light such as intensity, phase, polarization, wavelength etc. is modulated within the fibre by the measurand. The light never goes out of the fibre. Different types of intrinsic sensors are,

(i) ***Intensity modulated sensor:*** It is based on the principle of the measurand modulates the intensity of the transmitted light through the fibre and these variations in intensity is measured using a detector situated at the output end of the fibre. Example: microbending sensor, the reflection type sensor like Fotonic sensor etc.

(ii) ***Phase modulated sensor:*** It is based on the principle of the phase change between the signal and reference beam. Due to the phase change an interference pattern is produced. These sensors have very high sensitivity. Example: The fibre optic sensors using Michelson interferometer, Mach-Zehnder interferometer, Fabry-Perot interferometer etc.

(iii) ***Polarization modulated sensor:*** It is based on the principle of the modulation of the polarization of light by many external variables and hence polarisation modulated fibre optic sensors can be used for the measurement of a range of parameters. The phenomena like optical activity, Faraday rotation, electrogyration, electro-optic effect, Kerr effect, photoelastic effect etc. change the polarization of the light.

(iv) ***Wavelength modulated sensor:*** It is based on the principle of the modulation of the wavelength of the light by the measurand. Example: Fibre Bragg Grating sensor.

In passive sensor, the modulation takes place outside the fibre and hence the fibre acts as a transmitting medium. The light has to leave the fibre and reach the sensing region outside and comes back to the fibre.

The physical parameters to be measured and the corresponding modulation effects taking place in optical fibres are listed in Table 9.1.

Table 9.1 Physical parameters measured using fibre optic sensors and their modulation effects

Physical parameters to be measured	*Modulation effects in optical fibres*
Temperature	Thermoluminescence
Pressure	Piezoelectric effect
Electric field	Electro-optic effect
Electric current	Electroluminescence
Magnetic field	Magneto-optic effect
Mechanical force	Stress birefringence
Density	Triboluminescence
Nuclear radiation	Radiation-induced luminescence

Let us discuss some of the fibre optic displacement and temperature sensors.

9.10.1 Fotonic Sensor

Principle

The fotonic sensor is based on the principle of intensity modulation of transmitted beam. The light is passed through one fibre and it is made to incident on the moving object. The light reflected from the moving object is transmitted by another fibre and it is sensed by the detector. The intensity of the light detected is directly proportional to the displacement of the object.

Construction

The fotonic sensor consists of a bundle of transmitting fibres coupled with laser source and a bundle of fibres coupled with the detector. The diagrammatic representation of this sensor is shown in Figure 9.27. The sensitivity of the sensor can be increased by varying the axis of the receiving and transmitting fibres with respect to the moving object.

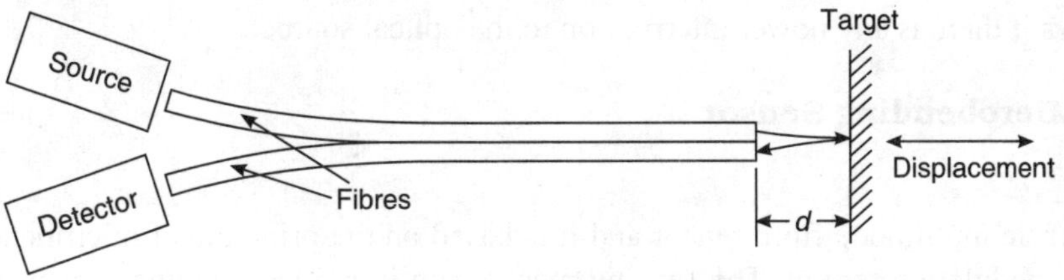

Figure 9.27 Fotonic sensor for displacement measurement.

Working

Light is transmitted through one fibre and it is made to fall on the target. The light reflected from the target is collected by the other fibre. The intensity of the light collected is a function of the distance between the fibre ends and the target. Therefore, the displacement of target is recorded in the optical detector. The sensitivity of this sensor is increased by placing the axes of these two fibres (one used for sending the signal and another used for receiving the signal) at an angle to one another with respect to the target. The main drawback of this intensity modulation sensor is the instability of the detector and the fibre cable, and inaccuracy occurs due to source.

9.10.2 Moire' Fringe Modulation Sensor

It is a multimode passive fibre sensor. The Moire' fringe modulation sensor consists of two gratings, in which one is fixed, whereas the other one is movable (Figure 9.28). The opaque lined grating produces dark Moire' fringes. The movable grating is moved transversely with respect to the other one. The transverse movement of one grating with respect to the other causes the fringes to move up or down. The displacement of the grating is measured by counting the number of fringes displaced.

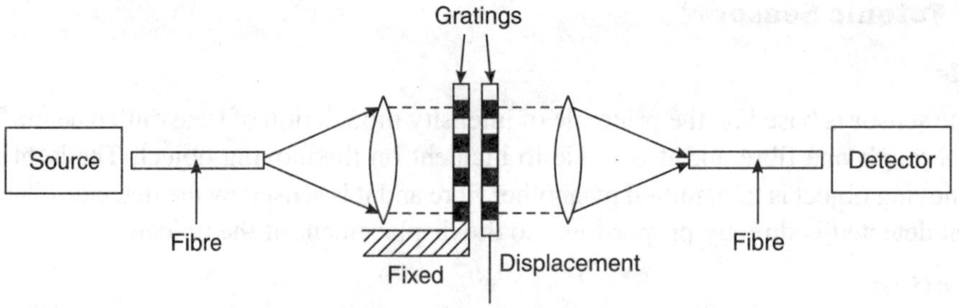

Figure 9.28 Moire' modulation sensor.

The fringe counting is independent of the instabilities produced in the intensity modulation sensors. However, the mechanical vibrations severely affect the measurement accuracy and it is difficult to eradicate. Another drawback of this sensor is the loss of count of interference fringes which occurs if there is any power interruption to the optical source.

9.10.3 Microbending Sensor

Principle

It is an intrinsic multimode fibre sensor and it is based on the principle of microbending of the fibre in the modulation region. The sensing mechanism in a microbending sensor is shown in Figure 9.29.

The deformation is produced in the fibre using a deformer on a small scale. This deformation causes the light to propagate through the cladding region where they are lost through radiation into the surroundings. If the spatial wavelength L of the fibre is correctly chosen, the radiation loss is very high and this offers very high sensitivity. By measuring the fluctuations in intensity of light emerging from core or cladding, one can measure the small deformation. If the deformation is produced by the measurement such as pressure, sound vibration, etc., one can measure them by measuring the fluctuations in the intensity of light emerging from the core or cladding. Thus monitoring of fibre core or cladding allows the detection of measurand. It is used to measure both pressure and displacement.

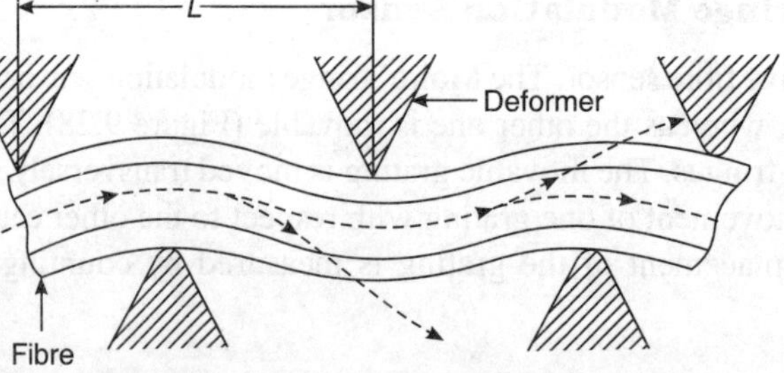

Figure 9.29 Fibre microbending sensor.

Theory

Consider that a periodic microbend is introduced as shown in Figure 9.30. The microbending loss depends upon numerical aperture, core size, core-to-cladding ratio and the spatial wavelength, L.

The spatial wavelength that corresponds to high microbending loss when all modes are coupled together is given by

$$L = \frac{2\pi a}{(2\Delta)^{1/2}} \tag{9.24}$$

where a is the fibre core radius and $\Delta = \frac{n_1^2 - n_2^2}{2n_1^2}$.

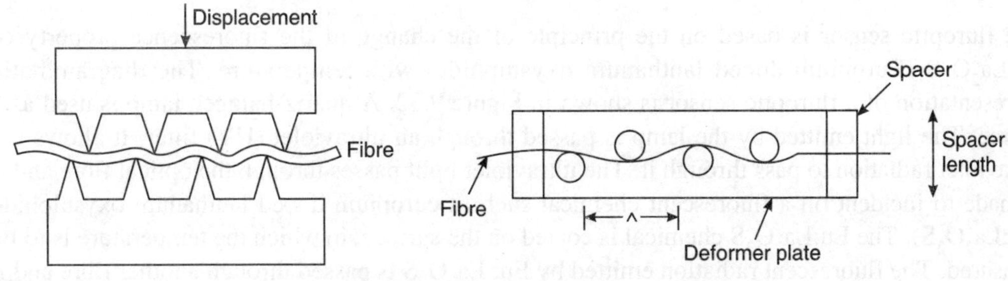

Figure 9.30 Microbending fibre sensor.

Here n_1 and n_2 are the refractive indices of the core and cladding respectively. Substituting $n_1 = 1.53$, $n_2 = 1.50$ and $a = 25$ μm, we get

$$L = \frac{2\pi a}{(2\Delta)^{1/2}} = \frac{2\pi(25\times10^{-6})}{2\times\left(\frac{1.53^2 - 1.5^2}{2\times1.5^2}\right)^{1/2}} = 0.7815 \text{ mm}$$

This calculation shows that the microbending loss is very high when the spatial wavelength, $L = 0.78$ mm.

Experimental measurement

The experimental arrangement used to measure displacement using microbend fibre sensor is shown in Figure 9.31. The laser beam produced by the laser source is made parallel by the lens L_1

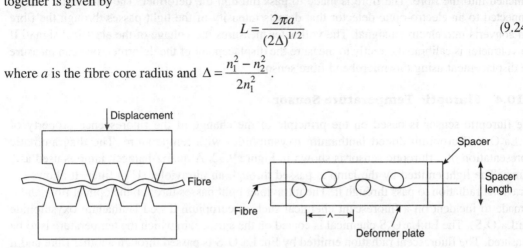

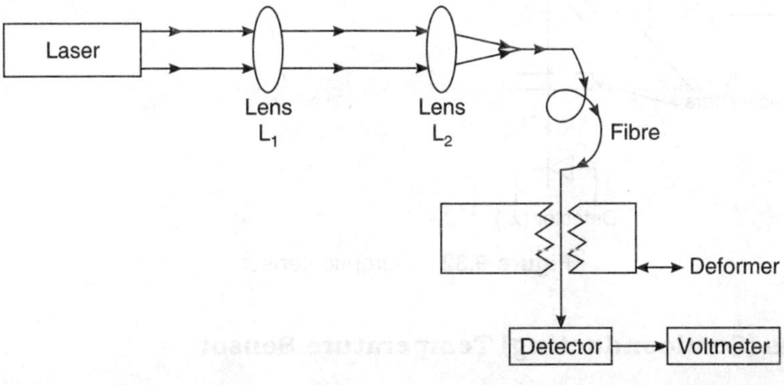

Figure 9.31 Experimental arrangement—microbend displacement/pressure sensor.

and the parallel beam is converged by the lens L_2. The laser source emerging from the lens L_2 is launched into the fibre. The fibre is made to pass through the deformer. The end of the fibre is connected to an electro-optic detector that detects intensity of the light passes through the fibre and converts into electrical signal. The voltmeter measures the voltage of the electrical signal. If the voltmeter is calibrated directly to measure the displacement of the deformer, one can measure the displacement using this microbend fibre sensor.

9.10.4 Fluroptic Temperature Sensor

The fluroptic sensor is based on the principle of the change of the fluorescence property of Eu:La_2O_2S (Europium doped lanthanum oxysulphide) with temperature. The diagrammatic representation of a fluroptic sensor is shown in Figure 9.32. A quartz-halogen lamp is used as a source. The light emitted by the lamp is passed through an ultraviolet (UV) filter. It allows the ultraviolet radiation to pass through it. The ultraviolet light passes through the optical fibre and it is made to incident on a fluorescent chemical such as europium doped lanthanum oxysulphide (Eu:La_2O_2S). The Eu:La_2O_2S chemical is coated on the surface, in which the temperature is to be measured. The fluorescent radiation emitted by Eu: La_2O_2S is passed through another fibre and it is made to incident on a beam splitter. The beam splitter splits the light into two sources. The intensity of these two splitted lights is measured by two different detectors. The ratio between these two intensities is used to find the temperature. A look-up table contains the ratio between these two intensities and the corresponding temperature. From the look-up table the temperature of the surface is determined.

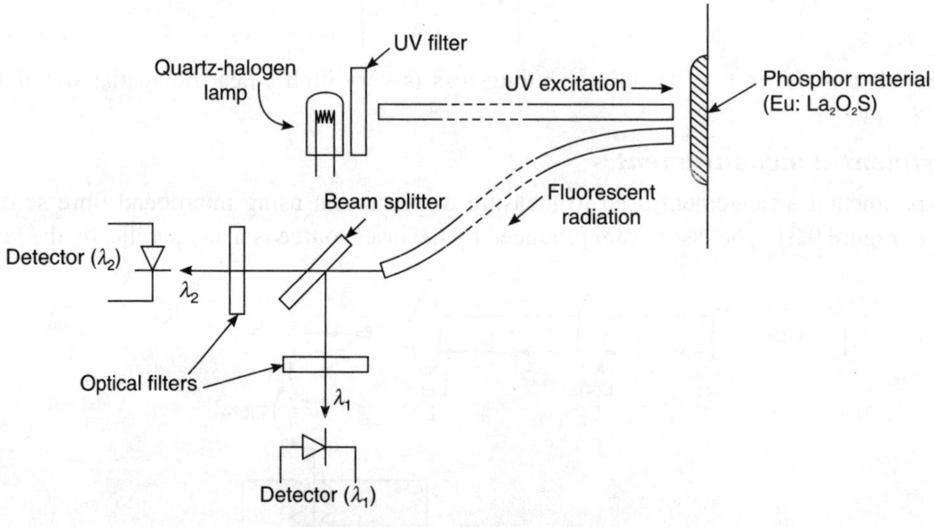

Figure 9.32 Fluroptic sensor.

9.10.5 GaAs (Semiconducting) Temperature Sensor

GaAs temperature sensor shown in Figure 9.33 is based on the principle of the change of the band gap of GaAs with temperature. The band gap energy of GaAs is 1.44 eV. The light having wavelength corresponding to the band gap energy is passed through one fibre. The light is passed through the GaAs semiconducting material that is kept at some higher temperature. The light emerging from GaAs is collected by another fibre and it is detected. The intensity of the detected light corresponds to the temperature of GaAs.

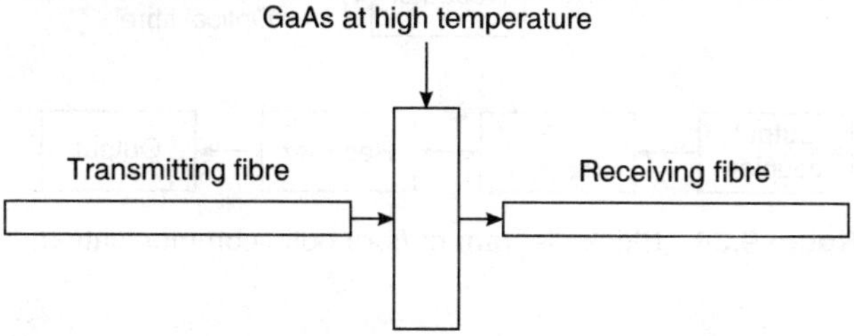

Figure 9.33 Semiconductor temperature sensor.

9.11 FIBRE OPTIC COMMUNICATION

One of the main advantages of optical fibre is communication. The generalized configuration of the fibre optic communication is shown in Figure 9.34.

The input signal may be sound, video signal or data. The non-electrical input is converted into an electrical signal using an input transducer. For example, a microphone is used to convert sound into electrical signal; a video camera is used to convert the video signal into electrical signal. The modulator is used to modulate the input signal by employing either analog modulation or digital modulation. If the information is in analog form it is converted into a digital form using a digital-to-analog converter. The optoelectronic sources (OE sources) are used to produce the electromagnetic radiations. Generally, LED (Light Emitting Diode) or ILD (Injection Laser Diode) is used as an optoelectronic source. The input channel coupler collects the light signal from the optoelectronic sources and sends it efficiently to the fibre cable. The signal is transmitted through the optical fibres. The function of the optical fibre is to transmit the signal from optoelectronic source to optoelectronic detector. Due to the attenuation and dispersion of the signal, it becomes weak after travelling a certain distance. Before this happens, the signals should be amplified and regenerated. The repeaters are used to regenerate or amplify the signal at an appropriate point along the length of the fibre. The output channel coupler is used to couple the optical signal from the optical fibre to the optoelectronic detector. The optoelectronic detector detects the optical signal and it converts the optical signal into the corresponding electrical signal. Generally, *p-i-n* diode or an avalanche photodiode is used as an optoelectronic detector. The electrical signal received from the detector is filtered and amplified by the receiver circuit. If the original signal is in digital form, it may be converted into an analog signal using digital-to-analog converter. Suitable output transducers are used to receive the output. If the input signal is a sound signal, a head phone/ear phone is used to convert the electrical signal into sound signal. If the input signal is a video signal, a television or any other device can be used to convert the electrical signal into a video signal.

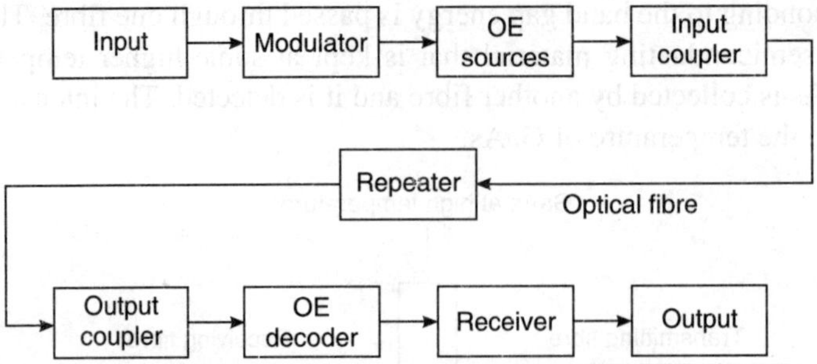

Figure 9.34 Block diagram of fibre optic communication.

9.11.1 Optical Sources

In optical communications, Light Emitting Diode (LED) or Laser Diode (LD) is used as optical sources. The optical source used in the fibre optic communication should have the following properties.

1. It should be able to couple the single mode fibre core of thickness 8.5 μm.
2. The signal should be easily modulated.
3. The optical source should be low cost, light weight, small size and high reliability.
4. It should provide high optical output.

Only the light emitting diodes and laser diodes fulfil many of these requirements.

Light emitting diode (LED)

A light emitting diode is a *p-n* junction device. The *p-n* junction is forward biased. The holes are added to the *p* region and the electrons are added to the *n* regions of the diode. These electrons and holes move towards the junction. During their movement towards the junction, they combine together and release the recombination energy in the form of light. Figure 9.35 illustrates this process.

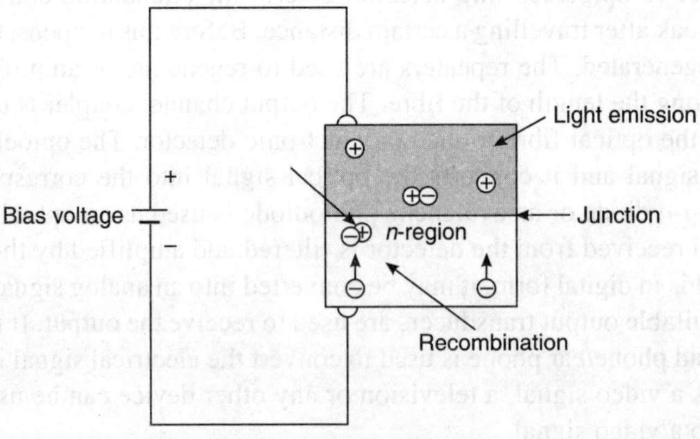

Figure 9.35 Basic LED structure.

The wavelength of the light emitted is given by

$$\lambda = \frac{hc}{E} \tag{9.25}$$

$$\lambda(nm) = \frac{1240}{E(\text{eV})} \tag{9.26}$$

where E is the band gap energy in eV and λ (in nanometer) is the wavelength of the light used.

LED materials

In an optical fibre the transmission loss is low at 850 nm, 1300 nm and 1550 nm. Therefore, the LED materials are selected such that they can emit these wavelengths. The GaAsP is used with plastic fibres at a wavelength of 660 nm. The optical fibres that operate at 800 to 900 nm are excited by GaAlAs ($\lambda_g \approx$ 800 to 900 nm) and GaAs ($\lambda_g \approx$ 900 nm). The InGaAsP with various compositions is used to emit wavelength from 1300 to 1550 nm.

LED geometry

The light emitted by the LED should be effectively coupled into the fibre core. LED emits lights in all directions and it is difficult to couple the light into the fibre because the fibre core is very thin. Two different structures of LED are used for this purpose. One is Burrus diode and another one is edge emitting diode.

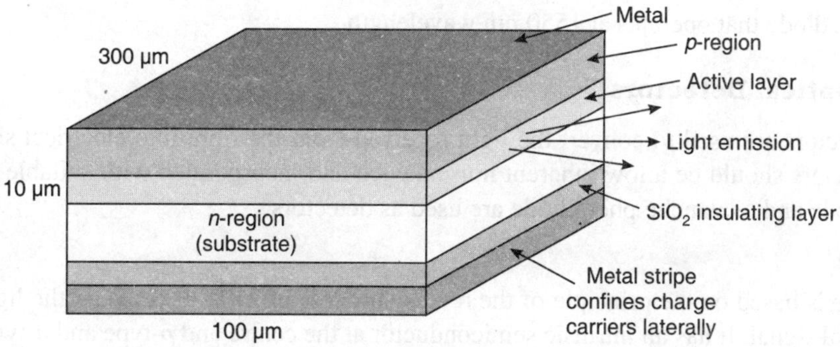

Figure 9.36 Edge emitting LED.

The Burrus diode has a small hole at the substrate and hence the light emission is confined. The fibres are directly inserted into these holes in the top of the device so as to collect the light.

The edge emitting diode has a small narrow active region for the emission of lights. The width of the emitting region is 10 μm and the light emitting zone is a few μm. Therefore, the structure confines the light emission and hence the light is coupled into the optical fibre. The structure of edge emitting diode is shown in Figure 9.36.

Laser diode

The laser diode is a *p-n* junction device. The p-type and n-type regions are heavily doped so as to produce population inversion. The dimension of the diode is 1 mm × 1 mm × 1 mm or even less than this value. The upper and bottom surface of the laser diode is used for ohmic contact and the back and front surfaces are used as cavity resonator and the remaining two surfaces are made as rough surfaces. The diode is forward biased with a voltage of $V = E_g/e$. The output wavelength from the diode is equal to, $\lambda_g = hc/E_g$. The basic structure of a laser diode is shown in Figure 9.37.

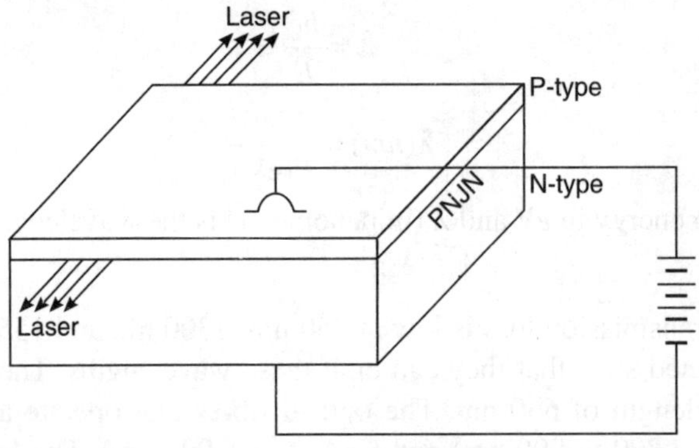

Figure 9.37 Laser diode.

Advances in laser technology

The older type of laser diodes is called edge emitting diode. It emits the light parallel to the boundaries. The latest technology uses the vertical cavity surface emitting laser diode (VCSEL, pronounced as vixel). VCSEL emits the laser beam perpendicular to the boundaries between layers of semiconductors. VCSEL is operating at 850 nm and 1300 nm wavelength range. It requires less current and emits narrow and more circular beam. The next challenging work is to produce low cost VCSEL diode that operates at 1550 nm wavelength.

9.11.2 Optical Detectors

Optical detectors are used to convert the light received from the fibre into electrical signal. The optical detectors should be a low inherent noise device and incorporated with suitable amplifier. The PIN diode and avalanche photodiode are used as detectors.

PIN diode

A PIN diode is based on the principle of the reverse process of LED. It converts the light energy into electrical signal. It has an intrinsic semiconductor at the centre and p-type and n-type regions at the end as shown in Figure 9.38. It is reverse biased (5–20 V). The reverse biasing is used to attract the charge carriers from the intrinsic regions.

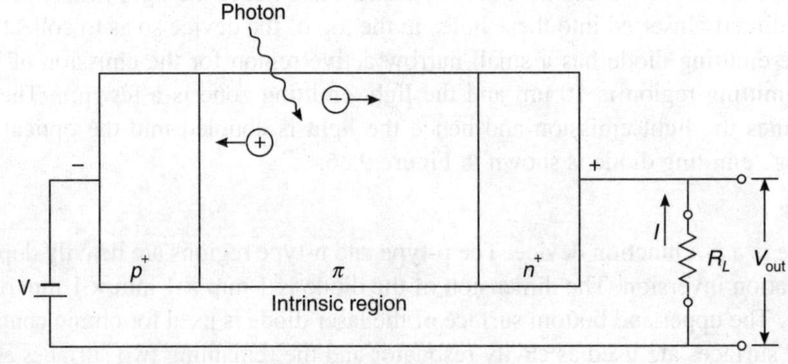

Figure 9.38 PIN diode.

When light is incident on the PIN diode, the intrinsic region receives more amount of light because of its large size. The photons incident on the intrinsic region produces electron-hole pair. The electron is raised from the valence band to the conduction band, leaving the hole. The electrons are attracted by the reverse biasing and hence move away from the junction. The movement of electrons in the conduction band creates the flow of charge and hence the light energy gets converted into electrical energy.

Avalanche photodiode

The avalanche photodiode is based on the principle of avalanche multiplication of the current. It consists of heavily doped p^+ and n^+ regions. The depletion region is lightly doped, almost intrinsic. The diode is reverse biased using 50–300 V. The light is made to incident on the depletion region. The incident light produces electron and hole pair. The electrons move towards the p region. Due to the strong reverse biasing, there is a depletion of charge carriers in the p region. The electrons in the p region undergo avalanche multiplication because of high reverse bias. The holes move towards the p^+ regions without producing further multiplication. The avalanche photodiode has better noise performance, because the carrier multiplication is limited to electrons only. The basic configuration of avalanche photodiode is shown in Figure 9.39.

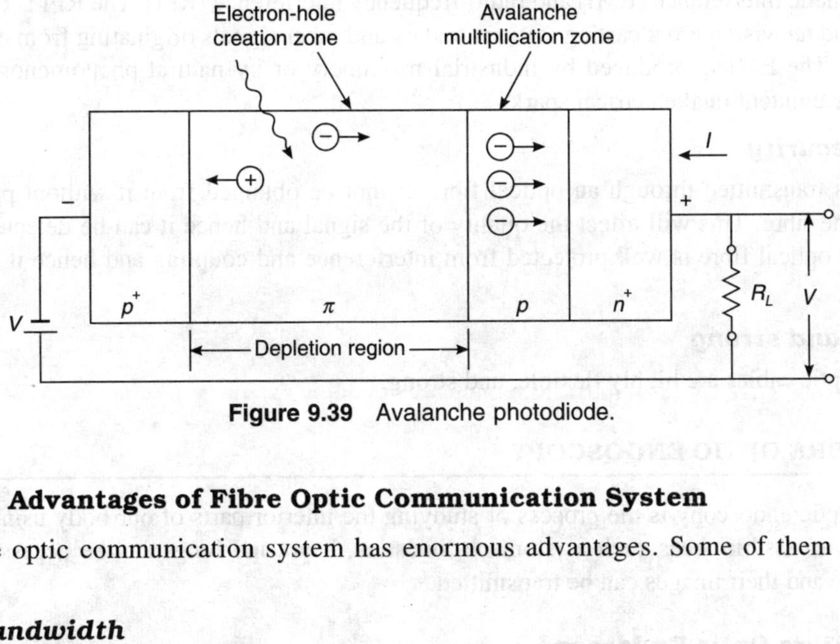

Figure 9.39 Avalanche photodiode.

9.11.3 Advantages of Fibre Optic Communication System

The fibre optic communication system has enormous advantages. Some of them are given below.

Wide bandwidth

The bandwidth of optical fibre communication is very large. For ordinary co-axial cable communication, the bandwidth is 500 MHz, whereas for the optical fibre communication has a bandwidth of 10^5 GHz. The information carrying capacity of optical fibres is also very large.

Electrical isolation

The optical fibres are prepared using glass and plastics. These two materials are very good electrical insulators. So, the fibre optic cables are electrically isolated.

Lack of cross-talk

Since the total internal reflection of light is used as the basic principle behind the fibre optic communication, there is no leakage of signal. Therefore, there is no cross-talk.

Low cost

The optical fibres are prepared using glass. The SiO_2 is the raw material used for the glass preparation. SiO_2 is easily available and hence the cost of fibre optic cables is very low.

Small size and light weight

The thickness of the optical fibres is nearly 250 μm (including core, cladding and sheath). Therefore, the size and hence the weight of the optical fibres is low.

Low transmission loss

The transmission loss of signal in optical fibre communication is typically around 0.2 dB/km and hence the transmission loss is low.

Immunity to electromagnetic interference

The optical fibres are prepared from glass and plastics. They are insulators and free from electromagnetic interference (EMI) and radio-frequency interference (RFI). The RFI is caused by the radio and television broadcasting stations, radars and other signals originating from electronic equipment. The EMI is produced by industrial machinery or by natural phenomenon such as lightning or unintentional electrical spark.

Signal security

The signals transmitted through an optical fibre cannot be obtained from it without physically intruding the fibre. This will affect the quality of the signal and hence it can be detected easily. Further the optical fibre is well protected from interference and coupling and hence it is highly secured.

Flexible and strong

The fibre optic cables are highly flexible, and strong.

9.12 FIBRE OPTIC ENDOSCOPY

The fibre optic endoscopy is the process of studying the interior parts of our body using optical fibres. The accessible areas such as stomach, intestine, heart and lungs can be accessed using optical fibre and their images can be transmitted.

9.12.1 Fibre Optic Endoscope

The fibre optic endoscope consists of a 10 mm diameter flexible shaft of length nearly 0.6 m to 1.8 m depending upon the applications as shown in Figure 9.40. The bottom portion of the shaft has a deflectable section of length nearly 5 cm to 8.5 cm, and a distal tip at the end. The distal tip consists of the following arrangements.

1. An irrigation channel to wash the objective lens by pumping water.
2. A coherent bundle of fibres to transmit the light from outside to the interior parts of the body.

3. Another bundle of fibres to transmit the reflected light from the interior parts of the body.
4. An operation channel to perform the task.

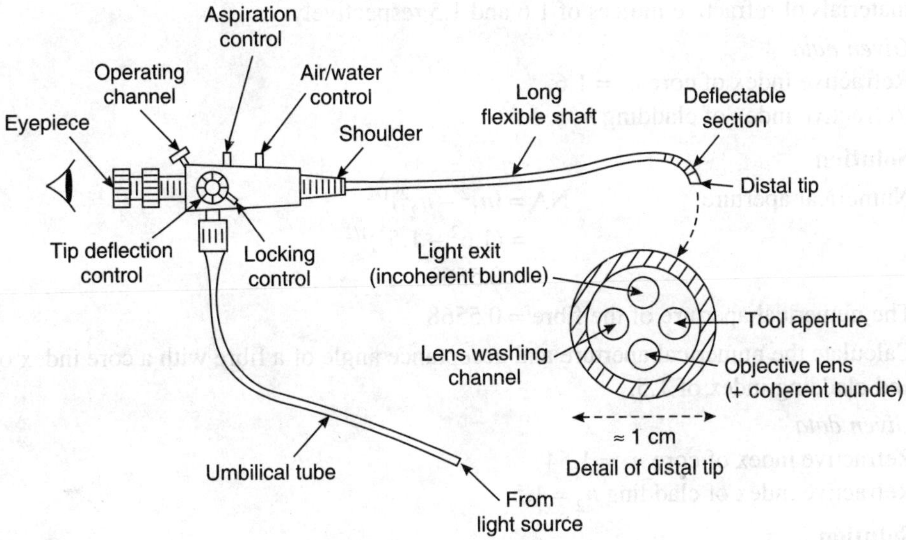

Figure 9.40 Fibre-optic endoscope.

The viewing end of the fibre optic endoscope consists of the following arrangements.

1. An eyepiece with camera attachment and focus control.
2. The distal tip control, capable of rotating it through 200°.
3. Operation channel valve that controls the entry of catheters, electrode, biopsy forceps and other flexible devices.
4. Valve control for application of water or air through irrigation channel.
5. A connection with umbilical tube for the control of light transmission from the source.

Working

The flexible shaft is inserted into our body and the distal tip is positioned correctly so as to study the interior organ of our body. If necessary, the objective lens is cleaned and the light is transmitted through one bundle of fibre. The light reflected from the interior parts of our body is transmitted by another bundle. The interior part of our body is viewed using the eyepiece or it may be photographed and then treatment is given suitably.

Advantages

1. Fibre optic endoscope is used for the examination of the gastrointestinal tract.
2. It is used for the study and treatment of ulcers, cancers, constriction, bleeding sites and so on.
3. The proportion of haemoglobin in blood is measured.

SOLVED PROBLEMS

9.1 Calculate the numerical aperture of an optical fibre whose core and cladding are made of materials of refractive indices of 1.6 and 1.5 respectively.

Given data

Refractive index of core $n_1 = 1.6$

Refractive index of cladding $n_2 = 1.5$

Solution

Numerical aperture

$$\text{NA} = (n_1^2 - n_2^2)^{1/2}$$
$$= (1.6^2 - 1.5^2)^{1/2}$$
$$= 0.5568$$

The numerical aperture of the fibre = 0.5568

9.2 Calculate the numerical aperture and acceptance angle of a fibre with a core index of 1.54 and cladding index of 1.5.

Given data

Refractive index of core $n_1 = 1.54$

Refractive index of cladding $n_2 = 1.5$

Solution

Numerical aperture

$$\text{NA} = (n_1^2 - n_2^2)^{1/2}$$
$$= (1.54^2 - 1.5^2)^{1/2}$$
$$= 0.3487$$

Acceptance angle

$$\alpha = \sin^{-1}\left[\frac{\sqrt{n_1^2 - n_2^2}}{n_0}\right]$$
$$= \sin^{-1}\left[\frac{\sqrt{1.54^2 - 1.5^2}}{1}\right]$$
$$= 20.408° = 20°24'30.78''$$

The numerical aperture of the fibre = 0.3487.

The acceptance angle = 20°24′30.78″

9.3 The refractive indices of the core and cladding are 1.6 and 1.49 respectively. Calculate the critical angle at the core–cladding interface.

Given data

Refractive index of core $n_1 = 1.6$

Refractive index of cladding $n_2 = 1.49$

Solution

Critical angle

$$\Phi_c = \sin^{-1}\left(\frac{n_2}{n_1}\right)$$

$$= \sin^{-1}\left(\frac{1.49}{1.6}\right)$$

$$= 68.63° = 68°37'49.85''$$

The critical angle of the fibre, $\Phi_c = 68°37'49.85''$

9.4 An optical fibre has a numerical aperture of 0.15 and a cladding refractive index of 1.55. Determine the acceptance angle of the fibre in water whose refractive index is 1.33.

Given data

Numerical aperture, NA = 0.15

Refractive index of cladding $n_2 = 1.55$

Refractive index of water, $n_0 = 1.33$

Solution

Numerical aperture

$$\text{NA} = (n_1^2 - n_2^2)^{1/2}$$

$$0.15 = (n_1^2 - 1.55^2)^{1/2}$$

$$0.15^2 = (n_1^2 - 1.55^2)$$

$$n_1 = (0.15^2 + 1.55^2)^{1/2} = 1.5572$$

Acceptance angle

$$\alpha = \sin^{-1}\left[\frac{\sqrt{n_1^2 - n_2^2}}{n_0}\right] = \sin^{-1}\left[\frac{\text{NA}}{n_0}\right]$$

$$= \sin^{-1}\left[\frac{0.15}{1.33}\right]$$

$$= 6.4757° = 6°28'32.55''$$

The refractive index of the core, $n_1 = 1.5572$

Acceptance angle = $6°28'32.55''$

9.5 A step index fibre has a numerical aperture of 0.26, core refractive index of 1.5 and a core diameter of 100 μm. Calculate the refractive index of the cladding.

Given data

Numerical aperture, NA = 0.26

Refractive index of core $n_1 = 1.5$

Core diameter $d = 100$ μm

Solution

Numerical aperture

$$\text{NA} = (n_1^2 - n_2^2)^{1/2}$$

$$0.26 = (1.5^2 - n_2^2)^{1/2}$$

$$n_2 = (1.5^2 - 0.26^2)^{1/2} = 1.477$$

The refractive index of the core = 1.477

9.6 Calculate the refractive indices of core and cladding material of the fibre from the following data: (i) numerical aperture is 0.26 and (ii) refractive index difference of the fibre (Δ) is 0.015.

Given data

Numerical aperture, NA = 0.26

Refractive index difference of the fibre Δ = 0.015

Solution

Numerical aperture

$$NA = (n_1^2 - n_2^2)^{1/2}$$

$$0.26 = (n_1^2 - n_2^2)^{1/2}$$

$$n_1^2 - n_2^2 = 0.0676$$

The refractive index difference of the fibre = 0.015

i.e.
$$\Delta = \frac{n_1^2 - n_2^2}{2n_1^2} = 0.015$$

Substituting the values of $n_1^2 - n_2^2$, we get

$$2n_1^2 = \frac{0.0676}{0.015}$$

$$n_1 = 1.5$$

Substituting the value of n_1 in the equation, $NA = (n_1^2 - n_2^2)^{1/2}$, we get $n_2 = 1.477$.

The refractive index of the core, $n_1 = 1.5$

The refractive index of the cladding, $n_2 = 1.477$

SHORT QUESTIONS

1. What is fibre optics?
2. What is the principle behind the fibre optics?
3. What is total internal reflection?
4. What is meant by critical angle?
5. Deduce an expression for critical angle.
6. Define acceptance angle.
7. What is numerical aperture?
8. What are the materials used for optical fibre preparation?
9. What is glass fibre?
10. What are plastic fibres?
11. What is step index fibre?
12. What is graded index fibre?
13. What is single mode fibre?
14. What is multimode fibre?

15. What do you mean by attenuation in optical fibres?
16. What is meant by modal dispersion?
17. What are the bending losses?
18. What is Fresnel's reflection loss?
19. Write about the loss of signal due to mismatch in core diameter.
20. Write about the loss of signals due to numerical aperture mismatch.
21. What do you mean by splicing of fibres?
22. What is fusion splicing?
23. Mention the advantages of fusion splicing.
24. What is mechanical splicing?
25. What are fibre optic sensors?
26. What is the principle of fluroptic sensor?
27. What is a fotonic sensor?
28. What is the principle of microbending sensor?
29. Write about GaAs temperature sensor.
30. Mention the advantages of fibre optic communications.
31. What is fibre optic endoscopy?
32. What are the advantages of fibre optic endoscopy?

DESCRIPTIVE TYPE QUESTIONS

1. What is acceptance angle? Derive a mathematical expression for acceptance angle.
2. What is numerical aperture? Derive an expression for the numerical aperture.
3. What are step index and graded index fibres? Explain the propagation of signals in step index and graded index fibres.
4. Describe with a suitable diagram about the double crucible method for the preparation of optical fibres.
5. Describe about the modified chemical vapour deposition and explain how optical fibres are prepared using this method.
6. Write an essay about the losses of signals in optical fibres.
7. Write an essay about the splicing of fibres.
8. What is a fibre optic sensor? Explain with neat diagram about the temperature and displacement sensor.
9. Describe the fibre optic communication system with suitable block diagram. What are the advantages of fibre optic communications?

PROBLEMS

1. Numerical aperture of optical fibre is 0.5, core refractive index is 1.54. Find refractive index of cladding.
2. An optical fibre has numerical aperture of 0.2 and a cladding refractive index of 1.59. Determine the acceptance angle for the fibre in water which has a refractive index of 1.33.
3. Calculate the numerical aperture, acceptance angle, and the critical angle of a fibre having a core refractive index of 1.50 and the cladding refractive index of 1.45.
4. The numerical aperture of an optical fibre is 0.6828 and its refractive index difference is 0.091. Determine the refractive indices of the core and cladding.
5. An optical fibre of core and cladding materials of refractive indices of 1.55 and 1.48 respectively is used in an experiment. Calculate the angle of cone ($2 \times \theta_a$) in which any light ray that enters should get propagated along the length of the fibre.
6. Calculate the light collecting capacity (numerical aperture) of the fibre whose core and cladding refractive indices are 1.55 and 1.5 respectively.

CHAPTER 10

CONDUCTING MATERIALS

10.1 INTRODUCTION

The conducting materials are those having a large number of conduction electrons. The electrons in the outermost orbit of an atom are loosely bounded. These loosely bounded electrons are easily removed from the parent atom with a small amount of energy. These electrons are called free electrons. The free electrons move freely here and there within the crystals and it is responsible for the most of the properties of metals. The properties of metals, such as electrical conduction, thermal conduction, etc., are due to these free electrons. These free electrons are called conduction electrons, since they are responsible for the conducting properties (electrical and thermal conduction) of metals.

This chapter deals with the electrical conduction, thermal conduction, density of states, number of electrons in a metal, etc.

10.2 CLASSICAL FREE ELECTRON THEORY OF METALS

This theory was proposed in 1900 by P. Drude and it was extended by H.A. Lorentz in 1909 and hence this theory is also known as **Drude–Lorentz theory**.

According to this theory, the free electrons are fully responsible for electrical conduction. If no electric field is applied, the free electrons move here and there and make collision with other free electrons or positive ion cores or wall of the container. The collision is elastic collision. The free electrons behave like gaseous molecules in a gaseous container. The gaseous molecules move here and there within the container and makes collision with other gaseous molecules and with the wall of the container. So, the free electrons are called **electron gas**. The velocity and the energy distribution of the free electron are governed by Maxwell–Boltzmann distribution function. The average velocity of the free electrons is zero, since they are not able to come out from the specimen.

Whenever an electric field is applied, the free electrons acquire some energy from the applied field and they are directed to move away from the field. The free electrons acquire some constant velocity due to the applied field and the velocity is known as **drift velocity**. The movement of the electrons in the absence of the electric field and during the application of the field is shown in Figure 10.1.

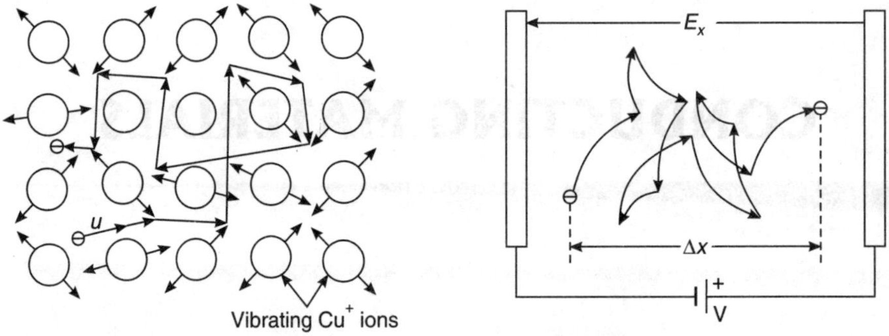

(a) In the absence of the electric field (b) In the presence of the electric field

Figure 10.1 Movement of the electrons.

10.2.1 Expression for the Electrical Conductivity of Metals

Electrical conductivity is defined as the flow of charge carriers in a material due to the applied electric field intensity. It is also defined as the reciprocal of the electrical resistivity. Its unit is $\Omega^{-1}\,m^{-1}$.

Consider that a specimen is subjected to an electric field of intensity E. Let m be the mass of the electron and e be the charge of the electron. The force acquired by the electron is $-eE$. The negative sign shows that the electron charge is negative. According to the Newton's law of motion, the force is given by

$$F = ma$$

Therefore, the equation of motion is written as

$$ma = -eE \tag{10.1}$$

Substituting $a = \dfrac{dv}{dt}$ in Eq. (10.1), we get

$$\frac{dv}{dt} = \frac{-eE}{m} \tag{10.2}$$

Integrating Eq. (10.2), we get

$$v = \frac{-eE}{m}t + C \tag{10.3}$$

where C is the constant of integration. The value of C can be evaluated by substituting the boundary condition, when $t = 0$, $<v> = 0$, we get

$$C = 0 \tag{10.4}$$

From Eq. (10.3) and Eq. (10.4), we can write

$$v = \frac{-eE}{m} t \tag{10.5}$$

The current density (J) is the current flowing through a unit area of cross section. It is given by

$$J = \frac{I}{A} \tag{10.6}$$

where I is the current flowing through the specimen and A is the area of cross section. The current density is also given by the relation

$$J = -nev_d \tag{10.7}$$

From Eq. (10.5) and Eq. (10.7), we get

$$J = \frac{ne^2 E}{m} t \tag{10.8}$$

From Ohm's law,

$$J = \sigma E \tag{10.9}$$

From Eq. (10.8) and Eq. (10.9), we get

$$\sigma = \frac{ne^2 t}{m} \tag{10.10}$$

From Eq. (10.10), we infer that the conductivity is directly proportional to the time t. It means that the conductivity increases, if we apply the electric field for a longer time. But in the actual measurement, the conductivity is found to be a constant. This discrepancy is due to the omission of collision of free electrons with other free electrons, positive ion core, and wall of the container. The force due to the collision is known as **frictional force**. The frictional force is given by $m <v>/\tau_r$, where τ_r is the relaxation time. The **relaxation time** is the average time taken by the electron between any two successive collisions in the presence of the electric field. Its value is in the order of 10^{-14} s. It is also known as mean free time. By introducing the frictional force, the equation of motion can be rewritten as

$$m\frac{dv}{dt} + \frac{m <v>}{\tau_r} = -eE \tag{10.11}$$

Dividing by m, we get

$$\frac{dv}{dt} + \frac{<v>}{\tau_r} = \frac{-eE}{m} \tag{10.12}$$

Multiplying by τ_r, we get

$$\tau_r \frac{dv}{dt} + <v> = \frac{-eE}{m}\tau_r \tag{10.13}$$

Rearranging Eq. (10.13), we get

$$\tau_r \frac{dv}{dt} = -\left[\frac{eE}{m}\tau_r + <v>\right] \tag{10.14}$$

i.e.

$$\frac{dv}{\left[\frac{eE}{m}\tau_r + \langle v \rangle\right]} = -\frac{dt}{\tau_r} \tag{10.15}$$

Integrating Eq. (10.15), we get

$$\ln\left[\frac{eE}{m}\tau_r + \langle v \rangle\right] = -\frac{t}{\tau_r} + \ln B \tag{10.16}$$

where ln B is the constant of integration. Eq. (10.16) can be written as

$$\left[\frac{eE}{m}\tau_r + \langle v \rangle\right] = Be^{-t/\tau_r} \tag{10.17}$$

Substituting the boundary condition, when $t = 0$, $\langle v \rangle = 0$, we get

$$B = \frac{eE}{m}\tau_r \tag{10.18}$$

From Eq. (10.17) and Eq. (10.18), the velocity of the electron is written as

$$\langle v \rangle = \frac{eE}{m}\tau_r\,[e^{-t/\tau_r} - 1] \tag{10.19}$$

Equation (10.19) gives the value of the velocity of the electron. The current density of the electron [from Eq. (10.19) and Eq. (10.7)] is written as

$$J = \frac{ne^2E}{m}\tau_r[1 - e^{-t/\tau_r}] \tag{10.20}$$

From Eq. (10.9) and Eq. (10.20), the conductivity can be written as

$$\sigma = \frac{ne^2}{m}\tau_r[1 - e^{-t/\tau_r}] \tag{10.21}$$

The conductivity increases exponentially only when $t < \tau_r$. If the field is applied continuously, the conductivity becomes constant, when $t > \tau_r$. This constant value of conductivity is known as **steady state conductivity**. The steady state conductivity is given by

$$\sigma = \frac{ne^2}{m}\tau_r \tag{10.22}$$

Similarly, from Eq. (10.19), the steady state velocity is given by

$$v_d = -\frac{eE}{m}\tau_r \tag{10.23}$$

Equation (10.23) is called the steady state velocity or **drift velocity** of the electrons. The negative sign shows that the electrons are moving in the opposite direction to that of the applied electric field.

(i) Resistivity (ρ)

The reciprocal of the electrical conductivity is known as electrical **resistivity**. It is represented by the letter ρ. Its unit is Ω m. From Eq. (10.22), ρ is given by

$$\rho = \frac{m}{ne^2\tau_r} \tag{10.24}$$

(ii) Mobility (μ)

The velocity of the free electrons per unit electric field intensity is known as **mobility**. It is represented by μ and it is given by

$$\mu = \frac{v}{E} = \frac{e\tau_r}{m} \tag{10.25}$$

Its unit is $m^2\ V^{-1}\ s^{-1}$.

(iii) Mean free path (λ)

The average distance travelled by an electron between any two successive collisions is known as **mean free path**. It is represented by the letter λ and it is given by

$$\lambda = v_d \tau_r \tag{10.26}$$

Its unit is m.

10.2.2 Thermal Conductivity of Metals

The term thermal conductivity represents the heat conducting property of the metal. It is defined as the ratio of the quantity of heat energy transferred through unit area of cross section in one second to the temperature gradient. It is given by

$$K = \frac{-Q}{\left(\frac{dT}{dx}\right)} \tag{10.27}$$

where the negative sign shows that the heat energy flows from the hot end to the cold end, Q is the quantity of heat energy flowing through unit area of cross section in one second and dT/dx is the temperature gradient. The unit for the thermal conductivity is $W\ m^{-1}\ K^{-1}$.

Generally, the thermal conductivity is produced due to the presence of the electron and the lattice vibration. The lattice vibration is called phonon. In the case of metals, since the electron concentration is very high, the thermal conductivity is only due to the presence of electron. For insulators, the thermal conductivity is only due to the lattice vibrations, since in insulators, all the states are completely occupied and hence no electron is available for electrical conduction.

Expression for the thermal conductivity of metals

Consider a metallic specimen AB of length 2λ, where λ is the mean free path of the electron as shown in Figure 10.2. Let the area of the cross-section of the metallic specimen AB is unity. Let the end 'A' is at a temperature T_1 and 'B' is at a temperature T_2. Let $T_1 > T_2$. Consider a plane C at a distance of λ from the end A. Let v_1 and v_2 be the velocities of the electron at the end A and B respectively.

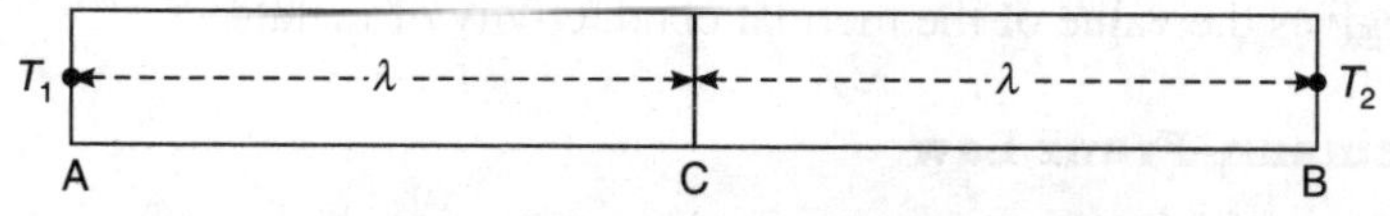

Figure 10.2 Thermal conductivity of metals.

The kinetic energy of the electron at A

$$\frac{1}{2}mv_1^2 = \frac{3}{2}kT_1 \tag{10.28}$$

The kinetic energy of the electron at B

$$\frac{1}{2}mv_2^2 = \frac{3}{2}kT_2 \tag{10.29}$$

The number of electrons crossing the plane C in one second per unit area of cross section

$$\frac{nv}{6} \tag{10.30}$$

The heat energy transferred from end A to B through unit area of cross section in one second

$$\frac{nv}{6} \times \frac{3}{2}kT_1 \tag{10.31}$$

The heat energy transferred from end B to A through unit area of cross section in one second

$$\frac{nv}{6} \times \frac{3}{2}kT_2 \tag{10.32}$$

Resultant energy transferred from one end to another end in one second through unit area of cross section, Q

$$\frac{nv}{6} \times \frac{3}{2}k(T_1 - T_2) \tag{10.33}$$

Temperature gradient

$$\frac{dT}{dx} = \frac{T_1 - T_2}{2\lambda} \tag{10.34}$$

The thermal conductivity is given by

$$K = \frac{\frac{nv}{6} \times \frac{3}{2}k(T_1 - T_2)}{\frac{(T_1 - T_2)}{2\lambda}} \tag{10.35}$$

$$K = \frac{1}{2}nvk\lambda \tag{10.36}$$

Substituting $\lambda = v\tau_r$ in Eq. (10.36), we get

$$K = \frac{1}{2}nv^2k\tau_r \tag{10.37}$$

Equation (10.37) gives the value of the thermal conductivity of metals.

10.2.3 Wiedemann–Franz Law

The **Wiedemann–Franz law** states that the ratio of the thermal conductivity to the electrical conductivity is directly proportional to the absolute temperature,

i.e.
$$\frac{K}{\sigma} \propto T$$
$$\frac{K}{\sigma T} = \text{constant} \tag{10.38}$$

The constant $\frac{K}{\sigma T}$ is known as **Lorentz number**. It is represented by the letter L,

i.e.
$$L = \frac{K}{\sigma T} \tag{10.39}$$

The experimental value of the Lorentz number is listed in Table 10.1 and its average value is equal to 2.3×10^{-8} W Ω K^{-2}.

Table 10.1 Electrical and thermal conductivities of selected metals measured at 20°C

Metal	$\sigma\,(10^7)\ \Omega^{-1}\ m^{-1}$	K W m^{-1} K^{-1}	$L(10^{-8})$ W Ω K^{-2}
Ag	6.15	423	2.45
Cu	5.82	387	2.37
Al	3.55	210	2.02
Na	2.10	135	2.18
Cd	1.30	102	2.64
Fe	1.00	67	2.31

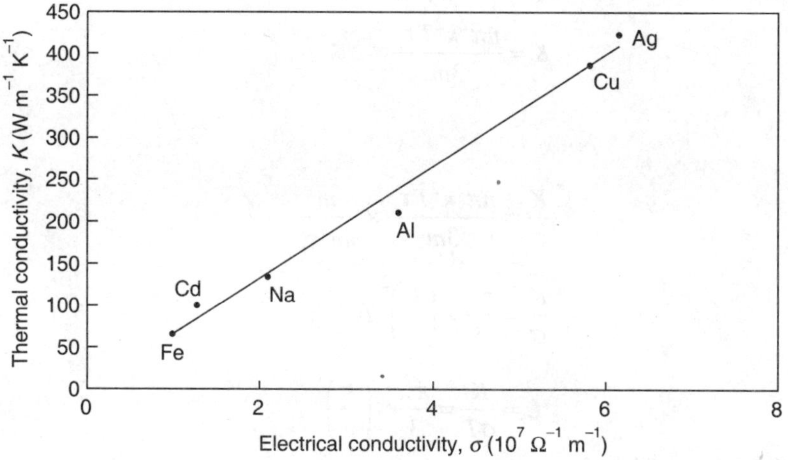

Figure 10.3 A plot between thermal conductivity and electrical conductivity.

If a graph is plotted for the data available in Table 10.1, a straight line is obtained as shown in Figure 10.3. The average value of $K/\sigma T$ obtained from Figure 10.3 is 2.28×10^{-8} W Ω K^{-2}.

Dividing the value of K by σ, we get

$$\frac{K}{\sigma} = \frac{\frac{1}{2} m v^2 k \tau_r}{\frac{n e^2 \tau_r}{m}} \tag{10.40}$$

$$\frac{K}{\sigma} = \frac{1}{2} mv^2 \frac{k}{e^2} \tag{10.41}$$

Substituting $\frac{1}{2} mv^2 = \frac{3}{2} kT$ in Eq. (10.41), we get

$$\frac{K}{\sigma} = \frac{3}{2} kT \frac{k}{e^2}$$

i.e.

$$\frac{K}{\sigma T} = \frac{3}{2} \frac{k^2}{e^2} \tag{10.42}$$

Substituting the values of k and e, we get

$$\frac{K}{\sigma T} = \frac{3}{2} \frac{(1.38 \times 10^{-23})^2}{(1.6 \times 10^{-19})^2}$$

$$\frac{K}{\sigma T} = 1.11 \times 10^{-8} \text{ W } \Omega \text{ K}^{-2} \tag{10.43}$$

This theoretical value is not in agreement with the experimental value of the Lorentz number. This shows that the classical free electron theory fails to give the correct mathematical expression for the thermal conductivity of metals. Therefore, the value of K is substituted from Quantum free electron theory. From Quantum free electron theory, the value of the thermal conductivity is given by

$$K = \frac{n\pi^2 k^2 T \tau_r}{3m} \tag{10.44}$$

The ratio $\frac{K}{\sigma}$ is

$$\frac{K}{\sigma} = \frac{n\pi^2 k^2 T \tau_r}{3m} \times \frac{m}{ne^2 \tau_r} \tag{10.45}$$

$$\frac{K}{\sigma} = \frac{\pi^2}{3} \times \left(\frac{k}{e}\right)^2 T$$

$$L = \frac{K}{\sigma T} = \frac{\pi^2}{3} \times \left(\frac{k}{e}\right)^2 \tag{10.46}$$

Substituting the values of k and e, we get

$$L = \frac{K}{\sigma T} = \frac{\pi^2}{3} \times \left(\frac{1.38 \times 10^{-23}}{1.6 \times 10^{-19}}\right)^2$$

$$L = 2.44 \times 10^{-8} \text{ W } \Omega \text{ K}^{-2} \tag{10.47}$$

This value is in good agreement with the experimental value of the Lorentz number. The Wiedemann–Franz law is not applicable at low temperature.

10.2.4 Advantages of Classical Free Electron Theory

The classical free electron theory has the following advantages.

1. It is used to derive a mathematical expression for the electrical conductivity of metals.
2. It is used to derive a mathematical expression for the thermal conductivity of metals.
3. It is used to verify Wiedemann–Franz law.
4. It is used to explain the optical properties of the materials.
5. It is used to verify ohm's law.

10.2.5 Drawbacks of Classical Free Electron Theory

The classical free electron theory has the following drawbacks.

1. According to the classical free electron theory, the Lorentz number is $L = \frac{K}{\sigma T} = \frac{3}{2}\frac{k^2}{e^2}$ $= 1.11 \times 10^{-8}$ W Ω K^{-2}. However, the experimental value of the Lorentz number is 2.3×10^{-8} W Ω K^{-2}.
2. In the case of metals, the Hall coefficient is negative because the charge carriers are electrons. It fails to explain why certain metals, e.g. Zn, exhibit positive values of Hall coefficient.
3. The electrical conductivity of a metal is inversely proportional to temperature. According to the free electron theory, $\sigma = \frac{ne^2}{m}\tau_r$, where τ_r is the relaxation time and it is temperature dependent. The relaxation time, τ_r is proportional to $T^{-1/2}$. Therefore, the classical free electron theory predicts the incorrect temperature dependence of σ.
4. The electronic specific heat of a metal is $C = \frac{3}{2}R$, where R is a universal constant. The experimental value of electronic specific heat is $C = 10^{-4}\,RT$. The experimental value shows that the electronic specific heat is temperature dependent, whereas the classical free electron theory says that it is temperature independent.
5. It fails to explain Compton effect, photoelectric effect, black body radiation, Zeeman effect.
6. It fails to give the correct mathematical expression for the thermal conductivity of metals.
7. It fails to explain the superconducting property and magnetic susceptibility.

10.3 QUANTUM CONCEPTS

The classical free electron theory fails to explain the black body radiations, photoelectric effect, Compton effect, etc. In order to explain the black body radiations, Max Planck proposed the Quantum theory of radiation in 1900. He introduced the quantum concepts in his theory. This concept was used by Einstein in 1905 to explain the photoelectric effect. Bohr used this concept in 1911 for his atom model. He partially succeeded in explaining the hydrogen spectral lines. Compton used this concept to explain the Compton scattering in X-rays. Therefore, the quantum concepts were accepted by the scientists during the earlier stage of the 20th century. The quantum concepts are:

1. The energy of the electromagnetic radiations is not continuous but discrete.
2. The electromagnetic radiation consists of small packet of energy called quanta. The energy of a quantum is $h\nu$. This discrete energy is also called a photon.
3. The absorption and emission of energy take place in the order of $h\nu$. Let E_1 and E_2 be the two energy levels. Let $E_2 - E_1 = h\nu$. The radiation absorbs energy in the order of $h\nu$ (i.e. $nh\nu$, where $n = 0, 1, 2, 3, \ldots$) and it gets excited to a higher energy level.
4. The emission of energy takes place if there is a transition from upper energy level to lower energy level and it emits energy in the order of $h\nu$ (i.e. $nh\nu$, where $n = 0, 1, 2, 3, \ldots$).

10.4 QUANTUM FREE ELECTRON THEORY

In order to rectify the drawbacks of the classical free electron theory, Sommerfield proposed the quantum free electron theory in 1928. This theory is based on the quantum concepts. This theory was proposed by making slight modification in the classical free electron theory and by retaining most of the postulates of the classical free electron theory.

According to the quantum free electron theory,

1. The free electrons are fully responsible for the electrical conduction.
2. The free electron is free to move anywhere within the crystal.
3. The loss of energy due to interaction of the free electron with the other free electron, or positive ion core or wall of the container is negligible.
4. The free electron behaves as a wave. It has to obey the quantum concepts.
5. The velocity and the energy distribution of the free electrons are governed by the Fermi–Dirac distribution function.

10.4.1 Advantages of Quantum Free Electron Theory

This theory explains most of the failures of the classical free electron theory.

1. It explains the specific heat capacity of materials.
2. It explains Compton effect, photoelectric effect, black body radiation, Zeeman effect.
3. It gives the correct mathematical expression for the thermal conductivity of metals.
4. It explains the superconducting property.

10.4.2 Drawbacks of Quantum Free Electron Theory

1. This theory fails to explain the positive value of Hall coefficients.
2. This theory fails to distinguish between the metal, semiconductor, and insulator.

10.5 FERMI–DIRAC DISTRIBUTION FUNCTION

The Fermi–Dirac statistics deals with the particles having half integral spin. The spin of the electron is $\frac{1}{2}$ and hence it is dealt with using Fermi–Dirac statistics and hence an electron is said to be a

Fermi particle or fermions. The Fermi–Dirac statistics is dealing with the velocity and energy distribution of the particles.

The Fermi-Dirac distribution function is given by

$$f(E) = \frac{1}{1 + \exp\left(\dfrac{E - E_F}{kT}\right)} \tag{10.48}$$

where $f(E)$ represents the probability that a quantum state is occupied by an electron, k is the Boltzmann's constant, T is the temperature in kelvin, E is the energy and E_F is the Fermi energy. Since $f(E)$ is a probability of finding an electron, its value lies between 0 and 1. The probability value 1 represents the presence of an electron. The probability value 0 represents the absence of an electron. If the probability value is 0.5, it means that there is 50% chance for the presence of the electron.

Let us find the value of $f(E)$, at different values of T and E.

Case (i) When $T = 0$ K and $E > E_F$

When $E > E_F$ and $T = 0$ K, the term $\exp\left(\dfrac{E - E_F}{kT}\right) = \infty$ and hence $f(E) = 0$.

Case (ii) When $T = 0$ K and $E < E_F$

When $E < E_F$ and $T = 0$ K, the term $\exp\left(\dfrac{E - E_F}{kT}\right) = e^{-\infty} = 0$ and hence $f(E) = 1$.

From these two cases, we infer that at 0 K, above the Fermi level all the states are not occupied by the electrons and below the Fermi level all the quantum states are occupied by the electrons. From these two cases one can define that *Fermi level is the maximum energy level occupied by the electrons at absolute zero degree kelvin. The Fermi level at absolute zero is also known as Fermi energy.*

Case (iii) When $T > 0$ K and $E = E_F$

When $E = E_F$ and $T > 0$ K, the term $\exp\left(\dfrac{E - E_F}{kT}\right) = e^0 = 1$ and hence $f(E) = \dfrac{1}{2}$. Thus from above, the term *Fermi level is defined as the energy level that has the probability of* $\dfrac{1}{2}$ *for the occupation of quantum states by the electrons.*

10.5.1 Variation of Fermi Function with Temperature

When the temperature is increased, some of the electrons are energized due to increase in temperature and get excited above the Fermi level. The quantum state that lies above the Fermi level also gets occupied by the electrons. The variation of the Fermi function with temperature is shown in Figure 10.4.

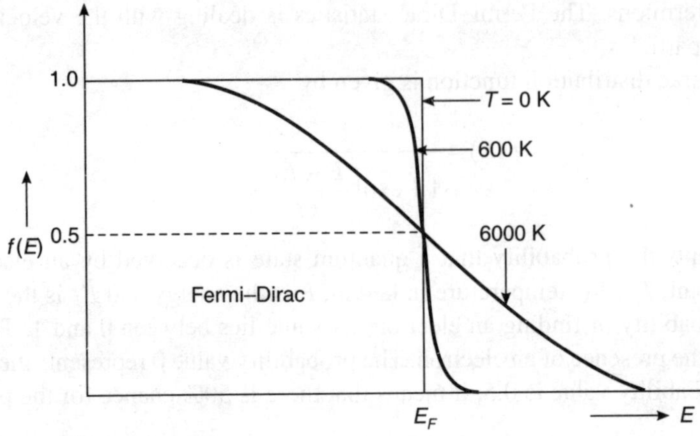

Figure 10.4 Variation of Fermi function with temperature.

10.6 DENSITY OF STATES

Density of states is the number of states available per unit volume between the energy levels E and $E + dE$.

$$\text{Density of states} = \frac{\text{Number of states available between the energy levels } E \text{ and } E + dE}{\text{Volume}} \tag{10.49}$$

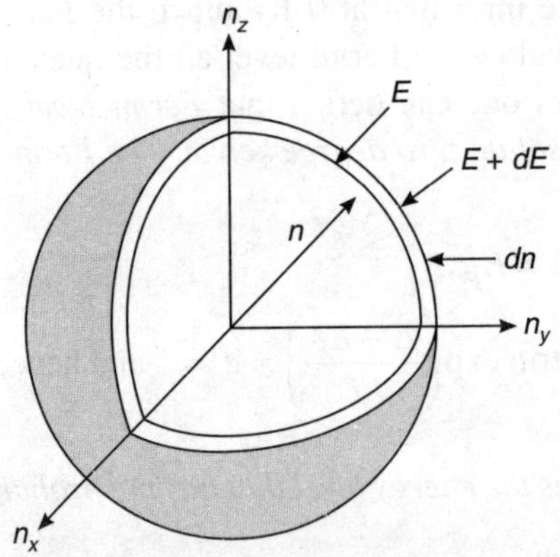

Figure 10.5 Density of states.

Let us construct a three-dimensional space of points using the quantum numbers n_x, n_y and n_z as shown in Figure 10.5. Let n be the quantum number in three-dimensional space of points, n_x, n_y and n_z such that

$$n^2 = n_x^2 + n_y^2 + n_z^2 \tag{10.50}$$

Let every point in the space represents a state, i.e. a unit cube contains exactly one state. The number of states available in any volume is equal to the numerical value of the volume. In a space of radius n, the number of states available is

$$dS = \frac{4}{3}\pi n^3 \tag{10.51}$$

Since only the positive values of the quantum numbers are allowed, the first octant of the cube is considered. Therefore, a factor $\frac{1}{8}$ is introduced.

The number of states available in the first octant is

$$dS = \frac{1}{8} \times \frac{4}{3}\pi n^3 \tag{10.52}$$

The energy and n are related using the relation,

$$E = \frac{n^2 h^2}{8mL^2} \tag{10.53}$$

where L is the width of the potential well.

The number of states available in the volume having energy less than E in the cube of width L is obtained by substituting the value of n given by Eq. (10.53) in Eq. (10.52), we get

$$dS = \frac{1}{8} \times \frac{4}{3}\pi n^3 = \frac{1}{6}\pi \left(\frac{8mL^2}{h^2}\right)^{3/2} E^{3/2} \tag{10.54}$$

$$dS = \frac{8}{6}\pi (2m)^{3/2} \frac{L^3}{h^3} E^{3/2} \tag{10.55}$$

The number of states available per unit volume is ($V = L^3$),

$$dS = \frac{8}{6}\pi (2m)^{3/2} \frac{1}{h^3} E^{3/2} \tag{10.56}$$

The number of energy states available in the energy interval dE can be obtained by differentiating Eq. (10.56),

$$\frac{dS}{dE} = \frac{8}{6}\pi (2m)^{3/2} \frac{1}{h^3} \times \frac{3}{2} E^{1/2} dE \tag{10.57}$$

Simplifying Eq. (10.57), we get

$$\frac{dS}{dE} = \frac{2\pi}{h^3}(2m)^{3/2} E^{1/2} dE \tag{10.58}$$

According to Pauli's exclusion principle, a state can occupy two electrons, one with spin up and another one with spin down. Therefore, the density of states available per unit volume in an energy interval dE is

$$N(E)dE = \frac{4\pi}{h^3}(2m)^{3/2} E^{1/2} dE \tag{10.59}$$

Expressing Eq. (10.59) in electron volt, we get

$$N(E)dE = \frac{4\pi}{h^3}(2m)^{3/2} E^{1/2} dE \times (1.6 \times 10^{-19})^{3/2} \tag{10.60}$$

Equation (10.60) can be written as

$$N(E)dE = \gamma E^{1/2}\, dE \tag{10.61}$$

where $\gamma = \dfrac{4\pi}{h^3}(2m)^{3/2}\,(1.6 \times 10^{-19})^{3/2} = 6.78 \times 10^{27}\ \text{m}^{-3}\ \text{eV}^{-3/2}$. The unit of γ is $\text{m}^{-3}\ \text{eV}^{-3/2}$.

Equation (10.59) or Eq. (10.60) gives the values of density of states. The variation of density of states with energy at different temperatures is shown in Figure 10.6. It represents the distribution of electrons at different temperatures. The number of electrons present in the material is given by the area under these curves. Since the areas of these curves are the same, the total number of free electrons present at different temperature in a metal is also the same.

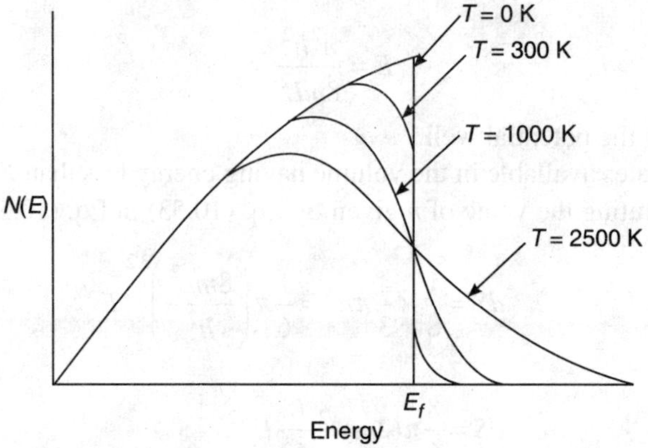

Figure 10.6 Variation of density of states with energy at different temperatures.

10.7 CARRIER CONCENTRATION IN METALS AND FERMI ENERGY AT 0 K

In the case of metals, the electrons are the charge carriers. The concentration of electrons in a metal is given by

$$n = \int_{-\infty}^{\infty} N(E)f(E)dE \tag{10.62}$$

where $N(E)$ is the number of electrons per unit volume and $f(E)$ is the probability that a quantum state is occupied by an electron. The product $N(E)f(E)$ represents the states occupied by electrons. Integrating $N(E)f(E)$ from $-\infty$ to ∞, gives the total number of electrons available in a metal. The value of $N(E)$ is

$$N(E) = \frac{4\pi}{h^3}(2m_e^*)^{3/2} E^{1/2} \tag{10.63}$$

and the value of $f(E)$ is

$$f(E) = \frac{1}{1 + \exp\left(\dfrac{E - E_F}{kT}\right)} \tag{10.64}$$

Substituting the values of $N(E)$ and $f(E)$ in Eq. (10.62), we get

$$n = \int_{-\infty}^{\infty} \frac{4\pi}{h^3}(2m_e^*)^{3/2} E^{1/2} \times \frac{1}{1+\exp\left(\frac{E-E_F}{kT}\right)} dE \tag{10.65}$$

Let us calculate the number of electrons at 0 K. At 0 K, the electrons occupy the energy levels up to the Fermi level. Therefore, the Fermi level is the maximum occupied energy level at 0 K and the minimum occupied energy level is 0 (zero). Let us denote the Fermi energy at 0 K as E_{F0}. Therefore, the integral varies from 0 to E_{F0}. The concentration of electrons at 0 K is

$$n = \int_{0}^{E_{F0}} \frac{4\pi}{h^3}(2m_e^*)^{3/2} E^{1/2} \times \frac{1}{1+\exp\left(\frac{E-E_F}{kT}\right)} dE \tag{10.66}$$

At 0 K, $E < E_F$ and hence $\exp\left(\frac{E-E_F}{kT}\right)$ is less than 1. Therefore, $\exp\left(\frac{E-E_F}{kT}\right)$ is negligible.

Equation (10.66) can be written as

$$n = \frac{4\pi}{h^3}(2m_e^*)^{3/2} \int_{0}^{E_{F0}} E^{1/2} dE \tag{10.67}$$

$$n = \frac{4\pi}{h^3}(2m_e^*)^{3/2} \frac{E_{F0}^{3/2}}{\frac{3}{2}} \tag{10.68}$$

$$n = \frac{8\pi}{3h^3}(2m_e^*)^{3/2} E_{F0}^{3/2} \tag{10.69}$$

Equation (10.69) gives the **concentration of electron in a metal** at 0 K.
Rearranging Eq. (10.69), we get

$$E_{F0} = \frac{h^2}{8m_e}\left(\frac{3n}{\pi}\right)^{2/3} \tag{10.70}$$

Equation (10.70) gives the expression for the **Fermi energy at 0 K**.

10.8 FERMI LEVEL IN A METAL, WHEN $T > 0$ K

The Fermi energy of the electron when $T > 0$ K is determined by finding the concentration of electrons. The maximum energy level occupied by the electron, when $T > 0$ K is infinity. The minimum energy level occupied by the electron when $T > 0$ K is 0 (zero). Therefore, the integral varies from 0 to ∞. The concentration of electrons, when $T > 0$ K is

$$n = \int_{0}^{\infty} \frac{4\pi}{h^3}(2m_e^*)^{3/2} E^{1/2} \times \frac{1}{1+\exp\left(\frac{E-E_F}{kT}\right)} dE \tag{10.71}$$

When $T > 0$ K, $E > E_F$ and hence $\exp\left(\frac{E-E_F}{kT}\right)$ is greater than 1. Therefore, the term 1 in the denominator is negligible.

$$n = \int_0^{\infty} \frac{4\pi}{h^3}(2m_e^*)^{3/2} E^{1/2} \times \exp\left(\frac{-(E-E_F)}{kT}\right) dE \tag{10.72}$$

Integrating Eq. (10.72) and rearranging the results, we get

$$E_F(T) = E_{F0}\left[1 - \frac{\pi^2}{12}\left(\frac{kT}{E_{F0}}\right)^2\right] \tag{10.73}$$

Equation (10.73) gives the variation of the Fermi level in a metal with temperature. Equation (10.73) shows that the Fermi energy decreases with the increase in temperature. However, the variation of the Fermi level is negligible, because the term $\left(\frac{kT}{E_{F0}}\right)^2$ is very low. Since the value of $kT \approx 0.025$ eV at 300 K and E_{F0} is in the order of few eV (say $E_{F0} = 5.32$ eV for Molybdenum), the value $\left(\frac{kT}{E_{F0}}\right)^2$ is negligible.

10.9 AVERAGE ENERGY OF AN ELECTRON IN A METAL

The average energy of an electron in a metal can be obtained using the relation,

$$\text{Average energy} = \frac{\text{Total energy of the electrons}}{\text{Total number of electrons}} \tag{10.74}$$

The total energy of the electrons is given by the product of the total number of electrons and the energy of an electron. Substituting the values of the total energy and the total number of electrons, we get

$$\text{Average energy} = \frac{\int_{-\infty}^{\infty} E \times N(E) f(E) dE}{n} \tag{10.75}$$

The values of $N(E)$ and $f(E)$ are given by Eq. (10.59) and Eq. (10.48) as

$$N(E) = \frac{4\pi}{h^3}(2m_e^*)^{3/2} E^{1/2} \tag{10.76}$$

$$f(E) = \frac{1}{1 + \exp\left(\frac{E-E_F}{kT}\right)} \tag{10.77}$$

Substituting the values of $N(E)$ and $f(E)$ in Eq. (10.75), we get

$$E_{AV} = \frac{1}{n}\int_{-\infty}^{\infty} E \times \frac{4\pi}{h^3}(2m_e^*)^{3/2} E^{1/2} \times \frac{1}{1 + \exp\left(\frac{E-E_F}{kT}\right)} dE \tag{10.78}$$

Equation (10.78) can be evaluated by assigning different values for the temperatures. Equation (10.78) is evaluated (i) at $T = 0$ K and (ii) $T > 0$ K.

Case (i) When $T = 0$ K

When $T = 0$ K, the integral varies from 0 to E_F, since 0 is the lowest energy and E_F is the highest energy occupied by the electrons. Let us represent the maximum energy occupied by the electron at 0 K as E_{F0}. Therefore, the integral varies from 0 to E_{F0}. Equation (10.78) can be written as

$$E_{AV} = \frac{1}{n}\int_0^{E_{F0}} E \times \frac{4\pi}{h^3}(2m_e^*)^{3/2} E^{1/2} \times \frac{1}{1+\exp\left(\dfrac{E-E_F}{kT}\right)} dE \tag{10.79}$$

At $T = 0$ K, $E < E_F$ and hence $\exp\left(\dfrac{E-E_F}{kT}\right)$ in the denominator of Eq. (10.79) is negligible. Therefore, Eq. (10.79) can be written as

$$E_{AV} = \frac{1}{n}\frac{4\pi}{h^3}(2m_e^*)^{3/2}\int_0^{E_{F0}} E^{3/2} dE \tag{10.80}$$

Integrating Eq. (10.80), we get

$$E_{AV} = \frac{1}{n}\frac{4\pi}{h^3}(2m_e^*)^{3/2}\frac{E_{F0}^{5/2}}{\dfrac{5}{2}} \tag{10.81}$$

Equation (10.81) can be written as

$$E_{AV} = \frac{1}{n} \times \frac{2}{5} \times \frac{4\pi}{h^3}(2m_e^*)^{3/2} E_{F0}^{3/2} \cdot E_{F0} \tag{10.82}$$

Substituting the values of E_{F0} in Eq. (10.82), we get

$$E_{AV} = \frac{1}{n} \times \frac{2}{5} \times \frac{4\pi}{h^3}(2m_e^*)^{3/2}\left(\frac{h^2}{8m_e}\left(\frac{3n}{\pi}\right)^{2/3}\right)^{3/2} \cdot E_{F0} \tag{10.83}$$

Simplifying Eq. (10.83), we get

$$E_{AV} = \frac{1}{n} \times \frac{2}{5} \times \frac{4\pi}{h^3}(2m_e^*)^{3/2}\frac{h^3}{8 \times (2m_e)^{3/2}}\frac{3n}{\pi} \cdot E_{F0} \tag{10.84}$$

$$E_{AV} = \frac{3}{5} \cdot E_{F0} \tag{10.85}$$

Equation (10.85) gives the value of the average energy of the electrons at 0 K.

Case (ii) When $T > 0$ K

When $T > 0$ K, the integral varies from 0 to ∞, since 0 is the lowest energy and ∞ is the highest energy occupied by the electrons. Therefore, the integral varies from 0 to ∞. Equation (10.78) can be written as

$$E_{AV} = \frac{1}{n}\int_0^{\infty} E \times \frac{4\pi}{h^3}(2m_e^*)^{3/2} E^{1/2} \times \frac{1}{1+\exp\left(\dfrac{E-E_F}{kT}\right)} dE \tag{10.86}$$

At $T > 0$ K, $E > E_F$ and hence $\exp\left(\frac{E-E_F}{kT}\right) > 1$, and hence the term 1 in the denominator of Eq. (10.86) is negligible. Therefore, Eq. (10.86) can be written as

$$E_{\text{AV}} = \frac{1}{n}\frac{4\pi}{h^3}(2m_e^*)^{3/2}\int_0^\infty E^{3/2}\exp\left(-\frac{E-E_F}{kT}\right)dE \qquad (10.87)$$

Integrating Eq. (10.87) and rearranging the results, we get

$$E_{\text{AV}}(T) = \frac{3}{5}E_{F0}\left[1+\frac{5\pi^2}{12}\left(\frac{kT}{E_{F0}}\right)^2\right] \qquad (10.88)$$

Equation (10.88) gives the variation of the average energy of the electron in a metal with temperature. Equation (10.88) shows that the average energy of the electron increases with the increase in temperature. However, the variation of the average energy is negligible, because the term $\left(\frac{kT}{E_{F0}}\right)^2$ is very low. Since the value of $kT \approx 0.025$ eV at 300 K and E_{F0} is in the order of few eV (say $E_{F0} = 5.32$ eV for Molybdenum) and hence $\left(\frac{kT}{E_{F0}}\right)^2$ is negligible.

10.10 SIGNIFICANCE OF FERMI ENERGY

The importance of the Fermi energy is explained using (i) contact potential in metals and (ii) Seebeck effect.

(i) Contact potential in metals

Consider two different metals such as platinum and molybdenum are joined together. The Fermi energy of platinum and molybdenum are 4.2 eV and 5.32 eV respectively. If these two metals are joined together, the Fermi level of these two metals lies in different positions, and hence there is a flow of electrons from molybdenum to platinum until the Fermi level of these two metals becomes equal. The positions of the Fermi levels at equilibrium and before the two metals are joined together is shown in Figure 10.7. At equilibrium, the Fermi level of these metals becomes equal. At equilibrium,

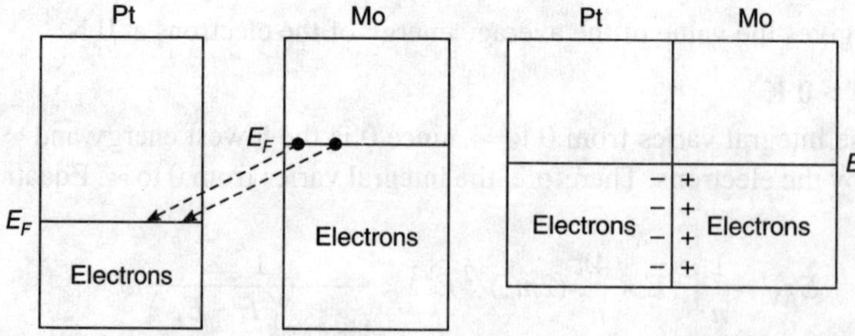

Figure 10.7 Contact potential-position of Fermi level when two metals are joined.

platinum becomes more negative since more number of electrons gets transferred to it and molybdenum becomes more positive due to deficiency of electrons.

(ii) Seebeck effect

Consider the ends of a metal which are at two different temperatures, say at 0°C and 100°C as shown in Figure 10.8. There will be a flow of heat from the hot end to the cold end. If the ends of the metals that lie at different temperatures are connected by the junctions made by the same wire, say copper wire as shown in Figure 10.9a, there is no flow of thermo emf because the Fermi level in the copper wires lies in the same level. One can measure the thermo emf by connecting ends of the metals by the junctions made by two different materials, say cromel and alumel, because the Fermi levels in these two materials lies at different positions. There will be a flow of electrons from a material having a high value of Fermi level to another material having a low value of Fermi level.

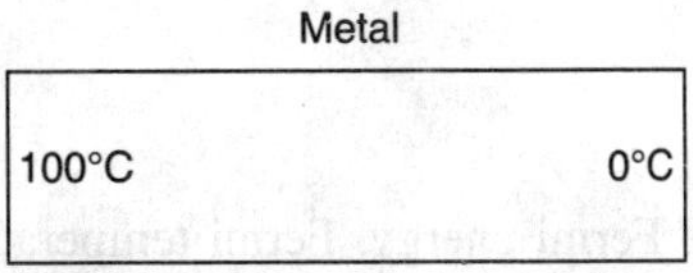

Figure 10.8 Seebeck effect.

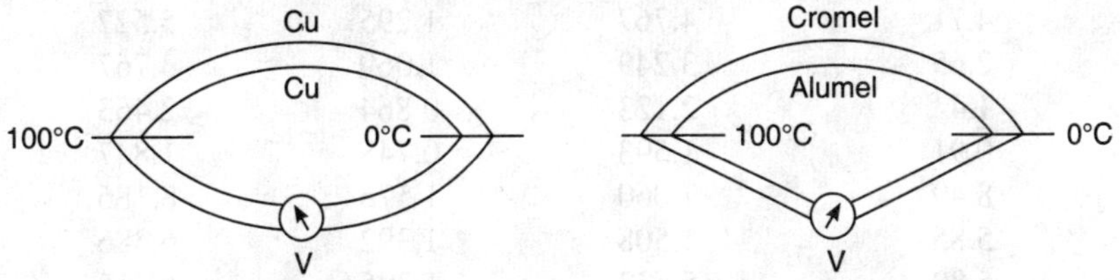

(a) Using two Cu wires (b) Two different wires

Figure 10.9 Measurement of thermo emf.

10.11 FERMI ENERGY, FERMI VELOCITY AND FERMI TEMPERATURE

10.11.1 Fermi Energy

Fermi energy is the maximum energy occupied by electron in a metal at absolute zero degree kelvin. It is also known as the Fermi level. It is given by the relation,

$$E_{F0} = \frac{h^2}{8m_e}\left(\frac{3n}{\pi}\right)^{2/3} \tag{10.89}$$

where E_{F0} is the Fermi energy at 0 K, n is the concentration of electrons, h is the Planck's constant and m_e is the mass of the electron.

10.11.2 Fermi Velocity

The velocity of the electron at the Fermi level is known as Fermi velocity. It is represented by v_F and it is given by

$$E_F = \frac{1}{2}mv_F^2$$

i.e.

$$v_F = \sqrt{\frac{2E_F}{m}} \tag{10.90}$$

10.11.3 Fermi Temperature

The ratio of the energy of the Fermi level of an assembly of fermions to the Boltzmann constant is known as the Fermi temperature. It is represented by the letter, T_F and it is given by

$$E_F = kT_F$$

$$T_F = \frac{E_F}{k} \tag{10.91}$$

The Fermi energy, Fermi temperature and Fermi velocity of some selected metals are listed in Table 10.2.

Table 10.2 Carrier concentration, Fermi energy, Fermi velocity, Fermi temperature and work function of some selected metals

Metals	*Concentration* $n(\times 10^{28}\ m^{-3})$	E_{F0} *in* eV	v_F *in* ms^{-1} $(\times 10^6)$	T_F *in* K $(\times 10^4)$	φ *in* eV
Li	4.71	4.767	1.295	5.527	2.48
Na	2.65	3.249	1.069	3.767	2.3
K	1.4	2.123	0.864	2.463	2.2
Cs	0.91	1.593	0.748	1.847	1.9
Cu	8.49	7.060	1.576	8.186	4.45
Ag	5.85	5.508	1.392	6.386	4.46
Au	5.89	5.533	1.395	6.415	4.9
Al	18.066	11.681	2.027	13.543	4.2

The Fermi energy, Fermi velocity and Fermi temperatures are calculated using Eq. (10.89), Eq. (10.90) and Eq. (10.91).

10.12 WORK FUNCTION OF A METAL

The minimum energy required to eject an electron at absolute zero is known as the work function of a metal. It is represented by φ. To liberate the electrons from the metal, the energy is given in the form of either light or heat. If the emission is by means of light, it is said to be photoelectric emission. If the emission of electron is by means of heat, then it is said to be thermionic emission.

At zero degree kelvin the maximum energy level occupied by the electron is known as the Fermi level. The top of the band of a metal is known as vacuum level. In order to free an electron, one has to supply the energy, which is equal to the difference between the Fermi energy and the vacuum level of the metal. This energy is known as work function of the metal.

For example, the Fermi energy of Li is 4.7 eV, and its work function is 2.5 eV. It is shown in Figure 10.10. The work function of selected metals is listed in Table 10.2.

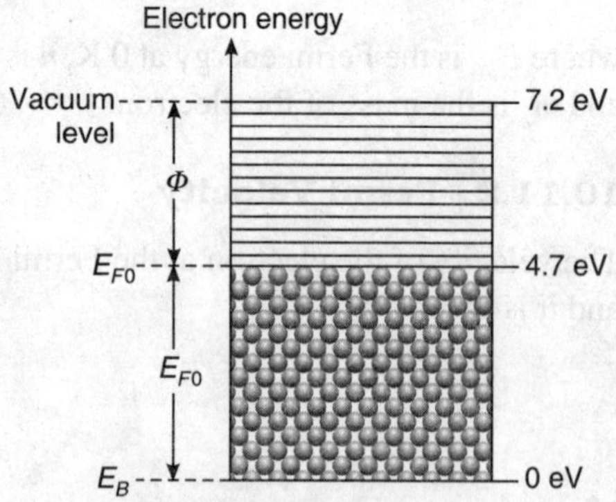

Figure 10.10 Work function of Li.

SOLVED PROBLEMS

10.1 Calculate the electrical resistivity of sodium at 0°C. It has 2.533×10^{28} electrons per unit volume and has a mean free time of 3.1×10^{-14} s.

Given data

Concentration of electrons = $2.533 \times 10^{28}\ m^{-3}$

Relaxation time = 3.1×10^{-14} s

Solution

Electrical conductivity $\sigma = \dfrac{ne^2\tau_r}{m}$

$$\rho = \frac{m}{ne^2\tau_r}$$

$$= \frac{9.1\times10^{-31}}{2.533\times10^{28}\times(1.6\times10^{-19})^2\times3.1\times10^{-14}}$$

$$= 4.527 \times 10^{-8}\ \Omega\ m.$$

The resistivity of sodium at 0°C is $4.527 \times 10^{-8}\ \Omega$ m.

10.2 If the slope of the graph between ln σ and $1/T$ is 3.75×10^3, find the band gap of the semiconductor.

Given data

$$\text{Slope} = 3.75 \times 10^3$$

Solution

Band gap

$$E_g = 2k \times \text{slope}$$

$$= \frac{2\times1.38\times10^{-23}\times3.75\times10^3}{1.6\times10^{-19}} = 0.647 \text{ eV}$$

The band gap of the semiconductor is 0.647 eV.

10.3 The occupied level of the electron at 989°C is above 0.5 eV of the Fermi level of a material. What is the probability of occupation of electrons at this temperature?

Given data

Temperature $T = 989°C = 1262$ K

$$E - E_F = 0.5 \text{ eV}$$

Solution

Fermi–Dirac distribution function,

$$f(E) = \frac{1}{1+\exp\left(\dfrac{E-E_F}{kT}\right)}$$

$$= \frac{1}{1+\exp\left(\dfrac{0.5\times 1.6\times 10^{-19}}{1.38\times 10^{-23}\times 1262}\right)}$$

$$= 0.01$$

The probability of occupation of electrons at 989°C is 0.01.

10.4 Calculate the drift velocity of the free electrons with a mobility of 0.0035 m² V⁻¹ s⁻¹ in copper for an electric field strength of 0.5 V m⁻¹.

Given data

Mobility of electrons $\mu_e = 0.0035$ m² V⁻¹ s⁻¹

Electric field strength $E = 0.5$ V m⁻¹

Solution

Drift velocity $v_d = \mu_e \times E$

$= 0.0035 \times 0.5 = 1.75 \times 10^{-3}$ m s⁻¹.

The drift velocity of the free electrons is 1.75×10^{-3} m s⁻¹.

10.5 Resistivity of copper is 1.73×10^{-8} Ω m. Its density is 8.92×10^3 kg/m³ and atomic weight is 63.5. Assuming classical laws, calculate the mobility of electrons.

Given data

Resistivity of copper $\rho = 1.73 \times 10^{-8}$ Ω m

Atomic weight = 63.5

Density = 8.92×10^3 kg m⁻³

Solution

The electrical conductivity is given by

$$\sigma = \frac{1}{\rho} = \frac{1}{1.73\times 10^{-8}} = 5.78 \times 10^7\ \Omega^{-1}\ \text{m}^{-1}$$

The concentration of carriers is given by

$$n = \frac{\text{Density} \times \text{Avogadro's constant}}{\text{Atomic weight}} = \frac{8.92\times 10^3 \times 6.022\times 10^{23}\times 10^3}{63.5}$$

$n = 8.459 \times 10^{28}$ m⁻³

Mobility of electrons is given by

$$\mu = \frac{\sigma}{ne} = \frac{5.78\times 10^7}{8.459\times 10^{28}\times 1.6\times 10^{-19}} = 4.270\times 10^{-3}\ \text{m}^2\ \text{V}^{-1}\ \text{s}^{-1}$$

The mobility of electrons is 4.270×10^{-3} m² V⁻¹ s⁻¹

10.6 The density of free electrons in aluminium is 18.1×10^{28} m⁻³. Calculate its Fermi energy at 0 K. ($h = 6.62 \times 10^{-34}$ J s, and $m_e = 9.1 \times 10^{-31}$ kg)

Given data

Concentration of electrons $n = 18.1 \times 10^{28}$ m⁻³

$h = 6.62 \times 10^{-34}$ J s

$m_e = 9.1 \times 10^{-31}$ kg

Solution

The Fermi energy at 0 K is given by

$$E_{F0} = \frac{h^2}{8m_e}\left(\frac{3n}{\pi}\right)^{2/3}$$

Substituting the values of h, n, m_e, we get

$$E_{F0} = \frac{(6.626\times10^{-34})^2}{8\times9.1\times10^{-31}}\left(\frac{3\times18.1\times10^{28}}{\pi}\right)^{2/3}$$

$$= 1.87 \times 10^{-18}\text{ J} = 11.69\text{ eV}$$

The Fermi energy of Al at 0 K is 11.69 eV.

10.7 Calculate the concentration of free electrons per unit volume of silver. The Fermi energy of its free electrons is 5.5 eV. (Given the value of Planck's constant, $h = 6.63 \times 10^{-34}$ J s and $m_e = 9.1 \times 10^{-31}$ kg)

Given data

Fermi energy $E_{F0} = 5.5$ eV

$h = 6.63 \times 10^{-34}$ J s

$m_e = 9.1 \times 10^{-31}$ kg

Solution

The Fermi energy at 0 K is given by

$$E_{F0} = \frac{h^2}{8m_e}\left(\frac{3n}{\pi}\right)^{2/3}$$

Rearranging the above equation, we get

$$n = (2m_eE_{F0})^{3/2} \times \frac{8\pi}{3h^3}$$

Substituting the values of h, m_e, E_{F0}, we get

$$n = (2 \times 9.1 \times 10^{-31} \times 5.5 \times 1.6 \times 10^{-19})^{3/2} \times \frac{8\pi}{3\times(6.626\times10^{-34})^3}$$

$$= 5.837 \times 10^{28}\text{ m}^{-3}$$

The concentration of free electrons per unit volume of silver is 5.837×10^{28} m^{-3}.

10.8 What is the probability of an electron being thermally promoted to the conduction band in silicon ($E_g = 1.07$ eV) at 25°C?

Given data

Energy gap of silicon $E_g = 1.07$ eV

Temperature $T = 298$ K

Solution

The probability of an electron being thermally promoted

$$f(E) = \frac{1}{1+\exp\left(\dfrac{E-E_F}{kT}\right)}$$

The electron has to be promoted to the conduction band. Therefore, $E = E_C$. Substituting, we get

$$f(E) = \frac{1}{1 + \exp\left(\frac{E_C - E_F}{kT}\right)} = \frac{1}{1 + \exp\left(\frac{1.07 \times 1.6 \times 10^{-19}}{1.38 \times 10^{-23} \times 298}\right)}$$

The Fermi level for intrinsic semiconductor lies at the middle of the band gap. Therefore,

$$(E_C - E_F) = \frac{E_g}{2}$$

$$f(E) = \frac{1}{1 + \exp\left(\frac{E_g}{2kT}\right)} = \frac{1}{1 + \exp\left(\frac{1.07 \times 1.6 \times 10^{-19}}{2 \times 1.38 \times 10^{-23} \times 298}\right)}$$

$$= 9.1226 \times 10^{-10}$$

The probability of an electron thermally excited to the conduction band at 25°C is 9.1226×10^{-10}.

10.9 Calculate the Fermi energy and Fermi temperature in a metal. The Fermi velocity of electrons in a metal is 0.86×10^6 ms^{-1}.

Given data

Fermi velocity $v_F = 0.86 \times 10^6$ m s^{-1}

Solution

The Fermi energy

$$E_F = \frac{1}{2} m v_F^2$$

$$= \frac{1}{2} \times 9.1 \times 10^{-31} \times (0.86 \times 10^6)^2$$

$$= 3.365 \times 10^{-19} \text{ J} = 2.10 \text{ eV.}$$

The Fermi energy

$$E_F = kT_F$$

The Fermi temperature

$$T_F = \frac{E_F}{k} = \frac{3.365 \times 10^{-19}}{1.38 \times 10^{-23}} = 2.438 \times 10^4 \text{ K}$$

The Fermi energy of the metal is 2.10 eV and the Fermi temperature is 2.438×10^4 K.

10.10 The electrical conductivity of copper at 27°C is 5.82×10^7 Ω^{-1} m^{-1}. Its thermal conductivity at 27°C is 387 W m^{-1} K^{-1}. Calculate the Lorentz number.

Given data

Electrical conductivity of copper = 5.82×10^7 Ω^{-1} m^{-1}

Thermal conductivity of copper = 387 W m^{-1} K^{-1}

Temperature = 27°C = 300 K

Solution

Lorentz number

$$L = \frac{K}{\sigma T} = \frac{387}{5.82 \times 10^7 \times 300} = 2.216 \times 10^{-8}\ \text{W}\ \Omega\ \text{K}^{-2}$$

The Lorentz number is 2.216×10^{-8} W Ω K^{-2}.

10.11 **The concentration of electrons in copper is $8.49 \times 10^{28}\ m^{-3}$ at 20°C. If the relaxation time of copper is 2.44×10^{-14} s at 20°C, calculate the electrical conductivity, thermal conductivity and hence the Lorentz number.**

Given data

Concentration of electrons in copper = $8.49 \times 10^{28}\ m^{-3}$

Relaxation time of electrons = 2.44×10^{-14} s

Temperature = 20°C = 293 K

Solution

Electrical conductivity

$$\sigma = \frac{ne^2\tau_r}{m}$$

$$= \frac{8.49 \times 10^{28} \times (1.6 \times 10^{-19})^2 \times 2.44 \times 10^{-14}}{9.1 \times 10^{-31}}$$

$$= 5.827 \times 10^7\ \Omega^{-1}\ m^{-1}$$

Thermal conductivity

$$K = \frac{n\pi^2k^2T\tau_r}{3m}$$

$$= \frac{8.49 \times 10^{28} \times \pi^2 \times (1.38 \times 10^{-23})^2 \times 293 \times 2.44 \times 10^{-14}}{3 \times 9.1 \times 10^{-31}}$$

$$= 417.89\ \text{W}\ m^{-1}\ K^{-1}$$

Lorentz number

$$L = \frac{K}{\sigma T} = \frac{417.89}{5.827 \times 10^7 \times 297} = 2.415 \times 10^{-8}\ \text{W}\ \Omega\ \text{K}^{-2}$$

The Lorentz number is 2.415×10^{-8} W Ω K^{-2}.

SHORT QUESTIONS

1. What is drift velocity?
2. What is meant by frictional force?
3. Define the term relaxation time.
4. Describe the classical free electron theory of metals.

5. Mention the drawbacks of the classical free electron theory.
6. Describe the advantages of the classical free electron theory.
7. Define the term electrical conductivity.
8. Define the term resistivity.
9. Define the term mobility of charge carriers.
10. Define the term mean free path of electron.
11. What do you mean by thermal conductivity of metals?
12. State Wiedemann–Franz law.
13. What is Lorentz number? What is its value?
14. What are the limitations of Wiedemann–Franz law?
15. Mention the quantum concepts.
16. Mention the postulates of quantum free electron theory.
17. Mention the advantages of quantum free electron theory.
18. What are the drawbacks of quantum free electron theory?
19. Write the Fermi–Dirac distribution function and explain the terms.
20. Define the terms Fermi level or Fermi energy.
21. Define the term Fermi velocity.
22. Define the term Fermi temperature.
23. Define density of states.
24. Define the term average energy of electrons.
25. Mention the significance of the Fermi energy.

DESCRIPTIVE TYPE QUESTIONS

1. Using the classical free electron theory, derive the mathematical expressions for the electrical conductivity and thermal conductivity of metals and hence deduce Wiedemann–Franz law.
2. What is density of states? Derive an expression for the density of states. Using density of states, obtain an expression for the carrier concentration in metals.
3. Describe the Fermi–Dirac distribution function and explain the effect of temperature on the Fermi function and hence derive an expression for the carrier concentration of metals.

PROBLEMS

1. The free electron density in sodium is 2.5×10^{28} m^{-3}. Calculate the Fermi energy of sodium in eV at 0 K. Given $h = 6.626 \times 10^{-34}$ J s and the electrons rest mass = 9.1×10^{-31} kg. The charge of the electron = 1.6×10^{-19} C.

2. The Fermi energy of cesium is 1.55 eV. Determine the conduction electron density.
3. Calculate the drift velocity of the free electrons (with a mobility of 3.5×10^{-3} m^2 V^{-1} s^{-1}) in copper for electric field strength of 0.5 V m^{-1}.
4. What is the probability of the electron being thermally promoted to the conduction band in diamond (E_g = 5.6 eV) at room temperature (25°C).
5. The density of copper is 8940 kg m^{-3} and its atomic weight is 63.54, calculate its Fermi energy at 0 K.
6. The Fermi energy of Cd is 7.47 eV. Calculate the Fermi velocity of electrons and Fermi temperature.
7. Calculate the concentration of free electrons per unit volume of copper. The Fermi energy of its free electrons is 7.06 eV. (Given the value of Planck's constant, $h = 6.626 \times 10^{-34}$ J s and $m_e = 9.1 \times 10^{-31}$ kg)
8. The electrical conductivity of silver at 20°C is 6.15×10^7 Ω^{-1} m^{-1}. Its thermal conductivity at 20°C is 423 W m^{-1} K^{-1}. Calculate the Lorentz number.
9. The concentration of electrons in silver is 5.85×10^{28} m^{-3} at 20°C. If the relaxation time of silver is 10^{-14} s at 20°C, calculate the electrical conductivity, thermal conductivity and hence the Lorentz number.
10. Copper has electrical conductivity at 300 K as 6.4×10^7 Ω^{-1} m^{-1}. Calculate the thermal conductivity of copper. Lorentz number, $L = 2.45 \times 10^{-8}$ W Ω K^{-2}.
11. The thermal conductivity of a metal is 123.92 W m^{-1} K^{-1}. Find the electrical conductivity and Lorentz number when the metal possesses the relaxation time of 10^{-14} s at 300 K. (Density of electron = 6×10^{28} m^{-3}).
12. For copper at 20°C, the electrical and thermal conductivities are 5.827×10^7 Ω^{-1} m^{-1} and 417 W m^{-1} K^{-1} respectively. Estimate the Lorentz number.

CHAPTER 11

SEMICONDUCTING MATERIALS

11.1 INTRODUCTION

Semiconductors are the materials having electrical conductivity in between the metals and insulating materials. They are mostly used in electronic components. After the invention of semiconducting components such as diodes, transistors, etc., the semiconductors have revolutionized the world since 1950s. The preparation of pure silicon and germanium semiconducting materials and the controlled doping of impurities into the semiconducting materials are the reasons for the fast development of the semiconducting usage. Thus, the study of the basic properties of the semiconducting materials is essential. This chapter deals with the carrier concentration of intrinsic and extrinsic semiconductors and their Fermi level, Hall effect, etc.

11.2 CLASSIFICATION OF SOLIDS

According to the band theory of solids, the materials are classified into metals, insulators and semiconductors. In the case of metals, the valence band and the conduction band overlap each other. The electrons are the charge carriers in metals. Due to the overlapping of the valence band and the conduction band, the electrons flow easily from valence band to the conduction band and hence the metals have high electrical conductivity.

In semiconductors, the valence band and the conduction band are separated by nearly 1 eV, whereas in the insulators the separation between the conduction band and the valence band is nearly 6 eV. The electron needs a very high amount of energy to cross the energy gap of the insulator and hence its conductivity is poor. The band structures for metals, insulators and semiconductors are shown in Figure 11.1.

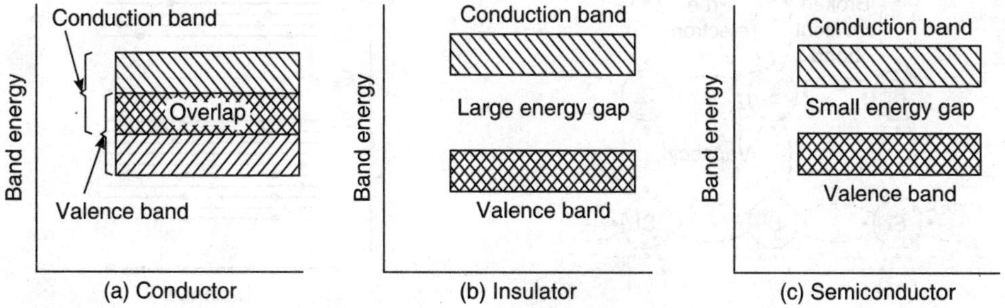

Figure 11.1 Band structures at 0 K.

11.3 CLASSIFICATION OF SEMICONDUCTORS

The semiconductors are classified into two types, namely (i) intrinsic semiconductors and (ii) extrinsic semiconductors.

11.3.1 Intrinsic Semiconductors

The intrinsic semiconductors are the pure form of semiconductors. The property of a pure semiconductor is controlled by the intrinsic property of the material and hence the pure semiconductors are called intrinsic semiconductors. The examples of intrinsic semiconductor are pure silicon and pure germanium crystals. Figures 11.2 and 11.3 depict the intrinsic semiconductors at 0 K and $T > 0$ K respectively.

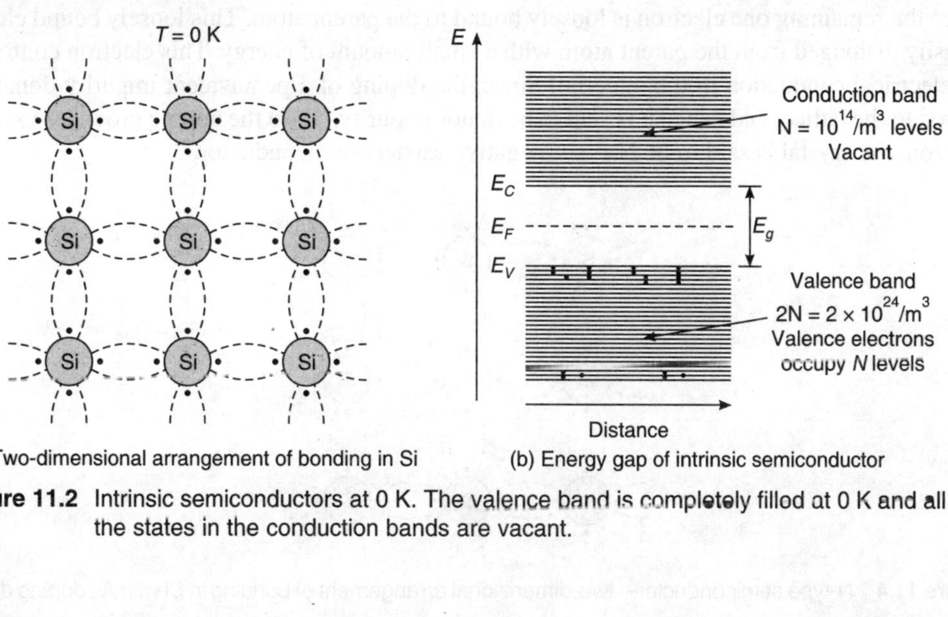

Figure 11.2 Intrinsic semiconductors at 0 K. The valence band is completely filled at 0 K and all the states in the conduction bands are vacant.

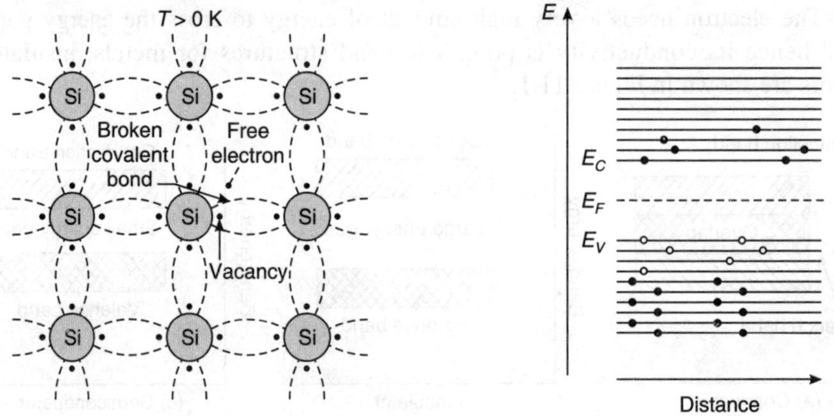

(a) Two-dimensional arrangement of bonding in Si (b) Energy gap of intrinsic semiconductor

Figure 11.3 Intrinsic semiconductors at $T > 0$ K. The valence band is not completely filled when $T > 0$ K and some of the states in the conduction bands are occupied.

11.3.2 Extrinsic Semiconductor

In electronic industry, the pure semiconductor is first prepared and then the impurities are added in a controlled manner, so as to obtain the desired property of the material. A doped or an impure semiconductor is known as extrinsic semiconductor. The property of the extrinsic semiconductor is controlled by the doped material. The extrinsic semiconductors are of two types. They are (i) N-type semiconductor and (ii) P-type semiconductor.

N-type semiconductor

Consider a pentavalent impurity such as P, Sb, As, doped with pure Si, as shown in Figure 11.4, four valence electrons of the impurity atom form covalent bondings due to sharing of electron and hence the remaining one electron is loosely bound to the parent atom. This loosely bound electron is easily dislodged from the parent atom with a small amount of energy. This electron contributes the electrical conduction in the material. Since, the doping of a pentavalent impurity donates an elctron to the lattice, and hence it is said to be donor impurity. Since the doping produces excess of electron, the crystal is said to be N-type (negative carrier) semiconductor.

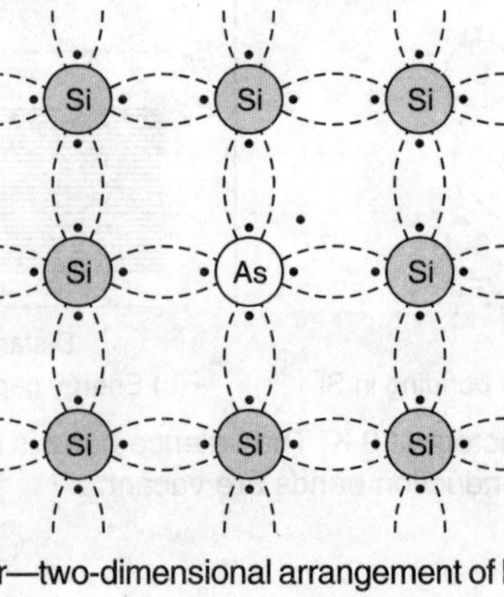

Figure 11.4 N-type semiconductor—two-dimensional arrangement of bonding in Si with As doping doping.

P-type semiconductor

The addition of trivalent impurities such as In, B, Al into pure Si, produces covelent bonding with three nearest atoms by sharing of electron and it is deficient of one electron as shown in Figure 11.5. The absence of one electron behaves as a positive charge carriers and hence it accepts an electron. This trivalent impurity is called acceptor impurity. The Si crystal doped with trivalent impurity is called P-type (positive carrier) material.

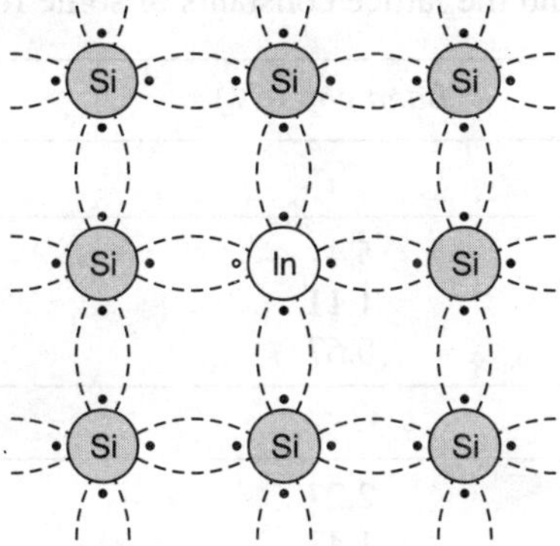

Figure 11.5 P-type semiconductor—two-dimensional arrangement of bonding in Si with In doping.

11.4 ELEMENTAL AND COMPOUND SEMICONDUCTOR

Silicon and germanium are the mostly studied semiconducting materials. They are said to be elemental semiconductors, since they are made up of only one type of atom. Si and Ge are tetravelent and they are found in the IV-B column of the periodic table. If a semiconductor consists of more than one element, then it is said to be compound semiconductor. They are prepared from the elements of the columns III and V or II and VI of the periodic table.

GaAs is a well-studied III-V compound. Its band gap is 1.4 eV and it offers higher mobility than Si and Ge. Generally, compound semiconductors have ionic characteristics. It is more difficult to break the bond in GaAs than in Si, because of the ionic bonding. Consider the semiconductors with one element is common, like GaP, GaAs and GaSb. It is found that the band gaps of these materials are 2.27 eV, 1.42 eV and 0.7 eV respectively. The increase in the atomic weight decreases the band gap of the material.

The atomic weight of these elements increases because of the increase in the atomic number of the other elements. The increase in the atomic number increases the charge (Ze) of the nucleus and these elements have more number of filled electronic inner shells. The valence electrons are farther from the nucleus and they are loosely bound. The bonding force between the atoms is less and hence the energy gap decreases.

ZnS, ZnSe and ZnTe are II-VI compounds. The band gaps of these materials are 3.67 eV, 2.58 eV and 2.26 eV. Their band gaps also decrease with the increase in the atomic weight. II-VI compounds are more ionic than III-V materials. But no II-VI compound can be made as both N-type and P-type. ZnS crystallizes in two different structures, namely (i) zinc blende structure and (ii) wurtzite structure. The bond length and density of these two structures are the same. The electrical properties of these structures are the same. The third nearest neighbour distance varies and hence they have different structures. Most of the III-V compounds also crystallizes in zinc blende structures, nitrides crystallizes in wurtzite structure.

GaAs and GaP are said to be binary semiconductors. GaAlAs is a ternary semiconductor. The addition of Al to GaAs increases the band gap. It is used in optoelectronic devices. The band gap and the lattice constants of some III-V and II-VI compounds are tabulated in Table 11.1.

Table 11.1 The band gap and the lattice constants of some III-V and II-VI compounds

Compound	*Band gap* (eV)	*Lattice spacings* (Å)
Group IV		
C	5.4	1.56
Si	1.11	5.43
Ge	0.67	5.66
Group III–V		
GaP	2.27	5.45
GaAs	1.42	5.65
GaSb	0.70	6.10
InP	1.27	5.80
InAs	0.36	6.06
InSb	0.17	6.48
Group II–VI		
ZnS	3.67	5.41
ZnSe	2.58	5.67
ZnTe	2.26	6.10
CdS	2.42	5.58
CdSe	1.74	6.05
CdTe	1.50	6.48

11.5 DIRECT AND INDIRECT BANDGAP SEMICONDUCTORS

In semiconductors, the separation of the valence band (VB) and the conduction band (CB) by a suitable energy is called bandgap. There are two types of bandgaps in semiconductors: (i) direct bandgap semiconductors and (ii) indirect bandgap semiconductors.

If a graph is drawn between energy E and wave vector k (or momentum, $p = \hbar k$, where $\hbar = h/2\pi = 1.054 \times 10^{-34}$ J-s and h is the Planck's constant), a plot as shown in Figure 11.6 is obtained. In some semiconductors, the lowest energy value in the conduction band and the highest energy value in the valence band have same momentum as shown in Figure 11.6(a). Such semiconductors are said to be direct bandgap semiconductors. The transition of electron from the lower energy level of the conduction band easily recombines with a hole in the valence band and the recombination energy is liberated. GaAs is one of the most studied direct bandgap semiconductor. In GaAs, the recombination energy is liberated in the form of light.

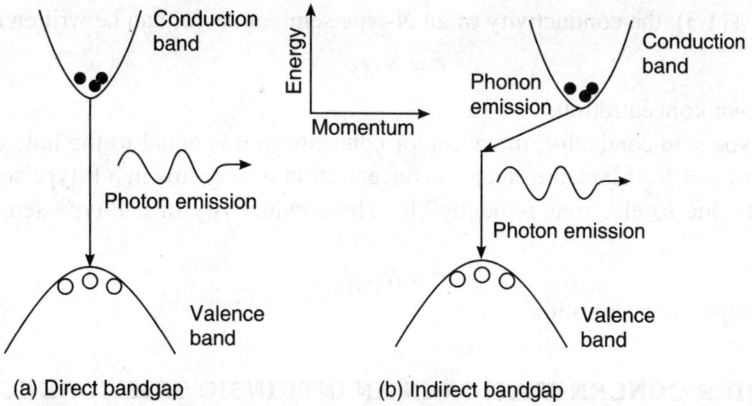

Figure 11.6 Semiconductor bandgap.

In certain semiconductors, the lowest energy of the conduction band and higher energy of the valence band have different momentum as shown in Figure 11.6(b). Such semiconductors are called indirect bandgap semiconductors. The recombination of electron and hole is not easily possible because the lowest energy value of the conduction band and highest energy value of the valence band have different momentum. In this case, the electron in the conduction band has to lose some of its energy by means of phonon emission and then it has to recombine with a hole in the valence band. Silicon is one of the examples for indirect bandgap semiconducting material. In silicon, the recombination energy is liberated in the form of heat.

11.6 CONDUCTIVITY OF A SEMICONDUCTING MATERIAL

The conductivity of a metal is given by

$$\sigma = \frac{ne^2\tau_r}{m} \tag{11.1}$$

and the mobility of electrons is given by

$$\mu = \frac{e\tau_r}{m} \tag{11.2}$$

Combining Eq. (11.1) and Eq. (11.2), the conductivity is given by $\sigma = ne\mu$. In the case of a semiconductor, since there are two charge carriers, electrons and holes, the conductivity for a semiconductor can be written as

$$\sigma = ne\mu_e + pe\mu_h \tag{11.3}$$

where the terms $ne\mu_e$ and $pe\mu_h$ are the conductivities due to electrons and holes respectively. In Eq. (11.3), n, e, p, μ_e and μ_h represent the concentrations of electrons, charge of electrons, concentration of hole, mobility of electrons and holes respectively.

For an intrinsic semiconductor, $n = p = n_i$. Equation (11.3) can be written as

$$\sigma = n_i e(\mu_e + \mu_h) \tag{11.4}$$

For an N-type semiconductor, $n = N_D$ and p is very small and hence the conductivity due to holes is negligible.

From Eq. (11.3), the conductivity of an N-type semiconductor can be written as

$$\sigma = N_D e \mu_e \tag{11.5}$$

where N_D is donor concentration.

For a P-type semiconductor, the acceptor concentration is equal to the hole concentration and it is given by $p = N_A$. Since the electron concentration is very low in a P-type semiconductor, the conductivity due to electrons is negligible. The conductivity of a P-type semiconductor is given by

$$\sigma = N_A e \mu_h \tag{11.6}$$

where N_A is acceptor concentration.

11.7 CARRIER CONCENTRATION IN AN INTRINSIC SEMICONDUCTOR

The carrier concentration in an intrinsic semiconductor is determined by finding the concentration of electrons in the conduction band and the concentration of holes in the valence band and by applying the law of mass action.

11.7.1 Concentration of Electrons in the Conduction Band

In a semiconductor, the conduction band is completely empty and the valence band is completely filled at 0 K. If the temperature becomes greater than 0 K, some of the electrons in valence band get excited into the conduction band. Let us calculate the number of electrons available in the conduction band, when $T > 0$ K. The electron concentration in the conduction band can be obtained using the equation,

$$n = \int_{-\infty}^{\infty} N(E) f(E)\, dE \tag{11.7}$$

where $N(E)$ is the number of electrons per unit volume and $f(E)$ is the probability that a quantum state is occupied by an electron. The values of $N(E)$ and $f(E)$ are

$$N(E) = \frac{4\pi}{h^3}(2m_e^*)^{3/2} E^{1/2} \tag{11.8}$$

and

$$f(E) = \frac{1}{1 + \exp\left(\frac{E - E_F}{kT}\right)} \tag{11.9}$$

where E_F is the Fermi energy and m_e^* is the effective mass of the electron. Substituting the values of $N(E)$ and $f(E)$ given by Eq. (11.8) and Eq. (11.9) in Eq. (11.7), we get

$$n = \int_{-\infty}^{\infty} \frac{4\pi}{h^3}(2m_e^*)^{3/2} E^{1/2} \times \frac{1}{1 + \exp\left(\frac{E - E_F}{kT}\right)} dE \tag{11.10}$$

The above integral varies from E_C to ∞, because the lower energy value of the conduction band is E_C and the upper energy value is ∞. The term E in Eq. (11.10) is replaced by $(E - E_c)$ because E_c is the lower energy value of the conduction band. Equation (11.10) can be written as

$$n = \frac{4\pi}{h^3}(2m_e^*)^{3/2}\int_{E_C}^{\infty}(E-E_C)^{1/2} \times \frac{1}{1+\exp\left(\frac{E-E_F}{kT}\right)}dE \tag{11.11}$$

Since $E > E_F$, the term $\exp\left(\frac{E-E_F}{kT}\right) > 1$. Therefore, neglecting 1, Eq. (11.11) can be written as

$$n = \frac{4\pi}{h^3}(2m_e^*)^{3/2}\int_{E_C}^{\infty}(E-E_C)^{1/2}\exp\left(-\frac{(E-E_F)}{kT}\right)dE \tag{11.12}$$

The above integral is solved by substituting $E - E_C = x$. Then $dE = dx$, the lower limit of this integral becomes 0 and the upper limit is ∞.

Equation (11.12) can be written as

$$n = \frac{4\pi}{h^3}(2m_e^*)^{3/2}\int_0^{\infty}x^{1/2}\exp\left(-\frac{(x+E_C-E_F)}{kT}\right)dx \tag{11.13}$$

Equation (11.13) can be written as

$$n = \frac{4\pi}{h^3}(2m_e^*)^{3/2}\int_0^{\infty}x^{1/2}\exp\left(-\frac{(E_C-E_F)}{kT}\right)\exp\left(\frac{-x}{kT}\right)dx \tag{11.14}$$

$$n = \frac{4\pi}{h^3}(2m_e^*)^{3/2}\exp\left(-\frac{(E_C-E_F)}{kT}\right)\int_0^{\infty}x^{1/2}\exp\left(\frac{-x}{kT}\right)dx \tag{11.15}$$

The integral $\int_0^{\infty}x^{1/2}\exp\left(\frac{-x}{kT}\right)dx$ can be evaluated by substituting $\frac{x}{kT} = y$.

Then $kT\,dy = dx$. There is no change in the upper and lower limit values,

i.e.
$$\int_0^{\infty}x^{1/2}\exp\left(\frac{-x}{kT}\right)dx = \int_0^{\infty}(ykT)^{1/2}\exp(-y)(kT)\,dy \tag{11.16}$$

Equation (11.16) can be written as

$$(kT)^{3/2}\int_0^{\infty}y^{1/2}\exp(-y)dy \tag{11.17}$$

The integral $\int_0^{\infty}y^{1/2}\exp(-y)dy$ can be solved using the Gamma function. The Gamma function is

$$\int_0^{\infty}y^{(n-1)}\exp(-y)dy = \Gamma(n) \tag{11.18}$$

The integral $\int_0^{\infty}y^{1/2}\exp(-y)dy$ can be rewritten as $\int_0^{\infty}y^{(3/2-1)}\exp(-y)dy$. From Eq. (11.18), the solution of $\int_0^{\infty}y^{(3/2-1)}\exp(-y)dy = \Gamma\left(\frac{3}{2}\right)$. Using the properties of the Gamma function, $\Gamma(n+1) = n\Gamma(n)$ and $\Gamma\left(\frac{1}{2}\right) = \sqrt{\pi}$, $\Gamma\left(\frac{3}{2}\right)$ can be written as $\Gamma\left(\frac{3}{2}\right) = \Gamma\left(\frac{1}{2}+1\right) = \frac{1}{2}\Gamma\left(\frac{1}{2}\right) = \frac{\sqrt{\pi}}{2}$.

Equation (11.15) can be written as

$$n = \frac{4\pi}{h^3}(2m_e^*)^{3/2}\exp\left(-\frac{(E_C-E_F)}{kT}\right)(kT)^{3/2}\frac{\sqrt{\pi}}{2} \tag{11.19}$$

or $$n = 2\left[\frac{2\pi m_e^* kT}{h^2}\right]^{3/2} \exp\left(-\frac{(E_C - E_F)}{kT}\right) \tag{11.20}$$

or $$n = N_C \exp\left(-\frac{(E_C - E_F)}{kT}\right) \tag{11.21}$$

where $N_C = 2\left[\frac{2\pi m_e^* kT}{h^2}\right]^{3/2}$

Equation (11.21) or Eq. (11.20) gives the electron concentration in the conduction band.

11.7.2 Concentration of Holes in the Valence Band

In a semiconductor, the conduction band is completely empty and the valence band is completely filled at 0 K. If the temperature becomes greater than 0 K, some of the electrons in valence band gets excited into the conduction band. Let us calculate the number of holes available in the valence band, when $T > 0$ K. The hole concentration in the valence band can be calculated using the equation,

$$p = \int_{-\infty}^{\infty} N(E)(1 - f(E))dE \tag{11.22}$$

where $N(E)$ is the number of electrons per unit volume and $[1 - f(E)]$ is the probability of the absence of the electron, i.e. the probability that a quantum state is occupied by a hole. The values of $N(E)$ and $[1 - f(E)]$ are

$$N(E) = \frac{4\pi}{h^3}(2m_h^*)^{3/2} E^{1/2} \tag{11.23}$$

and $$1 - f(E) = 1 - \frac{1}{1 + \exp\left(\frac{E - E_F}{kT}\right)} \tag{11.24}$$

where E_F is the Fermi energy and m_h^* is the effective mass of the hole. Equation (11.24) can be written as

$$1 - f(E) = \frac{1 + \exp\left(\frac{E - E_F}{kT}\right) - 1}{1 + \exp\left(\frac{E - E_F}{kT}\right)}$$

$$1 - f(E) = \frac{\exp\left(\frac{E - E_F}{kT}\right)}{1 + \exp\left(\frac{E - E_F}{kT}\right)}$$

Since $E < E_F$, the term $\exp\left(\frac{E - E_F}{kT}\right)$ in the denominator is less than 1. So, it is negligible. Therefore,

$$1 - f(E) = \exp\left(\frac{E - E_F}{kT}\right) \tag{11.25}$$

Substituting the values of $N(E)$ and $f(E)$ respectively given by Eq. (11.23) and Eq. (11.25) in Eq. (11.22), we get

$$p = \int_{-\infty}^{E_V} \frac{4\pi}{h^3}(2m_h^*)^{3/2} E^{1/2} \exp\left(\frac{E - E_F}{kT}\right) dE \tag{11.26}$$

The above integral varies from $-\infty$ to E_V, because the lower energy value of the valence band is $-\infty$ and the upper energy value is E_V. The term E in Eq. (11.26) is replaced by $(E_V - E)$ because E_V is the highest energy value in the valence band. Equation (11.26) can be rewritten as

$$p = \frac{4\pi}{h^3}(2m_e^*)^{3/2} \int_{-\infty}^{E_V} (E_V - E)^{1/2} \times \exp\left(\frac{E - E_F}{kT}\right) dE \tag{11.27}$$

Consider the integral in Eq. (11.27), i.e.

$$\int_{-\infty}^{E_V} (E_V - E)^{1/2} \exp\left(-\frac{(E_F - E)}{kT}\right) dE \tag{11.28}$$

The above integral is evaluated by substituting $E_V - E = x$. Then $-dE = dx$. The lower limit of this integral becomes ∞ and the upper limit is 0.

The integral in Eq. (11.28) can be written as

$$\int_{\infty}^{0} x^{1/2} \exp\left(-\frac{(E_F - (E_V - x))}{kT}\right)(-dx) \tag{11.29}$$

The integral in Eq. (11.29) can be written as

$$\int_{0}^{\infty} x^{1/2} \exp\left(-\frac{(E_F - E_V)}{kT}\right) \exp\left(\frac{-x}{kT}\right) dx \tag{11.30}$$

$$= \exp\left(-\frac{(E_F - E_V)}{kT}\right) \int_{0}^{\infty} x^{1/2} \exp\left(\frac{-x}{kT}\right) dx \tag{11.31}$$

The integral $\int_0^\infty x^{1/2} \exp\left(\frac{-x}{kT}\right) dx$ can be evaluated by substituting $\frac{x}{kT} = y$.

Then $kT\,dy = dx$. There is no change in the upper and lower limits values, i.e.

$$\int_0^\infty x^{1/2} \exp\left(\frac{-x}{kT}\right) dx = \int_0^\infty (ykT)^{1/2} \exp(-y)(kT)dy \tag{11.32}$$

Equation (11.32) can be written as

$$(kT)^{3/2} \int_0^\infty y^{1/2} \exp(-y)dy \tag{11.33}$$

The integral $\int_0^\infty y^{1/2} \exp(-y)dy$ can be solved by using the Gamma function. The Gamma function is

$$\int_0^\infty y^{(n-1)} \exp(-y)dy = \Gamma(n) \tag{11.34}$$

The integral $\int_0^\infty y^{1/2} \exp(-y)dy$ can be written as $\int_0^\infty y^{(3/2-1)} \exp(-y)dy$. From Eq. (11.34), the solution of $\int_0^\infty y^{(3/2-1)} \exp(-y)dy = \Gamma\left(\frac{3}{2}\right)$. Using the properties of the Gamma function, $\Gamma(n+1) = n\Gamma(n)$ and $\Gamma\left(\frac{1}{2}\right) = \sqrt{\pi}$, $\Gamma\left(\frac{3}{2}\right)$ can be written as $\Gamma\left(\frac{3}{2}\right) = \Gamma\left(\frac{1}{2}+1\right) = \frac{1}{2}\Gamma\left(\frac{1}{2}\right) = \frac{\sqrt{\pi}}{2}$.

Equation (11.27) can be written as

$$p = \frac{4\pi}{h^3}(2m_h^*)^{3/2} \exp\left(\frac{(E_V - E_F)}{kT}\right)(kT)^{3/2}\frac{\sqrt{\pi}}{2} \tag{11.35}$$

or

$$p = 2\left[\frac{2\pi m_h^* kT}{h^2}\right]^{3/2} \exp\left(\frac{(E_V - E_F)}{kT}\right) \tag{11.36}$$

or

$$p = N_V \exp\left(\frac{(E_V - E_F)}{kT}\right) \tag{11.37}$$

where $N_V = 2\left[\frac{2\pi m_h^* kT}{h^2}\right]^{3/2}$

Equation (11.36) or Eq. (11.37) gives the hole concentration in the valence band.

11.7.3 Law of Mass Action

The law of mass action states that the product of the electron concentration in the conduction band and the hole concentration in the valence band is equal to the square of the intrinsic carrier concentration of the material. Let n_i, n and p be the intrinsic concentrtaion of the material, the electron concentration in the conduction band and the hole concentration in the valence band respectively, then

$$n_i^2 = n \cdot p \tag{11.38}$$

Equation (11.38) is known as the law of mass action.

11.7.4 Intrinsic Carrier Concentration

From the law of mass action, $n_i^2 = n \cdot p$
Substituting the values of n and p given by Eq. (11.21) and Eq. (11.37), we get

$$n_i^2 = N_C N_V \exp\left[\frac{E_F - E_C}{kT}\right]\exp\left(\frac{(E_V - E_F)}{kT}\right) \tag{11.39}$$

$$n_i^2 = N_C N_V \exp\left[\frac{-(E_C - E_V)}{kT}\right] \tag{11.40}$$

Since $E_C - E_V = E_g$, the band gap of the material, Eq. (11.40) can be written as

$$n_i^2 = N_C N_V \exp\left[\frac{-E_g}{kT}\right] \tag{11.41}$$

Substituting the values of N_C and N_V in Eq. (11.41), we get

$$n_i^2 = 2\left[\frac{2\pi m_e^* kT}{h^2}\right]^{3/2} \times 2\left[\frac{2\pi m_h^* kT}{h^2}\right]^{3/2} \exp\left[\frac{-E_g}{kT}\right] \tag{11.42}$$

$$n_i^2 = 4\left[\frac{2\pi kT}{h^2}\right]^3 \times [m_e^* m_h^*]^{3/2} \exp\left[\frac{-E_g}{kT}\right] \tag{11.43}$$

Multiplying the numerator and denominator of Eq. (11.43) by m^3, we get

$$n_i^2 = 4\left[\frac{2\pi kmT}{h^2}\right]^3 \times \left[\frac{m_e^* m_h^*}{m^2}\right]^{3/2} \exp\left[\frac{-E_g}{kT}\right] \tag{11.44}$$

The value of n_i can be written as

$$n_i = 2\left[\frac{2\pi kmT}{h^2}\right]^{3/2} \times \left[\frac{m_e^* m_h^*}{m^2}\right]^{3/4} \exp\left[\frac{-E_g}{2kT}\right] \tag{11.45}$$

$$n_i = 2\left[\frac{2\pi km}{h^2}\right]^{3/2} \times \left[\frac{m_e^* m_h^*}{m^2}\right]^{3/4} T^{3/2} \exp\left[\frac{-E_g}{2kT}\right] \tag{11.46}$$

Equation (11.46) can be written as

$$n_i = AT^{3/2} \exp\left[\frac{-E_g}{2kT}\right] \tag{11.47}$$

where $A = 2\left[\frac{2\pi km}{h^2}\right]^{3/2} \times \left[\frac{m_e^* m_h^*}{m^2}\right]^{3/4} = 4.82 \times 10^{21} \times \left[\frac{m_e^* m_h^*}{m^2}\right]^{3/4}$

Equation (11.46) gives the intrinsic carrier concentration in semiconductor.

11.8 FERMI LEVEL IN AN INTRINSIC SEMICONDUCTOR

In an intrinsic semiconductor, the concentration of hole in the valence band is equal to the concentration of electron in the conduction band,

i.e.
$$n = p \tag{11.48}$$

Substituting the values of n and p, we get

$$N_C \exp\left(\frac{(E_F - E_C)}{kT}\right) = N_V \exp\left(\frac{(E_V - E_F)}{kT}\right) \tag{11.49}$$

$$\frac{N_C}{N_V} = \exp\left[\frac{(E_V - E_F - E_F + E_C)}{kT}\right] \tag{11.50}$$

$$\frac{N_C}{N_V} = \exp\left(\frac{(E_V + E_C - 2E_F)}{kT}\right)$$

Taking log on both sides, we get

$$\ln\frac{N_C}{N_V} = \frac{E_V + E_C - 2E_F}{kT} \tag{11.51}$$

$$E_C + E_V - 2E_F = kT\ln\frac{N_C}{N_V}$$

i.e.

$$E_F = \frac{E_C + E_V}{2} - \frac{kT}{2}\ln\frac{N_C}{N_V} \tag{11.52}$$

Now

$$\frac{N_C}{N_V} = \frac{2\left(\dfrac{2\pi m_e^* kT}{h^2}\right)^{3/2}}{2\left(\dfrac{2\pi m_h^* kT}{h^2}\right)^{3/2}} = \left(\frac{m_e^*}{m_h^*}\right)^{3/2} \tag{11.53}$$

From Eq. (11.52) and Eq. (11.53), we get

$$E_F = \frac{E_C + E_V}{2} - \frac{3kT}{4}\ln\left(\frac{m_e^*}{m_h^*}\right) \tag{11.54}$$

Equation (11.52) or Eq. (11.54) gives the value of the Fermi level in intrinsic semiconductor.

11.9 VARIATION OF FERMI LEVEL IN AN INTRINSIC SEMICONDUCTOR WITH TEMPERATURE

Case (i) When $T = 0$ K

When the temperature is 0 K, Eq. (11.54) becomes

$$E_F = \frac{E_C + E_V}{2} \tag{11.55}$$

The Fermi level lies in between the conduction band and the valence band, i.e. the Fermi level lies at the middle of the band gap (Figure 11.7).

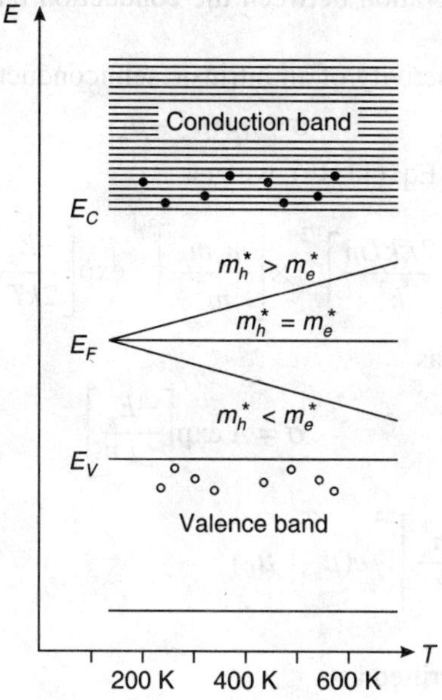

Figure 11.7 Variation of Fermi level in an intrinsic semiconductor.

Case (ii) When $T > 0$ K and $m_e^* = m_h^*$

When $T > 0$ K and $m_e^* = m_h^*$, Eq. (11.54) becomes

$$E_F = \frac{E_C + E_V}{2} \tag{11.56}$$

The Fermi level lies in between the conduction band and the valence band, i.e. the Fermi level lies at the middle of the band gap.

Case (iii) When $T > 0$ K and $m_e^* > m_h^*$

When $T > 0$ K and $m_e^* > m_h^*$, from (11.54), we infer that the Fermi level decreases with the increase in temperature.

Case (iv) When $T > 0$ K and $m_e^* < m_h^*$

When $T > 0$ K and $m_e^* < m_h^*$, $\ln\left(\frac{m_e^*}{m_h^*}\right)$ becomes negative. From (11.54), we infer that the Fermi level increases with the increase in temperature.

The variation of Fermi level in an intrinsic semiconductor with temperature is shown in Figure 11.7. In most of the semiconductors, $m_e^* \approx m_h^*$. Therefore, the increase or decrease of the Fermi level with temperature in an intrinsic semiconductor is negligible.

11.10 EXPRESSION FOR THE BAND GAP OF A SEMICONDUCTOR

The band gap is the energy seperation between the conduction band and the valence band of a semiconducting material.

From Eq. (11.4), the conductivity of an intrinsic semiconductor is given by

$$\sigma = n_i e(\mu_e + \mu_h) \tag{11.57}$$

Substituting the value of n_i from Eq. (11.45), we get

$$\sigma = 2\left[\frac{2\pi kTm}{h^2}\right]^{3/2} \times \left[\frac{m_e^* m_h^*}{m^2}\right]^{3/4} \exp\left[\frac{-E_g}{2kT}\right] e(\mu_e + \mu_h) \tag{11.58}$$

Equation (11.58) can be written as

$$\sigma = A \exp\left[\frac{-E_g}{2kT}\right] \tag{11.59}$$

where $A = 2\left[\frac{2\pi kTm}{h^2}\right]^{3/2} \times \left[\frac{m_e^* m_h^*}{m^2}\right]^{3/4} e(\mu_e + \mu_h)$

As $\sigma = \frac{1}{\rho}$, Eq. (11.59) can be written as

$$\rho = B \exp\left[\frac{E_g}{2kT}\right] \tag{11.60}$$

where B = A^{-1}. Using the equation, $\rho = \frac{RA}{d}$, Eq. (11.60) can be written as

$$R = B\frac{d}{A}\exp\left[\frac{E_g}{2kT}\right]$$

$$R = C \exp\left[\frac{E_g}{2kT}\right] \tag{11.61}$$

where $C = \frac{(Bd)}{A}$.

Taking log on both sides of Eq. (11.61), we get

$$\ln R = \ln C + \frac{E_g}{(2kT)} \tag{11.62}$$

The band gap is given by

$$E_g = 2kT(\ln R - \ln C) \tag{11.63}$$

Equation (11.63) gives an expression of the band gap of a semiconducting material. Equation (11.62) is in the form of a straight line. Comparing Eq. (11.62) with the straight line equation, $y =$

$mx + c$, we get $y = \ln R$, $m = \frac{E_g}{(2k)}$, $c = \ln C$, and $x = \frac{1}{T}$. By taking $\ln R$ in the y-axis and $\frac{1}{T}$ in the x-axis, if a graph is plotted between $\ln R$ and $\frac{1}{T}$, a straight line is obtained as shown in Figure 11.8.

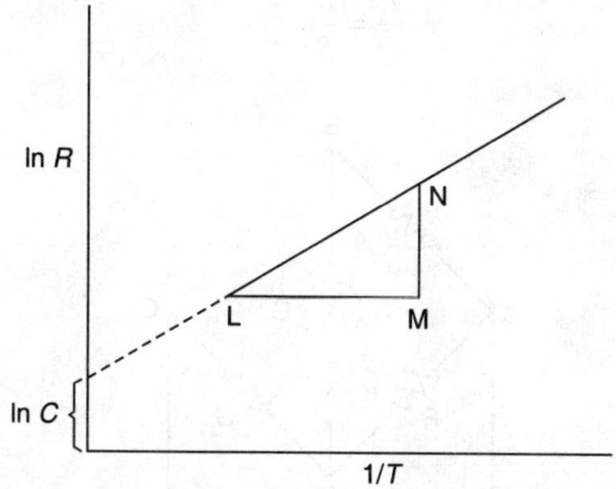

Figure 11.8 Variation of ln R with $1/T$ for a semiconductor.

By finding the slope of the straight line, the band gap of the semiconductor is determined using the relation,

$$E_g = 2k \times \text{slope of the straight line drawn between } \ln R \text{ and } \frac{1}{T} \qquad (11.64)$$

Using Eq. (11.64), the band gap of a semiconductor is determined.

11.11 EXPERIMENTAL DETERMINATION OF THE BAND GAP OF A SEMICONDUCTOR

The band gap of a semiconductor is determined by measuring its resistances at different temperatures and a graph is plotted between $\frac{1}{T}$ (T should be in kelvin) and $\ln R$. The band gap of the semiconductor is given by Eq. (11.64) as

$$E_g = 2\,k \times \text{slope of the straight line drawn between } \ln R \text{ and } \frac{1}{T}$$

where k is the Boltzmann's constant and it is equal to 1.38×10^{-23} J K^{-1}.

11.11.1 Measurement of Resistances at Different Temperatures

The resistance of a given semiconductor (thermister) is measured at different temperature either using post office box or Carey Foster bridge or using a simple circuit. Let us discuss about the measurement of resistances using post office box.

11.11.2 Principle of Post Office Box

A post office box is based on the principle of the Wheatstone's bridge network. The Wheatstone's bridge network has resistances in four arms, a galvanometer, a battery and keys as shown in Figure 11.9. The balanced condition in Wheatstone's bridge network is given by

$$\frac{P}{Q} = \frac{R}{S} \tag{11.65}$$

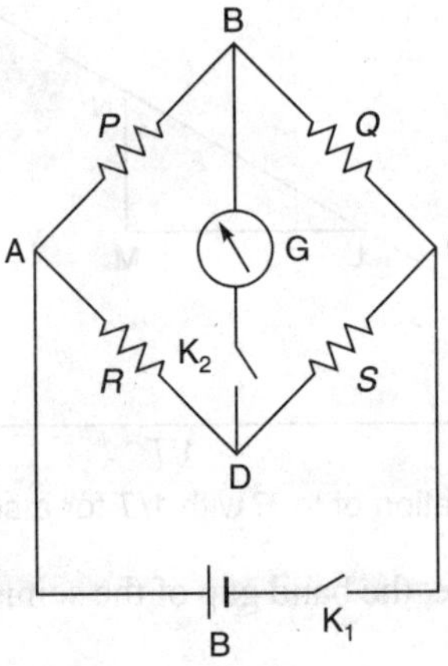

Figure 11.9 Wheatstone's bridge network.

11.11.3 Description of Post Office Box

A post office box consists of resistances *P*, *Q*, *R* either in the form of plug type or dial type and provisions for connecting the galvanometer, battery and the unkonwn resistance, *S*. The *P*-arm has 1 Ω, 10 Ω, 100 Ω, 1000 Ω resistances. The *Q*-arm also has 1 Ω, 10 Ω, 100 Ω, 1000 Ω resistances. The *R*-arm has reistances ranging from 1 Ω to 5000 Ω including an infinte resistance. The unknown resistance is connected in the *S*-arm.

11.11.4 Determination of Resistances Using Post Office Box

The circuit connection is made by connecting a battery, keys, a galvanometer, the unknown resistance (thermister) as shown in Figure 11.10. In order to check the circuit connection, introduce 10 Ω in the *P*-arm and introduce 10 Ω in the *Q*-arm. Introduce infinite resistance in the *R*-arm. By pressing the keys, the deflection in the galvanometer is noted. Introduce 0 Ω in the *R*-arm. Then the key is pressed and the deflection in the galvnometer is again noted. If opposite side deflection is obtained, the circuit connection is correct. Otherwise, the circuit has to be checked.

Initially the thermister is kept at room temperature. Its resistance is determined by introducing 10 Ω, 100 Ω, 1000 Ω in the *P*-arm as follows. To find the value of the unknown resistance, introduce 10 Ω in the *P*-arm and introduce 10 Ω in the *Q*-arm. The circuit is brought into the balanced condition by introducing suitable resistances in the *R*-arm. The unknown resistance value is, $R = S$, since $\frac{P}{Q} = \frac{10}{10} = 1$. The values are listed in Table 11.2.

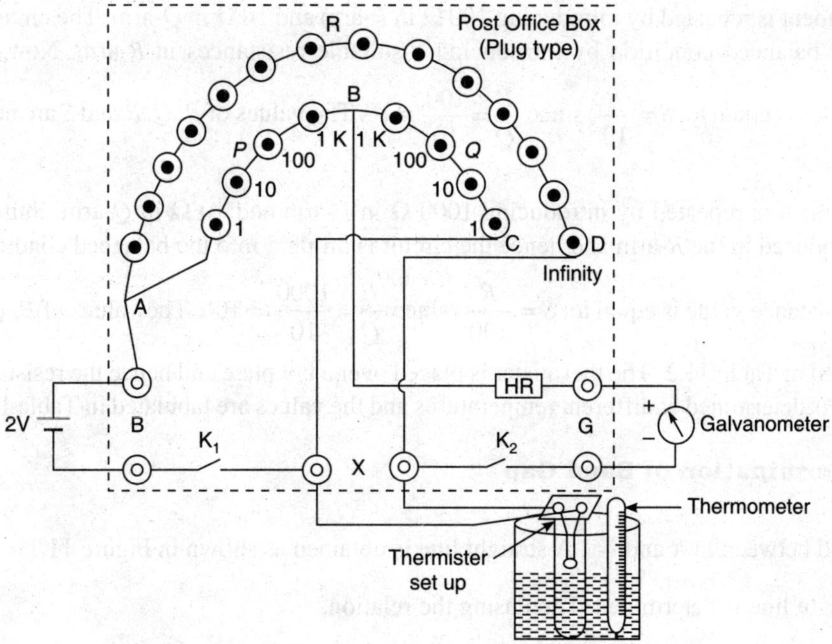

Figure 11.10 Detailed circuit connection in the post office box.

Table 11.2 Determination of resistance of a semiconductor at different temperatures

Temperature in °C	*Resistance in the first arm, P in Ω*	*Resistance in the second arm, Q in Ω*	*Resistance in the third arm, R in Ω*	*Unknown resistance, S = Q/P × R*
Room temperature, T = ... °C	10	10		
	100			
	1000			
40°C	10	10		
	100			
	1000			
45°C	10	10		
	100			
	1000			
50°C	10	10		
	100			
	1000			
55°C	10	10		
	100			
	1000			
60°C	10	10		
	100			
	1000			

The experiment is repeated by introducing 100 Ω in *P*-arm and 10 Ω in *Q*-arm. The circuit is brought into the balanced condition by introducing a suitable resistances in *R*-arm. Now, the unkonwn resistance is equal to, $S = \frac{R}{10}$, since $\frac{P}{Q} = \frac{100}{10} = 10$. The values of *P*, *Q*, *R* and *S* are noted in Table 11.2.

The experiment is repeated by introducing 1000 Ω in *P*-arm and 10 Ω in *Q*-arm. Suitable resistance is introduced in the *R*-arm and hence the circuit is brought into the balanced condition. The unkonwn resistance value is equal to, $S = \frac{R}{100}$, since $\frac{P}{Q} = \frac{1000}{10} = 100$. The values of *P*, *Q*, *R* and *S* are tabulated in Table 11.2. The thermister is placed over a hot plate and hence the resistance of the thermister is determined at different temperatures and the values are tabulated in Table 11.2.

11.11.5 Determination of Band Gap

A graph is plotted between ln *R* and $\frac{1}{T}$. A straight line is obtained as shown in Figure 11.11. The slope of the straight line is determined. Then using the relation,

$$E_g = 2k \times \text{slope of the straight line between } \ln R \text{ and } \frac{1}{T} \qquad (11.66)$$

The band gap value is converted into eV using the conversion factor, 1 eV = 1.6×10^{-19} J.

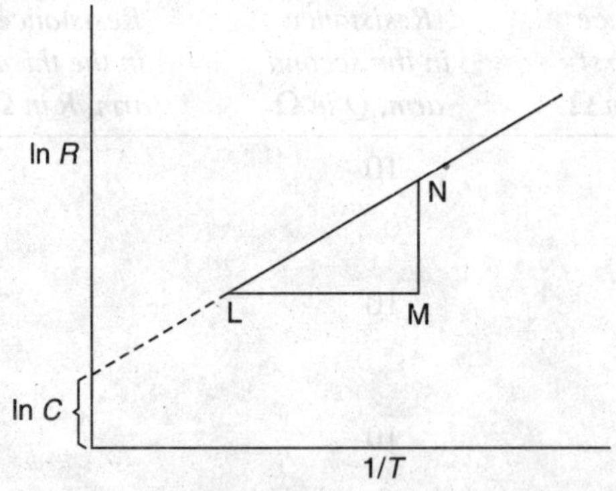

Figure 11.11 A plot between ln *R* and 1/*T*.

11.12 EXTRINSIC SEMICONDUCTOR

The impure semiconductor is said to be extrinsic semiconductor. There are two types of extrinsic semiconductors, namely (i) N-type semiconductor and (ii) P-type semiconductor. Doping of pentavalent impurity to Si converts the Si crystal into N-type semiconducting materials and the doping of trivalent impurities to Si makes the Si crystal as a P-type material.

The energy level of this donor impurities lies just below the conduction band of the semiconductor. This energy level is said to be donor level. It is represented by E_D. The donor level is shown in Figure 11.12(a). In a P-type material, the energy level of this acceptor impurity lies just above the valence band. This energy level is called acceptor level and it is represented as E_A. It is shown in Figure 11.12(b).

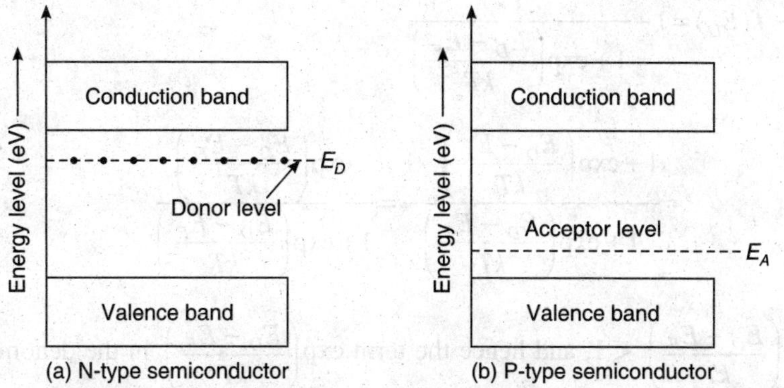

Figure 11.12 Energy bands.

The energy difference between donor energy level and the conduction band in an N-type Si and Ge semiconductor and the energy difference between the acceptor level and the valence band in a P-type Si and Ge semiconductor are listed in Table 11.3.

Table 11.3 Difference between donor energy level and the conduction band in an N-type Si and Ge semiconductor and the energy difference between the acceptor level and the valence band in a P-type Si and Ge semiconductor

	Impurities	*Germanium* (eV)	*Silicon* (eV)
Donor	Sb	0.0096	0.039
	P	0.0120	0.045
	As	0.0127	0.045
Acceptor	In	0.0112	0.160
	Ga	0.0108	0.065
	B	0.0104	0.045
	Al	0.0102	0.057

11.13 CARRIER CONCENTRATION OF AN N-TYPE SEMICONDUCTOR

The number of electrons present in the conduction band of the N-type semiconductor is given by the sum of the number of holes in the valence band and the number of ionized donor impurities. In mathematical form, the number of electrons in the conduction band is given by

$$n = N_h + N_D^+ \tag{11.67}$$

where N_h is the number of holes in the valence band and N_D^+ is the number of ionized donor impurities.

The number of ionized donor impurity is given by

$$N_D^+ = N_D[1 - f(E_D)] \tag{11.68}$$

where N_D is the number of donor impurities and $f(E_D)$ is the probability of finding an electron in the donor level and $[1 - f(E_D)]$ is the probability of finding a hole in the donor level. The value of $[1 - f(E_D)]$ is determined as follows:

$$1 - f(E_D) = 1 - \frac{1}{1 + \exp\left(\frac{E_D - E_F}{kT}\right)} \tag{11.69}$$

$$= \frac{1 + \exp\left(\frac{E_D - E_F}{kT}\right) - 1}{1 + \exp\left(\frac{E_D - E_F}{kT}\right)} = \frac{\exp\left(\frac{E_D - E_F}{kT}\right)}{1 + \exp\left(\frac{E_D - E_F}{kT}\right)} \tag{11.70}$$

Since $E_D < E_F$, $\exp\left(\frac{E_D - E_F}{kT}\right) < 1$, and hence the term $\exp\left(\frac{E_D - E_F}{kT}\right)$ in the denominator is negligible. The value of $[1 - f(E_D)]$ is written as

$$1 - f(E) = \exp\left(\frac{E_D - E_F}{kT}\right) \tag{11.71}$$

Therefore,
$$N_D^+ = N_D \exp\left(\frac{E_D - E_F}{kT}\right) \tag{11.72}$$

At $T = 0$ K, all the states in the valence band are occupied by the electrons and hence, N_h is equal to zero. Therefore, $n = N_D^+$.

$$n = N_D \exp\left(\frac{E_D - E_F}{kT}\right) \tag{11.73}$$

The concentration of electrons, in the conduction band is given by

$$n = N_C \exp\left(\frac{E_F - E_C}{kT}\right) \tag{11.74}$$

From Eq. (11.73) and Eq. (11.74), we get

$$N_C \exp\left(\frac{E_F - E_C}{kT}\right) = N_D \exp\left(\frac{E_D - E_F}{kT}\right) \tag{11.75}$$

$$\frac{N_C}{N_D} = \exp\left(\frac{-E_F + E_C}{kT}\right) \exp\left(\frac{E_D - E_F}{kT}\right) \tag{11.76}$$

$$\frac{N_C}{N_D} = \exp\left(\frac{-2E_F + E_C + E_D}{kT}\right) \tag{11.77}$$

Taking log on both sides of Eq. (11.77), we get

$$\frac{-2E_F + E_C + E_D}{kT} = \ln \frac{N_C}{N_D} \tag{11.78}$$

Rearranging Eq. (11.78), we get

$$E_F = \frac{E_C + E_D}{2} - \frac{kT}{2} \ln \frac{N_C}{N_D} \tag{11.79}$$

Equation (11.79) gives the equation for the Fermi level in an N-type semiconductor.

The concentration of electrons, in the conduction band is given by

$$n = N_C \exp\left(\frac{E_F - E_C}{kT}\right) \tag{11.80}$$

Substituting the value of E_F given by Eq. (11.79) in Eq. (11.80), we get

$$n = N_C \exp\left(\frac{E_C + E_D}{2kT} - \frac{E_C}{kT} - \frac{1}{2} \ln \frac{N_C}{N_D}\right) \tag{11.81}$$

$$n = N_C \exp\left(\frac{E_D - E_C}{2kT}\right)\left(\frac{N_D}{N_C}\right)^{1/2} \tag{11.82}$$

$$n = (N_C N_D)^{1/2} \exp\left(\frac{E_D - E_C}{2kT}\right) \tag{11.83}$$

Substituting the value of $N_C = 2\left(\frac{2\pi m_e^* kT}{h^2}\right)^{3/2}$ in Eq. (11.83), we get

$$n = (2N_D)^{1/2} \left(\frac{2\pi m_e^* kT}{h^2}\right)^{3/4} \exp\left(\frac{E_D - E_C}{2kT}\right) \tag{11.84}$$

Equation (11.84) gives the carrier concentration of electrons in an *N*-type material.

11.14 VARIATION OF FERMI LEVEL IN AN N-TYPE SEMICONDUCTOR WITH TEMPERATURE

11.14.1 Variation of Fermi Level with Temperature

For an N-type semiconductor, the expression for the Fermi level is given by

$$E_F = \frac{E_C + E_D}{2} - \frac{kT}{2} \ln \frac{N_C}{N_D} \tag{11.85}$$

(i) At $T = 0$ K

When the temperature is 0 K, Eq. (11.85) can be written as

$$E_F = \frac{E_C + E_D}{2} \tag{11.86}$$

It indicates that at 0 K, the Fermi level of an N-type semiconductor lies at the middle of the donor energy level and the lowest energy of the conduction band.

(ii) When $T > 0$ K

When temperature increases, the term, $\frac{kT}{2}\ln\frac{N_C}{N_D}$, also increases. From Eq. (11.85), we infer that the Fermi level decreases with the increase in temperature as shown in Figure 11.13. At a high temperature, the Fermi level lies at the middle of the band gap. This indicates that at high temperatures more number of electrons are dislodged from valence band and get excited to the conduction band. The number of holes in valence band is nearly equal to the number of electrons in the conduction band and hence an N-type semiconductor behaves as an intrinsic semiconductor at very high temperature.

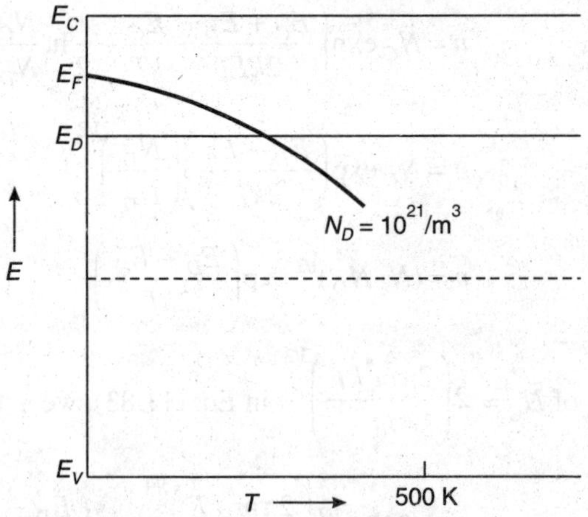

Figure 11.13 Variation of Fermi level with temperature in an N-type semiconductor.

11.14.2 Variation of Fermi Level with Concentration

If the donor concentration (N_D) is increased, the decrease of the Fermi level is minimized and hence it moves up as shown in Figure 11.14.

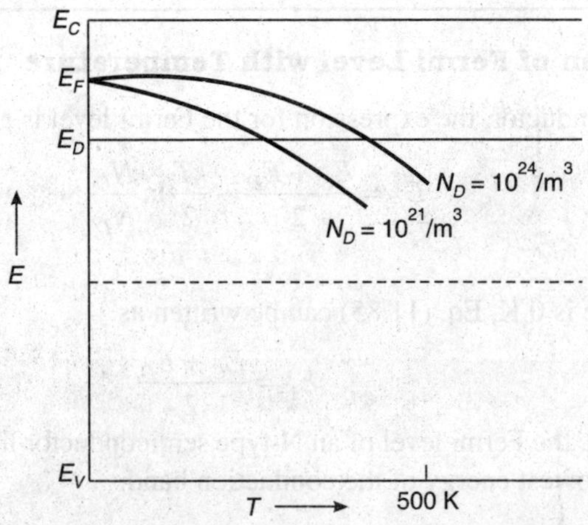

Figure 11.14 Variation of Fermi level with temperature and concentration in an N-type semiconductor.

11.15 P-TYPE SEMICONDUCTOR

The number of holes present in the valence band of the P-type semiconductor is given by the sum of the number of electrons in the conduction band and the number of ionized acceptor impurities. In mathematical form, the number of holes in the valence band is given by

$$p = N_e + N_A^+ \tag{11.87}$$

where N_e is the number of electrons in the conduction band and N_A^+ is the number of ionized acceptor impurities. The number of ionized acceptor impurity is given by

$$N_A^+ = N_A f(E_A) \tag{11.88}$$

where N_A is the number of acceptor impurities and $f(E_A)$ is the probability of finding an electron in the acceptor level. The value of $f(E_A)$ is determined as follows:

$$f(E_A) = \frac{1}{1 + \exp\left(\dfrac{E_A - E_F}{kT}\right)} \tag{11.89}$$

Since $E_A > E_F$, $\exp\left(\dfrac{E_A - E_F}{kT}\right) > 1$, and hence the term 1 in the denominator is negligible. The value of $f(E_A)$ is written as

$$f(E_A) = \exp\left(-\frac{E_A - E_F}{kT}\right) \tag{11.90}$$

Therefore,

$$N_A^+ = N_A \exp\left(\frac{E_F - E_A}{kT}\right) \tag{11.91}$$

At $T = 0$ K, all the states in the conduction band is empty and hence, N_e is equal to zero. Therefore, $p = N_A^+$.

$$p = N_A \exp\left(\frac{E_F - E_A}{kT}\right) \tag{11.92}$$

The concentration of holes in the valence band is given by

$$p = N_V \exp\left(\frac{E_V - E_F}{kT}\right) \tag{11.93}$$

From Eq. (11.92) and Eq. (11.93), we get

$$N_V \exp\left(\frac{E_V - E_F}{kT}\right) = N_A \exp\left(\frac{E_F - E_A}{kT}\right) \tag{11.94}$$

$$\frac{N_A}{N_V} = \exp\left(\frac{E_V - E_F}{kT}\right) \exp\left(-\frac{E_F - E_A}{kT}\right) \tag{11.95}$$

$$\frac{N_A}{N_V} = \exp\left(\frac{-2E_F + E_V + E_A}{kT}\right) \tag{11.96}$$

Taking log on both sides of Eq. (11.96), we get

$$\frac{-2E_F + E_V + E_A}{kT} = \ln\frac{N_A}{N_V} \tag{11.97}$$

Rearranging Eq. (11.97), we get

$$E_F = \frac{E_V + E_A}{2} - \frac{kT}{2}\ln\frac{N_A}{N_V} \tag{11.98}$$

Equation (11.98) gives the equation for the Fermi level in a P-type semiconductor.

The concentration of holes in the valence band is given by

$$p = N_V \exp\left(\frac{E_V - E_F}{kT}\right) \tag{11.99}$$

Substituting the value of E_F given by Eq. (11.98) in Eq. (11.99), we get

$$p = N_V \exp\left(\frac{E_V}{kT} - \frac{E_V + E_A}{2kT} + \frac{1}{2}\ln\frac{N_A}{N_V}\right) \tag{11.100}$$

$$p = N_V \exp\left(\frac{E_V - E_A}{2kT}\right)\left(\frac{N_A}{N_V}\right)^{1/2} \tag{11.101}$$

$$p = (N_V N_A)^{1/2} \exp\left(\frac{E_V - E_A}{2kT}\right) \tag{11.102}$$

Substituting the value of $N_V = 2\left(\frac{2\pi m_h^* kT}{h^2}\right)^{3/2}$ in Eq. (11.102), we get

$$p = (2N_A)^{1/2}\left(\frac{2\pi m_h^* kT}{h^2}\right)^{3/4} \exp\left(\frac{E_V - E_A}{2kT}\right) \tag{11.103}$$

Equation (11.103) gives the carrier concentration of holes in a P-type material.

11.16 VARIATION OF FERMI LEVEL WITH TEMPERATURE AND CONCENTRATION

11.16.1 Variation of Fermi Level with Temperature

The expression for the Fermi level of a P-type semiconductor is given by

$$E_F = \frac{E_V + E_A}{2} - \frac{kT}{2}\ln\frac{N_A}{N_V} \tag{11.104}$$

or

$$E_F = \frac{E_V + E_A}{2} + \frac{kT}{2}\ln\frac{N_V}{N_A} \tag{11.105}$$

(i) At $T = 0$ K

Substituting $T = 0$ K, in Eq. (11.105), we get

$$E_F = \frac{E_V + E_A}{2} \tag{11.106}$$

It represents that the Fermi level lies in between the valence band and acceptor level at 0 K for a P-type semiconducting material.

(ii) When $T > 0$ K

If temperature increases, the term $\frac{kT}{2}\ln\frac{N_V}{N_A}$ increases and hence, the Fermi level increases, as shown in Figure 11.15. At high temperature, the Fermi level lies at the middle of the energy gap. It shows that a P-type semiconductor becomes an intrinsic semiconductor at high temperature.

At high temperature, more number of electrons get dislodged from the valence band and excited into the conduction band. The number of electrons in the conduction band and the number of holes in the valence band are nearly equal at very high temperature and hence the material becomes an intrinsic semiconductor.

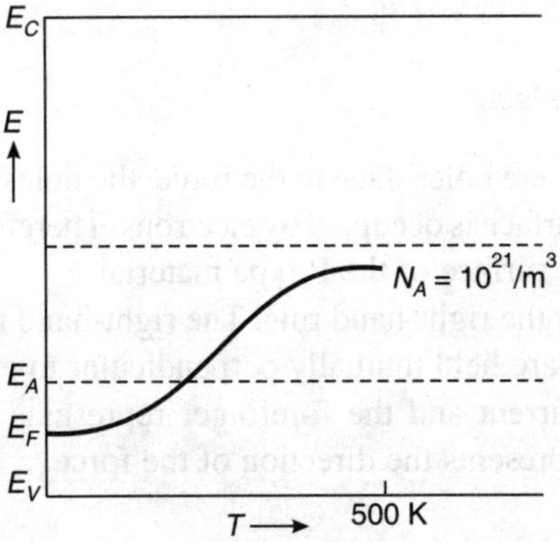

Figure 11.15 Variation of Fermi level of a P-type semiconductor with the temperature.

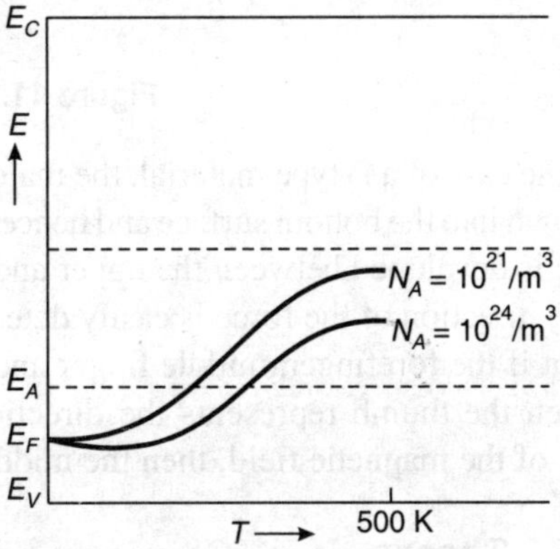

Figure 11.16 Variation of E_F in P-type semiconductor with temperature and carrier concentration.

11.16.2 Variation of Fermi Level with Concentration for a P-type Semiconductor

If the acceptor concentration is increased, the increase in E_F is minimized and hence the Fermi level moves down as shown in Figure 11.16.

11.17 HALL EFFECT

Consider that a current, I_X is applied to a semiconducting or a metal specimen along the X-direction and a magnetic field B_Z is applied to the specimen along the Z-direction and then a voltage is developed along the negative Y-direction as shown in Figure 11.17(a). This phenomenon is known as **Hall effect**. The voltage developed is known as **Hall voltage**.

The origin of the Hall voltage is easily explained. Due to the application of both current and the magnetic field perpendicularly, a downward force will be developed along a mutually perpendicular direction. As a result of that force, the charge carriers are forced down into the bottom surface. In the case of a metal, the electrons are the charge carriers. So, in metals, the electrons are forced down into the bottom surface and hence the upper surface is less negative and a voltage is developed between the upper and the bottom surface of the specimen.

In the N-type semiconductor, since the electrons are the charge carriers, the electrons are forced down into the bottom surface and hence the upper surface is occupied by the holes. Therefore, a potential difference is appearing between the upper and the bottom surface.

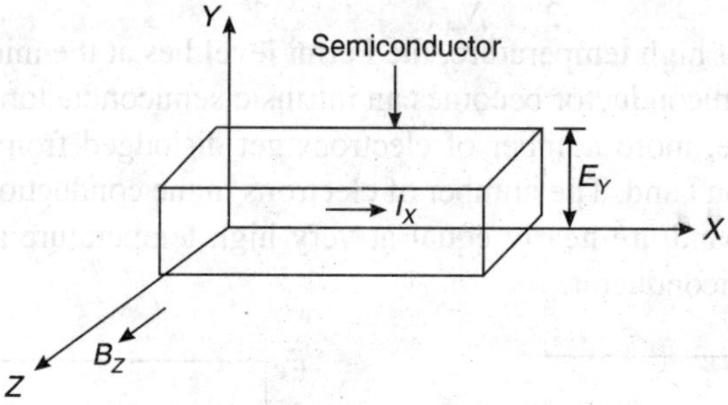

Figure 11.17(a) Hall effect.

In the case of a P-type material, the majority carriers are holes. Due to the force, the holes are forced down into the bottom surface and hence the upper surface is occupied by electrons. Therefore, a voltage is developed between the upper and the bottom surface of the P-type material.

The direction of the force is easily determined from the right-hand rule. The right-hand rule states that if the forefinger, middle finger and the thumb are held mutually perpendicular to each other, then the thumb represents the direction of the current and the forefinger represents the direction of the magnetic field, then the middle finger represents the direction of the force.

11.17.1 Theory

Let I_X be the current applied along the X-direction and B_Z be the magnetic field applied along the Z-direction, then the force due to the magnetic field is $B_Z e v_X$ and the force due to the electric field is eE_H, where E_H is the Hall electric field intensity. At equilibrium,

$$eE_H = B_Z e v_X \tag{11.107}$$

Cancelling the common terms, we get

$$E_H = B_Z v_X \tag{11.108}$$

The current density is given by

$$J = nev_X \tag{11.109}$$

From Eq. (11.108) and Eq. (11.109), we get

$$E_H = B_Z \frac{J}{ne} \tag{11.110}$$

The current density is also given by

$$J = \frac{I}{A} \tag{11.111}$$

From Eq. (11.110) and Eq. (11.111), we get

$$E_H = B_Z \frac{I}{neA} \tag{11.112}$$

Substituting $E_H = \frac{V_H}{d}$ in Eq. (11.112), we get

$$\frac{V_H}{d} = B_Z \frac{I}{neA} \tag{11.113}$$

By introducing the term, the Hall coefficient, R_H, where $R_H = \frac{1}{ne}$, Eq. (11.113) may be written as

$$\frac{V_H}{d} = \frac{B_Z I R_H}{A} \tag{11.114}$$

By substituting $A = w \times d$, (area of cross section = width × thickness), Eq. (11.114) can be written as

$$\frac{V_H}{d} = \frac{B_Z I R_H}{wd}$$

i.e.

$$V_H = \frac{B_Z I R_H}{w} \tag{11.115}$$

From Eq. (11.115), the Hall coefficient is given by

$$R_H = \frac{V_H w}{B_Z I} \tag{11.116}$$

The conductivity is given by, $\sigma = ne\mu$. From this equation, $\mu = R_H \sigma$. (11.117)

Equation (11.115) and Eq. (11.116) are derived by assuming that the velocity of the electrons is constant. But due to thermal agitation the velocity of the electrons is not a constant. It is randomly distributed. So, a correction factor $3\pi/8$ is introduced in the Hall coefficient. Therefore, R_H can be written as

$$R_H = \frac{3\pi}{8}\left(\frac{1}{ne}\right) \tag{11.118}$$

The term, charge density is defined as, $\rho = ne$. Therefore, Eq. (11.118) can be written as

$$R_H = \frac{3\pi}{8}\left(\frac{1}{\rho}\right) \tag{11.119}$$

For an N-type material, since the charge is negative, R_H is also negative and for a P-type material, R_H is positive.

Hall angle

The resultant electric field intensity in a semiconductor specimen is the vector sum of E_H and E_X. It is at an angle of θ_H, where θ_H is called the Hall angle [Figure 11.17(b)].

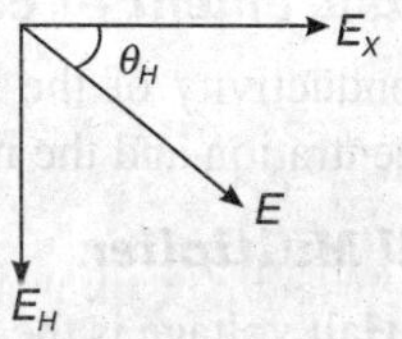

Figure 11.17(b) Hall angle.

The Hall angle θ_H is given by

$$\tan\theta_H = \frac{E_H}{E_X} = \frac{V_H}{d \times E_X}$$

Substituting the value of V_H from Eq. (11.115), we get

$$\tan\theta_H = \frac{B_Z I_X R_H}{w \times d \times E_X} = \frac{B_Z J_X R_H}{E_X} = \frac{B_Z \sigma E_X R_H}{E_X} = B_Z \sigma R_H = \mu B_Z$$

The Hall angle, θ_H is given by, $\theta_H = \tan^{-1}(\mu B_Z)$.

11.17.2 Experimental Measurement of Hall Voltage

A semiconducting specimen of width w, length l and thickness t is placed between the pole pieces of an electromagnet in such a way that the magnetic field is applied along the Z-direction. Apply a current of I ampere along the X-direction of the specimen. The Hall voltage is measured between the upper and the lower surface of the specimen as shown in Figure 11.18. The experiment is repeated for different values of I and B and hence the corresponding value of V is measured. The average value of V is taken as the Hall voltage and hence the value of R_H is determined from Eq. (11.116). The experiment is also repeated by varying the temperature of the specimen and hence the Hall voltage is measured at different temperatures.

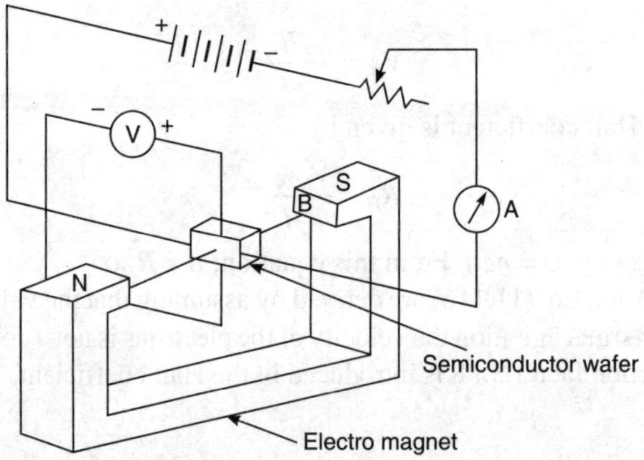

Figure 11.18 Experimental measurement—Hall effect.

11.17.3 Uses of Hall Effect

Semiconductor type

It is used to find out whether the given semiconductor is N-type or P-type.

Measurement of Hall voltage, Hall angle

It is used to measure the Hall voltage, Hall angle, and the Hall coefficient.

Measurement of carrier concentration and mobility

If conductivity of the specimen is also simultaneously measured, it is used to find the carrier concentration and the mobility of the charge carriers.

Hall Multiplier

The Hall voltage is the product of two input quantities, namely the current and the magnetic field. Using this principle, the Hall effect device is used as a multiplier.

Hall effect sensor

It is used as a magnetic field sensor. Using the Hall effect devices, the magnetic field ranging from 1 μT to 1 T is sensed. By using the required semiconducting materials, the Hall effect devices are used to sense a magnetic field as low as the earth's magnetic field (50 μT) and some high value of magnetic field such as 1 T.

SOLVED PROBLEMS

11.1 If the mobilities of electrons and holes in an intrinsic semiconductor at 300 K are 0.36 and 0.14 $m^2 V^{-1} s^{-1}$ respectively, calculate the number of charge carriers (Given that the conductivity is 2.2 $\Omega^{-1} m^{-1}$).

Given data

Mobility of the electrons $\mu_e = 0.36\ m^2 V^{-1} s^{-1}$

Mobility of the holes $\mu_h = 0.14\ m^2 V^{-1} s^{-1}$

Conductivity $\sigma = 2.2\ \Omega^{-1} m^{-1}$

Temperature $T = 300$ K

Solution

Electrical conductivity

$$\sigma = n_i e(\mu_e + \mu_h)$$

$$n_i = \frac{\sigma}{e(\mu_e + \mu_h)} = \frac{2.2}{1.6 \times 10^{-19} \times (0.36 + 0.14)}$$

$$= 2.75 \times 10^{19}\ m^{-3}$$

The carrier concentration of an intrinsic semiconductor = $2.75 \times 10^{19}\ m^3$.

11.2 The conductivity of a semiconductor at 20°C is 250 $\Omega^{-1} m^{-1}$ and at 100°C is 1100 $\Omega^{-1}m^{-1}$. What is its band gap, E_g?

Given data

Conductivity at 20°C = 250 $\Omega^{-1} m^{-1}$

Conductivity at 100°C = 1100 $\Omega^{-1} m^{-1}$

Solution

Electrical conductivity $\sigma = n_i e(\mu_e + \mu_h)$

where n_i is intrinsic concentration of carriers and it is given by

$$n_i = 2\left[\frac{2\pi kTm}{h^2}\right]^{3/2} \times \left[\frac{m_e^* m_h^*}{m^2}\right]^{3/4} \exp\left[\frac{-E_g}{2kT}\right]$$

The intrinsic carrier concentration at 20°C is given by

$$n_{i20} = 2\left[\frac{2\pi km \times 293}{h^2}\right]^{3/2} \times \left[\frac{m_e^* m_h^*}{m^2}\right]^{3/4} \exp\left[\frac{-E_g}{2k \times 293}\right] \quad (11.120)$$

The intrinsic carrier concentration at 100°C is given by

$$n_{i100} = 2\left[\frac{2\pi km \times 373}{h^2}\right]^{3/2} \times \left[\frac{m_e^* m_h^*}{m^2}\right]^{3/4} \exp\left[\frac{-E_g}{2k \times 373}\right] \quad (11.121)$$

Dividing Eq. (11.120) by Eq. (11.121), we get

$$\frac{n_{i20}}{n_{i100}} = \left[\frac{293}{373}\right]^{3/2} \times \frac{\exp\left(-\frac{E_g}{2 \times k \times 293}\right)}{\exp\left(-\frac{E_g}{2 \times k \times 373}\right)} \quad (11.122)$$

Rearranging Eq. (11.122), we get

$$\frac{n_{i20}}{n_{i100}} = \left[\frac{293}{373}\right]^{3/2} \times \exp\left(-\frac{E_g}{2 \times k}\left(\frac{1}{293} - \frac{1}{373}\right)\right) \quad (11.123)$$

The ratio between the σ_{i20} and σ_{i100} is given by

$$\frac{\sigma_{i20}}{\sigma_{i100}} = \frac{n_{i20}e(\mu_e + \mu_h)}{n_{i100}e(\mu_e + \mu_h)} = \frac{n_{i20}}{n_{i100}} \quad (11.124)$$

From Eq. (11.123) and Eq. (11.124), we get

$$\frac{\sigma_{i20}}{\sigma_{i100}} = \frac{n_{i20}}{n_{i100}} = \left[\frac{293}{373}\right]^{3/2} \times \exp\left(-\frac{E_g}{2 \times k}\left(\frac{1}{293} - \frac{1}{373}\right)\right) \quad (11.125)$$

$$\exp\left(-\frac{E_g}{2 \times k}\left(\frac{1}{293} - \frac{1}{373}\right)\right) = \frac{\sigma_{i20}}{\sigma_{i100}} \times \left(\frac{373}{293}\right)^{3/2} \quad (11.126)$$

Taking natural log on both sides

$$-\frac{E_g}{2 \times k}\left(\frac{1}{293} - \frac{1}{373}\right) = \ln\left(\frac{\sigma_{i20}}{\sigma_{i100}} \times \left(\frac{373}{293}\right)^{3/2}\right)$$

$$E_g = 2 \times k \times \left(\frac{1}{373} - \frac{1}{293}\right)^{-1} \times \ln\left(\frac{\sigma_{i20}}{\sigma_{i100}} \times \left(\frac{373}{293}\right)^{3/2}\right) \quad (11.127)$$

Substituting the values in Eq. (11.127), we get

$$E_g = 2 \times 1.38 \times 10^{-23} \times \left(\frac{1}{373} - \frac{1}{293}\right)^{-1} \times \ln\left(\frac{250}{1100} \times \left(\frac{373}{293}\right)^{3/2}\right) \quad (11.128)$$

$$E_g = 4.221 \times 10^{-20} \text{ J} = 0.2638 \text{ eV}$$

The band gap of the semiconductor is 0.2638 eV.

11.3 A silicon plate of thickness 1 mm, breadth 10 mm and length 100 mm is placed in a magnetic field of 0.5 Wb m^{-2} acting perpendicular to its thickness. If 10^{-2} A current flows along its length, calculate the Hall voltage developed if the Hall coefficient is 3.66×10^{-4} m^3/C.

Given data
Current $I = 10^{-2}$ A
length $l = 100$ mm
thickness $d = 1$ mm
breadth $b = 10$ mm
magnetic field $B = 0.5$ Wb m^{-2}
Hall coefficient $R_H = 3.66 \times 10^{-4}$ m^3/C

Solution

Hall voltage, $$V_H = \frac{BIR_H}{w}$$

$$= \frac{0.5 \times 10^{-2} \times 3.66 \times 10^{-4}}{10 \times 10^{-3}} = 1.83 \times 10^{-4} \text{ V}$$

The Hall voltage is 1.83×10^{-4} V.

11.4 Find the concentration of holes and electrons in N-type silicon at 300 K, if the conductivity is 300 S/cm. Also, find these values for P-type silicon. Given that for silicon, at 300 K, $n_i = 1.5 \times 10^{10}$ cm^{-3}, μ_e =1300 cm^2/Vs and μ_h = 500 cm^2/Vs.

Given data
Conductivity $\sigma = 300$ S/cm $= 3 \times 10^4$ S/m
temperature $T = 300$ K

$$n_i = 1.5 \times 10^{10} \text{ cm}^{-3},$$
$$\mu_e = 1300 \text{ cm}^2/\text{Vs}$$
$$\mu_h = 500 \text{ cm}^2/\text{Vs}$$

Solution

(i) For N-type material, $\sigma = N_D e \mu_e$

The concentration of electron is given by

$$N_D = \frac{\sigma}{e\mu_e}$$

$$N_D = \frac{3 \times 10^4}{1.6 \times 10^{-19} \times 0.13}$$

$$= 1.4423 \times 10^{24} \text{ m}^{-3}$$

The hole concentration is, $p \cdot N_D = n_i^2$

$$p = \frac{n_i^2}{N_D} = \frac{(1.5 \times 10^{16})^2}{1.4423 \times 10^{24}} = 1.56 \times 10^8 \text{ m}^{-3}.$$

(ii) For P-type material, $\sigma = N_A e \mu_h$

The concentration of hole is given by

$$N_A = \frac{\sigma}{e\mu_h}$$

$$N_A = \frac{3\times10^4}{1.6\times10^{-19}\times0.05}$$

$$= 3.75 \times 10^{24}\ \text{m}^{-3}$$

The electron concentration is, $n \cdot N_A = n_i^2$

$$n = \frac{n_i^2}{N_A} = \frac{(1.5\times10^{16})^2}{3.75\times10^{24}} = 6\times10^7\ \text{m}^{-3}$$

For N-type semiconductor

(i) The concentration of electron = 1.4423×10^{24} m^{-3}

(ii) The concentration of hole = 1.56×10^8 m^{-3}

For P-type semiconductor

(i) The concentration of hole = 3.75×10^{24} m^{-3}

(ii) The concentration of electron = 6×10^7 m^{-3}.

11.5 The Hall coefficient of a semiconductor was obtained as -3.68×10^{-5} m^3/C. What is the type of charge carriers? Also calculate the carrier concentration. (electron charge = 1.6×10^{-19} C)

Given data

Hall coefficient $R_H = -3.68 \times 10^{-5}$ m^3/C

Electron charge $e = 1.6 \times 10^{-19}$ C

Solution

Since the Hall coefficient is negative, the charge carriers of the semiconductors are electrons. The Hall coefficient is given by

$$R_H = \frac{3\pi}{8}\left(\frac{1}{ne}\right)$$

i.e.
$$n = \frac{3\pi}{8R_He} = \frac{3\pi}{8\times3.68\times10^{-5}\times1.6\times10^{-19}}$$

$$n = 2.0008 \times 10^{23}\ \text{m}^{-3}$$

The carrier concentration is = 2×10^{23} m^{-3}.

11.6 The energy gap of two intrinsic semiconductors A and B are 0.36eV and 0.72 eV respectively. Compare the intrinsic carrier density of A to B at 300 K. (Given $m_h^* = m_e^* = 9 \times 10^{-31}$ kg and $2\,kT = 0.052$ eV)

Given data

Energy gap of the first material $E_{g1} = 0.36$ eV

Energy gap of the second material $E_{g2} = 0.72$ eV

Temperature $T = 300$ K

$$m_h^* = m_e^* = 9 \times 10^{-31}\ \text{kg}$$

$$2\,kT = 0.052\ \text{eV}$$

Solution

The intrinsic concentration of carriers is given by

$$n_i = 2\left[\frac{2\pi kTm}{h^2}\right]^{3/2} \times \left[\frac{m_e^* m_h^*}{m^2}\right]^{3/4} \exp\left[\frac{-E_g}{2kT}\right]$$

The intrinsic carrier concentration for the materials *A* and *B* can be written as

$$n_i(A) = 2\left[\frac{2\pi kTm}{h^2}\right]^{3/2} \times \left[\frac{m_e^* m_h^*}{m^2}\right]^{3/4} \exp\left[\frac{-0.36\text{ eV}}{2kT}\right] \quad (11.129)$$

$$n_i(B) = 2\left[\frac{2\pi kTm}{h^2}\right]^{3/2} \times \left[\frac{m_e^* m_h^*}{m^2}\right]^{3/4} \exp\left[\frac{-0.72\text{ eV}}{2kT}\right] \quad (11.130)$$

Dividing Eq. (11.129) by Eq. (11.130), we get

$$\frac{n_i(A)}{n_i(B)} = \exp\left[\frac{-0.36\text{ eV}}{2kT}\right]\exp\left[\frac{0.72\text{ eV}}{2kT}\right]$$

$$\frac{n_i(A)}{n_i(B)} = \exp\left[\frac{-0.36\text{ eV}}{0.052\text{ eV}}\right]\exp\left[\frac{0.72\text{ eV}}{0.052\text{ eV}}\right]$$

$$\frac{n_i(A)}{n_i(B)} = \exp\left[\frac{0.36\text{ eV}}{0.052\text{ eV}}\right] = 1.0154 \times 10^3$$

The ratio of the intrinsic carrier densities of the materials *A* and *B* is 1.0154×10^3.

11.7 In N-type semiconductor, the concentration of electrons = 2×10^{22} m^{-3}. Its electrical conductivity =112 Ω^{-1} m^{-1}. Calculate the mobility of the electrons.

Given data

Concentration of electrons = 2×10^{22} m^{-3}

Conductivity = 112 Ω^{-1} m^{-1}

Solution

The mobility of electrons is given by

$$\mu = \frac{\sigma}{N_D e} = \frac{112}{2 \times 10^{22} \times 1.6 \times 10^{-19}} = 0.035\text{ m}^2\text{ V}^{-1}\text{ s}^{-1}$$

The mobility of the electrons = 0.035 m^2 V^{-1} s^{-1}.

11.8 A sample of N-type silicon has donor concentration of 10^{22} atoms m^{-3}. Find the Hall voltage in a sample with thickness = 500 μm. Area of cross section = 2.5×10^{-3} cm^{-2}, current = 1 A and magnetic field (B_Z) = 10^{-5} Wb cm^{-2}.

Given data

Thickness t = 500 μm

Area of cross section A = 2.5×10^{-3} cm^{-2} = 2.5×10^{-7} m^{-2}

Current I = 1 A

Magnetic field B = 10^{-5} Wb cm^{-2} = 0.1 Wb m^{-2}

Donor concentration N_D = 10^{22} m^{-3}

Solution

The Hall coefficient $R_H = \frac{3\pi}{8}\left(\frac{1}{ne}\right)$

Hall voltage $V_H = \frac{B_Z I_X \cdot R_H}{w} = \frac{B_Z I_X}{w} \times \frac{3\pi}{8}\left(\frac{1}{ne}\right)$

Substituting the values, we get

$$V_H = \frac{0.1 \times 1}{500 \times 10^{-6}} \times \frac{3\pi}{8}\left(\frac{1}{10^{22} \times 1.6 \times 10^{-19}}\right) = 0.1472 \text{ V}$$

The Hall Voltage in the sample = 0.1472 V.

11.9 In an intrinsic semiconductor, the energy gap is 1.2 eV. What is the ratio between its conductivity at 600 K and that of 300 K? (Given 1 eV = 1.602×10^{-19} J).

Given data

Energy gap E_g = 1.2 eV
Temperature T_1 = 300 K
Temperature T_2 = 600 K

$$1 \text{ eV} = 1.602 \times 10^{-19} \text{ J}$$

Solution

The conductivity is given by

$$\sigma = n_i e(\mu_e + \mu_h) \tag{11.131}$$

The conductivity at 300 K is given by

$$\sigma_{300} = n_{i300} e(\mu_e + \mu_h) \tag{11.132}$$

The conductivity at 600 K given by

$$\sigma_{600} = n_{i600} e(\mu_e + \mu_h) \tag{11.133}$$

Dividing Eq. (11.133) by Eq. (11.132), we get

$$\frac{\sigma_{600}}{\sigma_{300}} = \frac{n_{i600}}{n_{i300}}$$

The intrinsic concentration of carriers is given by

$$n_i = 2\left[\frac{2\pi kTm}{h^2}\right]^{3/2} \times \left[\frac{m_e^* m_h^*}{m^2}\right]^{3/4} \exp\left[\frac{-E_g}{2kT}\right] \tag{11.134}$$

$$\frac{n_{i600}}{n_{i300}} = \left(\frac{600}{300}\right)^{3/2} \times \exp\left[\frac{-1.2 \times 1.6 \times 10^{-19}}{2 \times 1.38 \times 10^{-23} \times 600}\right] \exp\left[\frac{1.2 \times 1.6 \times 10^{-19}}{2 \times 1.38 \times 10^{-23} \times 300}\right]$$

$$\frac{n_{i600}}{n_{i300}} = \left(\frac{600}{300}\right)^{3/2} \times \exp\left[\frac{+1.2 \times 1.6 \times 10^{-19}}{2 \times 1.38 \times 10^{-23}} \times \left(\frac{1}{300} - \frac{1}{600}\right)\right] = 3.0679 \times 10^5$$

The ratio between the conductivity of the material at 600 K and 300 K is 3.0679×10^5.

11.10 For an intrinsic GaAs, the room temperature electrical conductivity is $10^{-6}\ \Omega^{-1}\ m^{-1}$. The electron and hole mobilities are 0.85 and 0.04 $m^2\ V^{-1}s^{-1}$ respectively. Calculate the intrinsic carrier concentration at room temperature.

Given data

Electrical conductivity $\sigma = 10^{-6}\ \Omega^{-1}\ m^{-1}$

Electron mobility $\mu_e = 0.85\ m^2\ V^{-1}\ s^{-1}$

Hole mobility $\mu_h = 0.04\ m^2\ V^{-1}\ s^{-1}$

Solution

The electrical conductivity is given by

$$\sigma = n_i e(\mu_e + \mu_h)$$

The intrinsic carrier concentration is given by

$$n_i = \frac{\sigma}{e(\mu_e + \mu_h)} = \frac{10^{-6}}{1.6 \times 10^{-19} \times (0.85 + 0.04)} = 7.022 \times 10^{12}\ m^{-3}$$

The intrinsic carrier concentration of GaAs is $7.022 \times 10^{12}\ m^{-3}$.

11.11 Find the density of impurity atoms that must be added to an intrinsic silicon crystal in order to convert it into (i) 10 Ω cm P-type Si and (ii) 10 Ω cm N-type Si. Calculate also the concentration of minority carriers in each other. For Si, $\mu_e = 1350\ cm^2\ V^{-1}\ s^{-1}$, $\mu_h = 480\ cm^2\ V^{-1}\ s^{-1}$ and $n_i = 1.5 \times 10^{10}\ cm^{-3}$.

Given data

Resistivity of P-type Si $\rho = 10\ \Omega\ cm = 0.1\ \Omega\ m$

Resistivity of N-type Si $\rho = 10\ \Omega\ cm = 0.1\ \Omega\ cm$

Electron mobility $\mu_e = 1350\ cm^2\ V^{-1}s^{-1} = 0.135\ m^2\ V^{-1}s^{-1}$

Hole mobility $\mu_h = 480\ cm^2V^{-1}s^{-1} = 0.048\ m^2V^{-1}s^{-1}$

Solution

(i) For P-type semiconductor

The electrical conductivity is given by

$$\sigma = \frac{1}{\rho} = \frac{1}{0.1} = 10\ \Omega^{-1}\ m^{-1}$$

The electrical conductivity for the P-type material is given by

$$\sigma = N_A e \mu_h$$

The acceptor concentration is

$$N_A = \frac{\sigma}{e\mu_h} = \frac{10}{1.6 \times 10^{-19} \times 0.048} = 1.302 \times 10^{21}\ m^{-3}$$

The concentration of the minority carriers

$$n = \frac{n_i^2}{N_A} = \frac{(1.5 \times 10^{16})^2}{1.302 \times 10^{21}} = 1.728 \times 10^{12}\ m^{-3}$$

(ii) For N-type semiconductor

The electrical conductivity is given by

$$\sigma = \frac{1}{\rho} = \frac{1}{0.1} = 10\ \Omega^{-1}\ \text{m}^{-1}$$

The electrical conductivity for the N-type material is given by

$$\sigma = N_D e \mu_e$$

The donor concentration is

$$N_D = \frac{\sigma}{e\mu_e} = \frac{10}{1.6\times10^{-19}\times0.135} = 4.629\times10^{20}\ \text{m}^{-3}$$

The concentration of the minority carriers

$$p = \frac{n_i^2}{N_D} = \frac{(1.5\times10^{16})^2}{4.629\times10^{20}} = 4.86\times10^{11}\ \text{m}^{-3}.$$

For N-type material

The donor concentration = 4.629×10^{20} m^{-3}

The minority carriers concentration = 4.86×10^{11} m^{-3}

For P-type material

The acceptor concentration = 1.302×10^{21} m^{-3}

The minority carriers concentration = 1.728×10^{11} m^{-3}

SHORT QUESTIONS

1. What is a semiconductor?
2. What is an intrinsic semiconductor?
3. What is an extrinsic semiconductor?
4. What is an N-type semiconductor?
5. What is a P-type semiconductor?
6. What is an elemental semiconductor?
7. What is a compound semiconductor?
8. Write the equation for the electrical conductivity of an intrinsic semiconductor and explain the terms.
9. State mass action law.
10. What is meant by the Fermi level?
11. Discuss the variation of the Fermi level in an intrinsic semiconductor.
12. What is meant by band gap of a semiconductor?
13. Define Hall effect.
14. Mention the applications of Hall effect.

15. Discuss the variation of the Fermi level of an N-type semiconductor with temperature.
16. Discuss the variation of the Fermi level of an N-type semiconductor with concentration.
17. Discuss the variation of the Fermi level of a P-type semiconductor with temperature.
18. Discuss the variation of the Fermi level of a P-type semiconductor with concentration.
19. Write an expression for the conductivity of an N-type semiconductor.
20. Write an expression for the conductivity of a P-type semiconductor.

DESCRIPTIVE TYPE QUESTIONS

1. Derive the mathematical expressions for the concentration of electrons in the conduction band and the concentration of holes in the valence band and hence obtain the intrinsic carrier concentration.
2. What is Hall effect? Derive an expression for the Hall voltage. Explain an experimental method used to measure the Hall coefficient of a specimen. What are the uses of Hall effect?
3. Derive a mathematical expression for the carrier concentration of an N-type semiconductor and hence derive the expression for the Fermi level. Explain the variation of the Fermi level of an N-type semiconductor with temperature and concentration.
4. Derive a mathematical expression for the carrier concentration of a P-type semiconductor and hence derive the expression for the Fermi level. Explain the variation of the Fermi level of a P-type semiconductor with temperature and concentration.

PROBLEMS

1. The resistivity of an intrinsic semiconductor is 4.5 Ω m at 20°C and 2 Ω m at 32°C. Find the energy band gap in eV (Boltzmann constant = 8.617×10^{-5} eV K^{-1}).
2. The donor density of an N-type Ge sample is 10^{21} m^{-3}. The sample is arranged in a Hall experiment having magnetic field of 0.5 T and the current density is 500 A m^{-2}. Find the Hall voltage if the sample is 3 mm wide.
3. Calculate the conductivity of an intrinsic Ge at 300 K using the following data: $n_i = 2.4 \times 10^{19}$ m^{-3}, $\mu_e = 0.39$ m^2 V^{-1} s^{-1} and $\mu_h = 0.19$ m^2 V^{-1} s^{-1}.
4. The Hall coefficient of a certain Si specimen is found to be $-7.35 \times 10^{-5} m^3$ C^{-1} from 100 to 400 K. Determine the nature of the material. If the electrical conductivity is found to be 200 Ω^{-1} m^{-1}, calculate the density and mobility of charge carriers.
5. The intrinsic carrier density at room temperature for germanium is 2.37×10^{19} m^{-3}. The mobilities of electron and holes are 0.38 and 0.18 m^2 V^{-1} s^{-1} respectively. Calculate the resistivity.

6. Calculate the conductivity of Ge at 300 K. Given that at 300 K, $n_i = 23 \times 10^{18}$ m^{-3}, $\mu_e = 0.364$ m^2 V^{-1} s^{-1} and $\mu_h = 0.19$ m^2 V^{-1} s^{-1}.
7. Calculate the intrinsic concentration of charge carriers at 300 K. (Given that $m_e^* = 0.12\, m_o$, $m_p^* = 0.028\, m_o$, $m_o = 9.1 \times 10^{-31}$ kg and the band gap energy for Ge = 0.67 eV).
8. Find the resistivity and resistance of an intrinsic Ge rod of 1 cm length, 1 mm width and 1 mm thickness at 300 K. Given for Ge, $n_i = 2.5 \times 10^{19}$ m^{-3}, $\mu_e = 0.39$ m^2 V^{-1} s^{-1} and $\mu_h = 0.19$ m^2 V^{-1} s^{-1} at 300 K.
9. If a sample of Si is doped with 3×10^{23} atoms of arsenic and 5×10^{23} atoms of boron, determine the electron concentration if the intrinsic charge carriers are 2×10^{16} m^{-3}.
10. The conductivity of Ge at 20°C is 2 Ω^{-1} m^{-1}. What is its conductivity at 40°C? $E_g = 0.72$ eV.

CHAPTER 12

MAGNETIC MATERIALS

12.1 INTRODUCTION

The materials that can be magnetized are called magnetic materials. The magnetic materials have a lot of applications such as magnetic memories, inductors, transformers, ferrites antennas, and loudspeakers, etc. Most of the engineering components are made up of permanent magnets. Even though five different types of magnetic materials are available, ferromagnetic and ferrimagnetic materials are mostly used for application purposes. Diamagnetic and paramagnetic materials are rarely used. This chapter deals with ferromagnetic and ferrimagnetic materials and their uses.

12.2 DEFINITION OF SOME FUNDAMENTAL TERMS

(i) Magnetic dipoles

Any two opposite poles separated by a suitable distance constitute a dipole. A magnet is said to be a **magnetic dipole**. It has north and south poles. The distance of separation is the length of the magnet.

(ii) Magnetic dipole moment

The term magnetic dipole moment is defined as follows:

(a) It is defined as the product of the magnetic pole strength and the length of the magnet. It is represented by the letter μ. Therefore, $\mu = m \cdot l$.

(b) Consider that a current of i ampere is passed through a coil of wire and A be the area of cross section of the wire, then the term **magnetic dipole moment** is defined as the product of the current and the area of cross section. That is, $\mu = i \cdot A$. The unit for the magnetic dipole moment is A m^2.

(c) Consider that a magnet is suspended in a magnetic field. The magnet experiences certain torque and hence it will rotate and rest at some position. The torque experienced by the magnet is given by, $\tau = \bar{\mu} \times \bar{B}$, where $\bar{\mu}$ is the magnetic moment and $\bar{B}$ is the magnetic flux density.

(iii) Magnetic lines of forces

The magnetic field in a magnetic material is studied by drawing the magnetic lines of forces. The magnetic lines of forces are also called magnetic flux. The magnetic lines of forces originate from north pole and end at south pole.

(iv) Magnetic flux density, *B*

The magnetic flux passing through the unit area of cross section is known as the **magnetic flux density**. It is represented by B. Its unit is Wb m^{-2}.

The magnetic flux density is given by $B = \dfrac{\phi}{A}$, where ϕ is the magnetic flux and A is the area of cross section.

The magnetic flux density is also given by

$$B = \mu H \tag{12.1}$$

The flux density in air or vacuum

$$B_0 = \mu_0 H \tag{12.2}$$

and

$$B = \mu_0(M + H) \tag{12.3}$$

where M, H, μ_0, μ are magnetization, magnetic field strength, permeability of free space and permeability of the medium respectively.

(v) Magnetic field strength, *H*

The magnetic field strength is the force experienced by a unit north pole placed in the magnetic field region. It is represented by H. Its unit is A m^{-1}.

(vi) Magnetization, *M*

The magnetic moment per unit volume is known as magnetization. Its unit is A m^{-1} and it is represented by M.

(vii) Magnetic susceptibility, χ

The ratio of the magnetization to the magnetic field strength is known as the magnetic susceptibility. It has no unit and it is represented by χ.

$$\chi = \frac{M}{H}$$

(viii) Magnetic relative permeability, μ_r

The ratio of the permeability of a medium to the permeability of a free space is known as magnetic relative permeability. It is represented by μ_r and it is given by $\mu_r = \mu/\mu_0$. Here, μ_0 is the permeability of free space and it is equal to, $\mu_0 = 4\pi \times 10^{-7}$ H m^{-1}.

The magnetic relative permeability is also defined as

$$\mu_r = \frac{B}{B_0} \tag{12.4}$$

where B and B_0 are the magnetic flux density in a medium and magnetic flux density in vacuum respectively.

(ix) Bohr magneton

The magnetic moment is expressed in the unit ampere square metre (A m^2). Since the magnetic moment of an atomic particle is very low, it is represented by another unit known as Bohr magneton. The value of one Bohr magneton is given by

$$1 \text{ Bohr magneton} = \frac{eh}{4\pi m} \tag{12.5}$$

Substituting the values of e, m and h, we get

$$1 \text{ Bohr magneton} = \frac{1.6 \times 10^{-19} \times 6.626 \times 10^{-34}}{4\pi \times 9.1 \times 10^{-31}}$$

$$= 9.27 \times 10^{-24} \text{ A m}^2$$

It is represented by β. The value of 1 Bohr magneton is equal to 9.27×10^{-24} A m^{-2}.

12.3 CLASSIFICATION OF THE MAGNETIC MATERIALS

The magnetic materials are broadly classified into two types. They are (i) those do not have permanent dipole moments and (ii) those having permanent dipole moments. The term **permanent dipole moment** means the presence of the dipole moment even in the absence of the magnetic field. It is represented by the letter μ_P.

12.3.1 Materials not having Permanent Dipole Moment

Diamagnetic material is an example of material which does not have permanent dipole moment.

12.3.2 Materials having Permanent Dipole Moment

Paramagnetic materials, ferromagnetic materials, antiferromagnetic materials and ferrimagnetic materials are examples of material having permanent dipole moment.

Diamagnetic material

Diamagnetic materials exhibit negative susceptibility. The relative permeability of diamagnetic material is slightly less than unity. When a diamagnetic material is placed in a magnetic field, the magnetization vector **M** is in opposite direction to the applied field, H. The negative susceptibility of a diamagnetic material is due to the repulsive force experienced by diamagnetic material with the applied magnetic field as shown in Figure 12.1. Consider a diamagnetic material placed in a non-uniform magnetic field. It experiences a force towards smaller fields. The diamagnetic property is due to the presence of closed shell or subshell in the material. Mostly the covalent and ionic crystals exhibit the diamagnetic property. The superconducting materials exhibit perfect diamagnetism ($\chi = -1$).

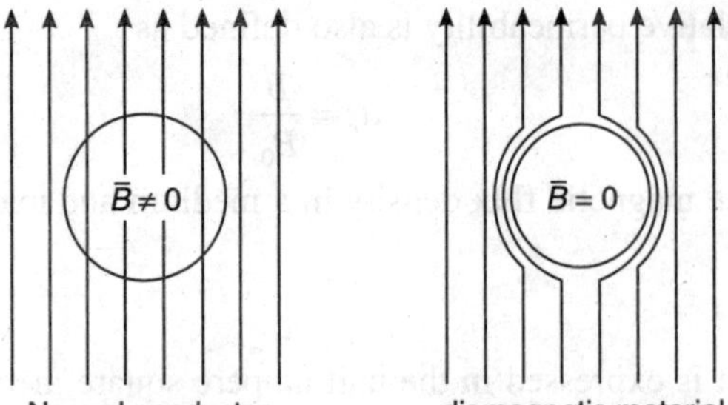

Figure 12.1 Diamagnetic property.

The examples of the diamagnetic materials are (i) the covalent metals such as Si, Ge, diamond, (ii) some metals such as copper, silver, gold, (iii) some ionic solids such as alkalai halides (iv) superconductors, and (v) organic materials such as polymers.

Paramagnetism

In a paramagnetic materials the dipoles are randomly oriented because of the random collisions of the molecules. The paramagnetic materials exhibit positive susceptibility. When a magnetic field is applied, the individual magnetic moment takes the alignment along the applied field as shown in Figure 12.2. The magnetization of a paramagnetic material increases with the increase in the applied field. The increase in temperature reduces the magnetization and it destroys the alignment of dipoles with the applied field.

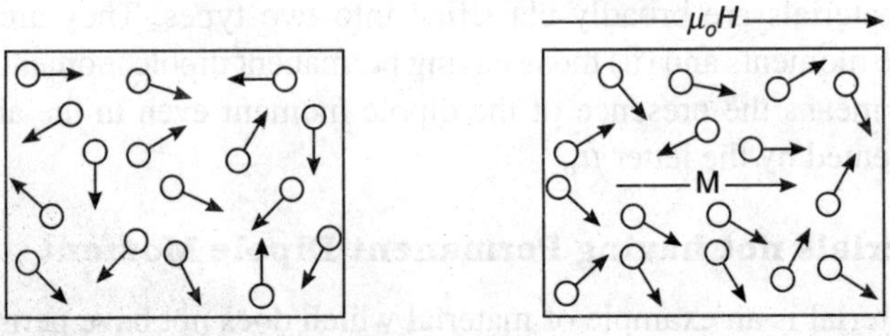

(a) Random orientation of dipoles in the absence of the field (b) The dipoles align towards the field

Figure 12.2 Paramagnetism.

Consider a paramagnetic material is placed in a non-uniform magnetic field. The paramagnetic material experiences a net force towards a greater field. The susceptibility of a paramagnetic material is given by

$$\chi = \frac{C}{T} \tag{12.6}$$

where C is the Curie constant and T is the temperature.

The examples of paramagnetic materials are Mg, gaseous and liquid oxygen, ferromagnetic material (Fe) at high temperature, antiferromagnetic material (Cr) at high temperature and ferrimagnetic material (Fe_3O_4) at high temperature.

Ferromagnetism

In a ferromagnetic material, all the dipoles are aligned parallel as shown in Figure 12.3. If a small value of magnetic field is applied, a large value of magnetization is produced. Ferromagnetic material has permanent dipole moment and the susceptibility is positive. The magnetization in a ferromagnetic material is non-linear and it becomes saturated, if a large value of magnetic field is applied.

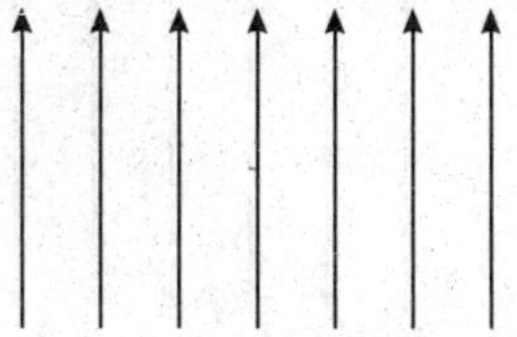

Figure 12.3 Alignment of dipoles in a ferromagnetic material.

A ferromagnetic material exhibits two different properties. It behaves as a ferromagnetic below a certain temperature known as ferromagnetic Curie temperature. Above that temperature, it behaves as a paramagnetic. In the ferromagnetic region, it exhibits a well-known curve known as hysteresis curve as shown in Figure 12.4.

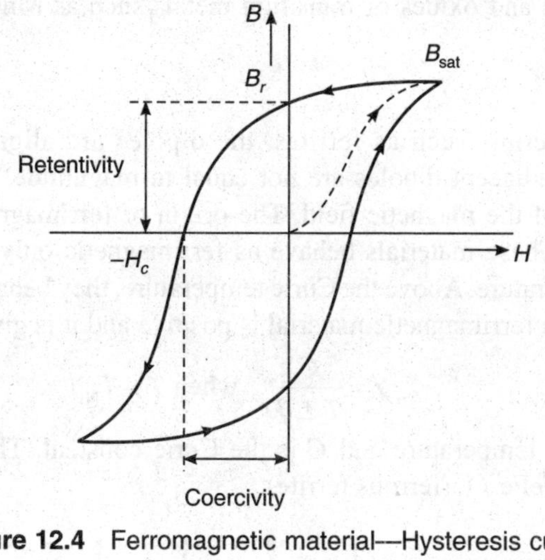

Figure 12.4 Ferromagnetic material—Hysteresis curve.

The susceptibility of a ferromagnetic material above the ferromagnetic Curie temperature, θ_f is given by

$$\chi = \frac{C}{T - \theta_f} \quad \text{when } T > \theta_f \tag{12.7}$$

where C is the Curie constant and θ_f is the ferromagnetic Curie temperature. The transition and rare earth metals such as Fe, Co, Ni, Gd, Dy are the examples of ferromagnetic material.

Antiferromagnetic materials

In an antiferromagnetic material, the dipoles are aligned antiparallel as shown in Figure 12.5. Antiferromagnetic material has positive value of susceptibility, but it is small. In the absence of the magnetic field, the magnetization produced by one dipole is cancelled by the other, because the magnitudes of the adjacent dipoles are the same. So, the antiferromagnetic materials do not have magnetization in the absence of the magnetic field. The antiparallel alignment of dipole is due to quantum mechanical exchange forces. The antiferromagnetism occurs only below certain temperatures called the Néel temperature, T_N. Above the Néel temperature, the material is paramagnetic.

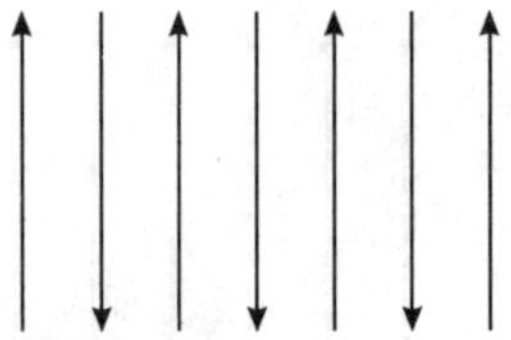

Figure 12.5 Antiferromagnetic material—alignment of dipoles.

The susceptibility of an antiferromagnetic material is given by

$$\chi = \frac{C}{T+\theta} \quad \text{when } T > T_N \tag{12.8}$$

where C is the Curie constant and T_N is the Néel temperature. The examples of antiferromagnetic materials are the salts and oxides of transition metals such as MnO, NiO, MnF_2, the transition metals like α-Cr, Mn.

Ferrimagnetism

In ferrimagnetic materials such as ferrites, the dipoles are aligned antiparallel as shown in Figure 12.6. But the adjacent dipoles are not equal in magnitude. So, they have magnetization even in the absence of the magnetic field. The origin of ferrimagnetism is due to the magnetic ordering of dipoles. These materials behave as ferrimagnetic only below a certain temperature known as Curie temperature. Above the Curie temperature, they behave as paramagnetic materials. The susceptibility of a ferrimagnetic material is positive and it is given by

$$\chi = \frac{C}{T \pm \theta} \quad \text{when } T > T_N \tag{12.9}$$

where T_N is the Néel temperature and C is the Curie constant. The examples of ferrimagnetic materials are Fe_3O_4, $FeFe_2O_4$ (ferrous ferrite).

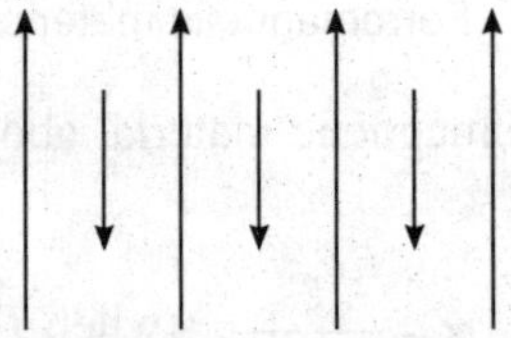

Figure 12.6 Ferrimagnetic material—alignment of dipoles.

12.4 ORIGIN OF PERMANENT DIPOLE MOMENT

The magnetic moment of a magnetic material is related to its angular momentum. If a magnetic material possesses angular momentum, then it has dipole moment in the absence of the magnetic field. The permanent dipole moment arises due to the following factors. They are

1. Orbital angular momentum of the electron,
2. Spin angular momentum of the electron, and
3. Nuclear magnetic moment.

12.4.1 Orbital Angular Momentum of the Electron

The orbital angular momentum of the electron arises due to the orbital motion of the electron. The relation between the orbital angular momentum and the magnetic moment is given by

$$\mu_m = -\frac{e}{2m} M_o \tag{12.10}$$

where μ_m is the magnetic moment, M_o is the orbital angular momentum of the electron, e is the charge of the electron and m is the mass of the electron.

An electron is represented by four quantum numbers. They are the principal quantum number, n, the orbital quantum number, l, the magnetic orbital quantum number m_l, and the magnetic spin quantum number, m_s. The principal quantum number takes the values of 1, 2, 3, 4, ..., etc. The orbital quantum number takes the values of 0, 1, 2, 3, 4, ..., $n - 1$. The m_l value varies from $-l$ to l including zero. The m_s value is either $+\frac{1}{2}$ or $-\frac{1}{2}$. The value of the orbital angular momentum is given by

$$M_o = \frac{h}{2\pi} m_l$$

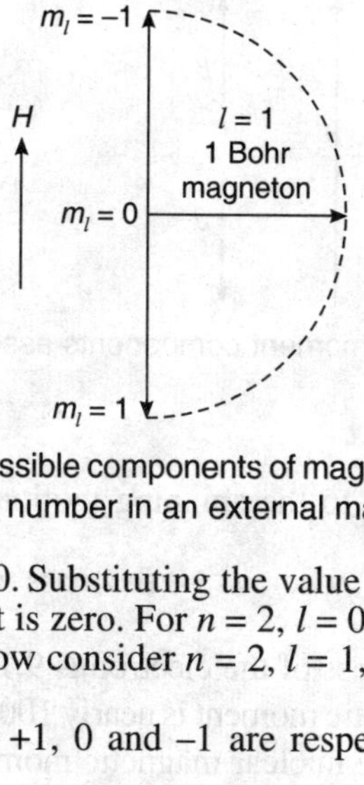

Figure 12.7 Illustration of three possible components of magnetic moment associated with orbital momentum quantum number in an external magnetic field.

Consider $n = 1$, $l = 0$ and $m_l = 0$. Substituting the value of $m_l = 0$, the angular momentum is zero and hence the magnetic moment is zero. For $n = 2$, $l = 0$, and $m_l = 0$, the angular momentum and the magnetic moment is zero. Now consider $n = 2$, $l = 1$, $m_l = +1$, 0 and -1. The value of the orbital angular momentum for $m_l = +1$, 0 and -1 are respectively given by, $M_o = \frac{h}{2\pi}$, 0 and $M_o = -\frac{h}{2\pi}$. The possible values of m_l in an external magnetic field is displayed in Figure 12.7.

The value of the magnetic moment corresponding to $n = 2$, $l = 1$, $m_l = +1$, 0 and -1 are $\mu_m = -\frac{eh}{4\pi m}$, 0 and $\mu_m = \frac{eh}{4\pi m}$. For a completely filled electronic energy states, the total angular momentum is zero. For an unfilled electronic energy states, the total angular momentum is not zero and hence the total magnetic moment is not equal to zero. For an electrical engineer, the iron group element with atomic numbers 21 to 28 are important.

12.4.2 Spin Angular Momentum of the Electron

The spin angular momentum arises due to the spinning motion of the electron. The spin angular momentum is given by $M_S = \frac{h}{2\pi} m_s$, where M_S is the spin angular momentum, m_s is the magnetic spin quantum number. The magnetic spin quantum number takes either $+\frac{1}{2}$ or $-\frac{1}{2}$. The spin angular momentum is given by $M_S = \frac{h}{4\pi}$, $M_S = -\frac{h}{4\pi}$. The relation between the magnetic moment and the spin angular momentum is given by $\mu_m = -\frac{e}{m} M_S$. Substituting the value of the spin angular momentum, we get, $\mu_m = -\frac{eh}{4\pi m}$, $\mu_m = \frac{eh}{4\pi m}$. Two possible magnetic moment components associated with electron spin is shown in Figure 12.8. For completely filled electronic states, the total magnetic moment is zero. For unfilled electronic states, the total magnetic moment is not equal to zero.

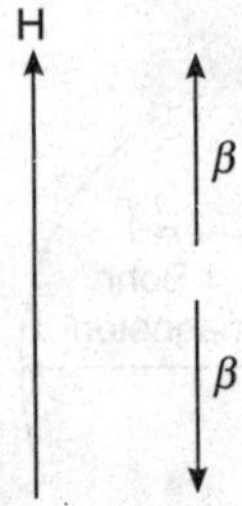

Figure 12.8 Two possible magnetic moment components associated with electron spin.

12.4.3 Nuclear Magnetic Moment

The nuclear magnetic moment arises due to the spinning motion of the nucleons. The nuclear magnetic moment is given by $\mu_m = \frac{eh}{4\pi m_N}$, where m_N is the mass of the nucleons. The mass of the nucleon is 1.67×10^{-27} kg, whereas the mass of the electron is 9.1×10^{-31} kg. Since, the mass of the nucleon is very large, the nuclear magnetic moment is nearly 1000 times lower than the magnetic moment due to the electrons and hence the nuclear magnetic moment is negligible.

12.5 FERROMAGNETIC MATERIALS

The dipoles of a ferromagnetic material are aligned parallel to each other. The parallel alignment produces magnetic moment even in the absence of the magnetic field.

12.5.1 Properties of Ferromagnetic Material

The ferromagnetic material behaves as a paramagnetic material above a certain temperature, known as ferromagnetic Curie temperature, θ_f and exhibits a well-known curve known as hysteresis curve below the ferromagnetic Curie temperature.

Paramagnetic behaviour of a ferromagnetic material

A ferromagnetic material behaves as a paramagnetic material above a certain temperature known as ferromagnetic Curie temperature, θ_f. If a graph is drawn between $1/\chi$ and T, a curve as shown in Figure 12.9 is obtained. In the case of a paramagnetic material, a straight line is obtained and it passes through the origin. Similar straight line behaviour is obtained for the ferromagnetic material. It indicates that the ferromagnetic material is behaving as a paramagnetic material, above the ferromagnetic Curie temperature. For the ferromagnetic material, nearer the X-axis, the straight line is slightly curved. The point at which the curve intersects the X-axis is known as ferromagnetic Curie temperature. If the line is extrapolated by identifying the straight line portion, it will meet the X-axis. The point of intersection of the extrapolated line at the X-axis is known as paramagnetic Curie temperature, θ. The paramagnetic and ferromagnetic Curie temperatures of some materials are listed in Table 12.1.

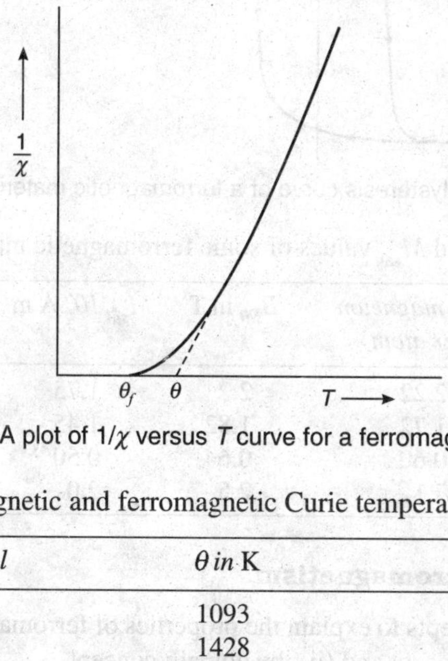

Figure 12.9 A plot of $1/\chi$ versus T curve for a ferromagnetic material.

Table 12.1 Paramagnetic and ferromagnetic Curie temperature of some materials

Material	θ in K	θ_f in K
Iron	1093	1043
Cobalt	1428	1393
Nickel	650	631

Hysteresis curve

Below the ferromagnetic Curie temperature, ($T < \theta_f$), ferromagnetic material exhibits a well-known curve known as hysteresis curve. If the magnetic field is increased gradually, the flux density increases and it becomes maximum. This maximum value of flux density is called saturated flux density (B_{sat}). If the field is reversed, the ferromagnetic material is found to have flux density even though the applied field becomes zero ($H = 0$). This property, the presence of flux density even in the absence of the magnetic field, is said to be retentivity or remanent flux density (B_r). The magnetization corresponding to the applied magnetic field is equal to zero ($H = 0$ and $M_r = B_r/\mu_0$) and is known as spontaneous magnetization. If the field is further reduced, the flux density becomes zero. The field required to bring the magnetic flux density into zero is called coercive field or coercivity ($-H_c$). If the field is further reduced, the flux density will become minimum. There is no further decrease in the value of the flux density beyond this value. If the field is increased, a closed loop as shown in Figure 12.10 is obtained. This closed loop between B and H is called hysteresis loop or B–H curve. The B_{sat} and M_{sat} values of some ferromagnetic materials are listed in Table 12.2.

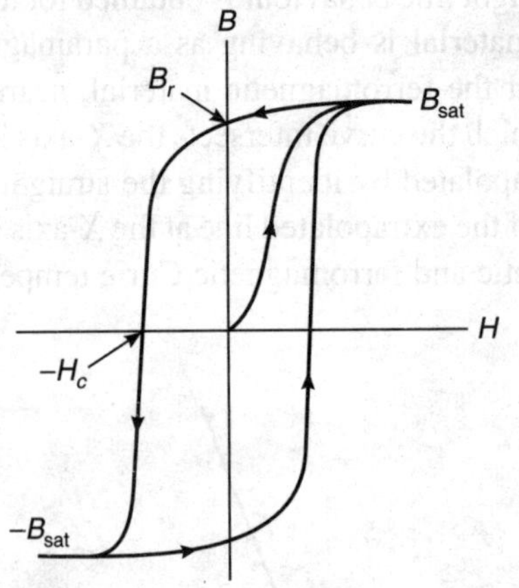

Figure 12.10 Hysteresis curve of a ferromagnetic material.

Table 12.2 B_{sat} and M_{sat} values of some ferromagnetic materials

Material	*Crystal structure*	*Bohr magneton per atom*	B_{sat} in T	M_{sat} 10^6 A m^{-1}	*Curie temperature, T_C in* K
Iron	BCC	2.22	2.2	1.75	1043
Cobalt	HCP	1.72	1.82	1.45	1393
Nickel	FCC	0.60	0.64	0.50	631
Gadolinium	HCP	7.1	2.5	2.0	289

12.5.2 Weiss Theory of Ferromagnetism

Weiss, in 1907, proposed two concepts to explain the properties of ferromagnetic materials. They are, namely (i) the internal field concept and (ii) the domain concept.

The magnetic field present at the location of a dipole is greater than the applied field. A dipole experiences the field due to the applied field and the field produced by the nearest neighbouring dipoles due to the interactions. The field produced by the interaction between the adjacent dipoles is known as internal field. The internal field is given by γM, where γ is the internal field constant and M is the magnetization. The internal field concept explains the spontaneous magnetization of ferromagnetic materials and the paramagnetic behaviour of ferromagnetic material.

A ferromagnetic material consists of a large number of localized regions called domains. The domain concept is used to explain the hysteresis property of the ferromagnetic material.

The Weiss theory explains most of the properties of ferromagnetic material. However, it has some drawbacks. According to Weiss theory, the ferromagnetic Curie temperature and the paramagnetic Curie temperatures are the same. But the experimental results indicate that these two values are different. The internal field constant, and the interaction energy (~kT) evaluated from Classical theory (Weiss theory) are 1000 times smaller than the actual value. This means that the interaction energy is due to the wave nature of electrons and hence they are quantum mechanical concepts. They are explained using quantum mechanics based on the exchange interaction concepts.

Magnetic domains

Consider that a ferromagnetic material is heated above its Curie temperature and cooled without applying the magnetic field, the magnetic domains are produced. The domains are the localized small region in which all the dipoles are aligned in one direction. A ferromagnetic material is found to have a number of domains and the dipoles of the adjacent domains are randomly oriented so that the resultant magnetization in the absence of the field is zero. The adjacent domains are separated by a region called domain wall or Bloch wall. The arrangement of domains in a polycrystalline iron sample is shown in Figure 12.11(a). If a magnetic field is applied to a ferromagnetic material, the domains that are parallel to the applied field increase in its size, whereas the size of the domains that are pointed in other directions to the applied field decreases.

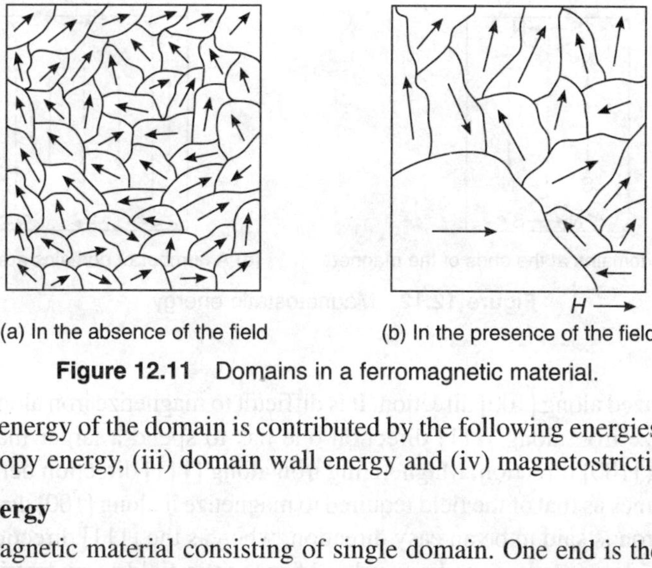

(a) In the absence of the field (b) In the presence of the field

Figure 12.11 Domains in a ferromagnetic material.

The internal energy of the domain is contributed by the following energies: (i) magnetostatic energy, (ii) anisotropy energy, (iii) domain wall energy and (iv) magnetostriction energy.

Magnetostatic energy

Consider a ferromagnetic material consisting of single domain. One end is the north pole of the magnet and the other end is the south pole of the magnet. The magnetic lines of forces originate from the north pole and they end at the south pole. The potential energy stored in a magnetic material is called magnetostatic energy. The potential energy of this material can be reduced by creating another domain. Consider these two domains are in the antiparallel direction. The region that separates these two domains is said to be domain wall or Bloch wall. The second domain is created by reducing the potential energy and it is rotated through an angle of 180° from the first domains. Similarly more number of domains are created by reducing the magnetostatic energy (potential energy). Figure 12.12(c) has four domains. These domains close the ends of the magnetic material with sideway domains and hence they are said to be closure domains. In Figure 12.12(d), the potential energy is further reduced and more number of domains are created. Thus, the creation of magnetic domains continues until the potential energy reduction in creating an additional domain is equal to the increase in potential energy for creating an additional wall. The specimen has the minimum potential energy and the net magnetization becomes zero.

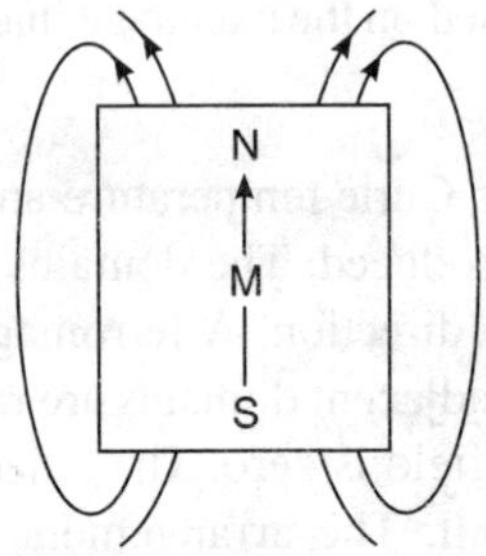

(a) A magnetized material with one domain

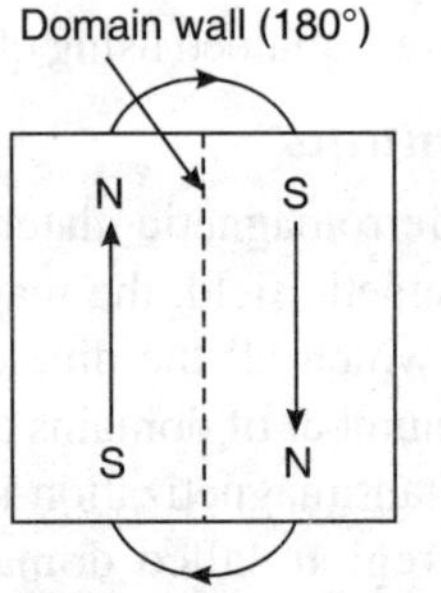

(b) Creation of two domains

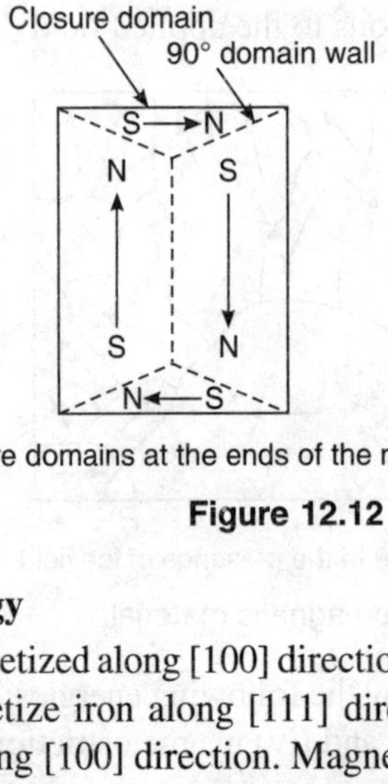

(c) Closure domains at the ends of the magnet

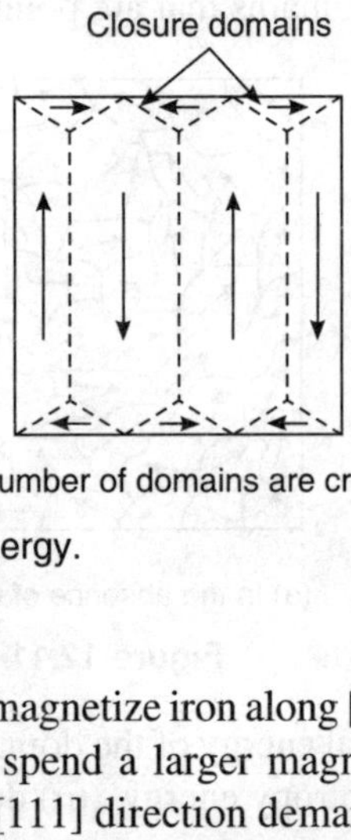

(d) A number of domains are created

Figure 12.12 Magnetostatic energy.

Anisotropy energy

Iron is easily magnetized along [100] direction. It is difficult to magnetize iron along [111] direction. In order to magnetize iron along [111] direction one has to spend a larger magnetic field than magnetizing it along [100] direction. Magnetizing iron along [111] direction demands a magnetic field of nearly four times as that of the field required to magnetize it along [100] direction. Therefore, [100] direction for iron is said to be an easy direction, whereas the [111] direction for iron is said to be hard direction. Iron needs a medium value of magnetic field to magnetize it along [110] direction. Therefore, [110] direction for iron is said to be medium direction. A plot of magnetization versus applied magnetic field for iron along [100], [110], and [111] directions are shown in Figure 12.13(a) and the directions [100], [110] and [111] for iron is shown in Figure 12.13(b).

The excess energy required to magnetize a material along a particular direction to its easy direction is known as anisotropy energy. It is denoted by K. The anisotropy energy for iron along [100] direction is zero and for the [111] direction, it is about 40 kJ m^{-3}. For nickel, [100] direction is the hard direction, [110] is the medium direction and [111] direction is the easy direction.

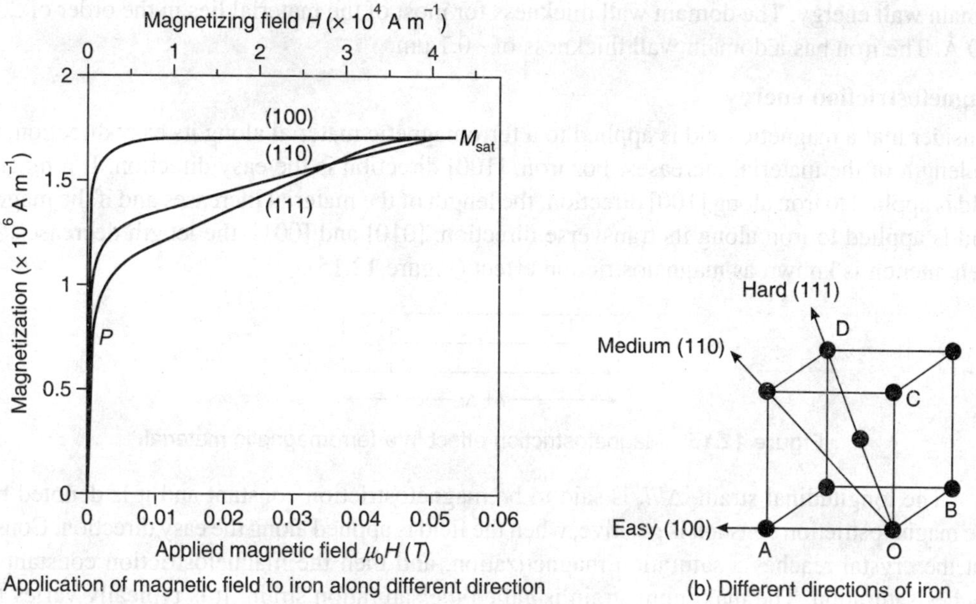

(a) Application of magnetic field to iron along different direction (b) Different directions of iron

Figure 12.13 Anisotrphy energy.

Bloch wall energy or domain wall energy

Consider a magnetic material with two domains. Consider that these two domains are antiparallel with each other as shown in Figure 12.14(a). The rotation of the domain wall does not take place abruptly. The spin magnetic moments rotates within the domain wall gradually. The exchange force and the anisotropy energy are responsible for the rotation within the domain wall. The exchange

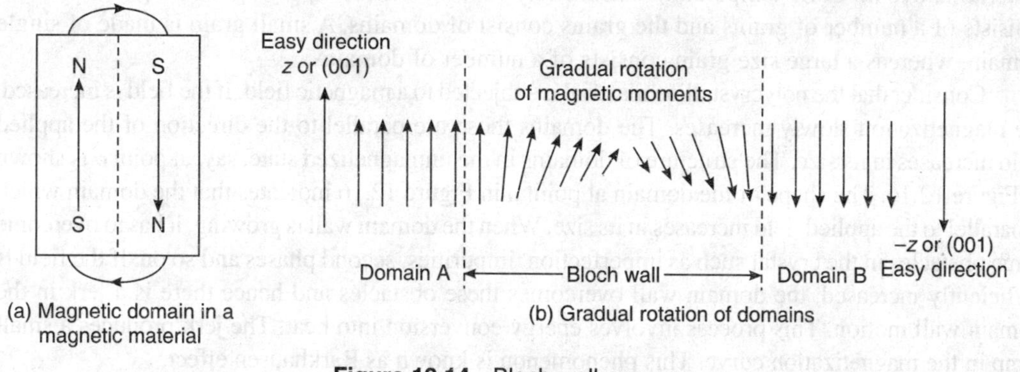

(a) Magnetic domain in a magnetic material (b) Gradual rotation of domains

Figure 12.14 Bloch wall energy.

force requires a very thick (infinitely thick) domain wall to achieve 180° rotation, whereas the anisotropy energy requires a spacing of one atomic scale for the rotation of domains through 180°. That is, the exchange force demands a very thick wall and the anisotropy energy demands a thin wall. Therefore, the domain wall has a thickness of an equilibrium value which minimizes the total potential energy, which is the sum of the exchange eenegy and the anisotropy energy. The minimum potential energy, which determines the domain wall thickness, is known as domain wall energy. The domain wall thickness for most of the material lies in the order of 200 to 300 Å. The iron has a domain wall thickness of ~0.1 μm.

Magnetostriction energy

Consider that a magnetic field is applied to a ferromagnetic material along its easy direction, then the length of the material increases. For iron, [100] direction is the easy direction. If a magnetic field is applied to iron along [100] direction, the length of the material increases and if the magnetic field is applied to iron along its transverse direction, [010] and [001], the length decreases. This phenomenon is known as magnetostriction effect (Figure 12.15).

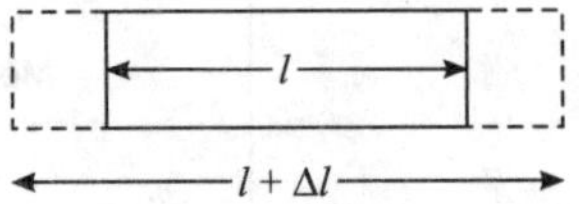

Figure 12.15 Magnetostriction effect in a ferromagnetic material.

The longitudinal strain, $\Delta l/l$, is said to be magnetostriction constant and it is denoted by λ. The magnetostriction constant is positive, when the field is applied along the easy direction. Consider that the crystal reaches a saturation magnetization, and then the magnetostriction constant also reaches saturation. The maximum strain is called the saturation strain. It is typically varies from 10^{-6} to 10^{-5}. The crystal lattice strain energy associated with magnetostriction is called magnetostriction energy.

The magnetostriction constant is negative for nickel and positive for iron along the easy direction. It may be controlled by suitably alloying. For 85% Ni and 15% Fe, it is zero.

Hysteresis curve (M versus H curve) for a polycrystalline material

A polycrystalline material consists of a number of grains. It is prepared by heating a ferromagnetic material above its Curie temperature and then by cooling without applying the magnetic field. It consists of a number of grains and the grains consist of domains. A small grain is made of single domain, whereas a large size grain consists of a number of domains.

Consider that the polycrystalline material is subjected to a magnetic field. If the field is increased, the magnetization slowly increases. The domains those are parallel to the direction of the applied field increases in its size. The structure of domains in the unmagnetized state, say, at point *o* is shown in Figure 12.16. The shape of the domain at point *a* in Figure 12.16 indicates that the domain which is parallel to the applied field increases in its size. When the domain wall is growing, it has to overcome some obstacles in the crystal such as imperfection, impurities, second phases and so on. If the field is sufficiently increased, the domain wall overcomes these obstacles and hence there is a jerk in the domain wall motion. This process involves energy conversion into heat. The jerk produces a small jump in the magnetization curve. This phenomenon is known as Barkhausen effect.

If the field is further increased, the magnetization increases. The domains those are parallel to the applied field increases whereas the domains that are pointing in other directions shrink. The domains shape at the points *o*, *a*, *b*, *c*, and *d* is shown in Figure 12.16. The curve *oabcd* is known as initial magnetization curve. At *c*, some grains are oriented along the direction of magnetization. At *d*, the entire specimen is made up of a single domain. If the field is further increased, then there is no increase in the size of the domain, which indicates that there is no further increase in the magnetization of the material.

If the field is slowly decreased, the size of the domain decreases and hence some new domains begin to grow. Even though the field is reduced to zero, the specimen has some value of magnetization indicating that the specimen does not regained its original domain structure as similar to its structure before the material gets magnetized (i.e. the structure at *o*). This property is said to be remanent or residual magnetization or retentivity (B_r). If the field is further applied in the reverse direction, the magnetization becomes zero, when the field is at $-H_C$. The field $-H_C$ is called coercive field or coercivity. The coercivity is the field required to bring the magnetization into zero. At $-H_C$, the material regains its actual domain structure.

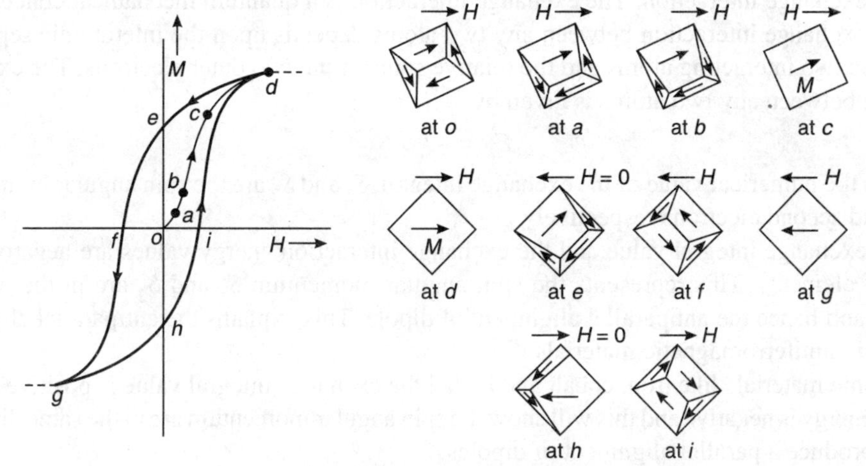

Figure 12.16 Hysteresis curve.

If the field is further decreased, the domains that is parallel to the applied field increases, whereas the domains that are in the other directions decreases. At *g*, the lowest value of magnetization occurs indicating that the specimen is fully occupied by a single domain. If the field is further decreased, there is no decrease in the magnetization value indicating that the entire specimen is fully occupied by a single domain. If the field is slowly increased, the curve takes the path *ghid* and then the curve is closed. This closed loop is called hysteresis loop.

12.5.3 Origin of Ferromagnetism and Heisenberg's Exchange Interaction

The ferromagnetic property is exhibited by transition elements such as iron, cobalt, and nickel at room temperature and rare earth elements like gadolinium and dysprosium. The ferromagnetic materials possess parallel alignment of dipoles. This parallel alignment of dipoles is not due to the magnetic force existing between any two dipoles. The reason is that the magnetic potential energy is very small and it is smaller than thermal energy.

The electronic configuration of iron is $1s^2$, $2s^2$, $2p^6$, $3s^2$, $3p^6$, $3d^6$, $4s^2$. For iron, the 3d subshell is an unfilled one. This 3d subshell has five orbitals. For iron, the six electrons present in the 3d subshell occupy the orbitals such that there are four unpaired electrons and two paired electrons as shown in Figure 12.17. These four unpaired electrons contribute a magnetic moment of 4β. This arrangement shows the parallel alignment of four unpaired electrons.

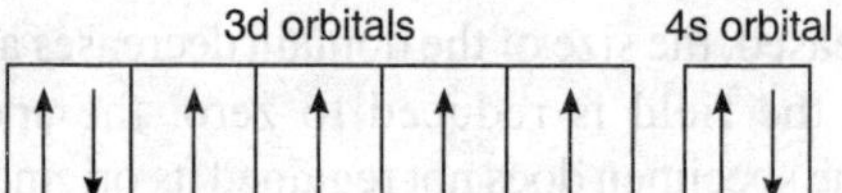

Figure 12.17 Orientation of electrons spins in 3d subshell in iron atom.

The parallel alignment of dipoles in iron is not due to the magnetic interaction. It is due to the Pauli's exclusion principle and electrostatic interaction energy. The Pauli's exclusion principle and electrostatic interaction energy are combined together and constitute a new kind of interaction known as exchange interaction. The exchange interaction is a quantum mechanical concept.

The exchange interaction between any two atoms depends upon the interatomic separation between the two interacting atoms and the relative spins of the two outer electrons. The exchange interaction between any two atoms is given by

$$E_{ex} = -J_e S_1 S_2 \tag{12.11}$$

where J_e is the numerical value of the exchange integral, S_1 and S_2 are the spin angular momenta of the first and second electrons respectively.

The exchange integral value and the exchange interaction energy values are negative for a number of elements. This represents the spin angular momentum S_1 and S_2 are in the opposite directions and hence the antiparallel alignment of dipole. This explains the antiparallel alignment of dipoles in antiferromagnetic materials.

In some materials like iron, cobalt and nickel the exchange integral value is positive and the exchange energy is negative and this will show the spin angular momentum are in the same direction. This will produce a parallel alignment of dipoles.

A plot between the exchange integral and the ratio of the interatomic separation to the radius of 3d orbital (r/r_d) is shown in Figure 12.18. For the transition metals like iron, cobalt, nickel and gadolinium the exchange integral is positive, whereas for manganese and chromium the exchange integral is negative. The positive value of the exchange integral represents the material is ferromagnetic and the negative exchange integral value represents the material is antiferromagnetic. In general, if the ratio, $r/r_d > 3$, the material is ferromagnetic, otherwise the material is antiferromagnetic. It should be noted that manganese is suitably alloyed so that $r/r_d > 3$, then it will become ferromagnetic.

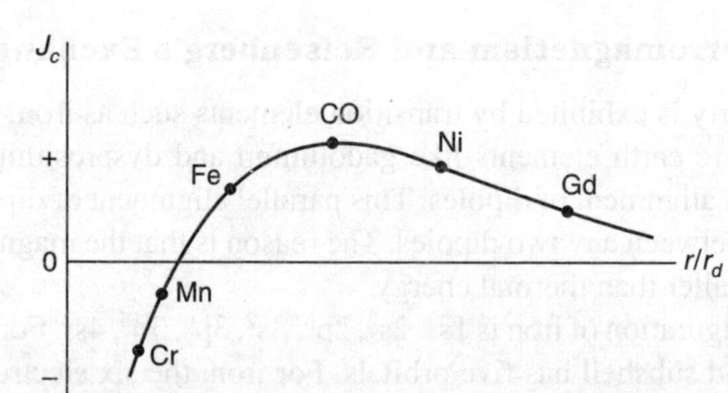

Figure 12.18 The exchange integral as a function of (r/r_d).

12.6 ANTIFERROMAGNETIC MATERIAL

In an antiferromagnetic material, the adjacent dipoles are aligned antiparallel. The dipoles are equal in magnitude and hence the magnetic moment produced by one dipole is cancelled by the other. Therefore, the resultant magnetic moment of antiferromagnetic materials is nearly zero in the absence of the magnetic field. The susceptibility of an antiferromagnetic material is given by

$$\chi = \frac{C}{T + \theta} \quad \text{where } T > T_N \tag{12.12}$$

where C is the Curie constant and T_N is the Nëel temperature.

For an antiferromagnetic material, if a graph is plotted between, χ and T, initially the susceptibility increases with the increase in temperature and it reaches a maximum value and then it decreases as shown in Figure 12.19. The temperature corresponding to this maximum value of χ is known as Nëel temperature, T_N.

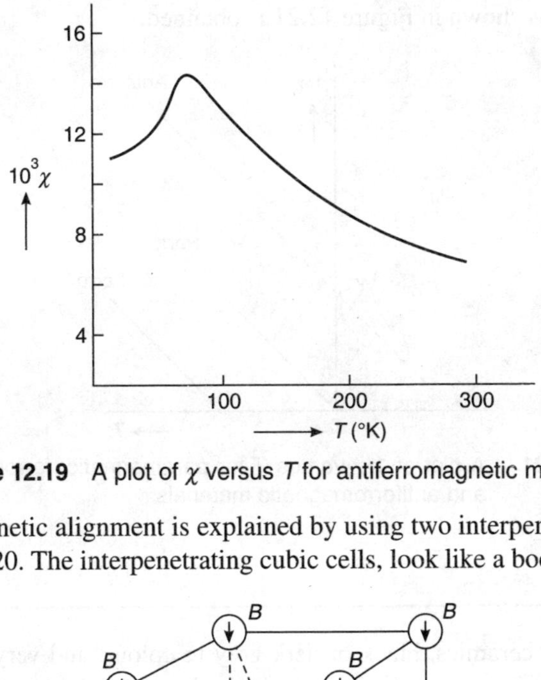

Figure 12.19 A plot of χ versus T for antiferromagnetic material.

The antiferromagnetic alignment is explained by using two interpenetrating cubic unit cells as shown in Figure 12.20. The interpenetrating cubic cells, look like a body centred unit cell. The

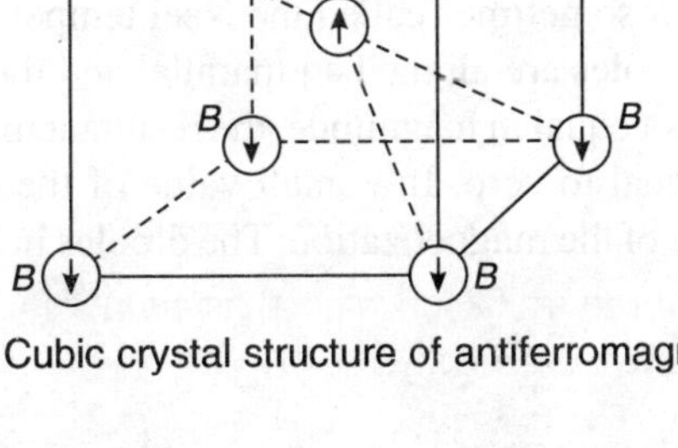

Figure 12.20 Cubic crystal structure of antiferromagnetic material.

corner atoms of the cubic unit cell are made of one type of atom and the body centred atom is made of another type of atoms. The dipoles of the corner atoms are pointed in the downward direction, whereas the dipoles of the body centred atom is pointed in the upward direction.

The susceptibility of paramagnetic, ferromagnetic and antiferromagnetic materials are given by

$$\chi = \frac{C}{T} \quad \text{for a paramagnetic material} \tag{12.13}$$

$$\chi = \frac{C}{T - \theta_f} \quad \text{where } T > \theta_f, \text{ for a ferrimagnetic material} \tag{12.14}$$

$$\chi = \frac{C}{T + \theta} \quad \text{where } T > T_N \text{ for an antiferromagnetic material} \tag{12.15}$$

If a graph is drawn between $\frac{1}{\chi}$ and T, for paramagnetic, ferromagnetic and antiferromagnetic materials, straight lines as shown in Figure 12.21 is obtained.

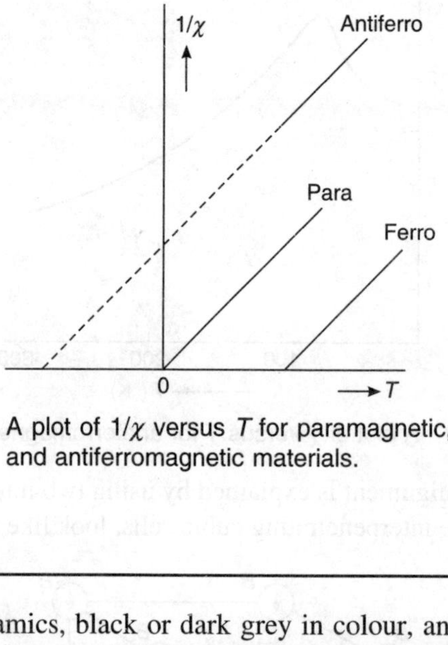

Figure 12.21 A plot of $1/\chi$ versus T for paramagnetic, ferromagnetic and antiferromagnetic materials.

12.7 FERRITES

Ferrites are mixed oxide ceramics, black or dark grey in colour, and very hard and brittle. The electrical resistivity of ferrites is very high, typically in the order of 10 to 10^{12} Ω cm. The ferrites have strong magnetic properties. The ferrites have well-defined magnetic transition temperature called the Curie temperature (or sometimes called the Néel temperature).

In ferrites the adjacent dipoles are aligned antiparallel and they are not equal in magnitude. Since the adjacent dipoles are not equal in magnitude, the resultant magnetic moment in the absence of the magnetic field is not equal to zero. If a small value of the magnetic field is applied to a ferrite, it produces a large value of the magnetization. The dipoles in ferrites are aligned antiparallel as shown in Figure 12.6.

12.7.1 Structure of Ferrites

There are three technically important ferrites, namely (i) spinels, (ii) garnets and (iii) hexaferrites. Spinels and garnets are magnetically soft, whereas the hexaferrite is hard. Let us discuss only the spinel structure.

Spinels

The chemical formula for the unsubstituted spinel is $M^{2+} Fe_2^{3+} O_4^{2-}$, where M^{2+} represents the metallic divalent ions such as Fe^{2+}, Mn^{2+}, Cd^{2+}, Zn^{2+}, Mg^{2+}, Cu^{2+}, Co^{2+}, etc. If the metallic ion M^{2+} is replaced by a ferrous ion (Fe^{2+}), then it is said to be a ferrous ferrite, $Fe^{2+} Fe_2^{3+} O_4^{2-}$. Similarly, if M^{2+} is replaced by Zn^{2+}, then it is said to be zinc ferrite, $Zn^{2+} Fe_2^{3+} O_4^{2-}$.

The crystal structure of spinels is cubic. There are 8 corner atoms in the cubic unit cell. Each and every corner atoms has one ferrite molecules. There are eight ferrite molecules in a ferrite unit cell. Therefore, there are 8 ferrous (Fe^{2+}) ions, 16 ferric (Fe^{3+}) ions and 32 oxygen (O^{2-}) ions in a ferrous ferrite unit cell. If oxygen atoms are considered alone, it constitutes a FCC structure. There are sixteen octahedral sites and eight tetrahedral sites in a spinel ferrite unit cell. A metallic ion surrounded by six oxygen atoms is called octahedral site. A metallic ion surrounded by four oxygen atoms is called tetrahedral site. So, in one spinel ferrite molecule there are two octahedral sites and one tetrahedral site.

Inverted spinel structure

Consider only one ferrite molecule. It has two octahedral sites and one tetrahedral site. Consider the octahedral and the tetrahedral sites as shown in Figure 12.22. In the case of a ferrous ferrite, since the ferrous ion is a magnetic material, it prefers to occupy the octahedral site. The remaining one octahedral site and the tetrahedral sites are occupied by the ferric ion. This type of structure is said to be inverted spinel structure. The ferrimagnetism arises only because of the Fe^{2+} ions which are coupled magnetically. There is no net moment arises from equal number of Fe^{3+} ions in the two sublattices (octahedral and tetrahedral sites).

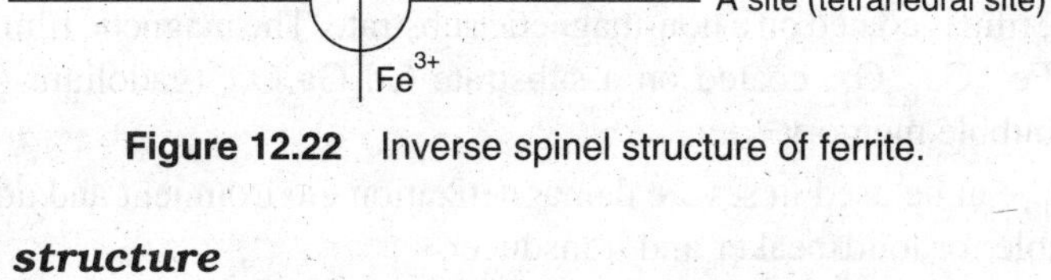

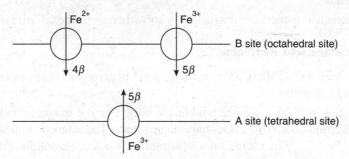

Figure 12.22 Inverse spinel structure of ferrite.

Normal spinel structure

Consider a zinc ferrite. Zinc is a non-magnetic material. The ferric ions prefers to occupy the octahedral site and the zinc ions occupy the tetrahedral site as shown in Figure 12.23. Therefore, the zinc ferrite is called normal spinel and it is not a ferrimagnetic material because all the Fe^{3+} ions are on octahedral site.

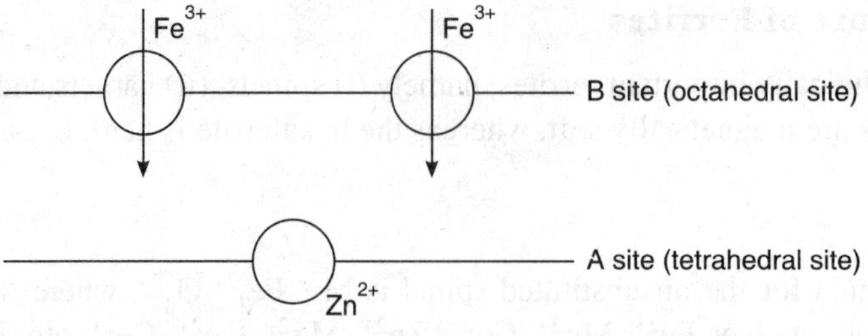

Figure 12.23 Normal spinel structure of ferrite.

Calculation of the magnetic moment

The magnetic moment of a ferrite molecule is calculated as follows. The ferrous ion has four unpaired electrons, whereas a ferric ion has five unpaired electrons. Consider that an unpaired electron produces a magnetic moment of one Bohr magneton (1 β), then the magnetic moment of a ferrous ion is 4 β and that of ferric ion is 5 β. If antiparallel alignment is considered, the total magnetic moment is 4 β. The experimental value of the magnetic moment of ferrous ferrite is 4.1 β. If all the dipoles are aligned parallel, the total magnetic moment is $4\beta + 5\beta + 5\beta = 14\beta$. This calculation confirms the antiparallel alignment of dipoles.

12.7.2 Properties of Ferrites

(a) Ferrites are metal oxide ceramics, black or grey in colour.
(b) Ferrites are very hard and brittle.
(c) The ferrites have strong magnetic properties due to ferrimagnetism.
(d) The magnetic dipoles are aligned antiparallel and they are not equal in magnitude.
(e) Ferrite has well-defined magnetic transition temperature called the Curie temperature.
(f) Ferrites have three technically important structures, namely spinels, garnets and hexaferrites.
(g) The spinels and garnets are magnetically soft whereas the hexaferrite is hard.

12.7.3 Applications of Ferrites

(a) Small toroids in MgMn, CuMn or LiNi spinel materials are used in large numbers as data storage materials.
(b) Garnets are used in magnetic bubble memory. For the magnetic bubble memory a magnetic, film is coated on a non-magnetic substrate. The magnetic film with composition $Y_{2.9}La_{0.1}Fe_{3.8}Ga_{1.2}O_{12}$ coated on a substrate $Gd_3Ga_5O_{12}$ (gadolium gallium garnet) is used for bubble memory.
(c) $BaFe_{12}O_{19}$ can be used in severe demagnetization environment and at low temperature. It is suitable for loudspeaker and transducer.
(d) $SrFe_{12}O_{19}$ can be used in dynamic applications with large external magnetic field. It is used in dynamic applications with large external demagnetization fields, e.g. motor stators, torque drives, etc.

(e) Ferrites are used for the preparation of microwave devices like circulators, isolators and phase shifter.

(f) Ferrites rods are used to produce low frequency ultrasonic waves using magnetostriction principles.

(g) Hard ferrites are used for the preparation of permanent magnets. They are used for the preparation of windscreen, wiper motor, loudspeaker, etc.

12.8 SOFT AND HARD MAGNETIC MATERIALS

The magnetic materials are classified into soft and hard magnetic materials based on their magnetic properties.

12.8.1 Soft Magnetic Material

The magnetic materials that are easy to magnetize and demagnetize are called soft magnetic materials. In a soft magnetic material, the domain wall move easily. Therefore, the application of a small value of magnetic field produces a large value of magnetization.

The soft magnetic materials are prepared so that they become very soft. The required composition of the materials is taken and they are heated above their melting points and the liquid solution is cooled slowly. The slow cooling of the melt makes them too soft and free from impurities.

Properties

(i) The B–H curve of a soft magnetic material is narrow and steep.

(ii) The hysteresis loop has a small area and hence the hysteresis power loss per cycle is low (Figure 12.24).

(iii) The resistivity of the soft magnetic material is very high and hence eddy current loss is low.

(iv) The soft magnetic material has high permeability and susceptibility.

(v) The soft magnetic materials have low coercivity.

(vi) The soft magnetic materials are free from impurities.

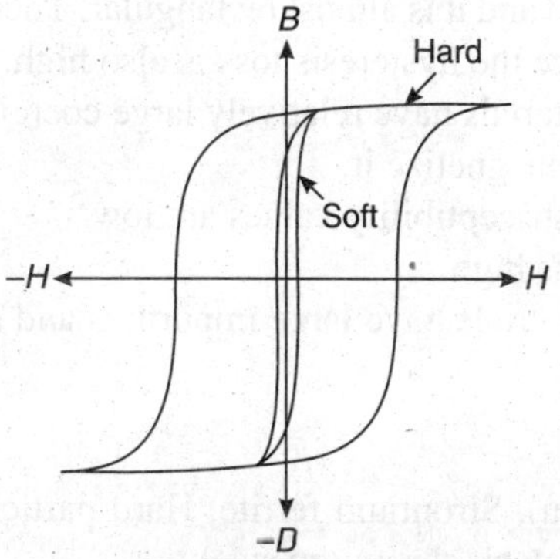

Figure 12.24 Hysteresis curve for soft and hard magnetic materials.

Examples of soft magnetic materials and uses

(a) Silicon iron (97%Fe–3%Si)

It possesses high resistivity and low eddy current losses. So, it is widely used in electrical machinery such as transformer.

(b) Supermalloy (79% Ni–16% Fe–5% Mo)

It possesses high permeability. It is used to prepare low loss electrical devices, e.g. transformers, magnetic amplifiers.

(c) Permalloy 78 (Ni–78% and Fe–22%)

It is used to prepare low loss electrical devices such as audio transformer, HF transformer, recording heads and filters.

(d) Glass metals (Fe-Si-B)

It is used to prepare low loss transformer core.

(e) Ferrites

(i) *Mn-Zn ferrites.* It has low conductivity and negligible eddy current losses. It is used in HF transformer, inductors (e.g. pot cores, recording heads)

(ii) *Mn ferrite and Mn-Zn ferrites.* The Mn ferrite such as $MnFe_2O_4$ and Mg-Zn ferrite $Mg_{(1-X)}Zn_XFe_2O_4$ are used for high frequency applications.

12.8.2 Hard Magnetic Materials

The magnetic materials that are difficult to magnetize and demagnetize them are called hard magnetic materials. Since the coercivity of the hard magnetic material is high, the material needs a large value of magnetic field to demagnetize it. The rotation of domain walls also needs very high magnetic field and the rotation of domains is difficult.

The materials are purposely made as hard as possible. Therefore, the materials are prepared by heating above their melting points and then they are suddenly cooled by quenching in a liquid. The impurities are purposely added to these materials so as to make them as hard as possible.

Properties

1. The *B–H* curve is broad and it is almost rectangular. Therefore, the area of the hysteresis curve is large and hence the hysteresis loss is also high.
2. The hard magnetic materials have relatively large coercivity. So they need a large value of magnetic field to demagnetize it.
3. The permeability and susceptibility values are low.
4. The eddy current loss is high.
5. The hard magnetic materials have large impurities and lattice defects. They have large magnetostatic energy.

Examples

NdFeB, Alnico (Fe-Al-Ni-Co-Cu), Strontium ferrite, Hard particles γ-Fe_2O_3, Rare earth cobalt, Sm_2Co_{15}, carbon steel, tungsten steel, chromium steel.

Applications

1. Alnico is used for wide range of permanent magnet applications.
2. Strontium ferrite is used for the preparation of loud speakers, telephone receiver, various toys, dc motor, starter motor.
3. γ-Fe_2O_3, is used for the magnetic coating in audio and video tapes, floppy disks.
4. NdFeB is used for a wide range of applications. Small motors, (e.g. in hand tools), and walkman equipment, DC motors, MRI body scanners, computer applications.
5. Carbon steel is used as magnets for toys, compass needle, latching relays, and certain types of metres.
6. Tungsten steel is used to prepare dc motors.
7. Chromium steel is used to prepare the best permanent magnets.

12.9 MAGNETIC RECORDING AND READING

12.9.1 Magnetic Recording

The magnetic materials are used for analog and digital recording of data. Let us consider the analog recording of data in an audio tape. The audio tape is a polymer-backed tape that has a magnetic coating over it. It uses a recording head to record the information on the audio tape. The recording head is a toroid type electromagnet and it has a small gap typically around 1 μm. The input signal is initially converted into a current signal. Usually a microphone is used to convert sound signal intq electrical signal. The electrical signal is passed through the coil of the electromagnet. When the current signal is passing through the coil of wire, it gets converted into magnetic field. The magnetic field, while passing through the electromagnet produces fringing magnetic field in the gap. Usually, the recording head touches the audio tape and hence the fringing magnetic field magnetizes the magnetic material coated on the tape. The fringing field changes according to the current signal and hence magnetization in the audio tape also changes. The electrical signal is recorded in the audio tape as a spatial magnetic pattern. As the tape advances the information are recorded in the audio tape continuously. This type of recording is said to be longitudinal recording (Figure 12.25).

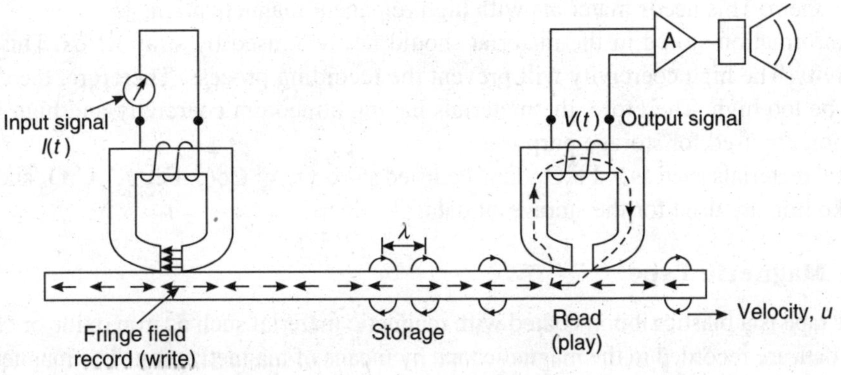

Figure 12.25 Magnetic recording and read out.

12.9.2 Magnetic Reading

The process of retrieving the data from the tape is known as reading. For reading of data the same recording head is used as reading head. The reading process is based on the Faraday's law of induction. During the reading process, the magnetic field stored in the tape passes through the reading head. A portion of the field penetrates through the core and flows around the whole core and hence it links the coil. Whenever the magnetic field is passing through the coil, it induces an emf. As the tape is moving with a constant speed, the play head produces continuous voltage signal. This induced emf is filtered and then amplified. The amplified voltage signal gets converted into sound using a head phone or ear phone.

Let f be the frequency of the spatial signal and u be the velocity of the tape, then the distance advanced by the tape is $\Delta x = u/f$. This Δx represents the spatial wavelength. The low spatial wavelength and greater frequency, f provide more number of storing of information.

12.9.3 Recording Head Material

A recording head material should produce the magnetization easily so as to follow the input signals. This property requires, the recording head should be a soft magnetic material. It should produce a strong fringing magnetic field at the gap so as to magnetize the material in the tape. It requires that the material should have low coercivity and large saturation magnetization.

The materials like permalloys (Ni-Fe alloys), sendust (Fe-Al-Si alloys), and some sintered soft ferrite (e.g. MnZn and NiZn ferrites) and the amorphous materials such as CoZrNb alloys are used for preparing the recording head.

12.10 MAGNETIC DATA STORAGE

The process of storing the data (audio or video) using the magnetic principle is known as magnetic data storage.

12.10.1 Magnetic Data Storage Materials

The materials used for storage purpose should retain the spatial magnetization pattern (information) recorded on them. This needs materials with high remanent magnetization, M_r.

The information stored in the material should not be erased by stray fields. This requires high coercivity. The high coercivity will prevent the recording process. Therefore, the coercivity should not be too high. Therefore, the materials having a medium coercivtiy and high remanent magnetization are used for storage purposes.

Typical materials such as γ-Fe_2O_3, Co-modified γ-Fe_2O_3, or Co(γ-Fe_2O_3), CrO_2 and metallic particles like iron are used for the storage of data.

12.10.2 Magnetic Tape

A magnetic tape is a plastic ribbon coated with magnetic material such as iron oxide or chromium oxide. The data are recorded in the magnetic tape by means of magnetization. The magnetized and non-magnetized regions are represented 1s and 0s. The data in a magnetic tape is erased and reused. The magnetic tapes with breadth 1/2 inch or 1/4 inch and length 50 to 2400 feet are available.

Data storage organization

The magnetic tape storage is divided into vertical columns, called frames and horizontal column called tracks or channels. The old version of magnetic tape has 7 tracks. The first six tracks were used for recording of data in BCD (binary coded decimal) code, and the seventh track is used for recording the parity bit or check bit. The parity bit is used to check whether there is a loss of any character or bit from a string of 6 bit from the input or output data during operation. Depending on the total number of bits is odd or even, the parity is called odd parity or even parity. If the total number of bit is odd then the parity bit is one so as to make even parity. To make odd parity, the parity bit is zero if the total number of bit is odd. In Figure 12.26, the A frame has odd parity and hence the parity bit becomes one so as to make the parity even.

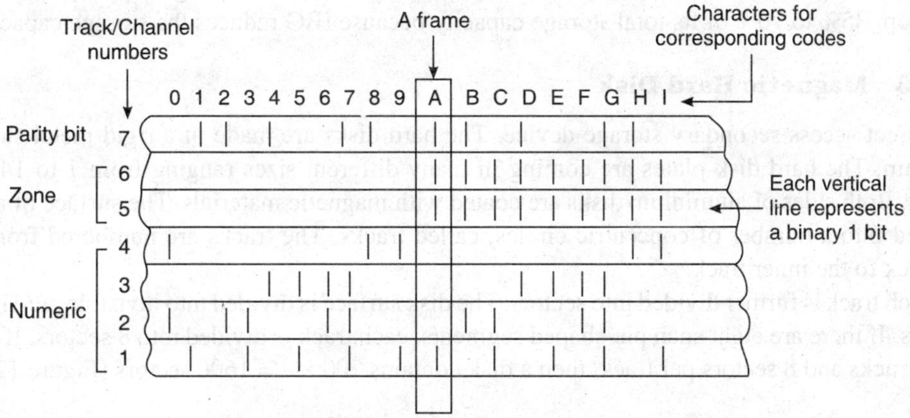

Figure 12.26 Data recording in a 7 track magnetic tape.

Another version of the magnetic tape has 9 tracks. The first eight tracks are used for recording the byte of data and the 9th track is used for recording the parity bit. In a 9 track tape, the data are recorded by nine different read/write heads. They record the data in nine parallel tracks.

In order to write the information in a tape, the tape is kept ready such that one end is wound in a spool and the other end is threaded manually in a take up spool. There is metal foil at the beginning of the tape (BOT), called a marker as shown in Figure 12.27. Whenever a write command is given, the data are recorded only after the tape moves with its full speed. During this time, the tape will move a distance of 0.6 inch. This distance is called inter block gap (IBG). After this gap, the data are recorded in a block. The length of the block is at least ten times as that of the IBG, so

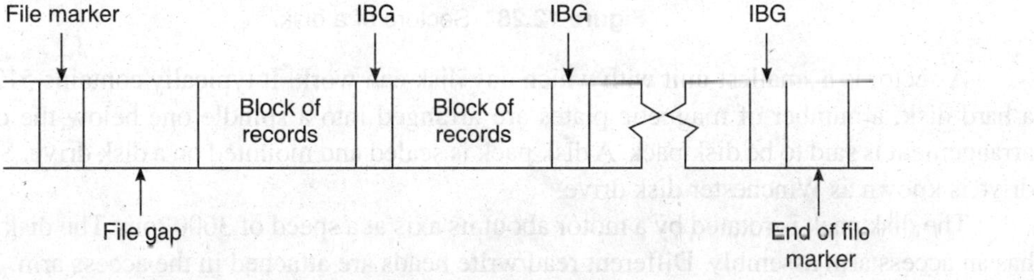

Figure 12.27 Data organization in a magnetic tape.

as to reduce the wastage of tape. After recording the data in a block, there is a gap called IBG. Again the data are recorded in another block. Similarly a number of blocks are written in a serial order and the end of data (EOD) is represented by a metal foil known as file marker. Now, the tape is rewound and kept for reading.

Since the data are recorded one-by-one in different blocks and there is no address to the data, the data are retrieved in the order in which they are written. For example, if any one wants to retrieve the data recorded on the last block, all the earlier data has to be read before reading the required data.

In older system, only 800 bpi (byte per inch) can be recorded, whereas the newer system has a recording density of 77,000 bpi. The storage capacity of a magnetic tape having a length of 2400 feet is $2400 \times 12 \times 512 \times 800 \times 2$ (for two sides) = 23 GB. The actual storage capacity of the tape varies from 35% to 70% of its total storage capacity because IBG reduces the storage capacity.

12.10.3 Magnetic Hard Disk

It is a direct access secondary storage device. The hard disks are made of a rigid metal such as aluminium. The hard disk plates are coming in many different sizes ranging from 1 to 14 inch diameter. Both sides of aluminium disks are coated with magnetic materials. The surface of a disk is divided into a number of concentric circles, called tracks. The tracks are numbered from the outer track to the inner track.

Each track is further divided into sectors. The disk surface is divided into invisible pie-shaped segments. If there are eight such pie-shaped segments, each track is divided into 8 sectors. If there are 200 tracks and 8 sectors per track, then a disk contains $200 \times 8 = 1600$ sectors (Figure 12.28).

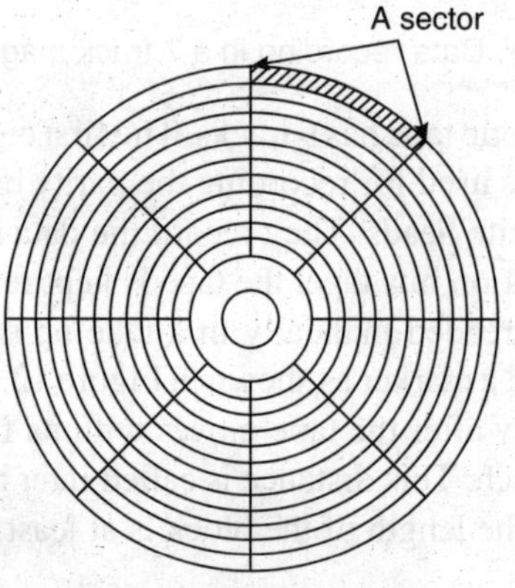

Figure 12.28 Sectors of a disk.

A sector is a smallest unit with which any disk can work. It typically contains 512 bytes. In a hard disk, a number of magnetic plates are arranged into a spindle one below the other. This arrangement is said to be disk pack. A disk pack is sealed and mounted on a disk drive. Such a disk drive is known as Winchester disk drive.

The disk pack is rotated by a motor about its axis at a speed of 3600 rpm. The disk drive also has an access arm assembly. Different read/write heads are attached in the access arm. The upper surface of the top disk and the lower surface of the bottom disk are generally not used because these surfaces may be easily scratched. For example, if a disk drive contains four disk platters as shown in figure 12.29, then there are six read/write heads, except the top surface of the disk, upper disk and the bottom surface of the lower disk. The read/write heads are arranged in the access arm assembly such that they can be moved simultaneously.

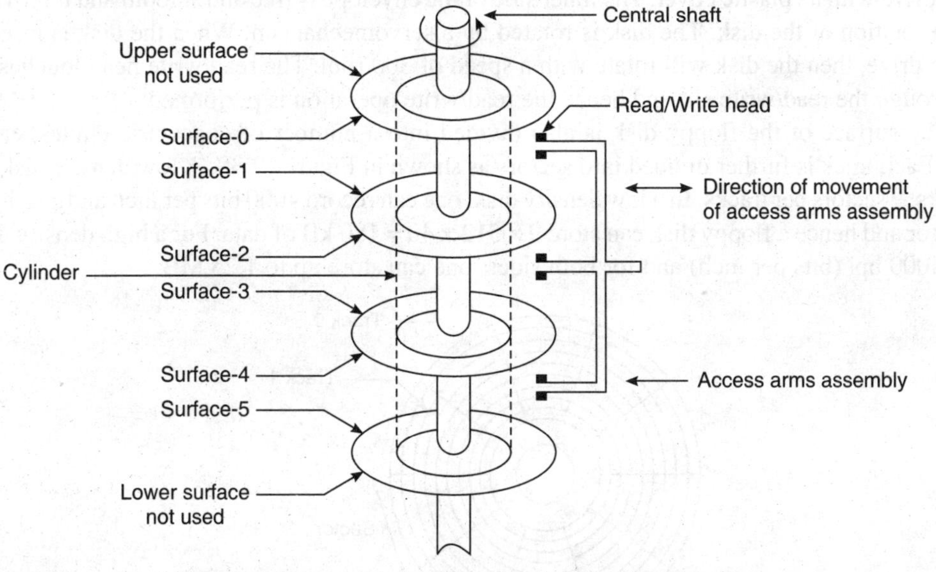

Figure 12.29 A hard disk pack containing four platters.

The fast access of data is achieved using a concept known as cylinder. A set of corresponding tracks in all the recording surfaces of a disk pack together form a cylinder. For, example, the 10th track of the entire recording surface constitutes the 10th cylinder. If there are 200 tracks in a disk, there are 200 cylinders. When a read command is given to the read/write head, all the read/write heads will assemble in a particular track, say, 20th track. In one revolution, the data stored on the 20th track of 0 (zero) surfaces are read. In the next revolution, the data stored in 20th track of 1st surface are read. Then the data stored in the 20th track of the 2nd surface are read. Similarly, up to the last surface the data are read in the same track. The disk address may consist of sector number, cylinder number and surface number.

The storage capacity of a hard disk having 5.25 inch diameter, 10 disk plates, 2655 tracks and 125 sectors per track and each sector can store 512 bytes per sector is $18 \times 2655 \times 125 \times 512 = 3 \times 10^9$ bytes $\approx$ 3 GB (3 Giga bytes).

12.10.4 Floppy Disk

A floppy disk is a round piece of flexible plastic, coated with magnetic material. It is covered by a square plastic container or vinyl jacket cover. The container protects the disk surface and it has a wiping action to remove the dust particle. Floppy disk can be bend and flexible and hence it is called floppy disk. It is also called floppies or diskettes.

There are two different types of floppies available. They are (i) 5.25 inch and (ii) 3.5 inch floppy disk.

5.25 inch floppy disk

It is a 5.25 inch circular flexible Mylar computer tape material. The magnetic oxide is coated on both sides of the disk. It is packaged in 5.25" square plastic envelope with a long slit for read/write access, a hole for index mark sensing and a hole for the hub. The floppy disk is inserted into the floppy drive with its plastic cover. The inner side of the envelope is free and smooth and it provides smooth rotation of the disk. The disk is rotated by a servomechanism. When the disk is inserted into the drive, then the disk will rotate with a speed of 300 rpm. The read/write head touches the disk through the read/write slit and hence the read/write operation is performed.

The surface of the floppy disk is also divided into a number of concentric circles, called tracks. Each track is further divided into sectors as shown in Figure 12.30. A low density disk has 40 tracks, 9 sectors per tracks. In a low density disk, one can record 4000 bits per inch and 512 bytes per sector and hence a floppy disk can store $9 \times 512 \times 40 = 180$ kB of data. For a high density disk, with 14000 bpi (bits per inch) and for both sides, one can store up to 1.25 MB.

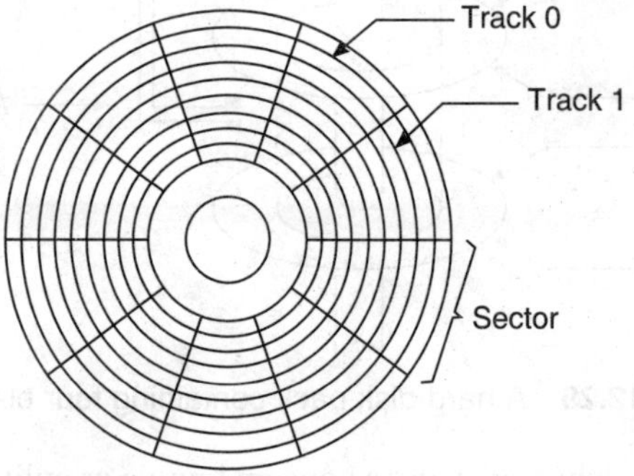

Figure 12.30 Sectors and tracks in floppy disk.

3.5 inch floppy disk

The 3.5 inch disk has a diameter of 3.5 inch as shown in Figure 12.31. It is encased in a 3.5 inch square, hard-plastic jacket cover. It has an opening for read/write head. The opening is covered by a sliding metal piece. When the disk is inserted into the floppy drive, the cover slides back to expose the opening. The read/write head will come into contact with the disk surface.

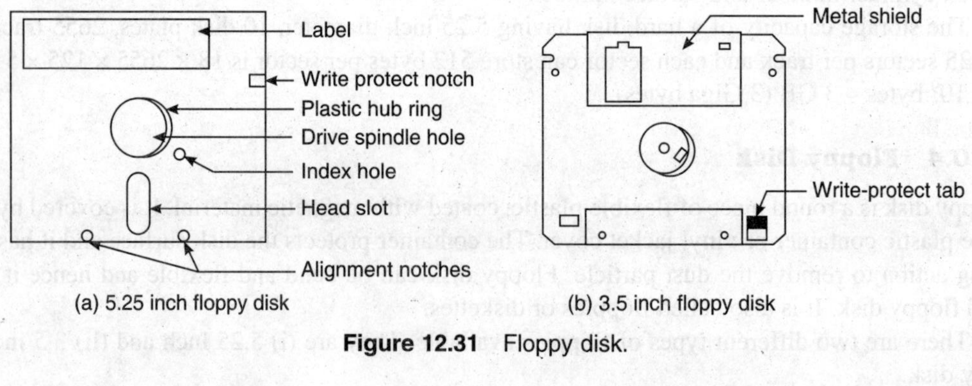

Figure 12.31 Floppy disk.

The 3.5 inch disk is available in three different capacities, namely (i) double density, (ii) high density and (iii) very high density. The double density, high density and very high density floppy disks can store up to 720 kB, 1.44 MB and 2.88 MB respectively.

12.11 MAGNETIC BUBBLE MEMORY

The magnetic bubble memory is a type of computer memory. It uses an epitaxially grown thin film of materials such as orthoferrite or garnet coated on a substrate, usually garnet. Garnet has a wavy domain structure as shown in Figure 12.32(a). If a single crystal plate (~μm thickness) with orthogonal easy direction is subjected to an increasing magnetic field (H_a) also normal to the plate, the antiparallel domains shrink as shown in Figure 12.32(b), until over a narrow range of (H_a), small cylindrical domains are formed [Figure 12.32(c)]. They are called magnetic bubbles. The diameter of the bubble is typically 2 to 5 μm. Each bubble can carry one bit of information. The plate should be free from imperfection, parallel and flat.

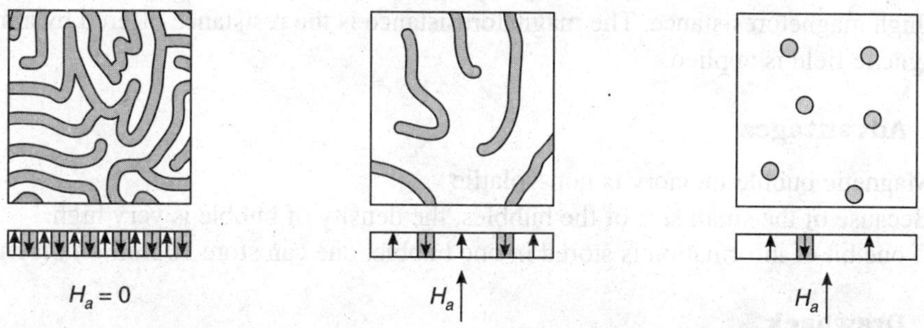

(a) Wavy domains in garnet (b) The domain shrinks due to field (c) Magnetic bubble formed

Figure 12.32 Domains in a thin plate of magnetic material with orthogonal easy direction.

The bubbles can be moved by applying the magnetic field. For moving the bubbles in a controlled manner along a particular direction, a pattern of small permalloy bars are crėated on the surface. These bars are produced using photoengraving.

The movement of the bubble from one place to the other can be achieved by applying magnetic field. Consider two permalloy bar as shown in Figure 12.33(a). Consider there is a magnetic bubble with its north pole upward in one of the bar (say, bar1). In the absence of the field, the permalloy bars are unmagnetized. If a field is applied, the permalloy bar1 gets magnetized and the bubble in the north pole of the magnet move towards south pole of the bar [Figure 12.33(b)]. In order to move the magnetic bubble from one bar to another bar, one has to change the direction of the field. Then the bubble moves to bar2 and the bar2 gets magnetized [Figure 12.33(c)].

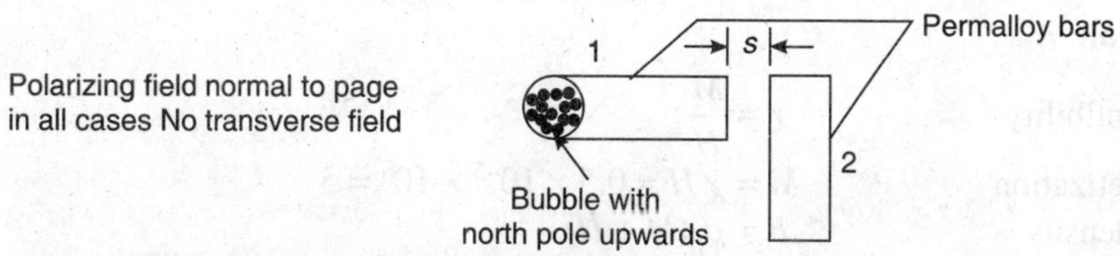

(a) The bubble is in the north pole of bar1

Figure 12.33 Movement of bubbles from one bar to another—magnetic bubble memory *(contd.)*

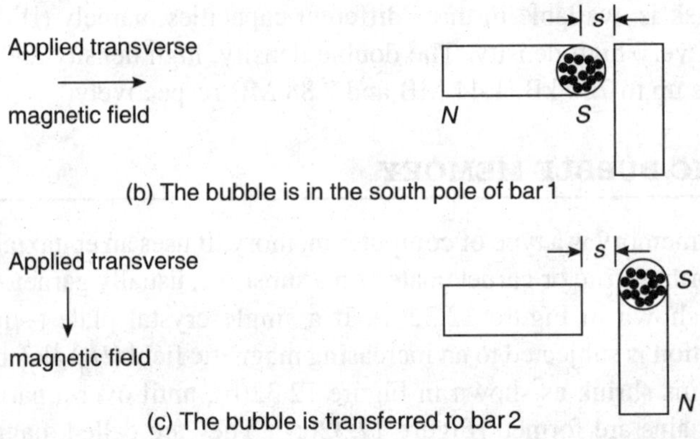

Figure 12.33 Movement of bubbles from one bar to another—magnetic bubble memory.

The bubbles can be detected by making to pass through a strip of Indium antimonide (InSb) which has high magnetoresistance. The magnetoresistance is the resistance offered by a material, when a magnetic field is applied.

12.11.1 Advantages

1. Magnetic bubble memory is non-volatile
2. Because of the small size of the bubbles, the density of bubble is very high.
3. If one bit of information is stored in one bubble, one can store 10 million bit cm^{-2}.

12.11.2 Drawback

1. The magnetic bubble memory is not a random access memory. Therefore, the information must be read serially. The achievable speed may be few hundred kbits s^{-1}.

SOLVED PROBLEMS

12.1 In a magnetic material, the field strength is found to be 10^6 A m^{-1}. If the magnetic susceptibility of the material is 0.5×10^{-5}, calculate the intensity of magnetization and the flux density of the material.

Given data

Magnetic field strength, $H = 10^6$ A m^{-1}

Magnetic susceptibility, $\chi = 0.5 \times 10^{-5}$

Solution

Susceptibility $\chi = \dfrac{M}{H}$

Magnetization $M = \chi H = 0.5 \times 10^{-5} \times 10^6 = 5$

Flux density $B = \mu_0(M + H)$

$$B = 4\pi \times 10^{-7}(5 + 10^6)$$
$$= 1.257 \text{ Wb m}^{-2}$$

Magnetization $M = 5 \text{ A m}^{-1}$
Flux density $B = 1.257 \text{ Wb m}^{-2}$

12.2 The saturation magnetic induction of nickel is 0.65 Wb m^{-2}. If the density of nickel is 8906 kg m^{-3} and its atomic weight is 58.7, calculate the magnetic moment of the nickel atom in Bohr magneton.

Given data
Saturation magnetic induction, $B = 0.65 \text{ Wb m}^{-2}$
Density $\rho = 8906 \text{ kg m}^{-3}$
Atomic weight of Ni $M_{at} = 58.7$

Solution

Number of atoms per m^{-3} $= \dfrac{\text{Density} \times \text{Avogadro's constant}}{\text{Atomic weight}}$

$$= \frac{8906 \times 6.022 \times 10^{23} \times 10^3}{58.7}$$
$$= 9.136 \times 10^{28} \text{ m}^{-3}$$

Magnetic moment, $\mu_m = \dfrac{B}{N\mu_0} = \dfrac{0.65}{9.136 \times 10^{28} \times 4\pi \times 10^{-7}}$

$$= 5.662 \times 10^{-24}$$
$$= \frac{5.662 \times 10^{-24}}{9.27 \times 10^{-24}} = 0.61\ \mu_B$$

Magnetic moment of nickel atom = 0.61 μ_B.

12.3 If a magnetic field of 1800 A m^{-1} produces a magnetic flux of 3×10^{-5} Wb in an iron bar of cross sectional area 0.2 cm^2, calculate permeability.

Given data
Magnetic field $H = 1800 \text{ A m}^{-1}$
Magnetic flux $\phi = 3 \times 10^{-5}$ Wb
Area of cross section $A = 0.2 \text{ cm}^2$

Solution

Magnetic flux density $B = \dfrac{\phi}{A} = \dfrac{3 \times 10^{-5}}{0.2 \times 10^{-4}} = 1.5 \text{ Wb m}^{-2}$

Magnetic flux density $B = \mu_0 \mu_r H$

Relative permeability $\mu_r = \dfrac{B}{\mu_0 H} = \dfrac{1.5}{4\pi \times 10^{-7} \times 1800}$

$$= 663.14$$

The permeability of the material is 663.14.

12.4 Calculate the saturation magnetization for Ni ferrite. The lattice parameter for the Ni ferrite is 0.835 nm and the magnetic moment per unit cell is 18.4 μ_B.

Given data

Magnetic moment $\mu = 18.4\ \mu_B$

Lattice parameter $a = 0.835$ nm

Solution

Magnetization

$$M = \frac{\text{Magnetic moment}}{\text{Volume}}$$

$$= \frac{18.4 \times 9.27 \times 10^{-24}}{(0.835 \times 10^{-9})^3} = 2.929 \times 10^5$$

Saturation magnetization = 2.929×10^5 A m^{-1}.

12.5 A magnetic field strength of 2×10^5 A m^{-1} is applied to a paramagnetic material with a relative permeability of 1.01. Calculate the value of B and M.

Given data

Magnetic field strength $H = 2 \times 10^5$ A m^{-1}

Relative permeability $\mu_r = 1.01$

Solution

Magnetic flux density, $B = \mu_0 \mu_r H$

$$= 4\pi \times 10^{-7} \times 1.01 \times 2 \times 10^5$$

$$= 0.2538 \text{ Wb m}^{-2}$$

Magnetic flux density, $B = \mu_0(M + H)$

Magnetization, $$M = \frac{B}{\mu_0} - H = \frac{4\pi \times 10^{-7} \times 1.01 \times 2 \times 10^5}{4\pi \times 10^{-7}} - 2 \times 10^5$$

$$= 2000 \text{ A m}^{-1}$$

Magnetic flux density, $B = 0.2538$ Wb m^{-2}

Magnetization, $M = 2000$ A m^{-1}

12.6 The magnetic material is subjected to a magnetic field of strength 500 A m^{-1}. If the magnetic susceptibility of the material is 1.2, calculate the magnetic flux density inside the material ($\mu_0 = 4\pi \times 10^{-7}$ H/m).

Given data

Magnetic field strength $H = 500$ A m^{-1}

Susceptibility $\chi = 1.2$

$\mu_0 = 4\pi \times 10^{-7}$ H/m

Solution

Susceptibility $$\chi = \frac{M}{H}$$

$$M = \chi H = 1.2 \times 500 = 600 \text{ A m}^{-1}$$

Magnetic flux density, $B = \mu_0(M + H)$

$= 4\pi \times 10^{-7} \times (600 + 500)$

$= 1.382 \times 10^{-3}$ Wb m^{-2}

Magnetic flux density, $B = 1.382 \times 10^{-3}$ Wb m^{-2}.

SHORT QUESTIONS

1. What are magnetic materials?
2. What is a magnetic dipole?
3. Define the term magnetic dipole moment.
4. What do you mean by magnetic lines of forces?
5. Define the term magnetic flux density.
6. Define magnetic field strength.
7. What is magnetization?
8. What is meant by magnetic susceptibility?
9. Define the term magnetic relative permeability.
10. What is Bohr magneton? What is its value?
11. Mention the five different types of magnetic materials.
12. What is a diamagnetic material? Mention any two properties of diamagnetic material.
13. What is a paramagnetic material? Mention any two properties of paramagnetic material.
14. What are ferromagnetic materials? Mention any two properties of ferromagnetic materials.
15. What are antiferromagnetic materials? Mention any two properties of antiferromagnetic materials.
16. What are ferrimagnetic materials? Mention the ferrimagnetic materials.
17. What are magnetic domains?
18. What is magnetostatic energy?
19. What is anisotropic energy?
20. What is domain wall energy?
21. What is magnetostriction energy?
22. What are ferrites?
23. Describe the spinel structure of ferrites.
24. Describe the inverse spinel structure of ferrites.
25. Write the properties of ferrites.
26. Write any two applications of ferrites.
27. What are soft magnetic materials?
28. Write the properties of soft magnetic materials.
29. Mention the uses of soft magnetic materials.
30. What are hard magnetic materials?

31. Mention the properties of hard magnetic materials.
32. Mention the applications of hard magnetic materials.
33. What are the materials used as recording head materials?
34. Mention the materials used as magnetic data storage.

DESCRIPTVE TYPE QUESTIONS

1. What are domains? Discuss the domain concept and hence explain the hysteresis curve.
2. What is meant by exchange energy? Explain the origin of ferromagnetism using Heisenberg's exchange interaction energy concept.
3. What are soft and hard magnetic materials? Mention the properties and applications of hard and soft magnetic materials.
4. What are antiferromagnetic materials? Explain the antiparallel alignment of dipoles in antiferromagnetic materials.
5. What are ferrites? Explain the structure of ferrites. Mention the properties and applications of ferrites.
6. Explain with neat sketch the process of magnetic recording and reading of data.
7. What are data storage materials? Explain the magnetic tape, floppy disks and hard disks.
8. What is magnetic bubble memory? Explain in detail the working of a magnetic bubble memory.

PROBLEMS

1. Magnetic field intensity of a paramagnetic material is 10^4 A m^{-1}. At room temperature, its susceptibility is 3.7×10^{-3}. Calculate the magnetization in the material.
2. In magnetic material the field strength is found to be 10^6 A m^{-1}. If the magnetic susceptibility of the material is 0.5×10^{-5}, calculate the intensity of magnetization and flux density in the material.
3. The magnetic susceptibility of copper is -0.5×10^{-5}. Calculate the magnetic moment per unit volume in copper when subjected to a field whose magnitude inside copper is 10^6 A m^{-1}.
4. The unit edge of Fe_3O_4 is about 0.8 nm and there are eight Fe^{++} atoms in the cell. Calculate the magnetization. For iron, the six outer electrons have five spins in one direction and the sixth in the other, giving a net moment of 4 Bohr magneton.
5. A paramagnetic material has a magnetic field intensity of 10^4 A m^{-1}. If the susceptibility of the material at room temperature is 3.7×10^3, calculate the magnetization and flux density of the material.
6. The saturation value of the magnetization of iron is 1.75×10^6 A m^{-1}. Given that iron has a body centred cubic structure with an elementary cube edge of 2.86 Å, calculate the average number of Bohr magnetons contributed to the magnetization per atom.

CHAPTER 13

SUPERCONDUCTING MATERIALS

13.1 INTRODUCTION

The electrical resistivity of a material decreases with the decrease in temperature. According to the classical free electron theory the electrical resistivity is inversely proportional to the relaxation time. The thermal vibration of atoms, molecules and ions decreases and hence the number of collisions between the conduction electrons with other constituent particles such as atom, ion, and molecules decreases. So the relaxation time increases and hence ρ decreases with the decrease in temperatures. Certain materials exhibit zero resistance, when they are cooled into the ultra low temperature and hence they are said to be superconductors. This chapter deals with the properties, theory and applications of superconductors.

13.2 OCCURRENCE OF SUPERCONDUCTIVITY

Helium gas was liquefied by Heike Kamerlingh Onnes in 1908. The boiling point of liquid helium is 4.2 K. He studied the properties of metals by lowering their temperatures using liquid helium. In 1911, he studied the electrical properties of mercury at very low temperatures. He found that the resistivity of mercury suddenly decreases nearly 10^5 times around 4.2 K as shown in Figure 13.1(a). The variation of resistivity of normal conductor and superconductor is shown in Figure 13.1(b). This drastic change of the resistivity of mercury around 4.2 K indicates that the mercury gets transformed from one conducting state to another. This property is said to be **superconducting properties**. The materials those are exhibiting the superconducting properties are said to be

superconductors. The temperatures at which a material gets transformed from one conducting state to another is said to be **transition temperature** or **critical temperature.** It is represented by the letter T_C.

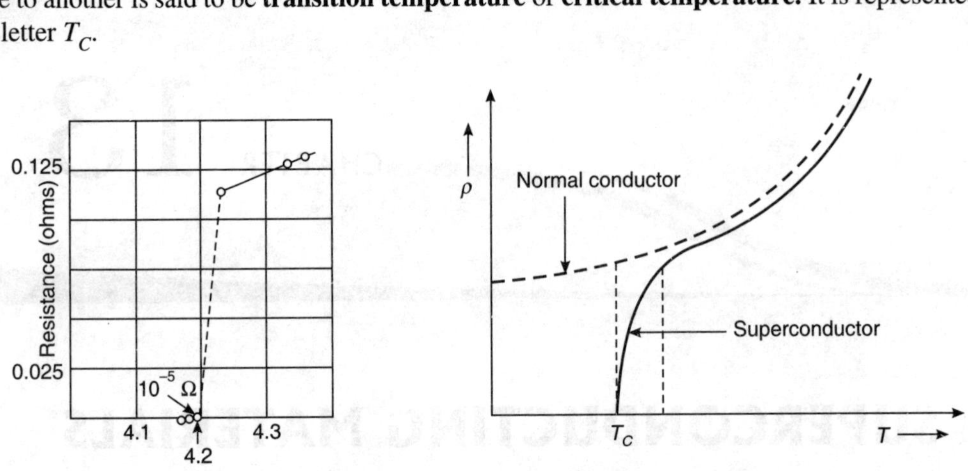

(a) Hg (b) Normal conductor and superconductors

Figure 13.1 Variation of resistivity versus temperature.

Table 12.1 Transition temperatures of some superconductors (*discovered in 2006, **discovered in 2009)

Materials	*Transition temperature, in* K	*Materials*	*Transition temperature, in* K
Elements		NbN	16.0
Al	1.175	Nb_3Sn	18.3
Hg α	4.15	Nb_3Al	18.9
Hg β	3.95	Nb_3Ge	23.0
In	3.41	Nb_3Au	11.5
Nb	9.46	La_3In	10.4
Zn	0.85	**Ceramics**	
Pb	7.196	$Bi_2Sr_2Ca_2Cu_3O_{10}$	110
Sn	3.72	$Bi_2Sr_2CaCu_2O_9$	110
Ti	0.4	$Bi_2Sr_2CaCu_2O_8$	91–92
La	4.88	$(Ca_{1-x}Sr_x)CuO_2$	110
Ta	4.47	$(Ba,Sr)CuO_2$	90
Cd	0.517	$(La,Sr)CuO_2$	42
Zr	0.61	$YBa_2Cu_3O_7$	90
Ga	1.083	$Tl_2Ba_2Ca_2Cu_3O_{10}$	127
Th_2	1.38	$TlBa_2CaCu_2O_7$	80
Compounds		$TlBa_2Ca_2Cu_3O_9$	105
V_3Ge	6.0	$TlBa_2Ca_3Cu_4O_{11}$	120
V_3Ga	14.2	$Hg_{12}Tl_3Ba_{30}Ca_{30}Cu_{45}O_{127}$	138*
V_3Si	17.4	$(Tl_4Ba)Ba_4Ca_2Cu_{10}O_y$	240**

After the invention of superconducting property in mercury, several materials were tested for superconductivity. Good electrical conductors at room temperatures such as gold, copper and silver are not good superconductors. The materials with high resistivity at room temperatures, generally found to be good superconductors.

The maximum transition temperature known to the scientists until the year 1985 is 23 K for Nb_3Ge. In January 1986, a copper oxide material (Lanthanum-Barium-Copper oxide) was reported to have transition temperatures of 35 K. Another material yttrium barium copper oxide (YBCO) was reported in February 1987 with a transition temperature of 92 K. After this invention, the materials having the transition temperatures above 77 K were studied. The high temperature superconductivity is mostly observed in Bi, Tl, Y, Hg compounds having CuO_2 layers. Recently, in May 2009, a compound $(Tl_4Ba)Ba_4Ca_2Cu_{10}O_y$ is found to have a transition temperature of 240 K. There is an increase in the transition temperature of the superconducting materials up to 100 K during the last four years (2006 to 2009). Generally the superconducting transition temperature varies from 0.001 K for Rh to 240 K for $(Tl_4Ba)Ba_4Ca_2Cu_{10}O_y$. The transition temperatures of some superconductors are listed in Table 13.1.

13.3 PROPERTIES OF SUPERCONDUCTORS

Electrical resistivity (ρ)

The electrical resistivity of the superconducting material is taken as zero, because the sensitivity of the equipments used to measure the resistivity is low and hence it is not possible to measure the resistivity of the superconducting materials. Gallop was able to predict the resistivity of a superconducting wire which is less than 10^{-26} Ω m using the lack of decay of a current circulating around a closed loop of superconducting wire. This value is nearly 10^{18} times less than the resistivity of copper at room temperature.

The scientists have studied the current flowing through a superconducting solenoid for a period of nearly three years. They found that there is no change in the current of the material. It represents the resistivity of the material is zero in the superconducting state. In a normal conducting material, if a small amount of current is applied, it will be destroyed within 10^{-12} s due to resistive loss (loss = i^2R).

Persistent current

Consider a small amount of current is applied to a superconducting ring. The superconducting current will keep on flowing through the ring without any changes in its value. This current is said to be persistent current. In a superconductor since there is no resistive heat loss (i^2R), the supercurrent will keep on flowing until the specimen is in the superconducting state.

Diamagnetic property

Consider a magnetic field applied to a normal conducting material. The magnetic lines of forces penetrate through the material. Consider that a normal conductor is cooled down to very low temperature for superconducting property. If it is cooled down below the critical temperature, then the magnetic lines of forces are ejected from the material. A diamagnetic material also repels the magnetic lines of forces. So, the ejection of magnetic lines of forces, when the superconducting material is cooled down is said to be the diamagnetic property. This property was first observed by Meissner and hence this property is also called *Meissner effect*.

The diamagnetic property as shown in Figure 13.2 is easily explained using the following mathematical treatment.

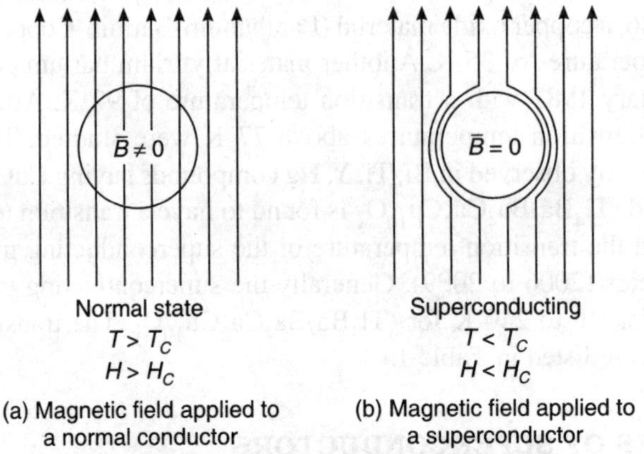

Figure 13.2 Diamagnetic property.

The magnetic flux density is given by

$$B = \mu_0(M + H)$$

For a superconducting material, $B = 0$. Substituting $B = 0$, we get

$$0 = \mu_0(M + H)$$

i.e.

$$M = -H$$

$$\chi = -\frac{M}{H} = -1 \qquad (13.1)$$

The negative value of the susceptibility shows the diamagnetic properties of the superconducting material.

Application of magnetic field

The superconducting materials exhibit the diamagnetic property only below the critical field. If the magnetic field is increased, the superconducting property of the material is destroyed, when the field is equal to or greater than the critical magnetic field. The minimum magnetic field required to destroy the superconducting property is known as the critical field (Figure 13.3). The field required to destroy the superconducting property is given by

$$H_C(T) = H_0\left(1 - \frac{T^2}{T_C^2}\right) \qquad (13.2)$$

where $H_C(T)$ is the critical field required to destroy the superconducting property at T K, T_C is the critical temperature, H_0 is the critical field required to destroy the superconducting property at 0 K.

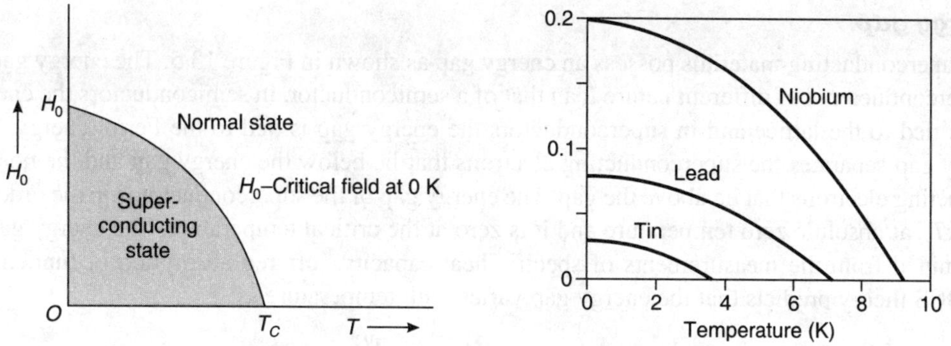

(a) Application of magnetic field to a superconductor

(b) Magnetic field versus temperature for Sn, Pb and Nb

Figure 13.3 Magnetic field versus temperature.

Application of current—Silsbee current

Consider a current of *i* ampere is applied to a superconducting coil of wire as shown in Figure 13.4. The application of the current induces a magnetic field and hence the superconducting property is destroyed. The critical current required to destroy the superconducting property is given by

$$i_C = 2\pi R H_C \tag{13.3}$$

where R is the radius of the superconducting wire, H_C is the critical magnetic field required to destroy the superconducting property. This property was discovered by Silsbee and hence Eq. (13.3) is known as Silsbee law.

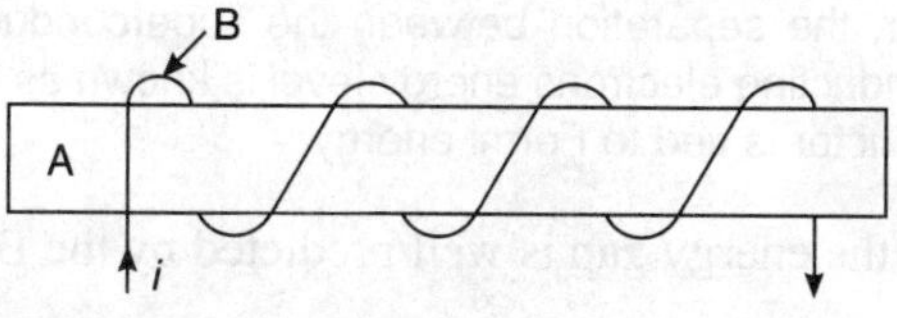

Figure 13.4 Application of current into a superconductor.

Effect of pressure

Cesium is a normal conductor at normal pressure. If the pressure of cesium is increased to 110 kbar after several phase transformation, it gets converted into a superconductor. The transition temperature of cesium at this pressure is 1.5 K. Silicon becomes a superconducting material at 165 kbar. Its transition temperature is 8.3 K.

Isotopic effect

The transition temperature of the superconducting materials varies with the mass numbers. It means that the presence of isotopes changes the transition temperature of the material. The mass number of mercury varies from 199.5 to 203.4. The transition temperature of mercury varies from 4.146 K to 4.185 K respectively. Maxwell showed that $T_C M^{\alpha}$ = constant. The value of α is found to be ½. This phenomenon is said to be isotopic effect.

Energy gap

The superconducting materials possess an energy gap as shown in Figure 13.5. The energy gap of a superconductor is of different nature than that of a semiconductor. In semiconductors the energy gap is tied to the lattice and in superconductors the energy gap is tied to the Fermi energy. The energy gap separates the superconducting electrons that lie below the energy gap and the normal conducting electrons that lie above the gap. The energy gap of the superconductor is in the order of ~3.5 kT_C at absolute zero temperature and it is zero at the critical temperature. The energy gap is determined from the measurements of specific heat capacity, infrared absorption or tunnelling. The BCS theory predicts that the energy gap varies with temperature as

$$E_g = 1.74 E_g(0) \left[1 - \frac{T}{T_C} \right]^{1/2} \tag{13.4}$$

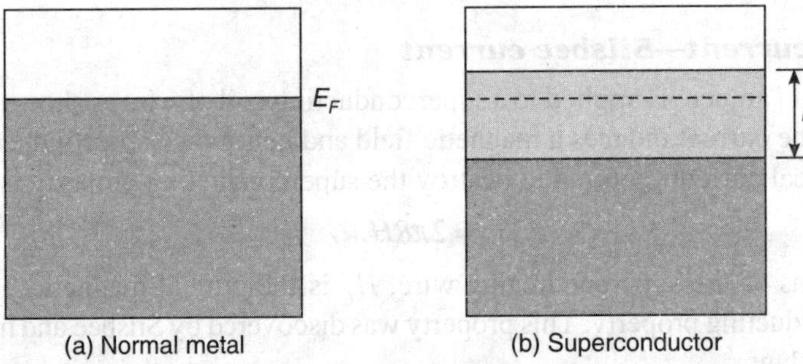

Figure 13.5 Energy gap (a) In semiconductor, the energy gap is tied to the lattice (b) For superconductor; the separation between the superconducting electrons energy level and normal conducting electrons energy level is known as energy gap. The energy gap of a superconductor is tied to Fermi energy.

The temperature variation of the energy gap is well predicted by the BCS theory (Figure 13.6).

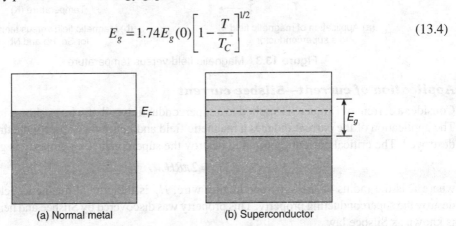

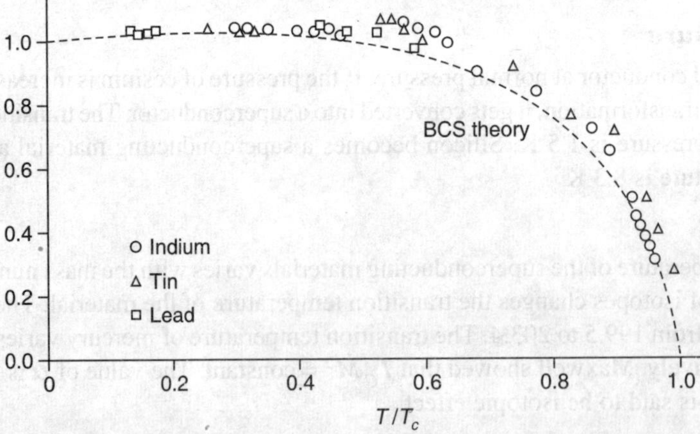

Figure 13.6 Variation of energy gap for Sn, In and Pb with temperature. The dotted line is drawn using BCS theory.

Specific heat capacity

The specific heat capacity of superconducting material has some discontinuity at the critical temperature. The specific heat capacity of tin is shown as a function of temperature in Figure 13.7. The critical temperature of tin is 3.72 K. A discontinuity of specific heat occurs at the critical temperature. The discontinuity is due to the absence of the magnetic field at $T = T_C$.

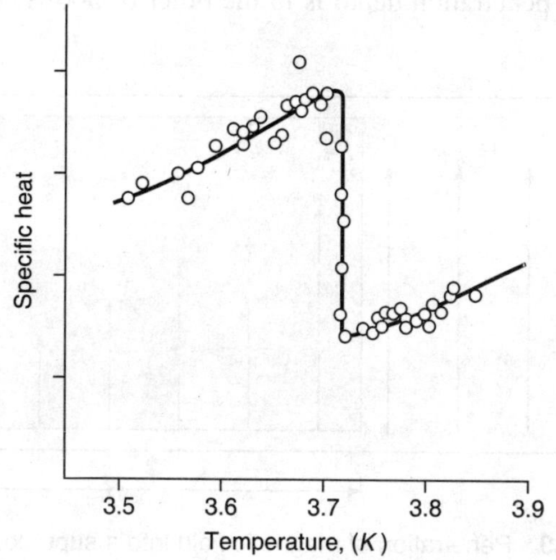

Figure 13.7 The variation of specific heat versus temperature for tin.

Entropy

The entropy is a measure of the disorder in a system. The entropy of aluminium is plotted against temperature. The entropy of the superconducting material is found to be lower than the normal conducting material. It indicates that in the superconducting states the electrons are in the more ordered state than the normal conducting state (Figure 13.8).

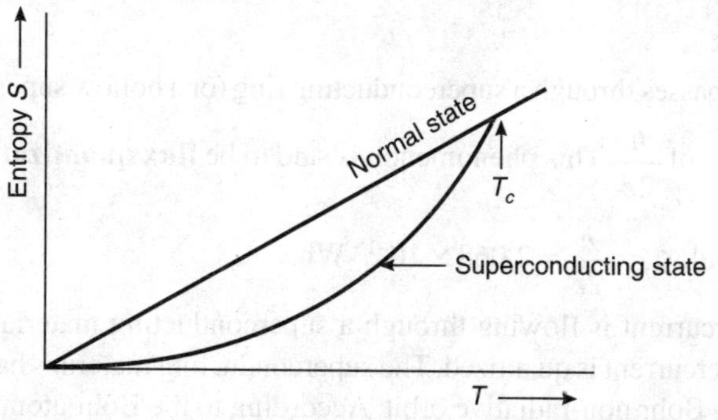

Figure 13.8 Entropy versus temperature for Al in the superconducting and normal conducting states.

Penetration depth

Consider a superconducting material is subjected to a magnetic field. The magnetic field does not drop to zero at the surface of the superconductors. It penetrates up to a small distance into the superconductors. The penetration of magnetic field within the superconductor varies exponentially (Figure 13.9)

$$H = H_0 e^{-x/\lambda} \tag{13.5}$$

where H_0 is the magnetic field at the boundary of the specimen, λ is called the penetration depth. The **penetration depth** is defined as the distance in which the field decreases by a factor $1/e$. For a pure superconductor, the penetration depth is in the order of 500 Å. It is also called **London penetration depth**.

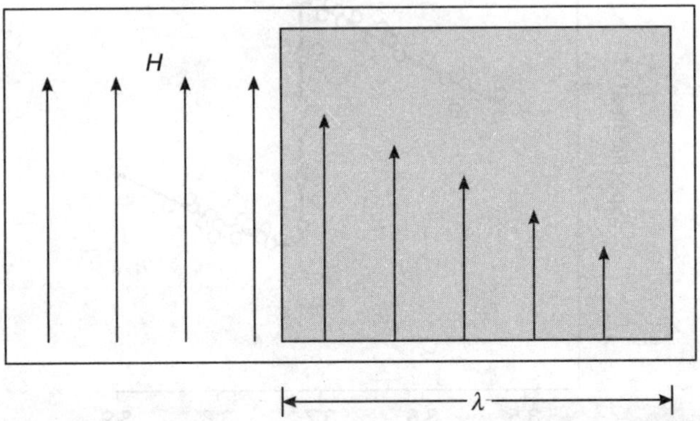

Figure 13.9 Penetration of magnetic field into a superconductor.

The penetration depth is also found to be temperature dependent. The temperature dependent of λ is

$$\lambda = \lambda_0 \left[1 - \frac{T^4}{T_C^4}\right]^{-1} \tag{13.6}$$

where λ_0 is the penetration depth at 0 K and λ is the penetration depth at a temperature T K.

Flux quantization

The magnetic flux that passes through a superconducting ring (or a hollow superconducting cylinder) is quantized in the order of $\frac{h}{2e}$. This phenomenon is said to be **flux quantization**. The value of the flux quantum, or fluxoid, $\phi_0 = \frac{h}{2e} = 2.068 \times 10^{-15}$ Wb.

Consider a supercurrent is flowing through a superconducting material. The magnetic flux associated with the supercurrent is quantized. The superconducting materials have persistent current. So, it is compared with Bohr non-radiative orbit. According to the Bohr atom model, the electrons orbits are quantized. As the Bohr's orbits are quantized, the supercurrent in a ring can also be considered as a macroscopic quantized orbit. The quantized orbit naturally produces a quantized magnetic field. The trapping of flux is shown in Figure 13.10(c).

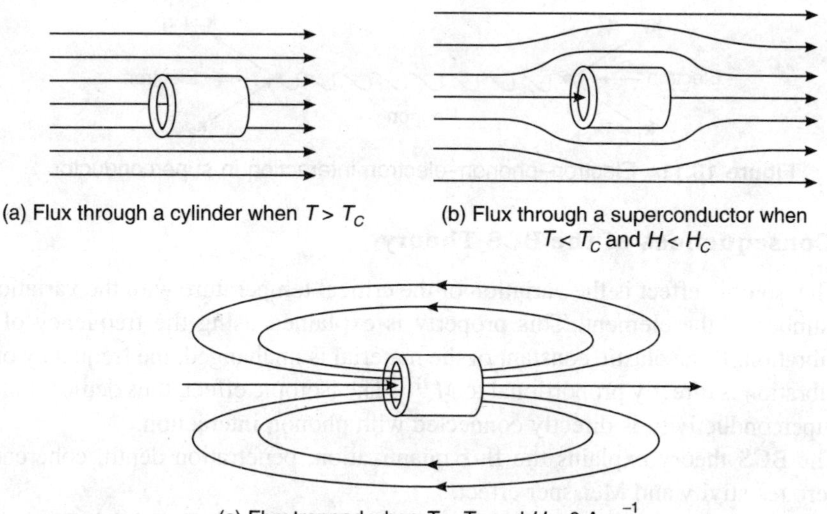
(a) Flux through a cylinder when $T > T_C$
(b) Flux through a superconductor when $T < T_C$ and $H < H_C$
(c) Flux trapped when $T < T_C$ and $H = 0$ A m^{-1}

Figure 13.10 Quantization of magnetic flux in superconductors.

13.4 BCS THEORY

This theory is a quantum mechanical theory and it was proposed by Bardeen, Cooper and Schrieffer in 1957. Some of the important concepts of this theory are given below.

According to this theory, two electrons with opposite spin and momentum pair up with one another due to a special type of attractive interaction. This pair of electrons is said to be superconducting pair or Cooper pair, named after the inventor. This pair is scattered, if there is sufficient energy so as to break up this pair of electrons. But in the superconducting state, the energy is not sufficient to break up this pair and hence this pair moves undeviated through the impurities. This explains the low resistivity of the superconductor.

There is an attractive interaction between two electrons. This interaction is caused by the positive ions in the crystal. When an electron is passing closer to a positive ion, the electron transfers some of its energy to the positive ion. Due to this, there is a modification in the vibration of the positive ion. If another electron passes closer to this deformed positive ion, it interacts with the positive ion. The energy is transferred from the positive ion to the second electron. It is an attractive interaction between two electrons and this would not arise if there is no positive ion.

In field theory, this interaction is called exchange of a virtual phonon, q between two electrons with wave vectors $\mathbf{k}_1$ and $\mathbf{k}_2$ (Figure 13.11). The electron with wave vector $\mathbf{k}_1$ interacts with a phonon moves with a reduced wave vector, $\mathbf{k}_1 - \mathbf{q}$ and transfers some of its energy to the phonon. If another electron of wave vector, $\mathbf{k}_2$ meets this phonon, then interacts with the phonon and gains some energy. Its wave vector is $\mathbf{k}_2 + \mathbf{q}$, i.e.

$$\mathbf{k}_1 - \mathbf{q} = \mathbf{k}_1' \text{ and } \mathbf{k}_2 + \mathbf{q} = \mathbf{k}_2' \quad (13.7)$$

The resultant wave vector is $\mathbf{k}_1 + \mathbf{k}_2 = \mathbf{k}_1' + \mathbf{k}_2'$ and the wave vector is conserved.

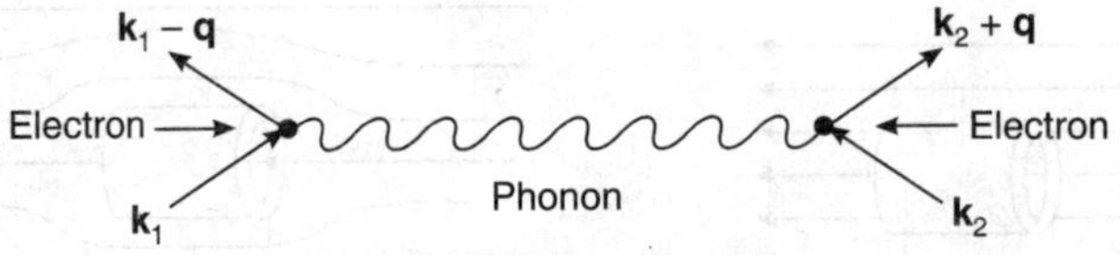

Figure 13.11 Electron–phonon–electron interaction in superconductor.

13.4.1 Consequences of the BCS Theory

1. The isotopic effect is the variation of the critical temperature with the variation of mass number of the element. This property is explained using the frequency of the ionic vibration. If the elastic constant of the material is unchanged, the frequency of the ionic vibration is directly proportional to $M^{-1/2}$. The isotopic effect, thus demonstrates that the superconductivity is directly connected with phonon interaction.
2. The BCS theory explains the flux quantization, penetration depth, coherence length, zero resistivity and Meissner effect.
3. The energy gap of a superconducting material is given by $E_g \approx 3.5\, k_B T_C$ at zero degree kelvin. At the critical temperature, the energy gap is zero.
4. The superconducting transition temperature of a metal or an alloy is given by

$$T_C = 1.14\theta_D \exp\left[-\frac{1}{UD(E_F)}\right] \tag{13.8}$$

where θ_D is the Debye temperature, U is the electron–lattice attractive interaction and $D(E_F)$ is the electron density of the orbital at the Fermi level.
5. The velocity of these two electrons is $-v$ and v respectively. The velocity of the pair of electron at the centre of mass is zero. The de Broglie wavelength of the pair of electrons is infinitely long. The charge of the pair of electrons is 2e.

13.5 JOSEPHSON JUNCTION

In 1962, B.D. Josephson predicted a number of remarkable phenomena about superconductivity, which are used to understand the superconducting properties.

13.5.1 DC Josephson Effect

An insulating material of thickness nearly 1 to 2 nm sandwiched between two different superconducting materials is known as Josephson device. In a Josephson device, a dc voltage is found to flow from a junction that has a higher density to the junction with lower density of superconducting electrons. This phenomenon is said to be DC Josephson effect. The dc current flowing through the device is equal to

$$J = J_0 \sin \delta \tag{13.9}$$

where δ is the phase difference and J_0 is a constant and it is the maximum current density through the insulator. The Josephson device is shown in Figure 13.12.

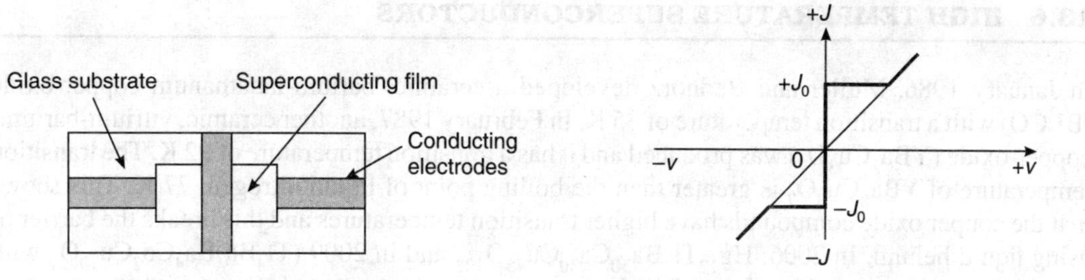

(a) Josephson device (b) V–I characteristics in Josephson junction

Figure 13.12 Josephson junction.

The pair of electrons present in a superconducting material is in the same phase. Whenever, a Josephson junction is formed by sandwiching an insulator in between two different superconductors, the superconducting electrons present in two different superconducting materials of a Josephson device need not be in the same phase. There will be a tunnelling of electrons from one superconducting material with higher electron density to another superconducting material with lower electron density. Due to this tunnelling of electrons, a dc voltage appears across the Josephson device, even though no field is applied there.

13.5.2 AC Josephson Effect

Consider a dc voltage, V applied to a Josephson device, then there will be a flow of ac voltage through the device. This phenomenon is known as AC Josephson effect. The Josephson current is given by

$$J = J_0 \sin\left(\delta(0) - \frac{qV}{\hbar}t\right) = J_0 \sin(\delta(0) - \omega t) \qquad (13.10)$$

where $\delta(0)$ is a constant and $\omega = \frac{qV}{\hbar}$ is the frequency of the ac signal. The application of a potential difference makes the Josephson current time dependent.

The photon energy of emission or absorption at the junction is $h\nu = qV = 2eV$

The frequency of the ac signal is

$$\omega = \frac{qV}{\hbar} \qquad (13.11)$$

i.e. $$\nu = \frac{qV}{h} = \frac{2eV}{h} = \frac{2 \times 1.602 \times 10^{-19} \times V}{6.626 \times 10^{-34}} = 483.55 \times 10^{12}\ V\,\text{Hz}$$

For an applied voltage of 1 μV, the frequency of the ac signal is

$$\nu = 483.55 \times 10^{12}\ V = 483.55 \times 10^{12} \times 10^{-6} = 483.55\ \text{MHz}$$

By measuring the frequency of the ac signal accurately, the value of $\frac{e}{h}$ and the photon energy, $2eV$, are accurately measured.

13.6 HIGH TEMPERATURE SUPERCONDUCTORS

In January 1986, Müller and Bednorz developed a ceramic, barium-lanathanum-copper-oxide (BLCO) with a transition temperature of 35 K. In February 1987, another ceramic, yttrium-barium-copper-oxide ($YBa_2Cu_3O_7$) was produced and it has a transition temperature of 92 K. The transition temperature of $YBa_2Cu_3O_7$ is greater than the boiling point of liquid nitrogen, 77 K. This shows that the copper oxide compounds have higher transition temperatures and this breaks the barrier of using liquid helium. In 2006, $Hg_{12}Tl_3Ba_{30}Ca_{30}Cu_{45}O_{127}$ and in 2009 $(Tl_4Ba)Ba_4Ca_2Cu_{10}O_y$ with transition temperatures 138 K and 240 K respectively were invented. Some high temperature superconducting materials and their transition temperatures are given in Table 13.1.

Conventional superconductors have their transition temperatures less than 23 K (Nb_3Ge). The superconductors those having the transition temperatures higher than the conventional superconductors are said to be high temperature superconductors (HTS).

Some of the important high temperature superconducting compounds are

1. Rare earth-based cuprates
2. Bi-based cuprates
3. Tl-based cuprates

13.6.1 Rare Earth-based Copper Oxide Compounds

The chemical formula for the rare earth modified compound is $MBa_2Cu_3O_{7-\delta}$, where M is a rare earth element such as Y, Nd, Sm, etc. The crystal structure of these compounds is oxygen defect modification of the perovskite structure, with nearly 1/3 of the oxygen vacant.

The crystal structure of YBCO is shown in Figure 13.13. It is an orthorhombic structure with cell constants, $a = 3.8227$ Å, $b = 3.8872$ Å and $c = 11.6802$ Å. The positive ion valencies for $YBa_2Cu_3O_{7-\delta}$ based on Y^{3+}, Ba^{2+}, Cu^{2+} are $3 \times 1 + 2 \times 2 + 3 \times 2 = 13$. The negative ion valencies for O^{2-} is $-2 \times (7 - \delta) = -14 + 2\delta$. The stoichiometric compound is $YBa_2Cu_3O_{6.5}$. The compound $YBa_2Cu_3O_6$ is an insulator with antiferromagnetic order. The increase in the oxygen above $O_{6.5}$ makes it as a non-magnetic metallic crystal. The crystal is superconducting above $O_{6.64}$. Due to doping in the ferromagnetic insulator, the excess charge carriers is hole. The $YBa_2Cu_3O_7$ compound is represented as Y-123.

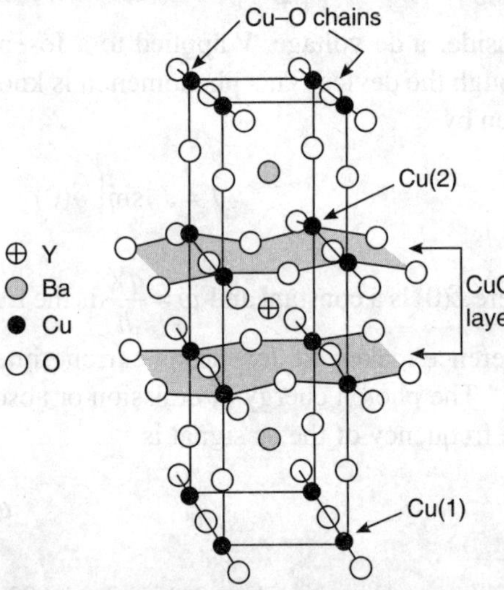

Figure 12.13 Crystal structure of $YBa_2Cu_3O_7$.

The YBCO crystal structure has CuO_2 sheets arranged in parallel along the *ab*-axes of the orthorhombic structure.

13.6.2 Bi-based and Tl-based Compounds

The Bi- and Tl-based copper oxide compounds are showing high temperature superconducting property. $Bi_2Sr_2Ca_2Cu_3O_{10}$ and $Tl_2Ba_2Ca_2Cu_3O_{10}$ are the examples for the Bi- and Tl-based compounds. They are represented by Bi-2223 and Tl-2223 respectively. The critical temperatures of these two compounds are 107 K and 127 K respectively. The superconducting property of these compounds is also controlled by the number of CuO_2 layers.

13.6.3 New Types of High T_C Superconductors

High critical temperature is also observed in some new materials, which do not have CuO_2 layers. Magnesium diboride (MgB_2) exhibits high temperature superconducting property. The crystal structure is shown in Figure 13.14. The boron atoms are arranged in two-dimensional hexagonal sheets within the cubic structure of magnesium. The critical temperature of magnesium diboride is 39 K.

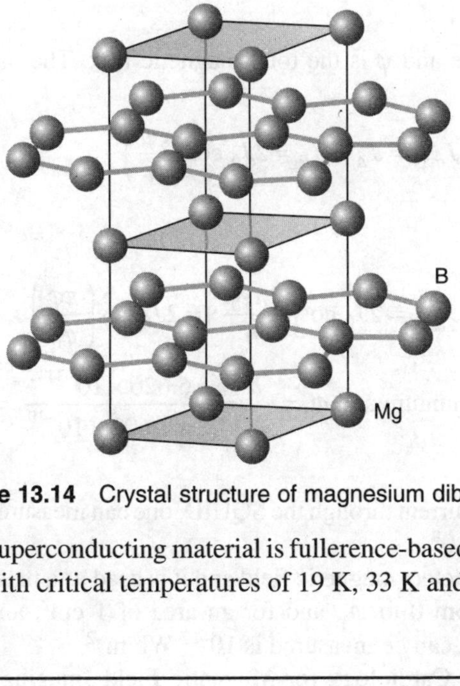

Figure 13.14 Crystal structure of magnesium diboride.

Another new type of superconducting material is fullerence-based compounds such as K_3C_{60}, Rb_3C_{60} and $Rb_{2.7}Tl_{2.3}C_{60}$ with critical temperatures of 19 K, 33 K and 42 K respectively.

13.7 APPLICATIONS

13.7.1 SQUID

SQUID is an acronym for **S**uperconducting **QU**antum **I**nterference **D**evice. It is a magnetometer. It is based on the principle of Josephson effect.

Consider two Josephson junctions A and B are connected parallel as shown in Figure 13.15. Let the current is applied through a common arm T_1 and a magnetic field is applied perpendicularly to this device. The current is splitted into two components and they can flow through the Josephson devices, A and B. It is similar to the splitting of light into two coherent sources in Young's double slit experiment. The current leaving the Josephson junctions (J_A and J_B) are combined together at the common arm T_2. It is similar to the two splitted beams of light are allowed to interfere with each other in Young's double slit experiment. The current flowing through the arm T_2 is

$$J = J_A + J_B = 2J_0 \sin(\delta\gamma_0) \cos\left(\frac{e\varphi}{\hbar}\right) \tag{13.12}$$

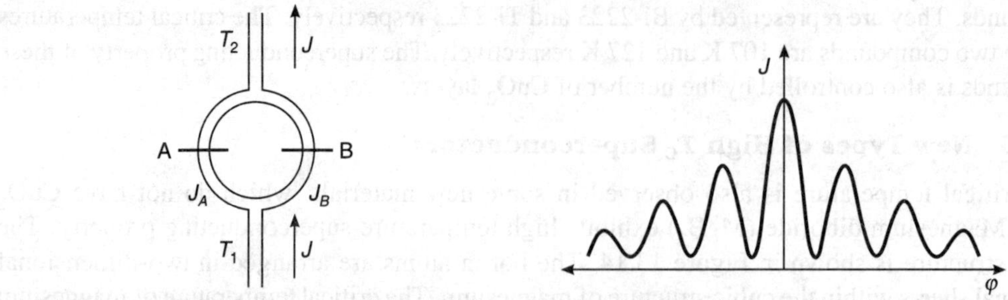

(a) Two Josephson junction connected parallel (b) Interference pattern produced between J and φ

Figure 13.15 SQUID. The fluxuation is due to the interference of the de Brogile waves inside the metals.

where $\delta\gamma_0$ is the initial phase and φ is the total magnetic flux. The maximum current will flow, when $\delta\gamma_0 = 90°$.

$$J_{max} = J_A + J_B = 2J_0 \cos\left(\frac{e\varphi}{\hbar}\right) \tag{13.13}$$

Substituting $\hbar = \frac{h}{2\pi}$, we get

$$J_{max} = 2J_0 \cos\left(\frac{2\pi e\varphi}{h}\right) = 2J_0 \cos\left(\frac{\pi\varphi}{\varphi_0}\right) \tag{13.14}$$

where φ_0 is known as flux quantum and $\varphi_0 = \frac{h}{2e} = \frac{6.626\times10^{-34}}{2\times1.602\times10^{-19}} = 2.07 \times 10^{-15}$ Wb.

Uses of SQUIDS

1. By measuring the current through the SQUID, one can measure the value of e/h accurately, correct up to 1 in 10^9.
2. SQUID is used to detect magnetic field and it is used as a magnetometer. For a complete cycle, φ varies from 0 to φ_0, and for an area of 1 cm^2, and an accuracy of 1%, the magnetic field that can be measured is 10^{-12} Wb m^{-2}.
3. SQUID is used in Cardiology for **M**agnetic **F**ield **I**maging (MFI), which detects the magnetic field of the heart for diagnosis and risk stratification.
4. In magnetogastrography, SQUID is used to record the weak magnetic fields of the stomach.
5. In magnetoencephalography (MEG), the measurements from an array of SQUIDs are used to make inferences about neural activity inside brains.
6. A novel application of SQUIDs is the magnetic marker monitoring method, which is used to trace the path of orally applied drugs.
7. MRI (magnetic resonance imaging) scanning using SQUID is used to measure the field in the microtesla region, whereas the ordinary MRI is used to measure the fields from one to several tesla.

13.7.2 Magnetic Levitation

The process of floating an object with the application of a magnetic field is called magnetic levitation. A diamagnetic material repels the magnetic field. Using this property, a diamagnetic material is made to float in air due to the application of a magnetic field. Since, the superconducting material behaves as a diamagnet, it is also made to float by applying a magnetic field. The magnetic levitation is based on the principle of Meissner effect. This property explains the zero resistivity and Meissner effect.

Uses of magnetic levitation

(a) Magnetic levitation principle is used to levitate vehicles. Magnetic-levitated trains are operated in Japan. Due to the levitation, there is no contact between the train and rail. These trains are called **Maglev trains**. There are two primary types of maglev technology:

- Electromagnetic suspension (EMS) uses the attractive magnetic force of a magnet beneath a rail to lift the train up.
- Electrodynamic suspension (EDS) uses a repulsive force between two magnetic fields to push the train away from the rail.

In the EMS systems, the train levitates above a steel rail while electromagnets, attached to the train, are oriented toward the rail from below. The electromagnets use feedback control to maintain a train at a constant distance from the track.

In electrodynamic suspension (EDS), both the rail and the train exert a magnetic field, and the train is levitated by the repulsive force between these magnetic fields. The magnetic field in the train is produced by either electromagnets or by an array of permanent magnets. The repulsive force in the track is created by an induced magnetic field in wires or other conducting strips in the track.

(b) The magnetic levitation is used in toy Maglev trains. The objects like golf balls, picture frames, clocks are suspended using magnetic levitation principle.

13.7.3 Cryotron

Consider two superconducting materials, A and B in the form of a rod and wire respectively. Let H_{CA} and H_{CB}, be the critical fields of the materials A and B respectively where $H_{CA} < H_{CB}$. Consider the wire, B is wound on the superconducting rod, A as shown in Figure 13.16. If a current of i ampere is passed through the wire, such that the magnetic field induced by the wire, H lies between H_{CA} and H_{CB}. Then the superconducting property of the rod A is destroyed and the contact is broken. This device acts as a switch. Such a device is called as cryotron.

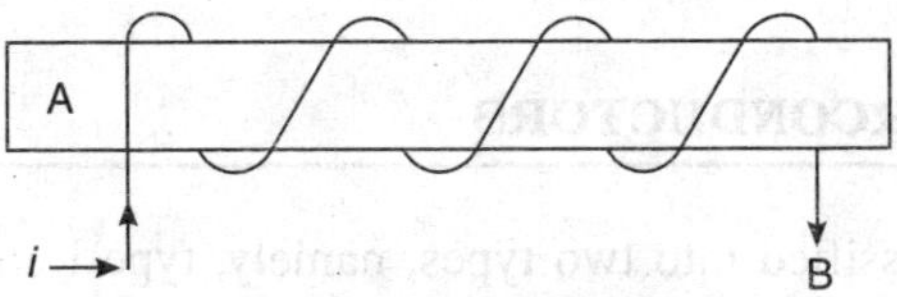

Figure 13.16 Cryotron.

13.7.4 Other Applications

Superconducting magnets

The superconducting materials are used to prepare magnets. A superconducting magnet of size 12 cm × 20 cm can produces a magnetic field of 20 T. To produce such amount of magnetic field, one can need electromagnets of large size and very high electric field to energise it.

Metrology

The accurate determination of the Planck's constant is possible, by measuring the voltage across the Josephson junction and the frequency of the ac signal and hence finding the value of flux quantum, $\phi_0 = \dfrac{h}{2e}$. From this value one can easily determine the value of the Planck's constant, h. The new value of h is changed to 6.626196×10^{-34} J s from its old value 6.62559×10^{-34} J s.

Radiation detector

The resistance of a superconducting material kept above the critical temperature is a function of temperature. If nuclear radiation is made to incident on a superconductor kept above the critical temperature, due to the heat produced by the incident radiation, the resistance changes. By measuring the change of resistance one can determine the intensity of the incident radiation.

Military applications

The Josephson device is used for the military applications to detect the nuclear submarines passing through underwater. The nuclear submarines passing through underwater cannot be detected by conventional methods using microwave, light or sound, because water is one of a good absorber for microwave, sound and light. The nuclear submarine passing through underwater produces a small perturbance in the earth magnetic field. By measuring the changes in the earths magnetic field using the SQUID magnetometer one can identify the nuclear submarines.

Transmission line

The superconducting transmission lines have no resistive losses ($i^2R = 0$), because the resistance of the superconducting materials is zero. Therefore, the superconducting materials are used as electric transmission lines.

Superconducting generators

The superconducting materials are used for fabricating generators. The superconducting generators are small in size and they produce more power than the conventional electric generators.

Switching circuits

Superconducting materials are used as switching devices such as relays.

13.8 TYPES OF SUPERCONDUCTORS

The superconductors are classified into two types, namely, type I and type II superconductors. Type I superconductors are called soft superconductors and type II superconductors are called hard superconductors.

13.8.1 Type I Superconductors

Type I superconductors behave as perfect diamagnetic materials. They perfectly obey the Meissner effect. Type I superconductors do not allow the magnetic lines of forces to penetrate through them, whenever the applied magnetic field is less than the critical field. If the applied magnetic field is greater than the critical field, the type I superconducting materials allow the magnetic field to pass through them. A plot between the applied magnetic field and the magnetization is shown in Figure 13.17(a). In Figure 13.17(a), the negative sign for magnetization indicates that the magnetization opposes the application of the magnetic field. The opposing force suddenly drops to zero at the critical field, H_C and hence the materials allow the magnetic field to pass through it. The superconducting materials behave as a normal conductor above the critical field. The simple metallic superconductors such as Al, Zn, Ti, Ga, Zr, Mo, W, In, Sn and Pb are examples of type I superconductors.

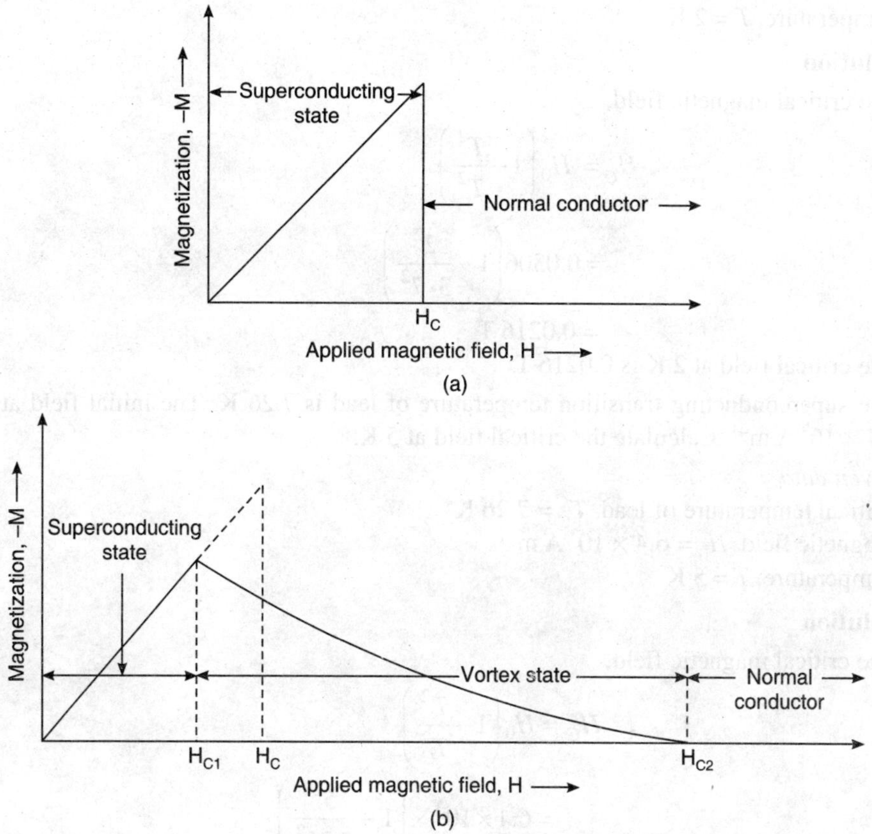

Figure 13.17 Types of superconductors: (a) Type I and (b) Type II superconductors.

13.8.2 Type II Superconductors

Type II superconductors do not perfectly obey the Meissner effect. They possess two critical fields, namely lower critical field, H_{C1} and the upper critical field, H_{C2}. The type II superconducting materials do not allow the magnetic lines of forces up to the lower critical field, H_{C1}. Above the lower critical field, H_{C1} the magnetic flux penetrates through the type II superconducting materials. Between H_{C1} and H_{C2}, type II superconductors behave as superconducting materials and at the same time the magnetic flux penetrates through the materials. After H_{C2}, they behave as normal conductors. Between H_{C1} and H_{C2} the materials is in the mixed state and it is commonly known as vortex state (Figure 13.17(b)). The alloys like NbTi, NbN, Nb_3Ge, Nb_2Si, PbMoS, Nb_3Sn, and V_3Sn are classified into type II superconductors.

SOLVED PROBLEMS

13.1 A superconducting tin has a critical temperature of 3.7 K at zero magnetic fields and a critical field of 0.0306 T at 0 K. Find the critical field at 2 K.

Given data

Critical temperature of tin, $T_C = 3.7$ K

Magnetic field, $H_0 = 0.0306$ T

Temperature, $T = 2$ K

Solution

The critical magnetic field,

$$H_C = H_0\left(1 - \frac{T^2}{T_C^2}\right)$$

$$= 0.0306\left(1 - \frac{2^2}{3.7^2}\right)$$

$$= 0.0216 \text{ T}$$

The critical field at 2 K is 0.0216 T.

13.2 The superconducting transition temperature of lead is 7.26 K. The initial field at 0 K is 6.4×10^3 A m^{-1}. Calculate the critical field at 5 K.

Given data

Critical temperature of lead, $T_C = 7.26$ K

Magnetic field, $H_0 = 6.4 \times 10^3$ A m^{-1}

Temperature, $T = 5$ K

Solution

The critical magnetic field,

$$H_C = H_0\left(1 - \frac{T^2}{T_C^2}\right)$$

$$= 6.4 \times 10^3 \times \left(1 - \frac{5^2}{7.26^2}\right)$$

$$= 3364.38 \text{ A m}^{-1}$$

The critical field at 5 K is 3364.38 T.

13.3 The superconducting transition temperature (T_C) for mercury with isotopic mass 199.5 is 4.185 K. Calculate the value of T_C when its mass changes to 203.4.

Given data

Critical temperature of Hg, $T_C = 4.185$ K

Atomic mass = 199.5

Atomic mass = 203.4

Solution

According to Maxwell's equation,

$$T_C M^{0.5} = \text{constant}$$

For the atomic mass, 199.5, the above equation is,

$$T_{C1} M_1^{0.5} = \text{constant}$$

For the atomic mass, 203.4, the above equation is,

$$T_{C2} M_2^{0.5} = \text{constant}$$

Equating,

$$T_{C1} M_1^{0.5} = T_{C2} M_2^{0.5}$$

i.e.

$$4.185 \times 199.5^{0.5} = T_{C2} \times 203.4^{0.5}$$

$$T_{C2} = \frac{4.185 \times 199.5^{0.5}}{203.4^{0.5}} = 4.14468 \text{ K}$$

The critical temperature of Hg with atomic mass, 203.4 is 4.14468 K.

13.4 Calculate the critical current density for 1 mm diameter wire of lead at 4.2 K. A parabolic dependence of H_C upon temperature may be assumed. Given: $T_C = 7.18$ K and $H_0 = 6.5 \times 10^4$ A m^{-1}.

Given data

Diameter of the wire = 1 mm = 1×10^{-3} m

Temperature, $T = 4.2$ K

Critical temperature, $T_C = 7.18$ K

Critical field, $H_C = 6.5 \times 10^4$ A m^{-1}

Solution

The critical magnetic field,

$$H_C = H_0\left(1 - \frac{T^2}{T_C^2}\right)$$

The critical magnetic field at 4.2 K,

$$H_C = 6.5 \times 10^4\left(1 - \frac{4.2^2}{7.18^2}\right)$$

$$= 4.276 \times 10^4 \text{ A m}^{-1}$$

Critical current,

$$i_C = 2\pi R H_C$$

$$= 2\pi \times 0.5 \times 10^{-3} \times 4.276 \times 10^4$$

$$= 134.33 \text{ A}$$

Critical current density,

$$J = \frac{I}{A} = \frac{I}{\pi r^2} = \frac{134.33}{\pi (0.5 \times 10^{-3})^2} \; 1.7103 \times 10^8 \text{ A m}^{-2}$$

The critical current density is 1.71×10^8 A m^{-2}.

13.5 A voltage of 6 μV is applied across a Josephson junction. What is the frequency of the radiation emitted by the junction?

Given data

Voltage applied across the junction = 6 μV

Solution

The frequency of the ac signal,

$$\nu = \frac{2eV}{h}$$

Substituting the values of e, V and h, we get

$$\nu = \frac{2 \times 1.6 \times 10^{-19} \times 6 \times 10^{-6}}{6.626 \times 10^{-34}}$$

$$= 2.897 \times 10^{9} \text{ Hz}$$

The frequency of the ac signal is 2.897×10^9 Hz.

13.6 The critical temperature of lead is 7.19 K. Calculate its band gap in the superconducting state.

Given data

The critical temperature of lead = 7.19 K

Solution

The energy gap of a superconductor,

$$E_g = 3.5\, k_B T_C$$

$$= 3.5 \times 1.38 \times 10^{-23} \times 7.19$$

$$= 3.47277 \times 10^{-22} \text{ J}$$

$$= 2.17 \times 10^{-3} \text{ eV}$$

The band gap of the superconducting lead is 2.17 meV.

SHORT QUESTIONS

1. What is superconductivity?
2. What is a superconductor?
3. What is meant by critical temperature?
4. What are high temperature superconductors?
5. What is Meissner effect?
6. What is isotopic effect?
7. What is the effect of magnetic field on the superconductors?
8. What is the effect of current on the superconductors?
9. Mention any two properties of superconductors.
10. What is SQUID?

11. What is cryotron?
12. What is magnetic levitation?
13. What are type I superconductors?
14. What are type II superconductors?
15. What is a Josephson junction?
16. What is dc Josephson effect?
17. What is ac Josephson effect?
18. Mention the applications of SQUID.
19. Mention the applications of superconductors.
20. Write about the energy gap of the superconductor.

DESCRIPTIVE TYPE QUESTIONS

1. What is superconductivity? Explain the properties of superconductors. Mention its uses. What are type I and type II superconductors?
2. Describe the BCS theory of superconductivity. What are the consequences of the BCS theory? Write an essay about high temperature superconductivity.
3. What is Josephson effect? Explain the dc and ac Josephson effects. What is SQUID? Describe SQUID and mention its uses.
4. Describe the following applications of superconductors in detail: (i) cryotron, (ii) SQUID and (iii) magnetic levitation.
5. Describe high T_C superconductors.

PROBLEMS

1. Two isotopes of lead of atomic mass 206 and 210 have T_C values of 7.193 K and 7.125 K respectively. Calculate the α value for lead.
2. Calculate the critical current through a long thin superconducting wire of radius 0.5 mm. The critical magnetic field is 7.2 kA/m.
3. The value of α is 0.5 and $T_C = 4.2$ K for Hg with mass number 202. Find T_C for Hg with mass number 200.
4. The critical field for V_3Ga is 1.4×10^5 A m^{-1} at 14 K and 4.2×10^5 A m^{-1} at 13 K. Determine the value of T_C and critical field at 0 K.
5. The critical temperature of NbN is 16 K. At 0 K, the critical field is 8×10^6 A m^{-1}. Calculate the critical field at 10 K.
6. Niobium has the critical temperature of 9.5 K. Calculate its band gap in the superconducting state.

CHAPTER 14

DIELECTRIC MATERIALS

14.1 INTRODUCTION

The dielectric materials are the insulating materials used to store the electrical energy. In insulators, all the states are completely occupied by electrons. Therefore, in insulators there is no free electrons available for electrical conduction, whereas a dielectric material has very few electrons for electrical conductivity and hence it has a dipole. The dielectric material has the interesting electrical properties because of the polarization produced by the applied electric field. So, the study of the theory, properties and applications of the dielectric material becomes essential. This chapter deals with the dielectric materials. Let us review some of the terms related to this topic.

14.2 DEFINITION OF SOME FUNDAMENTAL TERMS

Electric dipoles

Any two opposite charges separated by a suitable distance constitute an electric dipole. The dipole is represented by an arrow, which starts from the negative charge and ends at the positive charge. A dipole is shown in Figure 14.1.

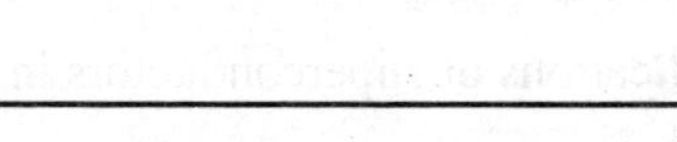

Figure 14.1 A dipole.

Electric dipole moment

The product of the charge and the distance of separation of a dipole is known as *electric dipole moment*. The electric dipole moment is given by

$$\mu = Q \cdot r \tag{14.1}$$

The unit for the electric dipole moment is coulomb metre (C m). For a system consisting of a number of charges, the dipole moment is given by

$$\mu = \sum_{i=1}^{n} Q_i \cdot r_i \tag{14.2}$$

where Q_i is the charge of the *i*th charged particle and r_i is the distance between the reference charge and the *i*th charge.

Electric lines of forces

The electric field in a dielectric material is studied by drawing the electric lines of forces. The electric lines of forces are also called *electric flux*. The electric lines of forces originate from positive charge and end at negative charge. The tangent drawn on the electric flux at a point gives the direction of the electric field at that point.

Electric flux density, D

The electric flux passing through the unit area of cross section is known as the **electric flux density**. It is represented by the letter *D*. Its unit is coulomb per square metre (C m^{-2}).

The electric flux density is given by, $D = \dfrac{\phi}{A}$, where ϕ is the electric flux and *A* is the area of cross section. From Gauss law, the flux emanating from a surface is equal to the total charge enclosed in that surface. Therefore, $D = \dfrac{Q}{4\pi r^2}$, where *Q* is the charge enclosed in the surface.

The electric flux density is also given by the following expressions:

$$D = \varepsilon E = \varepsilon_0 \varepsilon_r E \tag{14.3}$$

and
$$D = \varepsilon_0 E + P \tag{14.4}$$

where *P* is the polarization, *E* is the electric filed intensity and ε_r is the dielectric constant and ε_0 is the permittivity of free space.

Electric field strength, E

The electric field strength is the force experienced by a unit positive charge placed in the electric field region. It is represented by the letter, *E*. Its unit is V m^{-1}.

The electric field strength is given by $E = \dfrac{F}{q}$. It is also called *electric field intensity* or *electric field*. The Coulomb force is given by $F = \dfrac{q_1 q_2}{4\pi\varepsilon r^2}$. The electric field intensity is given by

$$E = \frac{q}{4\pi\varepsilon r^2}.$$

Polarization, P

The electric dipole moment per unit volume is known as *polarization*. Its unit is C m^{-2} and it is represented by the letter *P*.

$$P = \frac{\text{Dipole moment}}{\text{Volume}} = \frac{\mu}{V} \tag{14.5}$$

Substituting the value of $\mu = \alpha E$ in Eq. (14.5), we get

$$P = \frac{\alpha E}{V} \tag{14.6}$$

If there are *N* dipoles per unit volume, then the above equation can be written as

$$P = N\alpha E \tag{14.7}$$

The polarization is given by the following mathematical equations

$$P = \varepsilon_0(\varepsilon_r - 1)E \tag{14.8}$$

Substituting, $\chi = (\varepsilon_r - 1)$, the above equation can be written as

$$P = \varepsilon_0 \chi E \tag{14.9}$$

Dielectric susceptibility, χ

The polarization of a dielectric material is found to be

$P \propto$ the applied electric field intensity,

$P \propto$ the permittivity of free space.

By introducing the proportionality constant, the polarization can be written as

$$P = \varepsilon_0 \chi E \tag{14.10}$$

From the above equation the dielectric susceptibility is defined as $\left(\frac{1}{\varepsilon_0 E}\right)$ times of the polarization. It has no unit and it is represented by the letter, χ.

$$\chi = \frac{P}{\varepsilon_0 E} = \frac{\varepsilon_0(\varepsilon_r - 1)E}{\varepsilon_0 E} = (\varepsilon_r - 1) \tag{14.11}$$

It is a measure of the extent up to which a material can be polarized by the application of electric field.

Dielectric constant, ε_r

The ratio of the permittivity of the medium to the permittivity of the free space is known as *dielectric constant* or *relative permittivity* of the medium. It is represented by the letter, ε_r and it is given by, $\varepsilon_r = \varepsilon/\varepsilon_0$. Here, ε_0 is the permeability of free space and it is equal to, $\varepsilon_0 = 8.854 \times 10^{-12}$ F m^{-1} and ε is the permittivity of the medium.

The dielectric constant is also defined as the ratio of the capacitance of the capacitor with the given dielectric material as a dielectric (*C*) to the capacitance of the capacitor with vacuum as a dielectric (C_0).

$$\varepsilon_r = \frac{C}{C_0} \tag{14.12}$$

It is a measure of the extent to which a material can be polarized by the application of the electric field.

Dielectric polarizability, α

The induced dipole moment of an atom per unit electric field intensity is known as *polarizability*. It is also defined as 1/*NE* times of polarization. It is represented by the letter α and its unit is F m^2.

$$\alpha = \frac{\mu_{\text{ind}}}{E} = \frac{P}{NE} \tag{14.13}$$

14.3 POLARIZATION IN A DIELECTRIC MATERIAL

Whenever an electric field is applied to a dielectric material, a displacement of electrons or ions or rotation of dipoles takes place. Due to this phenomenon the dipoles are created and hence an induced dipole moment is produced. The induced dipole moment per unit volume is known as *polarization*. It is represented by the unit C m^{-2}. There are four different types of polarizations. They are as follows:

1. Electronic polarization
2. Ionic polarization
3. Orientation polarization and
4. Space-charge polarization

14.3.1 Electronic Polarization

Consider that the dielectric material is subjected to electric field intensity, *E*. A dielectric material consists of a number of atoms. Consider only one atom in a dielectric material as shown in Figure 14.2. Let *Z* be the atomic number and *e* be the charge of an electron. The charge of the nucleus is *Ze*. Consider that the electrons are revolving in the extra-nuclear space. Consider that the electrons charge is uniformly distributed throughout the atom. Due to the application of the electric field, the nucleus moves away from the field and the electron moves towards the field. Due to the movement of the electrons and the nucleus of the atom, there is a displacement and hence there is an induced dipole moment. This induced dipole moment per unit volume gives the induced polarization. This induced polarization is known as *electronic polarization*, since this polarization is produced by the displacement of electrons.

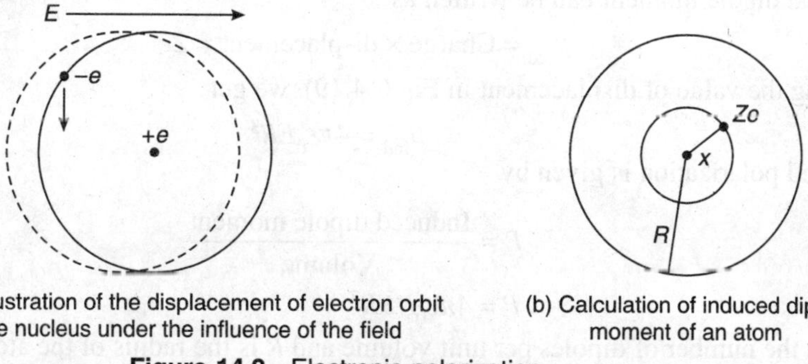

(a) Schematic illustration of the displacement of electron orbit relative to the nucleus under the influence of the field

(b) Calculation of induced dipole moment of an atom

Figure 14.2 Electronic polarization.

Consider Z is the atomic number, Ze is the charge of the nucleus and E is the electric field intensity applied to the atom.

The force due to the electric field,

$$F = ZeE \tag{14.14}$$

Consider that the electron is slightly displaced towards the field and the nucleus is displaced away from the field due to the application of the field as shown in Figure 14.2(b). Let x be the distance between the electron and the nucleus. Draw a sphere by taking the nucleus of the atom as the centre and x as the radius. From Gauss law, one can easily prove that the Coulomb's attractive force exists between the charge enclosed in the inner sphere and the nucleus. There is no force acting between the charge enclosed in the outer sphere and the nucleus.

$$\text{The charge enclosed in the inner sphere} = \frac{\frac{4}{3}\pi x^3}{\frac{4}{3}\pi R^3} Ze = \frac{x^3}{R^3} Ze \tag{14.15}$$

where x is the distance between the electron and the nucleus, R is the radius of the atom and Ze is the charge of the nucleus.

The Coulomb's attractive force between the nucleus and the electron cloud enclosed in the inner sphere

$$\frac{Ze\left(-\frac{x^3}{R^3} Ze\right)}{4\pi\varepsilon_0 x^2} \tag{14.16}$$

At equilibrium, the force due to the electric field and the Coulomb's attractive force are equal. Equating these two forces, we get

$$ZeE = \frac{Ze\left(\frac{x^3}{R^3} Ze\right)}{4\pi\varepsilon_0 x^2} \tag{14.17}$$

From Eq. (14.17), the value of x can be written as

$$x = \frac{4\pi\varepsilon_0 ER^3}{Ze} \tag{14.18}$$

The induced dipole moment can be written as

$$\mu_{\text{ind}} = \text{Charge} \times \text{displacement} = Ze \cdot x \tag{14.19}$$

Substituting the value of displacement in Eq. (14.19), we get

$$\mu_{\text{ind}} = 4\pi\varepsilon_0 ER^3 \tag{14.20}$$

The induced polarization is given by

$$P = \frac{\text{Induced dipole moment}}{\text{Volume}} \tag{14.21}$$

$$P = 4\pi\varepsilon_0 NER^3 \tag{14.22}$$

where N is the number of dipoles per unit volume and R is the radius of the atom.

This polarization is produced by the displacement of the electron and hence it is called *electronic polarization*. The electronic polarizability is given by

$$\alpha_e = \frac{\mu_{\text{ind}}}{E} = \frac{P}{NE}$$

i.e.
$$\alpha_e = 4\pi\varepsilon_0 R^3 \tag{14.23}$$

Equation (14.23) gives the value of electronic polarizability.

The polarization is also given by

$$P = \varepsilon_0(\varepsilon_r - 1)E \tag{14.24}$$

Equating Eq. (14.24) and Eq. (14.22), we get

$$\varepsilon_0(\varepsilon_r - 1)E = 4\pi\varepsilon_0 NER^3$$

i.e.
$$\varepsilon_r = 1 + 4\pi NR^3 \tag{14.25}$$

Equation (14.25) gives the equation for the dielectric constant of the material. For argon, $N = 2.7 \times 10^{25}$ m^{-3}, $\varepsilon_r = 1.0000192$. Substituting these values in Eq. (14.25), we get $R = 0.384$ Å. The value of the atomic radius obtained from Eq. (14.25) is in good agreement with the experimental results. Even though this model is a crude model, one cannot omit this model because it gives the correct value of the atomic radius.

14.3.2 Ionic Polarization

The ionic polarization is produced in ionic substances such as NaCl, KBr, KCl, etc. Consider an electric field is applied to an ionic crystal. The positive ions move away from the applied field, whereas the negative ions move towards the electric field. This movement of positive ion and negative ion produces displacement and hence an induced dipole moment is created. The induced dipole moment per unit volume is known as *ionic polarization* (Figure 14.3).

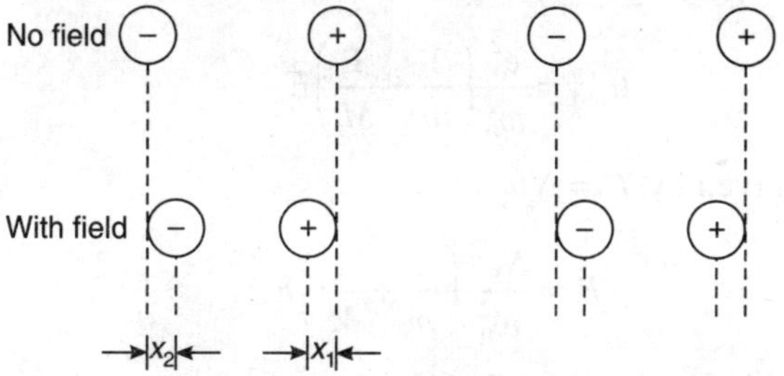

Figure 14.3 Ionic polarization.

Let x_1 and x_2 be the distances moved by the positive ion and negative ion respectively. The induced dipole moment is given by

$$\mu_{\text{ind}} = e(x_1 + x_2) \tag{14.26}$$

where x_1 and x_2 are the distances moved by the positive ion and negative ion respectively. Let E be the electric field intensity applied to the ionic crystal. The force acquired by the positive ion due to the application of the electric field is given by

$$F = eE \tag{14.27}$$

The force due to the electric field is directly proportional to the displacement, i.e.

$$F \propto x_1 \tag{14.28}$$

By introducing the proportionality constant, Eq. (14.28) can be written as

$$F = \beta_1 x_1 \tag{14.29}$$

The proportionality constant β_1 is equal to $m\omega_0{}^2$. Therefore, Eq. (14.29) can be written as

$$F = m\omega_0^2 x_1 \tag{14.30}$$

From Eq. (14.27) and Eq. (14.30), we get

$$eE = m\omega_0^2 x_1 \tag{14.31}$$

From Eq. (14.31), we get

$$x_1 = \frac{eE}{m\omega_0^2} \tag{14.32}$$

By using the similar arguments for the negative ion, one can write

$$F = eE$$

and

$$F \propto x_2$$

$$F = \beta_2 x_2$$

i.e.

$$F = M\omega_0^2 x_2 \tag{14.33}$$

$$eE = M\omega_0^2 x_2$$

where M is the mass of the negative ion and ω_0 is the angular frequency.

$$x_2 = \frac{eE}{M\omega_0^2} \tag{14.34}$$

Substituting the values of x_1 and x_2 in Eq. (14.26), we get

$$\mu_{\text{ind}} = \frac{e^2}{\omega_0^2}\left(\frac{1}{m} + \frac{1}{M}\right)E \tag{14.35}$$

The ionic polarization is given by, $P_i = N\mu_{\text{ind}}$

$$P_i = \frac{Ne^2}{\omega_0^2}\left(\frac{1}{m} + \frac{1}{M}\right)E \tag{14.36}$$

Comparing Eq. (14.36) with the equation, $P = N\alpha_i E$, we get

$$\alpha_i = \frac{e^2}{\omega_0^2}\left(\frac{1}{m} + \frac{1}{M}\right) \tag{14.37}$$

Equation (14.37) gives the expression for the ionic polarizability.

14.3.3 Orientation Polarization

The orientation polarization occurs in polar molecules. The liquid dielectric materials those having dipole moment even in the absence of the field are known as *polar dielectrics*. Nitrobenzene, H_2O, HCl, etc. are the examples of the polar dielectrics. The presence of the dipole moment even in the absence of the electric field is called *permanent dipole moment*. In the case of the water molecules, due to the arrangement of its molecular structure, it possesses dipole moment even in the absence of the electric field. So, it is a polar dielectric material. The molecules, those are not having any permanent dipole moment in the absence of the electric field, are known as *non-polar dielectrics*. The example of non-polar dielectrics is CO_2 molecule.

The polarization produced in the case of a polar molecule due to the application of an electric field is known as *orientation polarization* (Figure 14.4).

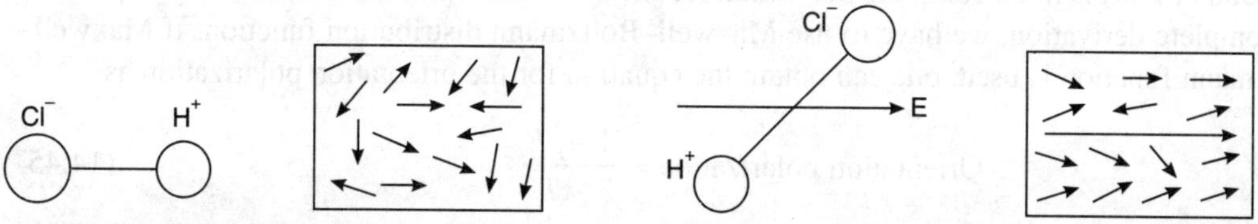

(a) HCl molecule (b) Random orientation of dipoles (c) Rotation of dipoles due to the applied field (d) Parallel alignment of dipoles with the electric field

Figure 14.4 Orientation polarization.

Consider a polar dielectric material such as HCl is subjected to an electric field. The dipole experiences certain force and hence it rotates and then it comes to rest at some particular position. The torque experienced by the dipole is given by

$$\tau = \bar{\mu}_P \times \bar{E} \tag{14.38}$$

$$\tau = \mu_P E \sin\theta \tag{14.39}$$

When the field is applied, the dipole will rotate and try to align parallel to the applied field. If the dipole is anti-parallel to the field, it has to rotate through 180° and if it is parallel to the applied field, the angle of rotation is 0.

$$\text{The maximum dipole energy} = \int_0^{\pi} \mu_P E \sin\theta\, d\theta \tag{14.40}$$

The above integral varies from 0 to π, because the angle of rotation of dipoles varies from 0 to π.

$$\text{The maximum dipole energy} = 2\mu_P E \tag{14.41}$$

The minimum dipole energy = 0 and hence the average dipole energy = $\mu_P E$.

$$\text{Orientation polarizaton} \propto \frac{\text{Average dipole energy}}{\text{Average thermal energy}} \tag{14.42}$$

The orientation polarization is found to be effective, if Eq. (14.42) is greater than one.

Substituting the values of average dipole energy and average thermal energy in Eq. (14.42), we get

$$\text{Orientation polarization} \propto \frac{\mu_P E}{\frac{5}{2}kT} \tag{14.43}$$

By introducing the proportionality constant, Eq. (14.43) can be written as

$$\text{Orientation polarization} = \text{Permanent dipole moment} \times \frac{\text{Average dipole moment}}{\text{Average thermal energy}}$$

$$\text{Orientation polarization} = \mu_P \times \frac{\mu_P E}{\frac{5}{2}kT}$$

$$\text{Orientation polarization} = \frac{2}{5}\frac{\mu_P^2 E}{kT} \tag{14.44}$$

Equation (14.44) is not a complete derivation for the orientation polarization. In order to perform the complete derivation, we have to use Maxwell–Boltzmann distribution function. If Maxwell–Boltzmann function is used, one can obtain the equation for the orientation polarization as

$$\text{Orientation polarization} = \frac{1}{3}\frac{\mu_P^2 E}{kT} \tag{14.45}$$

Equation (14.45) gives the expression for the orientation polarization produced by one dipole in polar molecules. If there is N number of dipoles per unit volume, then the total orientation polarization is given by

$$\text{Orientation polarization for } N \text{ number of dipoles per m}^{-3} = \frac{1}{3}\frac{N\mu_P^2 E}{kT} \tag{14.46}$$

From Eq. (14.46), we infer that the orientation polarization is temperature dependent.

14.3.4 Space-charge Polarization

Consider that a container has an electrolyte solution and two electrodes. Before, the application of the electric field, the positive ions and negative ions are not separated in the electrolyte solution. Consider that the electrodes are connected to a battery. The negative charges (electrons) get accumulated in the positive electrode and the positive charges get accumulated in the negative electrode. The separation of positive and negative charges produces an induced dipole moment and hence induced polarization. The induced polarization produced by the separation of positive and negative charges in an electrolyte due to the application of an electric field is known as *space-charge polarization*. The space-charge polarization is very small and negligible (Figure 14.5).

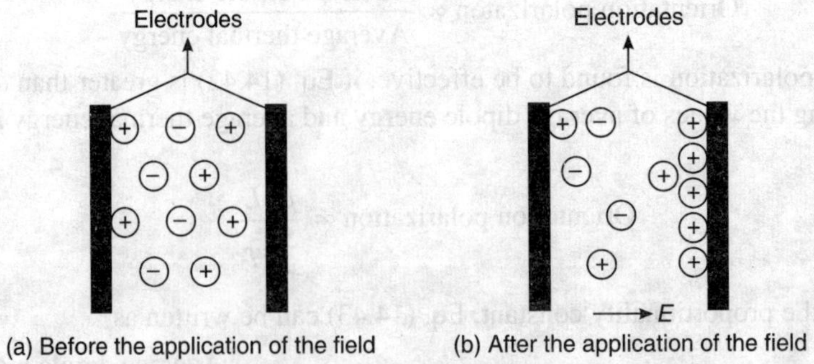

Figure 14.5 Space-charge polarization.

14.3.5 Total Polarization

The total polarization is the sum of the electronic, ionic, orientation and space-charge polarization. The space-charge polarization is very small and it is not well-defined and hence it is negligible. Therefore, the total polarization is the sum of the electronic, ionic and orientation polarizations.

$$\text{Total polarization, } P = P_e + P_i + P_o \tag{14.47}$$

Substituting the values of P_e, P_i and P_o, we get

$$P = NE\left(4\pi\varepsilon_0 R^3 + \frac{e^2}{\omega_0^2}\left(\frac{1}{m} + \frac{1}{M}\right) + \frac{1}{3}\frac{\mu_P^2}{kT}\right) \tag{14.48}$$

Equation (14.48) gives the value of the total polarization in a dielectric material.

14.4 LOCAL FIELD OR INTERNAL FIELD IN A SOLID DIELECTRIC MATERIAL

14.4.1 Local Field in a Solid Dielectric Material

Consider a dielectric material is subjected to an electric field of electric field intensity E. The electric field present at the location of an atom is not equal to the applied field. It is greater than the applied field. In a dielectric material, the atom experiences an induced field by the nearest neighbouring atoms in addition to the applied field. The field present at the location of an atom of a dielectric material is known as *internal field* or *local field*. The internal field is the sum of the applied field and the induced field.

Internal field = applied field + induced field

Consider an array of atoms as shown in Figure 14.6. Let a be the distance between any two adjacent atoms. Each and every atom acts as a dipole. Consider that the atoms are represented by A, B_1, B_2, C_1, C_2, etc. Consider the atom A. The field present on the atom A is equal to the applied field plus the field induced by the nearest neighbouring atoms.

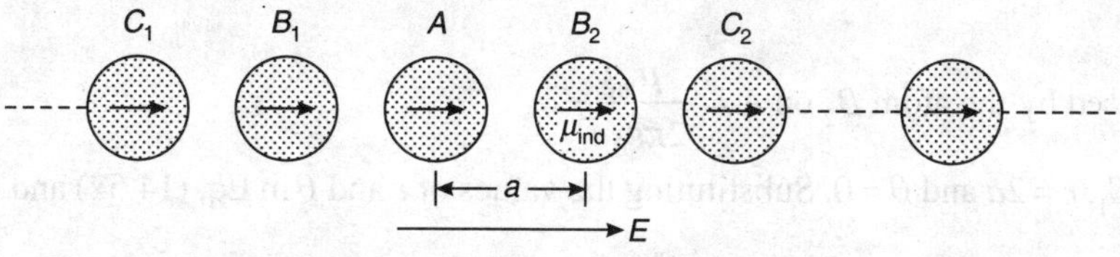

Figure 14.6 Linear chain of atoms—calculation of internal field.

The potential induced by a dipole at a point P as shown in Figure 14.7 is given by

$$V(r,\theta) = \frac{\mu\cos\theta}{4\pi\varepsilon_0 r^2} \tag{14.49}$$

The field present at the point P can be written as two components. They are the r component of the field and θ component of the field. They are given by

$$E_r = -\frac{\partial V}{\partial r} \tag{14.50}$$

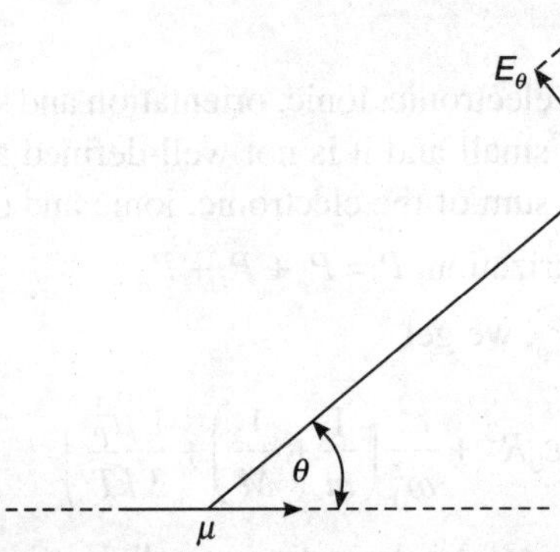

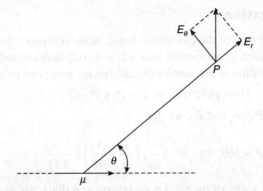

Figure 14.7 Field induced by a dipole at the point *P*.

$$E_\theta = -\frac{1}{r}\frac{\partial V}{\partial \theta} \tag{14.51}$$

The r and θ component of the fields are determined by differentiating Eq. (14.49).

$$E_r = \frac{\mu \cos\theta}{2\pi\varepsilon_0 r^3} \tag{14.52}$$

$$E_\theta = \frac{\mu \sin\theta}{4\pi\varepsilon_0 r^3} \tag{14.53}$$

The field induced by the nearest atoms on the atom A can be determined from Eq. (14.52) and Eq. (14.53) by substituting the values of r and θ. For the atom B_1, $r = a$ and $\theta = 0$. Substituting the values of r and θ in Eq. (14.52) and Eq. (14.53), we get

The field induced by the atom B_1 on $A = \dfrac{\mu_{\text{ind}}}{2\pi\varepsilon_0 a^3}$

For the atom B_2, $r = a$ and $\theta = 0$. Substituting the values of r and θ in Eq. (14.52) and Eq. (14.53), we get

The field induced by the atom B_2 on $A = \dfrac{\mu_{\text{ind}}}{2\pi\varepsilon_0 a^3}$

For the atom C_1, $r = 2a$ and $\theta = 0$. Substituting the values of r and θ in Eq. (14.52) and Eq. (14.53), we get

The field induced by the atom C_1 on $A = \dfrac{\mu_{\text{ind}}}{2\pi\varepsilon_0 (2a)^3}$

For the atom C_2, $r = 2a$ and $\theta = 0$. Substituting the values of r and θ in Eq. (14.52) and Eq. (14.53), we get

The field induced by the atom C_2 on $A = \dfrac{\mu_{\text{ind}}}{2\pi\varepsilon_0 (2a)^3}$

$$\text{Total field present at the atom } A = E + 2\times\frac{\mu_{\text{ind}}}{2\pi\varepsilon_0 a^3} + 2\times\frac{\mu_{\text{ind}}}{2\pi\varepsilon_0 (2a)^3} + 2\times\frac{\mu_{\text{ind}}}{2\pi\varepsilon_0 (3a)^3} + \cdots \tag{14.54}$$

$$= E + \frac{\mu_{\text{ind}}}{\pi\varepsilon_0 a^3} + \frac{\mu_{\text{ind}}}{\pi\varepsilon_0 (2a)^3} + \frac{\mu_{\text{ind}}}{\pi\varepsilon_0 (3a)^3} + \cdots \quad (14.55)$$

$$= E + \frac{\mu_{\text{ind}}}{\pi\varepsilon_0 a^3} \times \left[1 + \frac{1}{2^3} + \frac{1}{3^3} + \frac{1}{4^3} + \cdots\right] \quad (14.56)$$

$$= E + \frac{\mu_{\text{ind}}}{\pi\varepsilon_0 a^3} \times \sum_{n=1}^{N} \frac{1}{n^3} \quad (14.57)$$

The sum of the series $\sum_{n=1}^{N} \frac{1}{n^3} = 1.2$. Substituting the value of $\sum_{n=1}^{N} \frac{1}{n^3}$ in Eq. (14.57), we get

$$\text{Total field on the atom } A = E + \frac{1.2 \times \mu_{\text{ind}}}{\pi\varepsilon_0 a^3} \quad (14.58)$$

By taking $P = \frac{\mu_{\text{ind}}}{a^3}$ and $\gamma = \frac{1.2}{\pi}$ Eq. (14.58) can be written as

$$E_i = E + \frac{\gamma P}{\varepsilon_0} \quad (14.59)$$

Equation (14.59) represents the internal field in a solid dielectric material. For a cubic structure, $V = a^3$ and $P = \frac{\mu_{\text{ind}}}{a^3}$, $\gamma = \frac{1.2}{\pi} = \frac{1}{3}$, Eq. (14.59) can be written as

$$E_i = E + \frac{P}{3\varepsilon_0} \quad (14.60)$$

where E_i is the internal field or local field present at the location of an atom. Equation (14.60) is known as *Lorentz internal field equation.*

14.4.2 Evaluation of Local Field for a Cubic Structure

Consider that a dielectric material is placed in an electric field. The polarizing field produced in an atom of the dielectric material is known as *local field* or *internal field*.

Consider a dielectric material is placed between the parallel plates *P*, *Q* of a capacitor. Consider an electric field *E* is applied to the dielectric. Let there is a spherical cavity (also known as *Lorentz cavity*) of radius *r*. Consider a dipole, *A*, at the centre of the dielectric material. Consider that the radius of the cavity is large compared to the size of the dipole. The space outside the cavity may be treated as continuous as far as the dipole is concerned.

The field acting at the location of the dipole is given by the sum of the following terms.

$$E_{\text{loc}} = E_0 + E_1 + E_2 + E_3 \quad (14.61)$$

where E_0 is the field due to the charged plates, E_1 is the field due to polarizing charges on the plane surfaces of the dielectric medium. It is also called the *depolarizing field*, E_2 is the field due to the polarizing charges lying on the surface of the Lorentz cavity. It is also known as *Lorentz field*. E_3 is the field due to all other dipoles lying within the sphere (cavity).

To find E_0 (field due to charged plates)

Let E be the macroscopic field within the dielectric, i.e. the applied field, then

$$D = \varepsilon_0 E + P \tag{14.62}$$

The relation between the electric flux density and the electric field due to the charged plates when there is no dielectric medium is present is given by

$$D = \varepsilon_0 E_0 \tag{14.63}$$

Equating these two equations, we get

$$\varepsilon_0 E_0 = \varepsilon_0 E + P$$

$$E_0 = E + \frac{P}{\varepsilon_0} \tag{14.64}$$

To find E_1 (depolarizing field)

This field is due to the polarization charges on the external surfaces. This will act opposite to the applied field. This will depend on the geometrical shape of the dielectrics. The value of the depolarizing field is given by

$$E_1 = -\frac{P}{\varepsilon_0} \tag{14.65}$$

To find E_2 (Lorentz field)

It is the field due to the polarizing charges lying on the surfaces of the cavity. It is also known as *Lorentz field*. Consider an enlarged view of the spherical cavity. Let r be the radius of the cavity. Consider a small surface S with an elemental area dA at an angle of θ with respect to the polarization direction as shown in Figure 14.8. The induced charge on the surface (S) of the cavity is given by the normal component of polarization multiplied by the surface area.

$$q = P \cos\theta \, dA \tag{14.66}$$

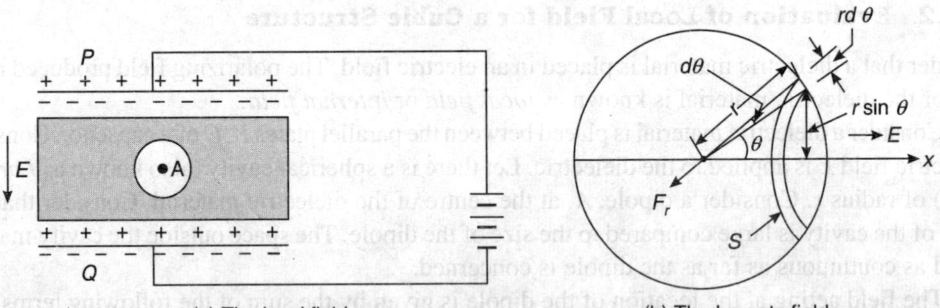

Figure 14.8 Local field in a solid dielectric.

According to Coulomb's law, the force between the charges on this surface area dA and the reference charge q is

$$F_r = \frac{q_1 q_2}{4\pi\varepsilon_0 r^2} = \frac{qP\cos\theta \, dA}{4\pi\varepsilon_0 r^2} \tag{14.67}$$

The area dA is calculated as follows. The thickness of the reference area dA is $rd\theta$. The circumference of the element between the angles θ and $\theta + d\theta$ is $2\pi r \sin\theta$. The surface area dA of the sphere between θ and $\theta + d\theta$ is $2\pi r^2 \sin\theta\, d\theta$. The force acting along the direction (x-axis) of the applied field is $F_x = F_r \cos\theta$.

Substituting the value of dA, we get

$$F_x = \frac{qP\cos^2\theta}{4\pi\varepsilon_0 r^2} 2\pi r^2 \sin\theta\, d\theta = \frac{qP}{2\varepsilon_0}\cos^2\theta \sin\theta\, d\theta \tag{14.68}$$

The Lorentz force can be evaluated by integrating the above equation from $\theta = 0$ to $\theta = \pi$.

$$F_x = \frac{qP}{2\varepsilon_0}\int_0^{\pi} \cos^2\theta \sin\theta\, d\theta \tag{14.69}$$

Substituting, $x = \cos\theta$ and $dx = -\sin\theta\, d\theta$, we get

$$F_x = \frac{qP}{2\varepsilon_0}\int_1^{-1} -x^2 dx = \frac{qP}{2\varepsilon_0}\int_{-1}^{1} x^2 dx$$

$$F_x = \frac{qP}{3\varepsilon_0} \tag{14.70}$$

The electric field (Lorentz field) intensity is

$$E_2 = \frac{F}{q} = \frac{P}{3\varepsilon_0} \tag{14.71}$$

Field due to the other dipoles in the cavity, E_3

The field E_3 is due to the other dipoles within the cavity. For a cubic structure, $E_3 = 0$.

Total field

The total field is $E = E_0 + E_1 + E_2 + E_3$

$$E_i = E + \frac{P}{\varepsilon_0} - \frac{P}{\varepsilon_0} + \frac{P}{3\varepsilon_0} + 0$$

$$E_i = E + \frac{P}{3\varepsilon_0} \tag{14.72}$$

This is known as *Lorentz relation*. The difference between the Maxwell field (E) and the Lorentz field, E_{local} is given below (Figure 14.9).

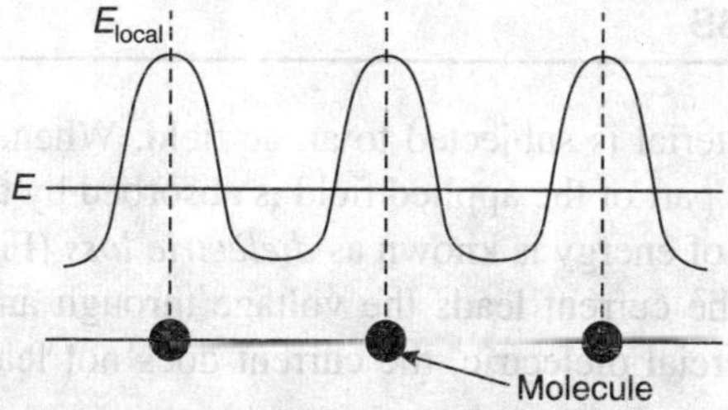

Figure 14.9 The difference between the Maxwell field E and the local field E_{local}.

The Maxwell field is the average field and it is macroscopic in nature, whereas the Lorentz field is the microscopic field and it is periodic in nature. The Lorentz field is quite large at the molecules, indicating that the molecules are more effectively polarized than they are under the average field.

14.5 CLAUSIUS–MOSOTTI EQUATION

The total polarization of a dielectric material is given by

$$\text{Total polarization, } P = P_e + P_i + P_o \tag{14.73}$$

For an elemental dielectric material, the ionic and the orientation polarizations are equal to zero. Therefore, the equation for the total polarization is given by

$$\text{Total polarization, } P = P_e = N\alpha_e E_i \tag{14.74}$$

where E_i is the internal field in a dielectric material. Substituting the values of E_i, we get

$$P = N\alpha_e\left(E + \frac{P}{3\varepsilon_0}\right) \tag{14.75}$$

The polarization is also given by

$$P = \varepsilon_0(\varepsilon_r - 1)E \tag{14.76}$$

Substituting the value of the polarization from Eq. (14.76) in Eq. (14.75), we get

$$\varepsilon_0(\varepsilon_r - 1)E = N\alpha_e\left(E + \frac{\varepsilon_0(\varepsilon_r - 1)E}{3\varepsilon_0}\right) \tag{14.77}$$

Simplifying Eq. (14.77), we get

$$\varepsilon_0(\varepsilon_r - 1) = N\alpha_e\left(1 + \frac{(\varepsilon_r - 1)}{3}\right) \tag{14.78}$$

Rearranging Eq. (14.78), we get

$$\frac{(\varepsilon_r - 1)}{(\varepsilon_r + 2)} = \frac{N\alpha_e}{3\varepsilon_0} \tag{14.79}$$

Equation (14.79) is known as Clausius–Mosotti equation. It is used to obtain the value of the dielectric constant of an elemental dielectric material.

14.6 DIELECTRIC LOSS

Consider that a dielectric material is subjected to an ac field. Whenever a dielectric material is subjected to an ac field, some part of the applied field is absorbed by the material and it is wasted in the form of heat. This loss of energy is known as *dielectric loss* (Figure 14.10).

In an ideal dielectric, the current leads the voltage through an angle of 90° as shown in Figure 14.10(a). In a commercial dielectric, the current does not lead the voltage through 90°.

There is an angle of lagging. This angle of lagging is known as *dielectric loss angle*. The power loss in a dielectric material is given by

$$P = VI \cos \theta \qquad (14.80)$$

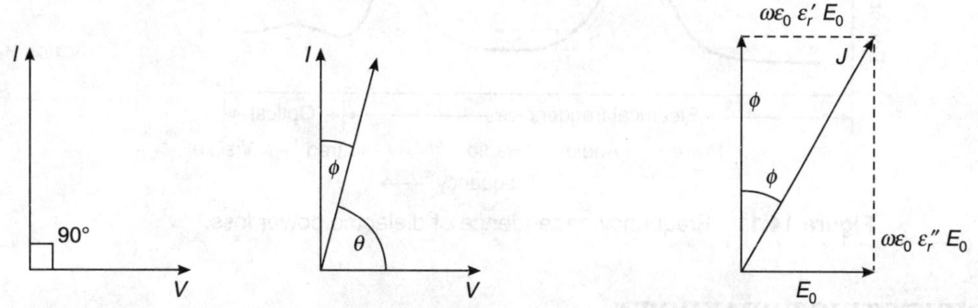

(a) In an ideal dielectric (b) In a commercial dielectric (c) Current density component with respect to field direction in a lossy dielectric material

Figure 14.10 Phase difference between current and voltage.

From Figure 14.10(b), $\theta = 90 - \phi$, where ϕ is the angle of lagging. It is also called *dielectric loss angle*. Substituting the value of θ in Eq. (14.80), we get

$$P = VI \cos (90 - \phi) \qquad (14.81)$$

i.e.

$$P = VI \sin \phi \qquad (14.82)$$

The current passing through a capacitor is, $I = \dfrac{V}{X_C}$, where X_C is the capacitive reactance and it is equal to $X_C = \dfrac{1}{j\omega C}$. Substituting the value of I and X_C, we get

$$P = j\omega\, CV^2 \sin \phi \qquad (14.83)$$

For smaller value of ϕ, $\sin \phi = \tan \phi$. Equation (14.83) can be written as

$$P = j\omega\, CV^2 \tan \phi \qquad (14.84)$$

where $\tan \phi$ is the power factor or loss tangent and it is given by

$$\tan \phi = \frac{\varepsilon_r''}{\varepsilon_r'} \qquad (14.85)$$

where ε_r' and ε_r'' are the real and imaginary part of the dielectric constant.

From Eq. (14.84), we infer that the dielectric loss is directly proportional to

1. the power factor of the dielectric material,
2. square of the applied voltage,
3. frequency of the applied voltage and,
4. the temperature and humidity.

The frequency dependent of the dielectric loss is shown in Figure 14.11. In optical frequency the dielectric loss is due to the optical absorption. At lower frequency the dielectric loss is duc to the dc resistivity. The dielectric losses at radio frequency are due to dipole rotation.

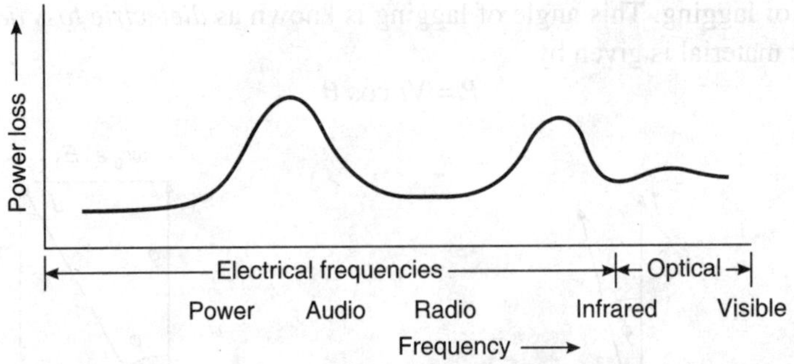

Figure 14.11 Frequency dependence of dielectric power loss.

14.7 DIELECTRIC BREAKDOWN

A dielectric material is used to store the electrical energy and it is also used to resist the flow of electric current, since it is an insulator. Whenever it is used as an insulating material, it has to resist the flow of current through it. It can withstand up to a maximum voltage known as *dielectric strength*. If the field is applied beyond this value, the dielectric material allows the electric current to flow through it. The failure of a dielectric material to resist the flow of current is known as *dielectric breakdown*. There are five different types of dielectric breakdown. They are

1. intrinsic breakdown
2. thermal breakdown
3. electrochemical breakdown
4. discharge breakdown and
5. defect breakdown

Intrinsic breakdown

Consider that a dielectric material is subjected to a very high voltage. The application of a high voltage to a dielectric material dislodges the electrons in the valence band and the dislodged electrons acquire some energy from the applied fields and hence these electrons are excited into the conduction band. The flow of these electrons in the conduction band constitutes an electric current. The flow of electric current in the conduction band indicates the failure of the dielectric material. This type of failure is known as *intrinsic breakdown*.

Whenever a high electric field is applied, the electrons get dislodged from the valence band. These dislodged electrons make collisions with the bond and hence they further dislodge some other electrons. If the process goes like this, a large number of electrons get dislodged simultaneously. These electrons get excited to the conduction band simultaneously and hence a large amount of electric field is produced. This phenomenon is known as *avalanche breakdown*.

Characteristics

1. It occurs when a large electric field is applied.
2. The breakdown occurs even at low temperatures.
3. It mostly occurs in thin samples.
4. It occurs within a short period of time.

Thermal breakdown

Consider that a dielectric material is subjected to an electric field. The application of the electric field generates heat in the material. The heat generated in the material should be dissipated from it. In some dielectric material, the generation of heat is higher than its dissipation. In that case, if the field is applied for a longer time, the temperature of the material increases. The increase in temperature brings the material into breakdown. This type of breakdown is known as thermal breakdown.

Characteristics

1. It occurs at very high temperature.
2. The breakdown takes place within few milliseconds.

Electrochemical breakdown

The application of electric current to a dielectric material for a longer time increases its temperature. The increase in the temperature of the material increases the mobility of the charge carriers. The increase in the mobility of the charge carriers induces the chemical reaction and hence the material is brought into breakdown due to the chemical reaction. This type of breakdown is called *electrochemical breakdown* or *chemical breakdown*.

Characteristics

1. It occurs due to the chemical reaction.
2. The electrochemical breakdown depends upon temperature.

Discharge breakdown

In certain dielectric materials, there is a possibility of the presence of occluded gas bubbles. If an electric field is applied to this dielectric material, the gaseous ions are more easily ionized than the solid state ions. Due to the ionization of the gaseous ions, ionization current flows through the material. This ionization current brings the material into breakdown. This type of breakdown is known as *discharge breakdown*.

Characteristics

1. It occurs due to the presence of gas bubbles.
2. It occurs even at a low voltage.

Defect breakdown

Consider that a dielectric material has some surface defects such as cracks. Due to the presence of crack, the moisture may get accumulated over the dielectric material. If an electric field is applied to this dielectric material, the material is easily brought into breakdown due to the presence of moisture. This type of breakdown is known as *defect breakdown*.

14.8 FREQUENCY DEPENDENCE OF POLARIZATION

Whenever a dielectric material is placed in an ac field, the polarization produced by it depends upon the frequency of the electric field applied. If the field is applied for a longer time, the increase in polarization for polar dielectrics with time can be written as

$$P_0(t) = P_0[1 - e^{-t/\tau_r}] \tag{14.86}$$

where P_0 is the maximum polarization attained by a dielectric material due to the prolong application of the field. The term, τ_r represents the relaxation time. It is the characteristics time that determines the sluggishness of the dipole response to an applied field. It relates the amount of polarization at a point in the dielectric to the field at that point.

The space-charge polarization is one of the slowest polarizations. It mostly occurs when the frequency is less than 10^2 Hz. It occurs at machine frequencies (50 to 60 Hz). The orientation polarization occurs when the frequency is less than 10^6 Hz. Generally it occurs in polar molecules between radio and microwave frequencies. It is slower than ionic and electronic polarization. But it is faster than space-charge polarization. The ionic polarization takes place in ionic crystals. It occurs when the frequency is less than 10^{13} Hz. This polarization requires a time period of 10^{-11} to 10^{-13} s. It is faster than space-charge and orientation polarization but it is slower than the electronic polarization. The electronic polarization is the fastest polarization. It occurs at optical frequencies. The frequency is in the range 10^{14} to 10^{16} Hz. It is quickly built between 10^{-14} and 10^{-16} s. The electronic contribution of the polarization is determined by measuring the refractive index at optical frequencies. The ionic and orientation polarizations are small at high frequencies because of the inertia of the ions and molecules (Figure 14.12).

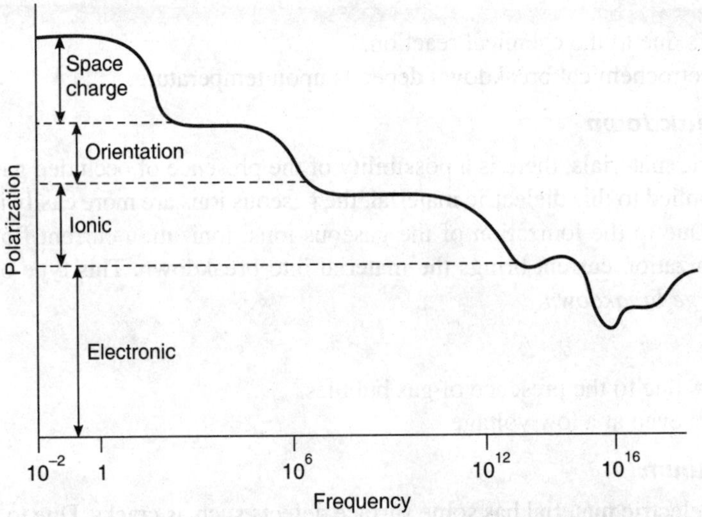

Figure 14.12 Frequency dependence of polarization.

14.9 TEMPERATURE DEPENDENCE OF POLARIZATION

The electronic and ionic polarizations are not dependent on temperature. Only the orientation polarization is dependent on temperature. The expression for the orientation polarization is given by,

$$P_o = \frac{1}{3}\frac{NE\mu_P^2}{k_B T}$$

The total polarization is also temperature dependent. If a curve is plotted between the total polarization and $1/T$, a straight line as shown in Figure 14.13 is obtained.

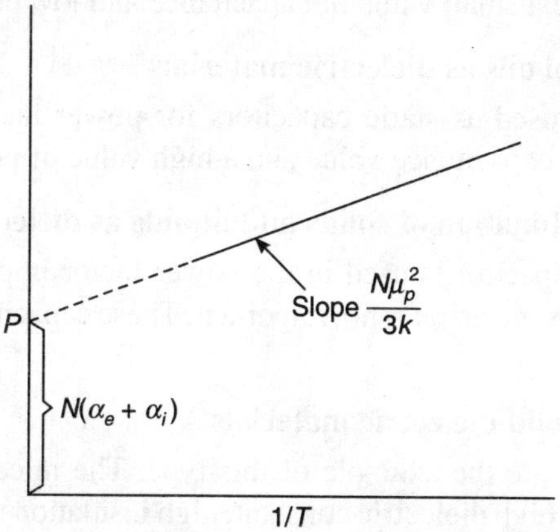

Figure 14.13 Temperature dependence of polarization.

A plot of polarization versus temperature for CH_4, CH_3Cl, CH_2Cl_2, CCl_4, $CHCl_3$ is shown in Figure 14.14. From this plot, one can determine the dipole moment. It is also possible to find the molecular structure.

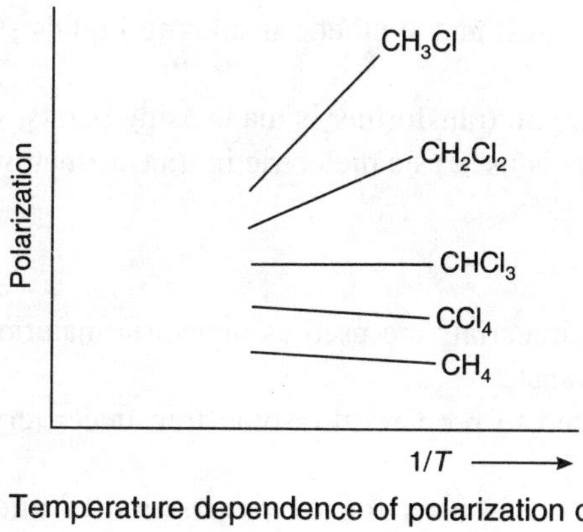

Figure 14.14 Temperature dependence of polarization of some liquids.

14.10 APPLICATIONS OF DIELECTRIC MATERIAL

In capacitors

The dielectric materials are used in capacitors to store the electrical energy. The capacitors are classified based on the dielectric materials used. The capacitors can be broadly classified into the following categories:

(i) Capacitors using vacuum, air, gas as dielectric materials,
(ii) Capacitors using mineral oils as dielectric materials,
(iii) Capacitors using a combination of solids and liquids as dielectric materials,
(iv) Capacitors using only solid dielectric materials.

(i) Capacitors using vacuum, air, gas as dielectric materials

The vacuum or air or gas capacitors are used in high frequency circuits like radios and for frequency measuring devices. They have a small value of capacitance and low energy loss.

(ii) Capacitors using mineral oils as dielectric materials

The oil-filled capacitors are used as static capacitors for power factor improvements in small installations. They have large capacitance value and a high value of power loss.

(iii) Capacitors using a combination of solids and liquids as dielectric materials

The oil impregnated paper capacitor is used in the power factor improvement device and radio circuit. They are used wherever precision is not important. These capacitors have a large capacitance value and are smaller in size.

(iv) Capacitors using only solid dielectric materials

Mica and ceramic capacitors are the example of this type. The mica capacitor is used to make standard capacitors. Mica has high dielectric constant, high insulation resistance and low dielectric loss. The dielectric constant of mica is not affected by temperature. High frequency power capacitors are prepared using class 1 ceramic dielectrics. The ceramic dielectrics are used to produce a capacitance value from 0.1 to 2 μF.

In transformer

1. In transformer, the liquid dielectric is used as a coolant and as a dielectric material. The mineral insulating liquids and synthetic insulating liquids such as askarels are used as dielectric medium.
2. The insulation coating in transformer is made using paints, varnish and some enamels.
3. Sulphur hexafluoride is used as a dielectric in transformer, waveguides, capacitors etc.

Other applications

Solid dielectrics

1. The solid dielectric materials are used as dielectric materials in capacitors, e.g. mica, ceramics, barium titanate.
2. Quartz crystal is used to prepare ultrasonic transducer, crystal oscillator, delay line, filters, etc.
3. Lead zirconate titanate is used for the preparation of microphone, spark generator, accelerometer, ear phones, etc.

Liquid dielectrics

1. The liquid dielectrics are used as a cooling medium in transformers and some electronic equipment. The liquid dielectrics are also used as a filling medium in capacitors, bushings, etc.
2. The liquid dielectrics are used as an insulating and arc-quenching medium in switchgear.
3. The liquid dielectrics are impregnated with solid materials such as paper, porous polymers and press board. These are used in transformers, switch gears, capacitors and cables.

Gaseous dielectrics

1. Compressed air is used as a dielectric insulation in air-blast circuit breaker. It is also used as arc-extinguishing medium.

2. Very high voltage equipments such as cyclotron and Van de Graff generator use high vacuum as insulator. Vacuum is also used as insulation in vacuum circuit breaker and contactors.
3. Silicone fluid (poly-dimethyl siloxanes) is used as alternative fire resistant insulating liquids.
4. Askarels (Polychlorinated biphenyls) is used as high permittivity ($\varepsilon_r = 3–6$) fire resistant insulating liquids.

14.11 FERROELECTRIC MATERIAL

In a ferroelectric material, the polarization is not a linear function of the applied electric field intensity. The ferroelectric materials exhibit hysteresis curve, when a curve is drawn between E and P. The hysteresis curve exhibited by a ferroelectric material is shown in Figure 14.15. If the electric field intensity is increased, the polarization increases and it reaches a maximum value (P_{max}). Beyond this maximum value of the polarization, if we increase the electric field there is no further increase in the polarization value. If the field is decreased, the polarization decreases. The polarization is not equal to zero, when the field becomes zero. The existence of polarization in the absence of the electric field is known as *remanent polarization.* In order to bring the polarization into zero, the field should be applied in the reverse direction. The field required to bring the polarization into zero is known as *coercive field* or *coercivity.* If the field is further applied in the reverse direction, the minimum value of the polarization (P_{min}) is obtained. There is no further decrease in the polarization value, beyond P_{min}. Now, if the field is gradually increased, the ferroelectric material traces the path as shown in Figure 14.15 and a closed plot known as a *hysteresis curve* is obtained.

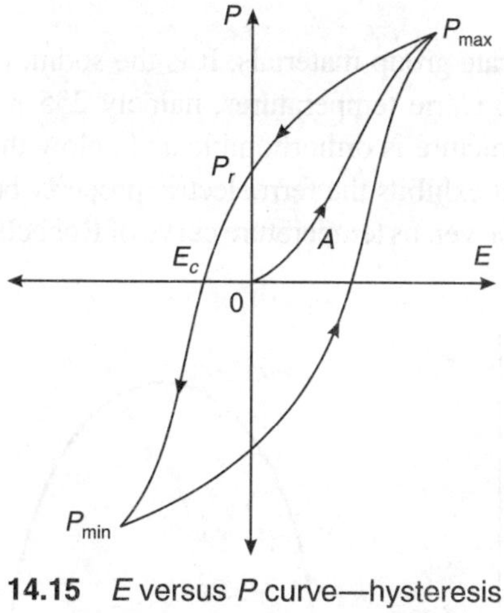

Figure 14.15 *E* versus *P* curve—hysteresis curve.

The ferroelectric property is explained using the domain concept. Similar to the ferromagnetic material, the ferroelectric material also has domain structure. Domains are small localized regions. The domains are needle-shaped and have a diameter of nearly 10^{-6} m. The ferroelectric domains are shown in Figure 14.16. Each and every domain acts as a dipole and they are pointed in random directions. If an electric field is applied to a ferroelectric material, the domain that is parallel to the applied field increases in its size and the domain that is opposite to the applied field decreases in its size. The polarization becomes maximum, when the entire material is made up of single domain. The presence of polarization even in the absence of the electric field is due to the domains present in the material are not able to return to their original position.

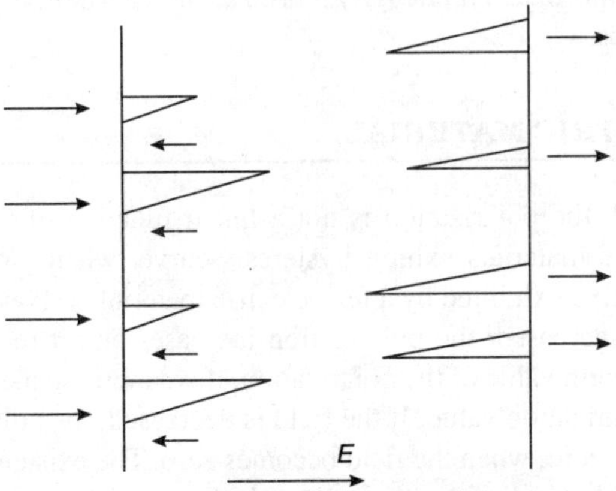

Figure 14.16 Ferroelectric domains.

14.11.1 Classification of Ferroelectric Materials

The ferroelectric materials are classified into three different groups depending on their chemical composition and structure. They are, (i) tartarate group, (ii) dihydrogen phosphate and arsenates alkalai metals and (iii) oxygen octahedron group.

Tartarate group

Rochelle salt is one of the tartarate group materials. It is the sodium potassium salt of tartaric acid ($NaKC_4H_4O_6.4H_2O$). It has two Curie temperatures, namely 255 K and 296 K. Above the upper Curie temperature, its crystal structure is orthorhombic and below the lower Curie temperature, its crystal structure is monoclinic. It exhibits the ferroelectric property between the Curie temperatures 255 K and 296 K. The polarization versus temperature curve of Rochelle salt is shown in Figure 14.17.

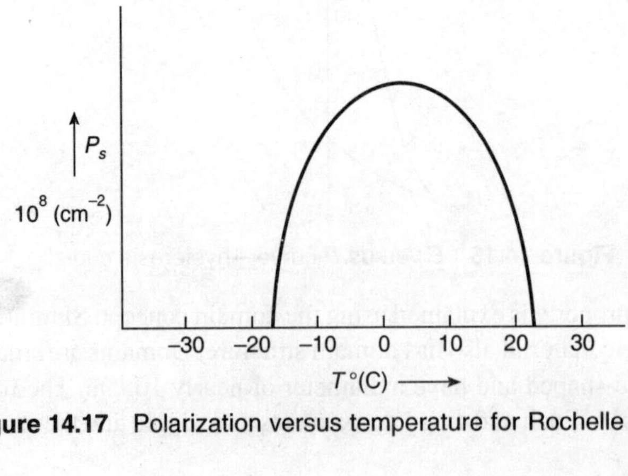

Figure 14.17 Polarization versus temperature for Rochelle salt.

Dihydrogen phosphate and arsenates alkali metals

The potassium dihydrogen phosphate (KH_2PO_4) belongs to this group. It exhibits only one Curie temperature, 123 K. It exhibits the ferroelectric property below the Curie temperature. Above the Curie temperature its crystal structure is tetragonal and below the Curie temperature, its structure is orthorhombic. The polarization versus temperature curve of potassium dihydrogen phosphate is shown in Figure 14.18.

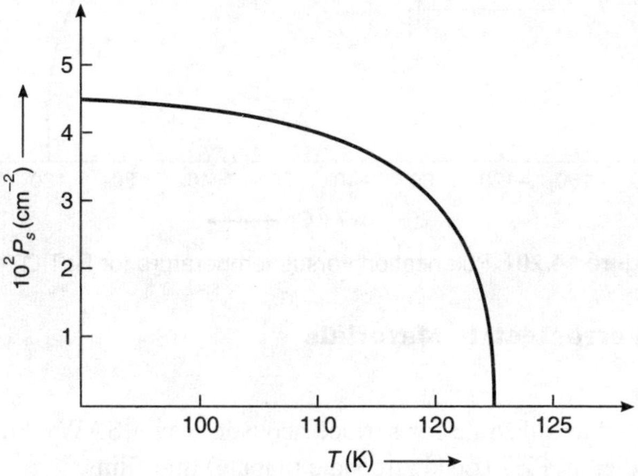

Figure 14.18 Polarization versus temperature for KH_2PO_4.

Oxygen octahedron group

Barium titanate ($BaTiO_3$) is an example of this group of materials. It has three Curie temperatures, namely 193 K, 278 K and 393 K. Above 393 K, it has cubic structure. Below 193 K, it has orthorhombic and between 193 K and 278 K, its crystal structure is tetragonal.

The cubic crystal structure is shown in Figure 14.19. The Ba^{2+} occupies the corner of the cube, O^{2-} occupies the face-centered position and Ti^{4+} occupies the body-centered position. The Ti^{4+} ion is considerably smaller in size than the space available in the oxygen octahedron. It brings ionic polarization due to the following reasons, (i) its charge is 4e, (ii) it will be displaced relatively long distance due to the application of the field. The polarization versus temperature curve of barium titanate is shown in Figure 14.20.

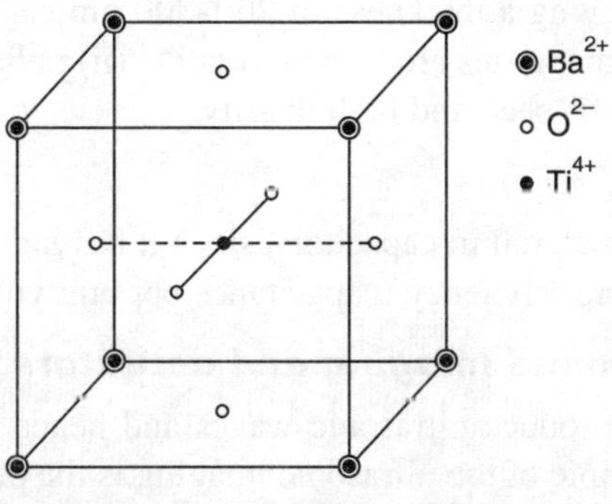

Figure 14.19 Cubic crystal structure of $BaTiO_4$.

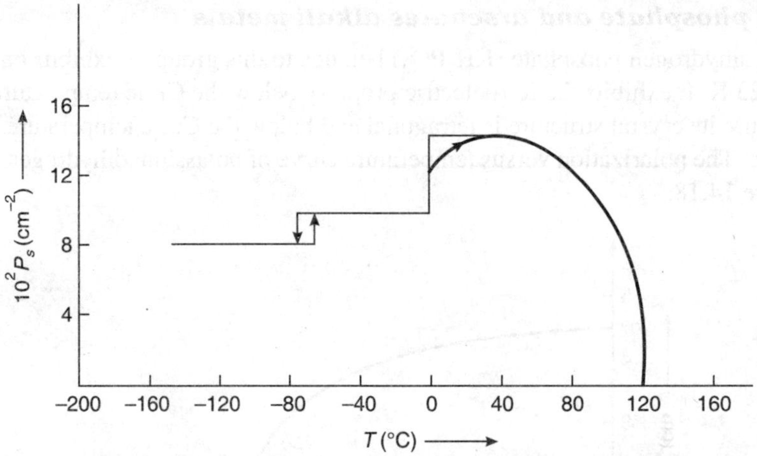

Figure 14.20 Polarization versus temperature for $BaTiO_4$.

14.11.2 Uses of Ferroelectric Materials

SAW devices

$LiNbO_3$, $LiTaO_3$ are used as substrate for surface acoustic wave (SAW). Surface acoustic wave propagation is also studied in PZT (Lead zirconate titanate) thin films.

Pyroelectric detector

$LiTaO_3$, $(Sr,Ba)Nb_2O_6$ are used for heat sensing applications. $PbTiO_3$, $(Pb,La)TiO_3$ and PZT (Lead zirconate titanate) are also studied for heat sensing applications.

Optical memory display

Lead-lanthanum-zirconate titanate (PLZT) thin film is used in optical display application.

Ferroelectric thin film waveguide

$LiNbO_3$, $Li(Nb,Ta)O_3$, Lead zirconate titanate (PZT), Lead lanthanum zirconate titanate (PLZT) are used in optical waveguide applications because of their large electro-optic coefficients.

Ferroelectric memories

Lead zirconate titanate (PZT) is used as ferroelectric memory because of its wide range of switching characteristics. The material having a thickness of 200–300 nm can be operated using 5 V. The advantages of the ferroelectric memories are (i) non-volatile, (ii) radiation hardened compatibility with CMOS and GaAs, (iii) high speed and high density.

Capacitors

$BaTiO_3$ is used as a dielectric material in capacitors, since it has high value of dielectric constant ($\varepsilon_r \approx 15{,}000$) and high volumetric efficiency (capacitance per unit volume).

Piezoelectrics for ultrasound imaging and actuators

The ferroelectrics are used to produce ultrasonic waves and hence the ultrasound image of the organs studied. The basic principle of the ultrasonic imaging is the pulse echo method. A pulse of ultrasonic wave is passed through the human body and it gets reflected back. The reflected signal is used to form the image of the interior parts of our body.

Thermistor

Due to the PTCR (positive temperature coefficient of resistances) properties in polycrystalline form, barium titanate is most often found used as a thermistor, e.g. in thermal switches.

Non-linear optics

Barium titanate is used as non-linear optics material.

Microphone

Barium titanate is used as a piezoelectric material for microphones and transducers.

SOLVED PROBLEMS

14.1 Calculate the relative dielectric constant of a barium titanate crystal, which when inserted in a parallel plate condenser of size 10 mm × 10 mm and distance of separation of 2 mm, gives a capacitance of 10^{-9} F. ($\varepsilon_0 = 8.854 \times 10^{-12}$ F m^{-1})

Given data

Area of the capacitor, $A = 10 \text{ mm} \times 10 \text{ mm} = 10 \times 10^{-3} \times 10 \times 10^{-3} \text{ m}^2$

Distance of separation, $d = 2 \text{ mm} = 2 \times 10^{-3}$ m

Capacitance, $C = 10^{-9}$ F

Solution

Capacitance of the capacitor, $C = \dfrac{\varepsilon_0 \varepsilon_r A}{d}$

$$\text{Dielectric constant } \varepsilon_r = \frac{Cd}{\varepsilon_0 A} = \frac{10^{-9} \times 2 \times 10^{-3}}{8.854 \times 10^{-12} \times 10^{-4}}$$

$$= \frac{2 \times 10^4}{8.854} = 2258.87$$

The dielectric constant of the material is 2258.87.

14.2 The dielectric constant of a helium gas at NTP is 1.0000684. Calculate the electronic polaraizability of the atoms if the gas contains 2.7×10^{25} atoms m^{-3}.

Given data

Dielectric constant of He gas $\varepsilon_r = 1.0000684$

Concentration of dipoles, $N = 2.7 \times 10^{25}$ m^{-3}

Solution

Polarization,
$$P = \varepsilon_0(\varepsilon_r - 1)E$$
$$= 8.854 \times 10^{-12} \times (1.0000684 - 1) \times E$$
$$= 6.056 \times 10^{-16} \times E \text{ C m}^{-2}$$

Electronic polarizability, $\alpha_e = \frac{P}{NE} = \frac{6.056 \times 10^{-16} \times E}{NE}$

$$= \frac{6.056 \times 10^{-16}}{2.7 \times 10^{25}} = 2.243 \times 10^{-41} \text{ F m}^2.$$

Electronic polarizability of He gas, $\alpha_e = 2.243 \times 10^{-41}$ F m^2.

14.3 Calculate the polarization produced in a dielectric medium of dielectric constant 6, when subjected to an electric field of 100 V m^{-1}.

Given data

Dielectric constant, $\varepsilon_r = 6$

Electric field intensity, $E = 100$ V m^{-1}

Solution

Polarization, $P = \varepsilon_0(\varepsilon_r - 1)E$

$$= 8.854 \times 10^{-12} \times (6 - 1) \times 100$$

$$= 4.427 \times 10^{-9} \text{ C m}^{-2}$$

Polarization, $P = 4.427 \times 10^{-9}$ C m^{-2}.

14.4 Calculate the electronic polarizability of neon. The radius of the neon atom is 0.158 nm.

Given data

Radius of neon $R = 0.158$ nm

Solution

Electronic polarizability $\alpha_e = 4\pi\varepsilon_0 R^3$

$$= 4\pi \times 8.854 \times 10^{-12} \times (0.158 \times 10^{-9})^3$$

$$= 4.388 \times 10^{-40} \text{ F m}^2$$

The electronic polarizability, is 4.388×10^{-40} F m^2.

14.5 Metal sheets and mica ($\varepsilon_r = 6$) of thickness 0.002 cm are given. Calculate the area of the metal sheet required to construct a capacitor of capacity 0.02 μF, $\varepsilon_0 = 8.854 \times 10^{-12}$ C^2 N^{-1} m^{-2}.

Given data

Thickness of mica = 0.002 cm

Thickness of metal sheets = 0.002 cm

Capacitance of the capacitor, $C = 0.02$ μF

$\varepsilon_0 = 8.854 \times 10^{-12}$ C^2 N^{-1} m^{-2}.

Solution

Capacitance of the capacitor, $C = \frac{\varepsilon_0 \varepsilon_r A}{d}$

$$0.02 \times 10^{-6} = \frac{8.854 \times 10^{-12} \times 6 \times A}{0.002 \times 10^{-2}}$$

$$A = \frac{0.02 \times 10^{-6} \times 0.002 \times 10^{-2}}{8.854 \times 10^{-12} \times 6}$$

$$= 7.529 \times 10^{-3} \text{ m}^2$$

The area of the metal sheet required = 7.529×10^{-3} m^2 = 75.29 cm^2.

14.6 A crystal is subjected to an electric field of 1000 Vm^{-1} and the resultant polarization is 4.3×10^{-8} C/m^2. Calculate the relative permittivity of the crystal.

Given data

Electric field $E = 1000$ V m^{-1}

Polarization $P = 4.3 \times 10^{-8}$ C m^{-2}

Solution

Polarization
$$P = \varepsilon_0(e_r - 1)E$$
$$4.3 \times 10^{-8} = 8.854 \times 10^{-12}(\varepsilon_r - 1) \times 1000$$
$$\varepsilon_r = 5.856$$

The dielectric constant is 5.856.

14.7 If the relative susceptibility of a material is 4.94 and the number of dipoles per unit volume is 10^{28} m^{-3}, calculate the polarizability of the material.

Given data

Relative susceptibility, $\chi = 4.94$

Number of dipoles per unit volume = 10^{28} m^{-3}

Solution

Polarization,
$$P = N\alpha E$$
Polarization,
$$P = \varepsilon_0 \chi E$$
Equating these two equations, we get
$$\varepsilon_0 \chi E = N\alpha E$$
$$\alpha = \frac{\varepsilon_0 \chi}{N} = \frac{8.854 \times 10^{-12} \times 4.94}{10^{28}} = 43.73876 \times 10^{-40} \text{ F m}^2$$

The polarizability of the material is 43.73876×10^{-40} F m^2.

SHORT QUESTIONS

1. What are dielectric materials?
2. What is an electric dipole?
3. Define the term electric dipole moment.
4. What are electric lines of forces?
5. Define electric flux density.
6. Define electric field strength.
7. Define the term polarization.
8. Define electric susceptibility.
9. What is dielectric constant?
10. Define electric polarizability.
11. What is electronic polarization?
12. What is ionic polarization?

13. What is orientation polarization?
14. What is space-charge polarization?
15. What do you mean by local field in a solid dielectric?
16. What is dielectric loss?
17. What is dielectric breakdown?
18. What is intrinsic breakdown?
19. What is thermal breakdown?
20. What is an electrochemical breakdown?
21. What is discharge breakdown?
22. What is defect breakdown?
23. Mention the applications of dielectric materials in capacitors.
24. Mention the applications of dielectric materials in transformer.
25. What are the ferroelectric materials?
26. Write about the different types of ferroelectric materials.
27. Write about the uses of ferroelectric material.
28. Write about the temperature dependence of polarization.

DESCRIPTIVE TYPE QUESTIONS

1. What do you mean by polarization in solid dielectrics? Discuss the different types of polarization mechanism and hence derive the mathematical expression for the total polarization.
2. What is internal field in solid dielectrics? Derive a mathematical expression for the internal field in solid dielectrics and hence deduce Clausius–Mosotti equation.
3. What is meant by dielectric breakdown? Explain the different types of dielectric breakdowns.
4. What are ferroelectric materials? Explain the properties of ferroelectric material and explain the different types of ferroelectric materials.
5. (i) What is dielectric loss? Discuss dielectric loss.
   (ii) Explain the frequency dependence of polarization in different types of ferroelectrics.

PROBLEMS

1. An insulating material is kept inside the parallel plate capacitor. The distance between the plates is 1 mm and the area of each plate is 10^4 mm^2. The resistivity of the dielectric is 10^{10} Ω m and its dielectric constant is 2.3. Calculate the dielectric loss per unit capacitance for dc voltage of 1 kV. {Hint: The dielectric loss per unit capacitance, $W_p = \frac{V^2}{R_p}\frac{1}{C}$}.

2. An elemental dielectric material has a relative dielectric constant of 12. It also contains 5×10^{28} atoms m^{-3}. Calculate its electronic polarizability, assuming Lorenz field.
3. What is the electric field between the parallel plates of a capacitor if the voltage applied between them is 2 V and the separation between the plates is 0.1 mm?
4. A monoatomic gas contains 3×10^{25} atoms m^{-3} at a certain temperature at one atomic pressure. The radius of the atom is 0.19 nm. What is the relative permittivity of the gas at the given pressure and temperature? What is the polarizability of the atom?
5. The following data refers to a dielectric material $\varepsilon_r = 4.94$ and $n^2 = 2.69$, where n is the index of refraction. Calculate the ratio of the electronic to ionic polarizability for the material.
6. A water molecule has a dipole moment of 6.2×10^{-30} C m. What is the polarization of a water drop of 0.1 cm radius polarized in the same direction?

CHAPTER 15

ADVANCED ENGINEERING MATERIALS

15.1 INTRODUCTION

A number of new engineering materials were discovered by the scientists. These materials have different properties and hence they can be used for different applications. In this section, the new engineering materials such as metallic glasses and shape memory alloys are discussed.

15.2 METALLIC GLASSES

Metallic glasses are a new type of metallic alloys, which have noncrystalline characteristics. The metallic glasses are also known as amorphous metals. Metallic glasses are prepared by the rapid solidification of liquid alloys (cooling rate in the order of 10^5 to 10^6 K s^{-1}). When the liquid alloys are quenched rapidly, it will suppress the growth of the crystalline phase and hence an amorphous metal is formed. The metallic glasses have some unusual electrical, mechanical, optical, magnetic and corrosion properties and hence they are used for different applications.

Glass is said to be a vitrified liquid. In order to form glass, the liquid alloys should be cooled below its thermodynamic melting point so as to prevent the formation of crystalline material. Due to this an amorphous material like glass is formed. Below the melting point an undercooled liquid undergoes the changes in viscosity and heat capacity. Some of the examples of the metallic glasses are $Au_{80}Si_{20}$, $Zr_{70}Pd_{30}$, $Pt_{75}P_{25}$, $Pd_{77}Cu_6Si_{17}$, etc.

15.2.1 Types of Metallic Glasses

The metallic glasses are classified into two classes. They are (i) metal-metalloid and (ii) metal-metal metallic glasses.

The metallic glasses consist of the base metals such as Fe, Ni, Co, Al, Mn, Cr and Cu together with metalloids such as B, C, Si, P, N, Ge and As are said to be metal–metalloid metallic glasses. The percentage of metalloid is about 15 to 30%.

The alloys that contain an early transition metals or rare earth element (e.g. Zr, Ti, Nb, Ta and so on) alloyed with late transition element like nickel, cobalt, iron and palladium and so on is said to be metal–metal metallic glasses. Both classes of alloys yielded metallic glasses that were stable at room temperature and above. The glass transition temperature and the glass crystallization temperature of some metal–metal and metal–metalloid metallic glasses are displayed in Table 15.1.

Table 15.1 Glass transition temperature and glass crystallization temperature of some metal–metal and metal-metalloid metallic glasses

Alloy	*Glass transition temperature, T_g in* K	*Glass crystallization temperature, T_C in* K
$Au_{81}Si_{19}$	292	320
$Au_{55}Pb_{22.5}Sb_{22.5}$	312.9	337.3
$Ni_{80}P_{20}$	622	640
$Pd_{80}P_{20}$	610	630
$Fe_{80}P_{13}C_7$	705	730
$Fe_{80}B_{20}$ (Metglass 2605)	>713	713
$Zr_{35}Cu_{65}$	781	815
$Zr_{50}Cu_{50}$	707	755.5
$Zr_{72}Ni_{28}$	642	671
$Zr_{60}Ni_{40}$	713	751
$Zr_{36}Ni_{64}$	834	864

15.2.2 Preparation of Metallic Glasses

The metallic glasses are prepared using a number of methods such as gun quenching, piston and anvil technique, melts spinning, thin film techniques like thermal evaporation (such as using electron beam guns), sputtering, deposition of metallic ions from an electrolytic solution (electroless and electrodeposition). The piston and anvil method and melt spinning techniques are discussed in this section.

Piston and anvil method

The experimental set-up for the piston and anvil technique is illustrated in Figure 15.1. The apparatus consists of an anvil A, which is supported by a pneumatic cushion, J. A piston, B is supported by a latch, F. The anvil and piston are kept inside a chassis, C. The required composition of the metal–metal or metal–metelloid is taken in a crucible, D. The crucible is heated by the heating element, *E*.

The sample is melted and a droplet is ejected and it is made to fall between the faces of anvil and piston. The piston is pneumatically accelerated and the droplet is struck by the face of the piston and it carries the droplet to the anvil. When the moving piston strike the anvil, the droplet gets spread into a thin layer. The droplet immediately solidifies by the conduction of heat to the anvil and the piston. Using this method, a cooling rate of 10^5 to 10^6 K s^{-1} is achieved. The sample obtained using this method is a foil of uniform thickness ranging from 30 μm to 50 μm. In Figure 15.1, G is a light source and H is a photocell and timing circuit, which is used to detect the flow of the droplet and hence the latch, F is released and the piston will move.

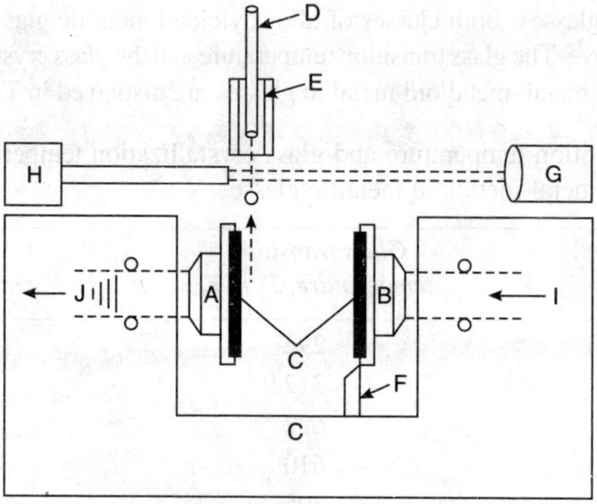

Figure 15.1 Piston and anvil technique.

Melt spinning

Melt spinning apparatus, shown in Figure 15.2, consists of a disc (rotating roller), generally made up of copper, is rotated at very high speed so as the rim will generate a speed of nearly 50 ms^{-1}.

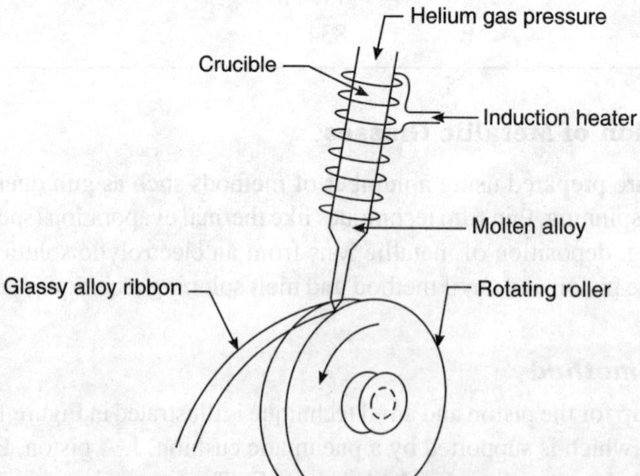

Figure 15.2 Melt spinning apparatus.

The required composition of the (metal–metal or metal–metalloid) material is taken in the crucible. The crucible has a nozzle at its base. The materials are heated slightly above (i.e. nearly 25° to 50°C) their melting point by an induction heater so as to get the liquid alloys. After the uniform composition is obtained, the liquid alloy is allowed to flow through the nozzle. For this purpose, an inert gas like argon or helium is passed into the crucible from its upper portion. When the liquid alloy is flowing out, it is made to fall on the rotating copper disc. Since, copper is a good conductor of heat, the liquid alloys solidifies when it comes into contact with the disc. The metallic glasses are formed continuously in the form of a ribbon.

Thermal evaporation

The metallic glasses are also produced using thermal evaporation of simple metals (Sn, Pb and so on) and the simple alloys of simple metals (Sn-Cu, Pb-Cu, and so on). The thermally evaporated metallic vapours are made to deposit on a cryogenically cooled substrate ($T < 10$ K). The ultralow temperature produces amorphous materials.

15.2.3 Properties of Metallic Glasses

Electronic property

1. The Hall coefficient of metallic glasses is found to be both positive and negative sign. The positive Hall coefficient is attributed to the hole-like conduction mechanism, whereas the negative Hall coefficient is interpreted due to electron-like conduction mechanism.
2. Metallic glasses have positive value of magnetoresistivity (the increase of the resistivity with the application of the magnetic field).
3. The thermopower of the metallic glasses varies linearly with the temperature. At low temperature, there is some anomalies in the thermopower of the metallic glasses has been observed and it is due to the electron–photon scattering and electron localization effect.
4. Metallic glasses exhibit electrical transport properties that are the characteristics of metals. The electrical resistivity varies from 50 μΩ cm to 250 μΩ cm at ambient temperatures.
5. For metals and alloys ρ decreases rapidly with the decrease in temperature. But for metallic glasses, ρ varies very little with temperature. The resistivity of some metallic glasses increases with temperature and in others decreases with temperature. The temperature coefficient of resistivity varies from -2×10^{-4} to 2×10^{-4}.

Magnetic property

1. The metallic glasses have ferromagnetic property. The metallic glass $Fe_{78}P_{12}C_{10}$ undergoes ferromagnetic transition at temperature near 400°C and develops a spontaneous magnetization.
2. The Curie temperature of the metallic glasses is somewhat lower than that of the crystalline transition metals. This reduction in the Curie temperature is due to the disorder in the local atomic environment in metallic glasses.
3. The saturation magnetic moment of ferromagnetic metallic glasses varies in a systematic way with the valency of the transition metal component, reaching a maximum value of 2 Bohr magneton per transition metals atom in Fe-Co base metal–metalloid glasses.
4. The saturation magnetizations are somewhat lower than those of the corresponding crystalline alloys. This reduction in the saturation magnetization is due to the disorder in the local atomic environment in metallic glasses.

5. The ferrous group metal–metalloid glasses have intrinsic magnetic anisotropy.
6. The ferrous group metal–metalloid glasses have very low intrinsic coercive force. This property has made the metallic glasses to be used for soft magnetic materials.
7. Since the metallic glasses are amorphous materials, the electrical conductivity of these materials is very poor and hence the eddy current loss is low. The combination of low coercivity and poor conductivity is ideal for applications in ac transformers.
8. The amorphous materials (metallic glasses) exhibit magnetostriction effect as the crystalline materials. The magnitude of the magnetostriction effect exhibited by the metallic glasses is comparable with those observed in crystalline materials.

Structural properties

The metallic glasses are amorphous materials. It has short range order of atomic ordering.

Thermodynamic property

1. The viscosity (η) of the undercooled liquid increases rapidly with falling temperatures. For a typical liquid, the viscosity is measured in centipoises. When the temperature of the liquid is reduced, the viscosity increases up to 10^{16} poise, a value generally taken to indicate the solid. The crystallization of metallic glasses gives the direct observation of the glass transition temperature (Figure 15.3).

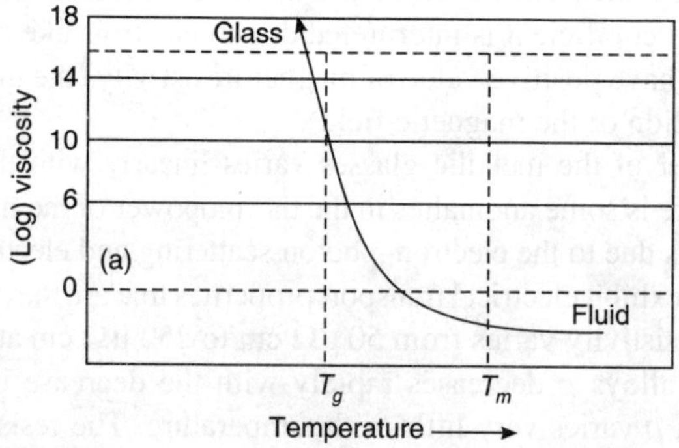

Figure 15.3 Variation of viscosity with temperature.

2. The heat capacity of the metallic glass shows an anomaly. The maximum value of the heat capacity (or the rate of change of viscosity is maximum) corresponds to the glass transition temperatures (Figure 15.4).

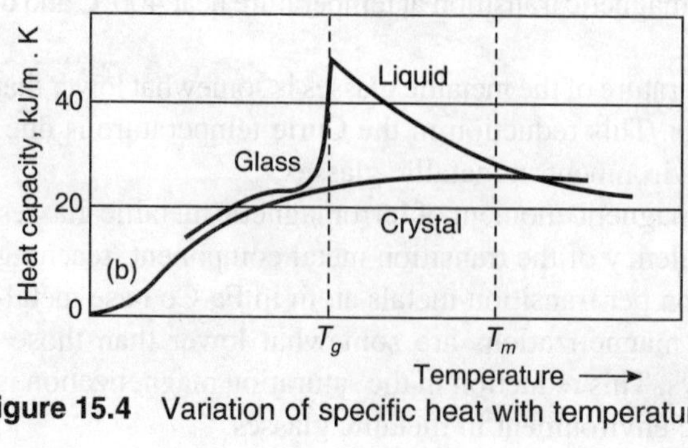

Figure 15.4 Variation of specific heat with temperature.

Superconducting property

1. The metallic glass $La_{80}Au_{20}$ exhibits superconducting property with 3.6 K as transition temperature. Metal–metalloid glasses based on ruthenium, rhenium and niobium were found to exhibit superconducting property.
2. Coherence length (ξ) of amorphous superconductor is low. It is in the order of 4 nm to 10 nm, but for crystalline material it is 50–100 nm. London penetration depth of amorphous superconductor is quite large. Amorphous superconductors are type II superconductors with small value of lower critical field (H_{C1}) and large value of upper critical field (H_{C2}).

Mechanical property

1. Metallic glasses undergo homogeneous and inhomogeneous deformation under an applied stress.
2. The elastic constants of metallic glasses are found to be similar to crystalline materials. The Young's modulus is somewhat smaller than the corresponding crystalline material. The bulk modulus is slightly reduced. The shear modulus is reduced by 15%.
3. Yield strength (σ_y) and hardness (H) are higher than the corresponding ductile crystalline solids. The yield strength and elastic constants of metallic glasses place them among the strongest known solids.

Chemical properties

1. Certain metallic glasses are found to have excellent corrosion resistance. For example, ferrous metal–metalloid glasses containing modest amount of Cr and Mo (for example, $Fe_{72}Cr_8P_{13}C_7$ and $Fe_{75}Mo_5P_{13}O_7$) are found to have excellent corrosion resistance.
2. Amorphous Fe-Ni metalloid alloys have been used for catalytic synthesis of hydrocarbons by hydrogenation of carbon monoxide.

15.2.4 Applications

Metallic glasses as transformer core materials

The metallic glasses made of ferrous amorphous alloys are soft magnetic materials with low coercivity and high permeability. It can be produced as a uniform sheet up to a thickness of 30–40 μm and width 1 m, essentially up to unlimited length. This property can be used as a core material in power distribution transformer. The distribution of electrical energy to the residence, stores, offices and small industries uses a small step-down transformer. The core of this transformer has to be frequently energized by a 50–60 Hz ac supply. This distribution transformer has a lifetime of nearly 25–40 years. If the core of these distribution transformers is prepared using metallic glasses, the eddy current losses and core losses are minimized and hence a lot of electrical energy is saved.

Coatings

The high yield strength and hardness of many metallic glasses and the observed corrosion resistance of Cr, Mo and other transition elements suggest possible applications in wear and corrosion resistance applications.

Reinforcing fibre

The high yield strength of metallic glasses has led to a number of applications in reinforcing fibres in composite materials.

Some refractory metal–metalloid glasses have tensile yield strength in excess of 7 GPa. Therefore they can be used to compete with the refractory materials.

The incorporations of a few percent volume of metallic glass with concrete increases the overall workload requirement to cause fracture by 100 times.

Golf club heads

Zirconium-based metallic glass is used to manufacture golf club heads.

Consumer products

Vitreloy (41.2% Zr, 13.8%Ti, 12.5%Cu, 10%Ni and 22.5%Be) can be used for watch case to replace Ni and other metals.

The sporting goods such as fishing equipment, hunting bows, guns, scuba gear, marine applications and bicycle frames are made using liquid metals.

Medical

Vitreloy is highly biocompatible and it is ideal for corrosion and wear-resistant medical applications. For example, DePuy Orthopaedics, Inc. is using the material in knee-replacement device.

Vitreloy is also used for pacemaker's castings.

Ophthalmic scalpel blades are prepared using Vitreloy.

Other applications

The high strength, hardness, fracture toughness and fatigue strength of bulk metallic glasses make them ideal for use as optical, die, tool and cutting materials.

By the addition of ceramic as second phase particle, the ductility can be imparted to the metallic glasses. This composite can be used as armour penetrator material. It is possible to envisage structural materials for use in aircraft frames, automobiles and medical implants.

The bulk metallic glasses (BMG) have large supercooled liquid regions in which the workability of the material is very high. This property has been exploited in friction welding of Pd-based bulk metallic glass.

The edged tools like knives and blades are prepared using Vitreloy.

15.3 SHAPE MEMORY ALLOYS

A group of metallic materials that demonstrate the ability to return to some previously defined shape or size when subjected to appropriate thermal procedure are known as **shape memory alloys (SMA)**. The shape memory alloys are also known as a **smart metal**, **memory alloy**, or **muscle wire.** The shape memory transformation was first observed by A. Ölander in 1932 in AuCd alloy. The transformation was seen in brass (copper–zinc) in 1938. However, the potential practical uses of shape memory alloys commence in 1962, after the discovery of shape memory effect in Ni-Ti by Buehler and co-workers. The alloys like Ag-Cd, Au-Cd, Cu-Al-Ni, Cu-Zn, Ni-Ti, Cu-Sn, Ni-Al, Cu-Zn-X (where X=Si, Sn, Al) exhibit the shape memory effects.

15.3.1 Two Different Phases

The shape memory alloys are found to have two different phases, namely (i) martensite phase and (ii) austenite phase. The martensite is a low temperature phase. It is a soft and easily deformable phase. The austenite phase is a stronger phase of shape memory alloy. It occurs at high temperature. The austenite phase has cubic structure. The martensite and austenite phases are shown in Figure 15.5.

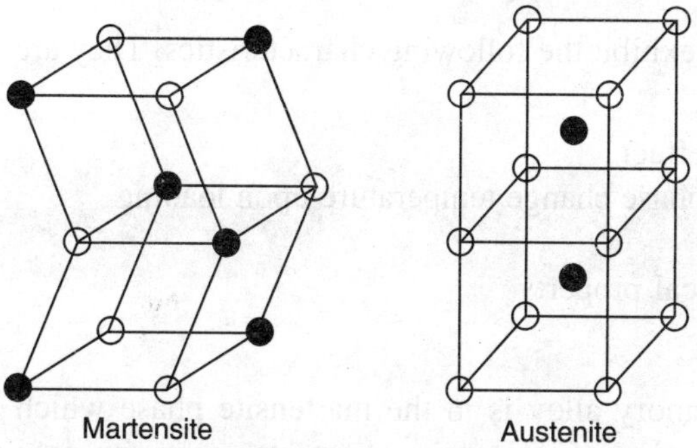

Figure 15.5 Martensite and austenite phase.

The macroscopic and the microscopic view of the shape memory alloys are shown in Figure 15.6. The macroscopic view of austenite and twinned martensite has no difference in its shape, whereas differences can be observed only in the microscopic scale. The austenite is a cubic, whereas the martensite is a twinned structure. Martensite is easily deformed by applying stress. The deformed martensite is shown in Figure 15.6(c).

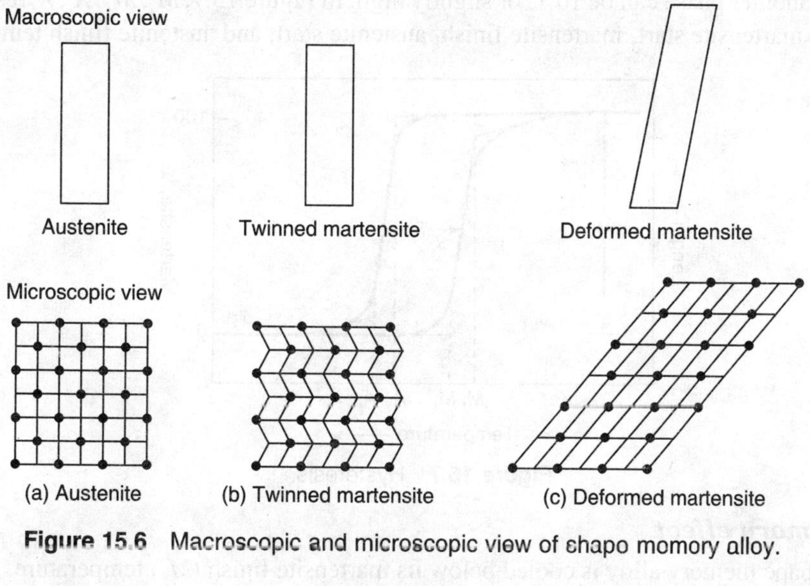

Figure 15.6 Macroscopic and microscopic view of chapo momory alloy.

15.3.2 Types of Shape Memory Alloys

There are two types of shape memory alloys. They are (i) one-way shape memory alloy and (ii) two-way shape memory alloy. Materials that exhibit the shape memory effect only upon heating is said to be **one-way shape memory** alloy. Materials that are exhibiting the shape memory effect both during heating and cooling are said to be **two-way shape memory alloy**.

15.3.3 Characteristics of Shape Memory Alloys

The shape memory alloys exhibit the following characteristics. They are

1. Hysteresis
2. Shape memory effect
3. Dependence of phase change temperature upon loading
4. Pseudoelasticity
5. Thermomechanical property

Hysteresis

Consider that a shape memory alloy is in the martensite phase which is subjected to thermal procedures. The transformation of martensite phase into austenite does not take place at single temperature. It takes place over a range of temperatures.

The transformation of martensite to austenite phase takes place somewhat at higher temperature during heating. During cooling the transformation from austenite to martensite takes place at some lower temperature. The transformation temperature from one phase to another, upon heating and cooling, do not overlap and it exhibits a curve known as hysteresis. The term hysteresis is defined as the difference between the temperature at which 50% transformed to austenite upon heating and 50% transformed to martensite upon cooling. The temperature difference for the transformation of one phase to another phase can be 10°C or slightly high. In Figure 15.7, M_s, M_f, A_s, A_f respectively represent the martensite start, martensite finish, austenite start, and austenite finish temperatures.

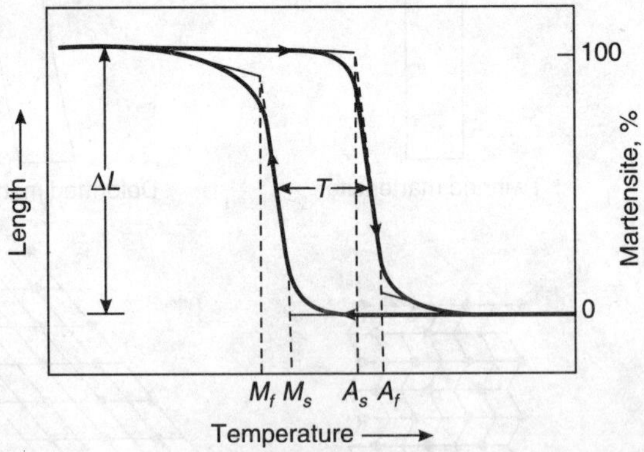

Figure 15.7 Hysteresis.

Shape memory effect

Consider a shape memory alloy is cooled below its martensite finish (M_f) temperature. The SMA is now in the martensite phase. If the SMA in the martensite phase is deformed by applying a stress, a deformed martensite is obtained. Upon heating, the deformed martensite gets converted into austenite. If the SMA is cooled in the austenite phase, twinned martensite phase is obtained. This behaviour of shape memory alloy is said to be shape memory effect (Figure 15.8).

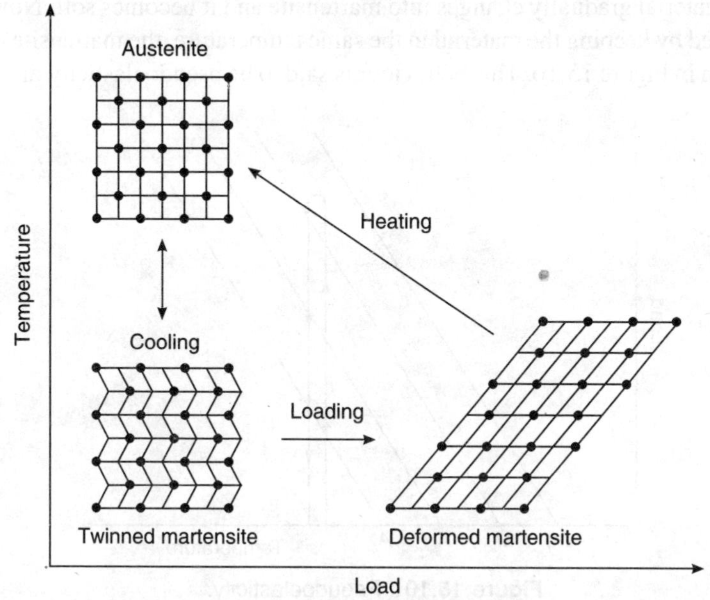

Figure 15.8 Shape memory effect.

The dependence of phase change temperature upon loading

Consider that M_s, M_f, A_s, A_f respectively represent the martensite start, martensite finish, austenite start, and austenite finish temperatures. These phase change temperatures vary linearly upon loading. The initial value of these four temperatures is also affected by the composition of the material (Figure 15.9).

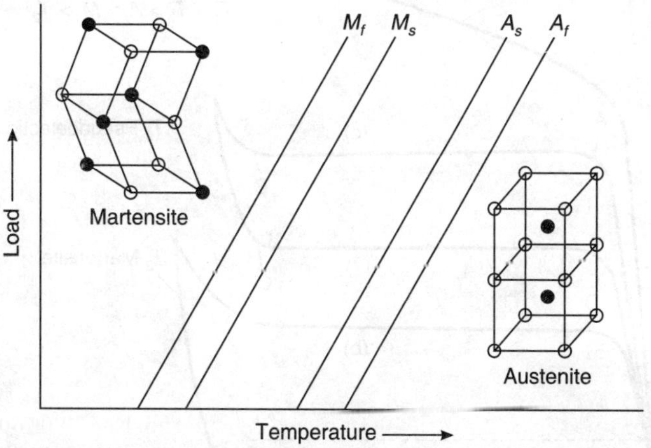

Figure 15.9 Dependence of phase change temperature upon loading.

Pseudoelasticity or super elasticity

Consider that a SMA is in the austenite phase at a temperature greater than A_f. Consider that a load is gradually increased by keeping the shape memory alloy in its austenite phase and its temperature is constant. The material gradually changes into martensite and it becomes soft. Now, if the load is gradually decreased by keeping the material in the same temperature, the martensite phase becomes austenite as shown in Figure 15.10. This behaviour is said to be pseudoelasticity or super elasticity.

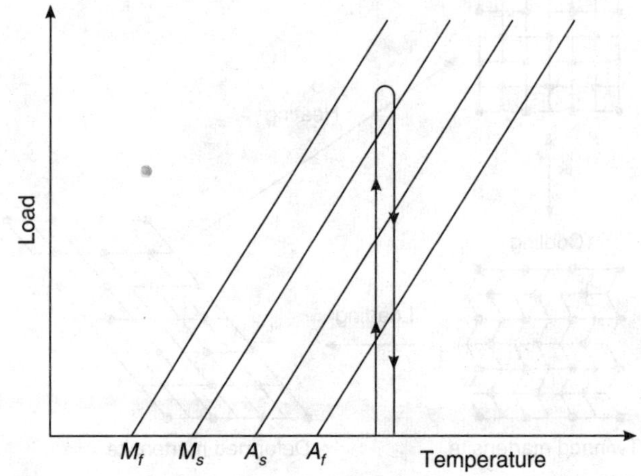

Figure 15.10 Pseudoelasticity.

Thermomechanical property

The mechanical properties of the SMAs vary greatly with temperature. Consider stress is applied and strain is measured by keeping the SMA at a temperature T_2, which is less than the martensite

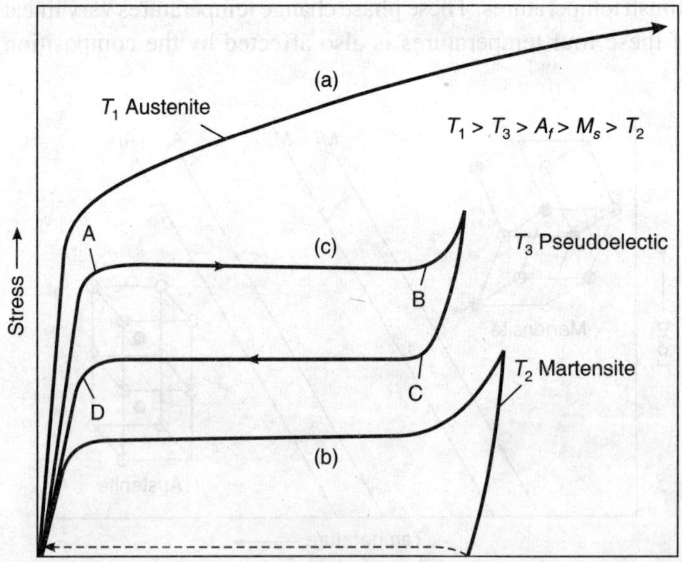

Figure 15.11 Thermomechanical property.

start (M_s) temperature. At very low stress the martensite phase is deformed as shown in Figure 15.11(b). If the stress is removed, initially the strain does not become zero. The dashed line represents the SMA is remembering its phase after removing the stress even though it is kept in the same temperature, T_2. Figure 15.11(a) indicates that the SMA is maintained at a constant temperature, T_1, which is greater than A_f. Upon loading the strain increases and the SMA does not return back to its original phase even after removing the stress by keeping it at a temperature T_1. This indicates that the SMA is in austenite phase and it is one of a rigid phase. Consider that the SMA is kept at a temperature T_3 in such a way that $T_1 > T_3 > A_f$. If stress is increased, a straight line portion, AB, as shown in Figure 15.11(c) is obtained. By keeping the SMA in the same temperature, if stress is decreased, it reverts to austenite at lower stress, as seen in line CD and the shape recovery occurs. This shows the elastic nature of SMA and this property is called pseudoelasticity.

15.3.4 Characterization of Shape Memory Alloys

The shape memory alloys are characterized using the following techniques. They are (i) differential scanning calorimeter, (ii) resistivity measurement, (iii) transformation of SMA with temperature by applying a constant stress and (iv) tensile test.

Differential scanning calorimeter

A small quantity of the sample of the material is heated and cooled through the transformation temperature range. The endotherm and exotherm peaks are identified and hence the heat absorbed or given off by the sample is measured. From this measurement the phase change is identified. This method is not an accurate one because it uses a small quantity (few grams) of the sample in the unstressed condition.

Resistivity measurement

The resistivity of the sample is measured during heating and cooling. The changes and the peaks obtained in the resistivity value in the transformation range are used to identify the phase change. However, the correlation between the measured resistivity changes and the mechanical properties or the measured phase changes has not always been successful.

The transformation of SMA with temperature by applying constant stress

In this method, a constant stress is applied to an appropriate sample of SMA. The temperature of SMA is varied and the corresponding value of strain is measured both during heating and cooling. A curve as shown in Figure 15.7 is obtained. From this measurement the value of martensite start (M_s), martensite finish (M_f), austenite start (A_s) and austenite finish (A_f) are determined. It gives slightly higher M_f, M_s, A_s, A_f values than DSC method, because in DSC method no stress is applied.

Tensile test

By applying tensile stress at different temperatures, the strain is measured. The temperature is varied across the transformation temperature range. From this measurement the transformation temperature values can be determined. This method is very imprecise. However, this method is used to measure the changes in properties of each phase due to work hardening or different heat treatment.

15.3.5 Commercial Shape Memory Alloys

The shape memory effect was observed in many alloys. But only two different types of alloys are mostly exploited for commercial applications. They are (i) Ni-Ti alloy (ii) Cu-based SMA. Let us discuss only about the Ni-Ti alloy in this section.

Ni-Ti alloy

The most common shape memory material is an alloy of nickel and titanium called Nitinol. It has very good electrical and mechanical properties, long fatigue life, and high corrosion resistance.

Synthesis

Solid Ni-Ti alloys are manufactured by a double vacuum melting process, to ensure the quality, purity and properties of the material. After the formulation of raw materials, the alloy is vacuum induction melted (1400°C). After the initial melting, the alloy transition temperature must be controlled due to the sensitivity of the transition temperature to small changes in the alloy composition. This is followed by vacuum arc remelting to improve the homogeneity and structure of the alloy. Double-melted ingots can be hot-worked (800°C) and cold-worked to a wide range of product sizes and shapes.

Properties of Ni-Ti alloy

1. Ni-Ti alloy is a binary, equiatomic intermetallic compound. It has a moderate solubility. The solubility allows alloying with many of the elements to modify its mechanical properties.
2. It exhibits ductility as comparable to most of the ordinary alloys.
3. Machining by turning or milling is very difficult except with special tools and practices.
4. Welding, brazing or soldering of the alloy is very difficult. Grinding and shearing or bunching can be done if the thickness is small.
5. The heat treatment to impart the desired property is often done at 500°C to 800°C, but it can be done at 300°C to 350°C, if sufficient time is allowed.

Some of the mechanical properties are displayed in Table 15.2.

Table 15.2 Mechanical properties of Ni-Ti alloy

Properties	*Values*
1. Melting temperatures	1300°C
2. Density	6.45 g cm^{-3}
3. Resistivity	
(i) Martensite	~70 μΩ cm
(ii) Austenite	~100 μΩ cm
4. Thermal conductivity	
(i) Martensite	8.5 W m^{-1} $°C^{-1}$
(ii) Austenite	18 W m^{-1} $°C^{-1}$
5. Corrosion resistance	Similar to 300 series stainless steel or Ti alloys

(Contd.)...

Table 15.2 Mechanical properties of Ni-Ti alloy

Properties	*Values*
6. Young's modulus	
(i) Martensite	~28–41 GPa
(ii) Austenite	~83 GPa
7. Yield strength	
(i) Martensite	70–140 MPa
(ii) Austenite	195–690 MPa
8. Ultimate tensile strength	895 MPa
9. Transformation temperatures	–200°C to 110°C
10. Latent heat of transformation	167 kJ/kg atom
11. Shape memory strain	8.5% maximum

15.3.6 Applications of Shape Memory Alloys

Superelastic applications

Eye glass frames use the superelastic Ni-Ti to absorb large deformations without damaging the frames.

Arch wires for orthodontic correction using N-Ti to give large rapid movement of teeth.

Guide wires for steering catheters into vessels in the body have been developed using Ni-Ti.

Force actuators

Cu-Zn-Al actuator is used to shut off toxic or flammable gas flow when fire occurs.

Cryofit hydraulic coupling

The cryofit fittings are manufactured as cylindrical sleeves slightly smaller than metal tubings they are to join. These diameters are expanded while in the martensite, and upon warming to austenite, their diameter shrinks and strongly holds the tube ends. This coupling produces a better joint and it is superior to a weld.

Blood clot filter

The Ni-Ti wire is shaped to anchor itself in a vein and catch passing clots. The part is chilled so it can be collapsed and inserted into the veins, then body heat is sufficient to turn the part in the functional shape.

Biocompatibility

Nitinol (Ni-Ti alloy) is highly biocompatible and has properties suitable for use in orthopaedic applications.

Temperature control system

Ni-Ti alloy can be used as a temperature control system, as it changes shape, it can activate a switch or a variable resistor to control the temperature.

Novelty products

Ni-Ti alloy is used to produce self-bending spoons which can be used by amateur and stage magician to demonstrate 'psychic' powers or as a practical joke, as the spoon will bend itself when used to stir tea, coffee, or any other warm liquid.

Antenna

Ni-Ti alloy is used in cell-phone technology as a retractable antenna due to its highly flexible and mechanical nature.

Tumour identification

Ni-Ti alloy can also be used as wires which are used to locate and mark breast tumours so that following surgery can be more exact.

15.4 NANOMATERIALS

The materials those having the particle size less than 100 nm are called nanomaterials. The value of 1 nm is equal to 10^{-9} m. When the particle size gets reduced, most of the physical properties of the nanomaterials get changed. Due to the change in the physical property, they have various applications that are different from those of the bulk materials, and hence the study of the nanomaterials becomes important.

The nanoscience and nanotechnology primarily deal with the synthesis, characterization, exploration and exploitation of nanostructured materials. The individual nanostructures include clusters, quantum dots, nanocrystals, nanowires and nanotubes. The collection of nanostructures involve arrays, assemblies and superlattices of the individual nanostructures.

15.4.1 Top-down and Bottom-up Process

The nanomaterials are synthesized using a number of methods. These methods are generally classified into two categories, namely

1. top-down process and
2. bottom-up process

Top-down process

To synthesise a nanomaterial, if a bulk material is used as a starting material then the method is known as top-down process. In the top-down process, a bulk material is crushed into fine particles using the process like mechanical alloying, laser ablation, sputtering, etc. The examples of the top-down process are

1. Ball milling
2. Laser ablation
3. Sputtering
4. Arc plasma
5. Electron beam evaporation
6. Photolithography, etc.

Bottom-up process

In some methods, the nanomaterials are prepared by arranging atom by atom. Due to the nucleation and growth, bigger size grains or a cluster of atoms having a size less than 100 nm are produced. The examples of the bottom-up process are

1. Chemical vapour deposition
2. Sol-gel method
3. Electrodeposition, etc.

These methods produce the nanomaterials due to some chemical reactions and hence these methods are called chemical methods. The schematic representation of these two processes is illustrated in Figure 15.12.

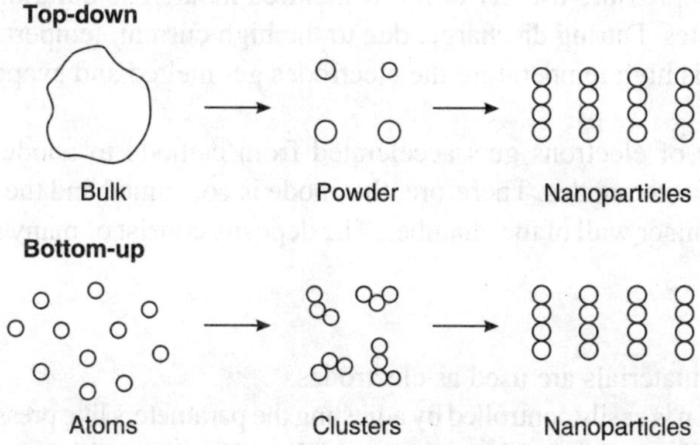

Figure 15.12 Top-down and bottom-up process.

Let us discuss the synthesis of nanomaterials using arc plasma, laser ablation, sol-gel method, electrodeposition, ball milling, and chemical vapour deposition.

15.4.2 Synthesis of Nanomaterials

Arc discharge

The experimental set up consists of a vacuum chamber, gas flow controls and two electrodes with power supply. The experimental arrangement is shown in Figure 15.13. The reactant materials are

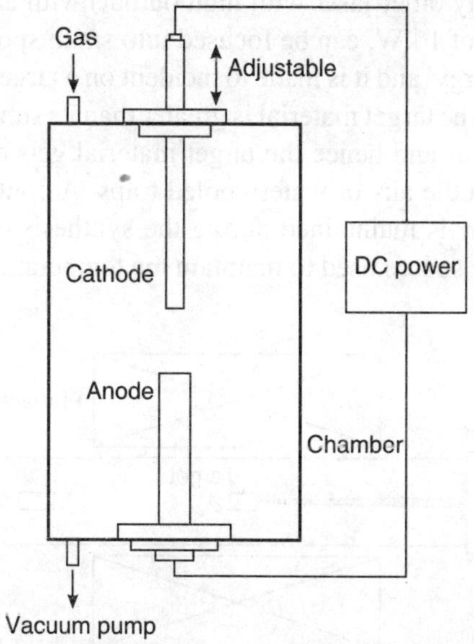

Figure 15.13 Arc discharge method.

compressed into rods and they are used as anode and cathode. The anode and cathode are separated by a distance of nearly 1 mm. A dc voltage of nearly 20 to 40 V with a current rating of 100 to 150 A is applied to the electrodes to generate an arc. The chamber is filled with inert gases such as argon or helium with a pressure usually of a few hundred mbar. The duration of the discharge is normally several minutes. During discharge, due to the high current, temperature around 3000°C is produced. Due to this high temperature the electrodes get melted and evaporate into clusters at atomic scale.

A large quantity of electrons gets accelerated from cathode to anode because of the arc discharge and collides with anodes. Therefore, the anode is consumed and the deposits are formed on the cathode and the inner wall of the chamber. The deposits consist of many nanosized structures including nanotubes.

Advantages

1. Conducting materials are used as electrodes.
2. The deposition is easily controlled by adjusting the parameters like pressure and temperature.
3. This method is used to produce nonmaterials in large quantities.
4. The nanomaterials with perfect structure are produced.
5. Both single wall and multiwall nanotubes can be produced.

Drawbacks

1. High temperature around 3000°C should be maintained in between the two electrodes.
2. Pressure around 400–700 mbar should be maintained.

Laser ablation technique

The laser ablation technique, shown in Figure 15.14, consists of a high intense laser beam, a cylindrical furnace, an air- or water-cooled metallic trap with filler, a vacuum pump and an inlet for the gas. A CO_2 laser (or any other laser with high output) with an output wavelength of 10.6 micrometer and power output of 1 kW, can be focused into small spot in the order of a mm using a lens produces a very high energy, and it is made to incident on a target material. If the temperature produced by the laser beam on the target material is greater than its sublimation point, the explosion of the target material may occur and hence the target material gets efface from the surface. The nanomaterials are collected on the air- or water-cooled traps. A continuous flow of nitrogen gas with a pressure of nearly 1 torr is maintained during the synthesis of the nanomaterials. A tube furnace is an additional energy source used to maintain the temperature of the target. This method is known as laser ablation.

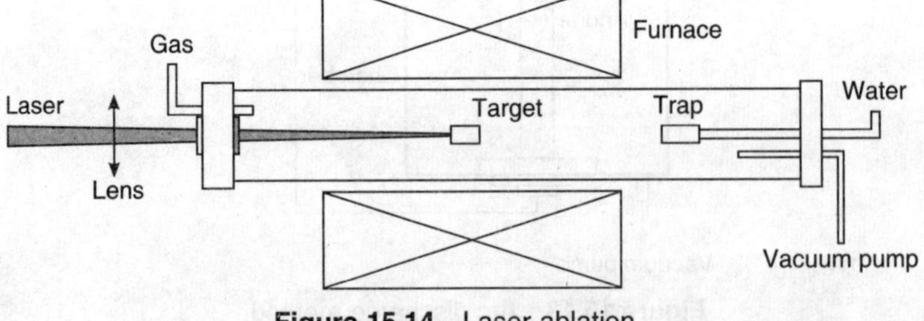

Figure 15.14 Laser ablation.

If the temperature produced by the laser beam on the target material is less than its sublimation point, the laser radiation heats the target material without the obvious ejection of the target material. This method is said to be laser heating. Both laser ablation and laser heating can be used for producing nanomaterials.

Advantages

1. The nanomaterials with perfect structures can be produced.
2. Both single wall and multiwall nanotubes can be produced.
3. The process is easily controllable by controlling the temperature, pressure and laser output power.

Drawbacks

1. High temperature should be maintained
2. Laser source with high output energy should be needed.
3. Only a small quantity of nanomaterials can be produced.

Ball milling and annealing

This method is a two-step process. The first step is ball milling at room temperature and another step is annealing at relatively low temperatures. These two processes correspond to separate nucleation and growth process.

A ball mill consists of a stainless steel container and several iron balls or silicon carbide balls. The material powder is taken in the ball milling chamber in addition to the several hundreds of silicon carbide balls. An external magnet introduces the pulling forces to the materials and hence it increases the milling energy. When the milling chamber rotates the balls, they drop to bottom due to the gravitational force and due to the magnetic force they reach the top of the chamber and hence the balls move continuously as shown in Figure 15.15. The silicon carbide balls impart very high amount of energy to the material powders and hence the powders get crushed. The milling is to be done nearly 100 h to 150 h to get uniform fine powders. After preparing fine powders, annealing is done at a suitable temperature for a certain period of time so as to get nanotubes.

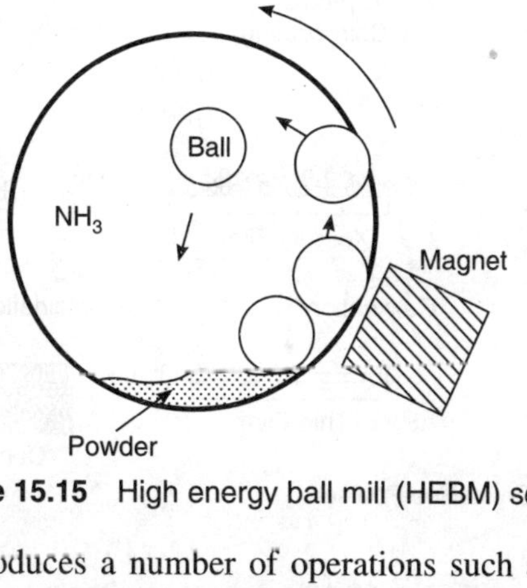

Figure 15.15 High energy ball mill (HEBM) set up.

The milling process produces a number of operations such as fracturing, grinding, high speed plastic deformation, cold welding, thermal shock, intimate mixing, etc. of the materials. In ball milling, the structural and the chemical changes are produced by the mechanical energy rather than the thermal energy at low temperature (room temperature).

Advantages

1. It is a well-defined, inexpensive process.
2. Nanopowders with particle size 2 to 20 nm in large quantities can be produced.

Disadvantages

1. The shape of the nanoparticles is irregular.
2. This method introduces crystal defects.
3. Introduction of impurities from balls and milling additives.

Sol-gel process

The sol-gel process is a wet-chemical method used to produce high purity oxide nanoparticles such as TiO_2, ZnO, etc. Suitable precursors in the form of nitrates/acetates/carbonates are taken and then dissolved in deionized water. The solutions are kept at suitable temperatures. The starting material is used to produce a colloidal suspension known as sol. Then a gelling agent like PVA (polyvinyl alcohol) is added. This will produce a gel. A thin film coating is made on a substrate such as glass or Ti or Ni sheets depending upon the requirements using sol-gel dip-coating or spin-coating. To make a uniform coating the pH, temperature and viscosity should be controlled. Then the film is annealed at suitable temperature and then studied.

Instead of forming a thin film coating, crystalline powders are also produced by suitably calcining the gel. The sequence of steps involved in a sol-gel process is displayed in Figure 15.16.

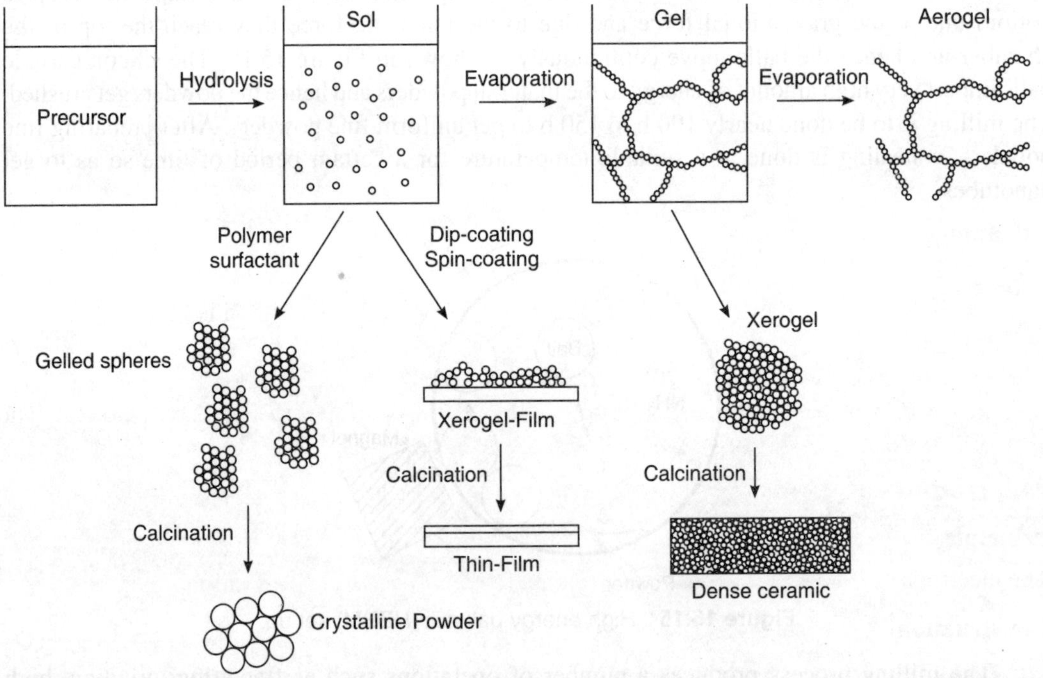

Figure 15.16 Sol-gel process.

Advantages

1. It is a low-cost method used to produce large quantities of nanopowders.
2. The sol-gel process can be done at low temperature.
3. This method produces homogeneous substances.

Chemical vapour deposition (CVD)

Chemical vapour deposition is a chemical method used to produce nanomaterials. The CVD apparatus consists of a quartz tube container, which is heated by a tubular furnace as shown in Figure 15.17.

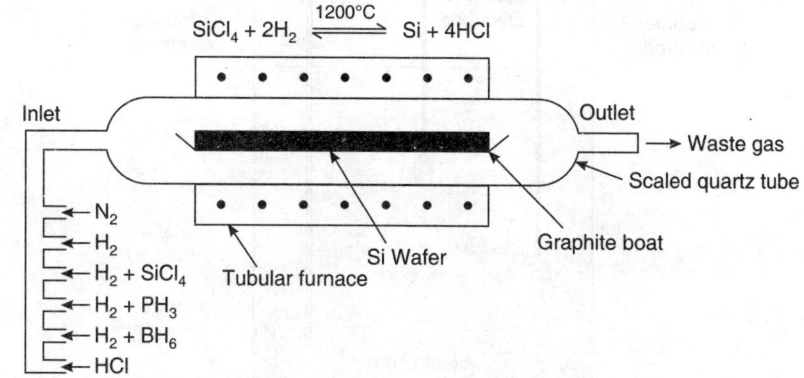

Figure 15.17 Chemical vapour deposition process.

A silicon substrate is kept inside the quartz container in a tungsten/graphite boat. The chamber is evacuated and an inert gas is passed into the chamber. The reactants are admitted into the chamber and the chamber is maintained at a suitable temperature. The coating is formed on the silicon substrate because of the chemical reaction. The unused gases flow through the outlet.

The chemical reaction, the hydrogen reduction of silicon tetrachloride, is used to produce the epitaxial growth of pure silicon. This reaction takes place at 1200°C. The chemical reaction involved is

$$SiCl_4 + 2H_2 \xrightarrow{1200^\circ C} Si + 4HCl$$

Advantages

1. CVD is a low-cost and high-yield method.
2. Compared to the other methods, the temperature of the deposition is low.
3. High purity (more than 99% pure) nanomaterials are produced using this method.
4. Both single wall and multiwall nanotubes can be produced.
5. The diameter of the nanotube can be controlled by controlling the thickness of the catalytic film.

Electrodeposition

Principle

The electrochemical reaction is the basic principle behind the electrodeposition process.

Construction

The circuit connection is made as shown in Figure 15.18. The electrodeposition set-up consists of a container, in which the electrolytes are taken in the required composition and pH. In electrodeposition process, three electrodes are used, namely working electrode, counter electrode and reference electrode. An inert electrode like Pt is used as counter electrode. The working electrode consists of conducting materials such as ITO (indium tin oxide)-coated glass plates or metals such as Ti or Ni. The standard calomel electrode is used as a reference electrode.

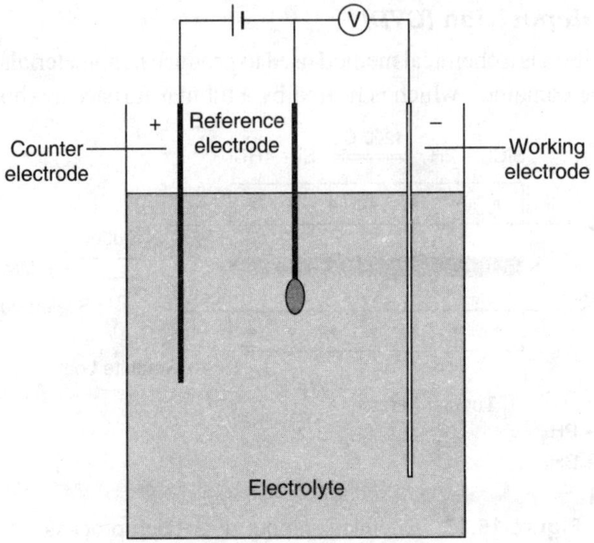

Figure 15.18 Electrodeposition process.

Working

The required electrolytes, with suitable compositions and pH, are taken in the electrodeposition chamber. In order to deposit the nanocrystalline material, a grain refiner should be added into the electrolyte. The set-up is kept in a constant temperature. Suitable potential difference (ac or dc) is applied between the counter electrode and the working electrode as shown in Figure 15.18. The deposition can be made on cathode by maintaining a constant potential difference between the standard calomel electrode and the working electrode. It is said to be potentiostatic method.

The deposition can also be made either in the cathode or anode. If the deposition is made in the cathode, then it is said to be cathodic deposition. If the deposition is made in the anode, then it is said to be anodic deposition. If the current through the circuit is made constant, then it is said to be galvanostatic method.

Advantages

1. It is one of the simplest and inexpensive methods.
2. The thickness of the film can be controlled by adjusting the deposition rate.

Drawbacks

1. Conducting electrodes are needed to produce electrodeposition.
2. After the deposition is made, the films should be annealed at high temperature. High temperature annealing is expensive.

15.4.3 Properties of nanomaterials

The physical, mechanical, optical and electronic properties of the nanomaterials differ from their bulk material. Some of the properties of the nanomaterials are given below:

Physical properties

Melting point

The melting temperature of gold in bulk form is around 1300 K. The melting temperature significantly decreases up to around 700 K, when the particle size is decreased around 10 to 20 nm. The variation of the melting temperature of gold with particle size is displayed in Figure 15.19. The melting temperature of CdS is slightly higher than 1700 K in the bulk form. The melting temperature of CdS reduces to 600 K, when its particle size is around 10 to 15 nm.

Interatomic distance

The interatomic distance gets reduced when there is a reduction in particle size.

Magnetism in nanoparticles

Sputtered alloy films of CoCu, FeCu, and CoAg exhibit giant magnetoresistance (GMR). GMR values as high as 55% at 4.2 K and 20% at room temperature have been observed.

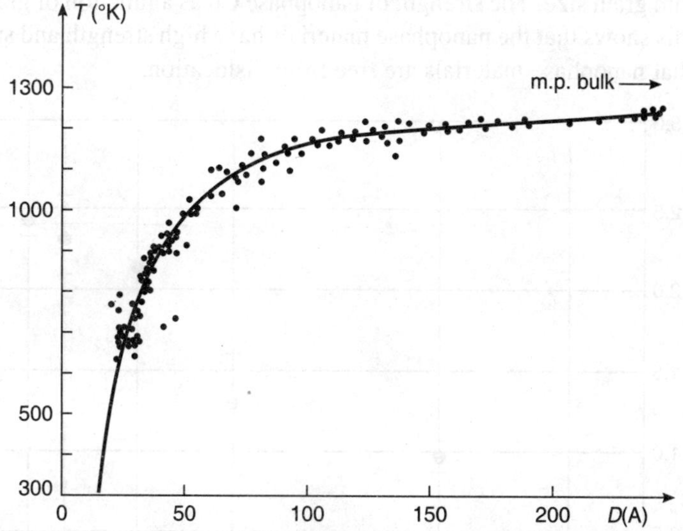

Figure 15.19 Variation of melting temperature of gold with particle size.

Magnetic moment

The magnetic moment of a nanomaterial increases with the reduction in coordination number. This shows that the smaller size particles are more magnetized. The magnetic moment of iron is greater by up to 30% when it is nanoscale than when it is in the bulk form. Clusters with small size particles are spontaneously magnetized. The variation of the magnetic moment with coordination number is displayed in Figure 15.20.

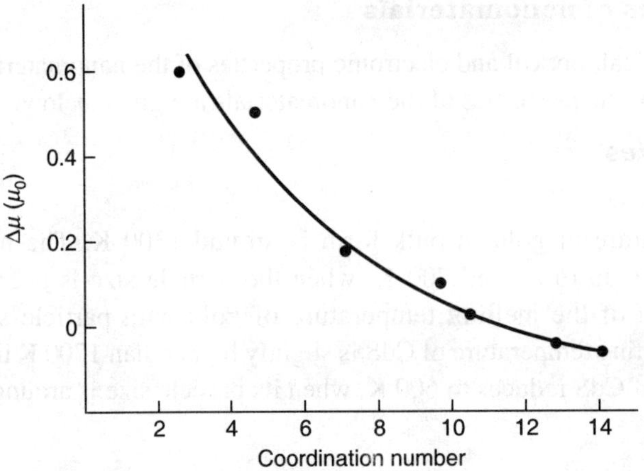

Figure 15.20 Variation of magnetic moment with coordination number.

Mechanical properties

The hardness of a nanomaterial increases with the reduction in particle size. The hardness of Cu gets increased to two times when its particle size is 50 nm and becomes 5 times harder than the bulk material at 6 nm grain size. The strength of nanophase Cu as a function of grain size is shown in Figure 15.21. This shows that the nanophase materials have high strength and super hardness. It is due to the fact that nanophase materials are free from dislocation.

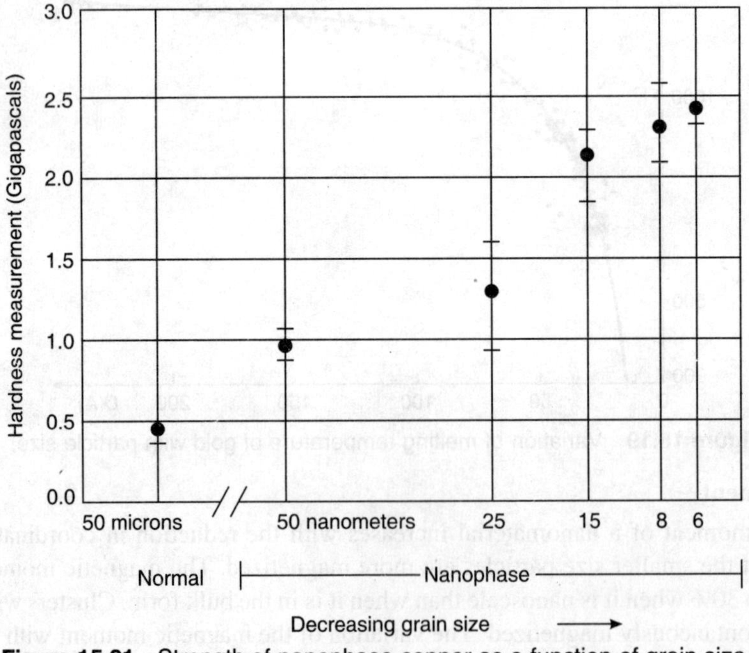

Figure 15.21 Strength of nanophase copper as a function of grain size.

Superelasticity

Superelasticity has been observed in nanophase materials.

Optical properties

Nanoclusters of different particle size exhibit different absorption spectra and they lie in the visible region. Due to this, the nanoscale gold solution with different particle size appears as orange, purple, red or greenish, as per the data available in the literature (Murray, R.W., et al., IBM J. Res. Dev., 45, 47, 2001). The absorption spectra of CdS nanoparticle is shown in Figure 15.22.

Some other properties

Metals show nonmetallic band gaps when the diameter of the nanocrystals is in the range of 1 to 2 nm. Hg clusters show a nonmetallic band gap that decreases with an increase in cluster size. Approximately 300 atoms appear to be necessary to close the gap.

Metal gold nanoparticles of 1 to 2 nm diameter exhibit unexpected catalytic activity.

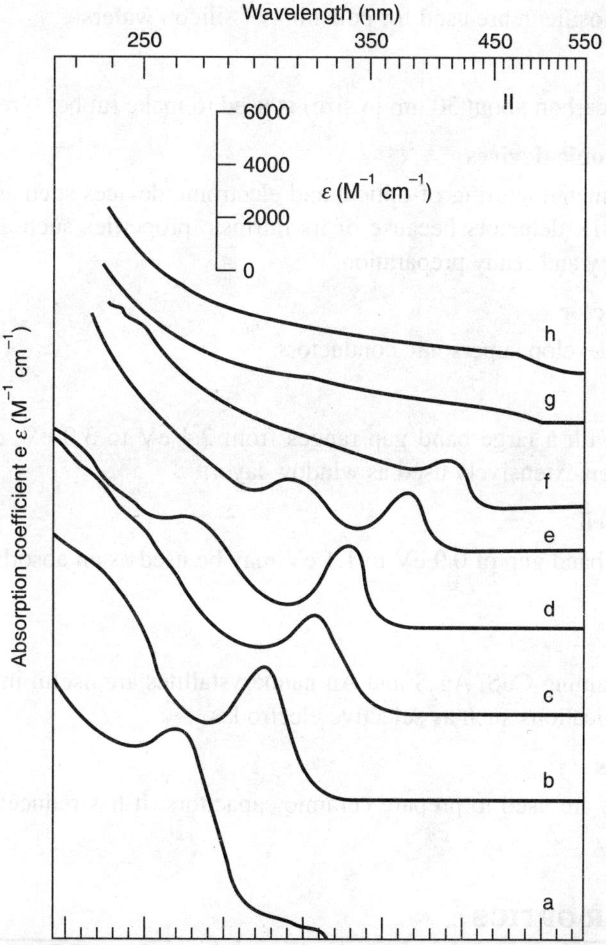

Figure 15.22 Electronic absorption spectra of CdS nanocrystals (a) 64 nm, (b) 72 nm, (c) 0.8 μm, (d) 0.93 μm, (e) 1.94 μm, (f) 2.8 μm and (g) 4.8 μm.

15.4.4 Uses of Nanomaterials

1. Liquid crystal display and cathode ray tube display

Nanophosphors are used to display colours in liquid crystal display and cathode ray tube.

2. Audio/video tape recorders and disk drives

Nano-iron oxide is used as a magnetic material in disk drives and audio/video tape recorder.

3. Blocking of UV rays

Nano-zinc oxide or nanotitania is used in many sunscreens to block harmful UV rays.

4. Catalytic converter

Nanoscale platinum is crucial for the operation of catalytic converters.

5. Polishing of silicon wafer

Nanoalumina and nanosilicon are used for polishing of silicon wafers.

6. Wear resistant

Carbon black (a nanocarbon about 30 nm in size) is used to make rubber tyres wear resistant.

7. Optical and electronic devices

Ag_2S is used for the manufacturing of optical and electronic devices such as photovoltaic cells, photoconductors and IR detectors because of its intrinsic properties such as narrow band gap, good chemical stability and ready preparation.

8. Supersonic conductor

Ag_2S is also used to develop supersonic conductors.

9. Optical window

Sulphides like CdS with a large band gap ranges from 2.4 eV to 3.0 eV, depending upon the particle size, have been extensively used as window layer.

10. Thin film solar cell

Ag_2S with an optical band gap of 0.9 eV to 1.1 eV may be used as an absorber in thin film solar cells.

11. Optical device

Silica thin films containing CuS, Ag_2S and Au nanocrystallites are useful in optical devices and nonconventional applications such as selective electrodes.

12. Electrical devices

$BaTiO_3$ nanopowders are used to prepare ceramic capacitors. It has reduced size and enhanced capacitances.

15.5 NONLINEAR OPTICS

Nonlinear optics is the study of the nonlinear optical properties exhibited by some crystals whenever a laser beam with high intensity is passed through it. When an ordinary light source is used, the optical properties such as refractive index and polarization of the crystal are represented by simple mathematical equations. These ordinary mathematical equations are not adequate whenever a laser beam is passed through the crystal. The laser beam is an electromagnetic radiation consisting of electric and magnetic vectors. The electric field produced by a laser beam varies from 10^6 V cm^{-1} to 10^7 V cm^{-1}. Since it can be focussed into a small area (A $\propto \lambda^2$) of spot size 10^{-12} m^2. The electric field produced by an ordinary light is in the order of 10 V cm^{-1}, whereas the internal field present in an atom is in the order of 10^9 V m^{-1}. The internal electric field is produced by the positive charge of the nucleus. Since the electric field produced by the laser beam is comparable to the internal field of the crystal, some crystals exhibit nonlinear optical properties.

The nonlinear optics is mainly useful in optical communication. Some well-known techniques such as mixing, heterodyning, and modulation can be done in optical frequencies. Therefore, the study of nonlinear optical materials becomes important. This chapter deals with the nonlinear optical properties, nonlinear optical materials and their applications.

15.5.1 Linear and Nonlinear Properties

Linear and nonlinear properties can be illustrated using the following example. Consider a spring is fixed at one end, and it is loaded at the other end as shown in Figure 15.23.

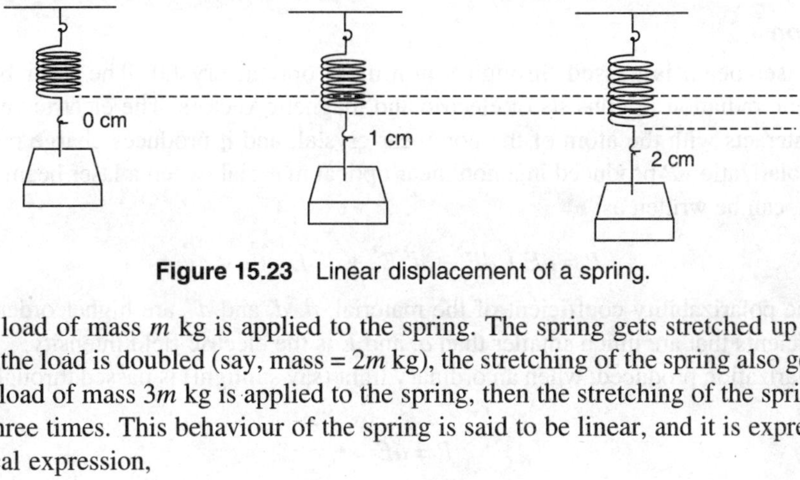

Figure 15.23 Linear displacement of a spring.

Let a load of mass m kg is applied to the spring. The spring gets stretched up to a certain distance. If the load is doubled (say, mass = $2m$ kg), the stretching of the spring also gets doubled. Consider a load of mass $3m$ kg is applied to the spring, then the stretching of the spring also gets increased three times. This behaviour of the spring is said to be linear, and it is expressed by the mathematical expression,

$$F = kx \tag{15.1}$$

where F is the force applied to the spring in the X-direction and k is the proportionality constant known as spring constant. This linear behaviour of the spring is shown in Figure 15.24. If the force is applied to the spring beyond certain limit, the stretching of the spring is not linear and hence the spring is said to be exhibiting nonlinear property. Equation (15.1) is not adequate to represent the nonlinear property of the spring. Equation (15.1) is written as:

$$F = kx + k'x^2 + k''x^3 + k'''x^4 + \cdots \tag{15.2}$$

where k', k'', k''' are the nonlinear coefficients, and they are very low as compared to the spring constant k. The nonlinear property of the spring is shown in Figure 15.24.

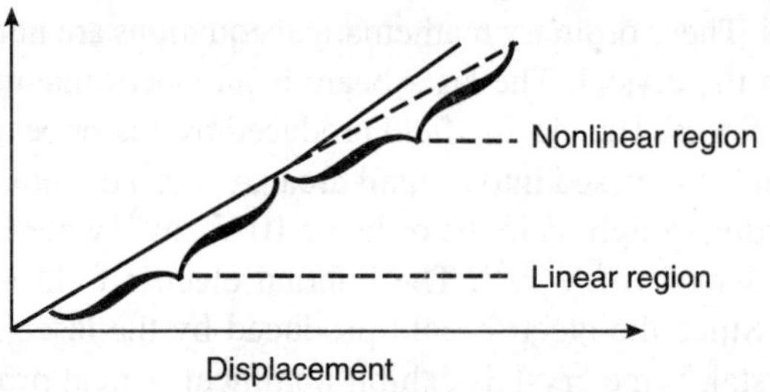

Figure 15.24 Linear and nonlinear behaviour of a spring.

Similarly, whenever an ordinary light beam is passed through some optical crystals, they exhibit linear property. These crystals exhibit nonlinear optical property, if a high intensity laser beam is used as a source.

15.5.2 Properties of Nonlinear Optical Materials

The nonlinear optical materials exhibit the properties such as polarization, frequency doubling, optical mixing, optical rectification, etc. These properties are discussed in this section:

Polarization

Consider a laser beam is passed through a nonlinear optical crystal. The laser beam is an electromagnetic radiation. It consists of electric and magnetic vectors. The electric vector of the laser beam interacts with the atom of the nonlinear crystal, and it produces charge polarization. The charge polarization P produced in a nonlinear optical material, when a laser beam is made to incident on it, can be written as:

$$P = aE + dE^2 + d'E^3 + d''E^4 + \cdots \tag{15.3}$$

where a is the polarizability coefficient of the material, d, d' and d'' are higher order nonlinear optical coefficients that are much smaller than a, and E is the electric field intensity.

The polarization produced, when an ordinary light (say sunlight) is passed through a crystal, is given by

$$P = aE \tag{15.4}$$

where a is the polarizability of the material, and E is the electric field intensity. Equation (15.4) is more accurate when an ordinary light is passed through the crystal. Whenever a laser beam is made to incident on a crystal, the electric field produced by the laser beam is comparable with the internal field of the crystal, and hence the polarization produced by the crystal becomes nonlinear. It is expressed by Eq. (15.3).

Frequency doubling or second harmonic generation (SHG)

The polarization produced, when light is passing through a nonlinear optical material, is given by

$$P = aE + dE^2 + d'E^3 + d''E^4 + \cdots \tag{15.5}$$

By neglecting the higher order terms and considering only the first two terms, Eq. (15.5) can be written as:

$$P = aE + dE^2 \tag{15.6}$$

The electric field produced by the electromagnetic radiation can be written as

$$E = E_0 \cos 2\pi\nu t = E_0 \cos \omega t \tag{15.7}$$

where E_0 is the amplitude and ω is the angular frequency. It is equated to the linear frequency using the relation, $\omega = 2\pi\nu$. Substituting Eq. (15.7) in Eq. (15.6), we get

$$P = a(E_0 \cos \omega t) + d(E_0 \cos \omega t)^2 \tag{15.8}$$

Using the trigonometric relation, $\cos^2\theta = \dfrac{1+\cos 2\theta}{2}$, Eq. (15.8) can be written as:

$$P = a(E_0 \cos \omega t) + \frac{dE_0{}^2}{2}\cos 2\omega t + \frac{dE_0{}^2}{2} \tag{15.9}$$

Equation (15.9) gives the resultant polarization when a light beam is passing through a nonlinear optical material. Equation (15.9) can be written as:

$$P = P(\omega) + P(2\omega) + \text{constant term}, \left(\frac{dE_0^2}{2}\right) \tag{15.10}$$

The first term $P(\omega)$ reveals that the charge polarization is a function of the angular frequency ω. The second term $P(2\omega)$ indicates that the charge polarization is a function of twice the angular frequency 2ω. This doubling of frequency is called frequency doubling or second harmonic generation (SHG). The third term $dE_0^2/2$ is independent of frequency. It provides DC charge polarization. These three charge polarizations are shown in Figure 15.25. Equation (15.10) shows that whenever light is passing through a nonlinear material, it produces two types of oscillating charge polarization, one with frequency ω and another with frequency 2ω in the optical frequency range (10^{14} Hz to 10^{16} Hz). This phenomenon was first observed in 1961 by Professor Peter Franken and some graduate students at the University of Michigan.

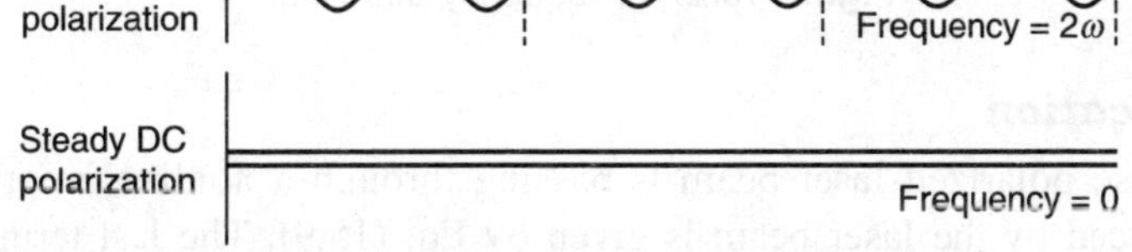

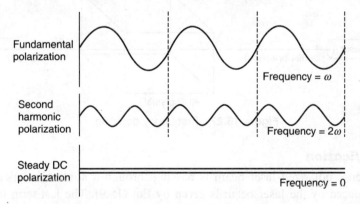

Figure 15.25 Frequency associated with space-charge polarization.

The experimental arrangement for observing the frequency doubling is shown in Figure 15.26. A Q-switched ruby laser emits a laser beam of wavelength 694.3 nm. The filter is used to remove the unwanted wavelength, and it passes the laser beam with wavelength 694.3 nm.

Lens L_1 converges the light into the NLO crystal. The emergent beam from the quartz crystal is made to pass through lens L_2 and then a quartz prism. The prism splits the incident light beam into two beams due to the dispersion of light. These two beams with wavelengths 694.3 nm and 347.2 nm are made to incident on a film (screen). The production of additional light beam with wavelength 347.2 nm shows that the wavelength of the incident light beam has halved or the frequency of the incident beam has doubled. This shows the frequency doubling property of the nonlinear materials.

The process of frequency doubling is shown in Figure 15.27. A laser beam emitted by ruby laser with wavelength 694 nm is passed through a potassium dihydrogen phosphate (KDP) crystal and the emergent beam from the KDP crystal has the wavelength of 347 nm. This shows the wavelength of the emergent beam has reduced into half of the wavelength of the incident beam or it shows the frequency doubling.

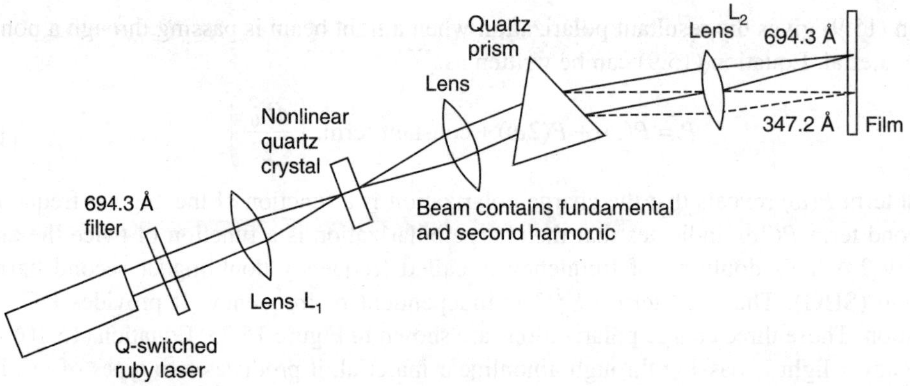

Figure 15.26 Second harmonic generation-experimental methods.

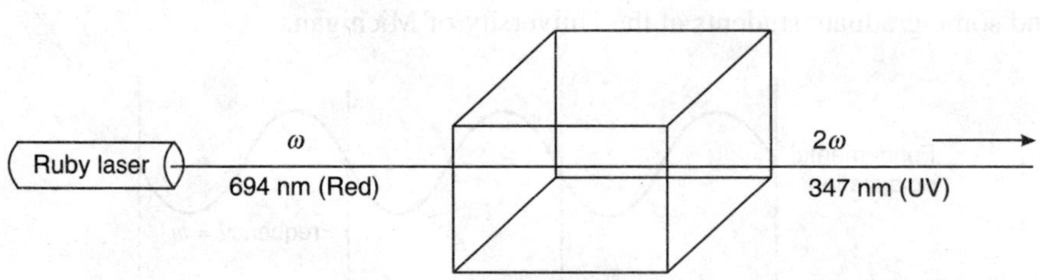

Figure 15.27 Frequency doubling.

Optical rectification

Consider an intense polarized laser beam is passing through a nonlinear optical crystal. The polarization produced by the laser beam is given by Eq. (15.9). The last term in Eq. (15.9) is independent of frequency, and it is a constant term. It is the DC signal produced whenever a laser beam is passed through a nonlinear optical material. This DC signal can be detected by connecting a voltmeter to the opposite sides of the nonlinear optical crystal. This process is called optical rectification, and it is shown in Figure 15.28. This phenomenon is similar to the rectification produced by a semiconductor diode when an AC signal is passed through it.

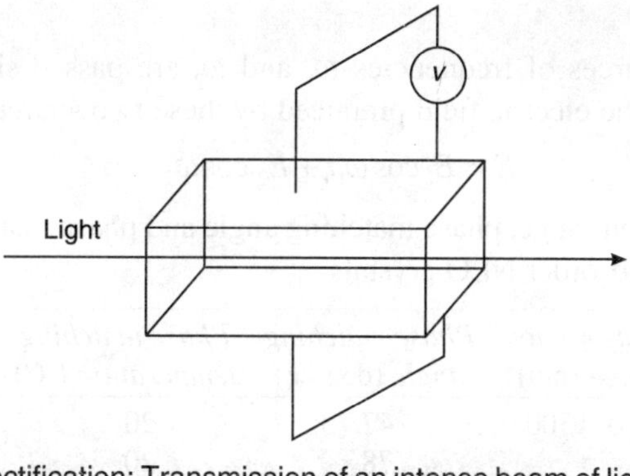

Figure 15.28 Optical rectification: Transmission of an intense beam of light through a NLO crystal produces a DC signal.

Phase matching

Consider a laser beam (say, light from a ruby laser $\lambda = 694.3$ nm) is passed through a NLO crystal such as potassium dihydrogen phosphate (KDP) at certain angle from the optic axis of the crystal, the incident beam ($\lambda = 694.3$ nm) and the second harmonic generation ($\lambda = 347.2$ nm) travels with the same speed. This phenomenon is known as phase matching or index-matching. Now, the crystal is said to be phase-matched or index matched. It is similar to the travelling of ordinary and extraordinary rays with same speed along the optic axis of a doubly refracting crystal. The ordinary and extraordinary rays in KDP crystals travel with same speed at $\theta \approx 50°$. The KDP crystal exhibits phase matching, and it is shown in Figure 15.29. In some NLO crystals, the phase matching can also be obtained by varying the temperature. The reason is that the extraordinary index is more temperature dependent than the ordinary index. The birefringence can be adjusted until the phase matching is obtained by varying the temperature. The transmission range, phase matching angle and phase matching temperature for some NLO materials are tabulated in Table 15.3.

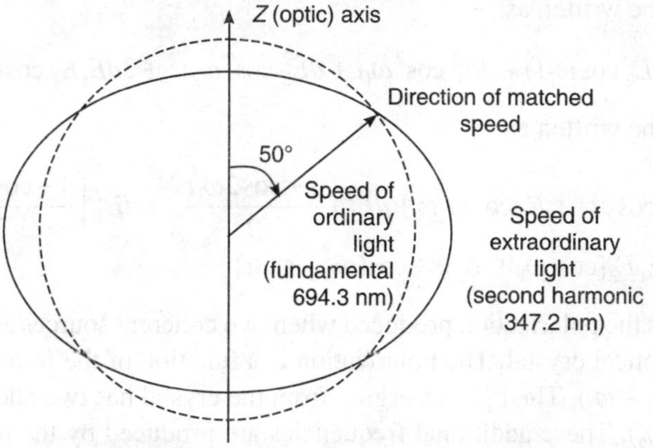

Figure 15.29 Velocities of ordinary and extraordinary rays in KDP crystal.

Optical mixing

Consider two coherent sources of frequencies ω_1 and ω_2 are passed simultaneously through a nonlinear optical crystal. The electric field produced by these two sources is given by

$$E = E_1 \cos\omega_1 t + E_2 \cos\omega_2 t \tag{15.11}$$

Table 15.3 Transmission range, phase matching angle and phase matching temperatures of some second-order NLO crystals

Material	*Symmetry*	*Transmission range* (nm)	*Phase matching angle* (degree)	*Phase matching temperature* (°C)	*Process*
KDP	$\bar{4}2m$	200–1500	47.73	20	527 nm → 1.054 μm
			78	20	2×532 nm
			90	70	2×532 nm
KD*P	$\bar{4}2m$	200–1500	86	20	2×532 nm
			90	40	2×532 nm
ADP	$\bar{4}2m$	200–1200	51	20	2×532 nm
			90	53.2	2×532 nm
β-BBO	R3	190–3000	21	20	2×694 nm
			48	20	2×532 nm
Urea	$\bar{4}2m$	200–1430	$\theta = 87$	20	266 nm + 1.064 μm

Neglecting the higher order terms, the polarization produced in the nonlinear crystal is given by

$$P = aE + dE^2 \tag{15.12}$$

Substituting the value of E from Eq. (15.11) in Eq. (15.12), we get

$$P = a(E_1 \cos\omega_1 t + E_2 \cos\omega_2 t) + d(E_1 \cos\omega_1 t + E_2 \cos\omega_2 t)^2 \tag{15.13}$$

Equation (15.13) can be written as:

$$P = a(E_1 \cos\omega_1 t + E_2 \cos\omega_2 t) + dE_1^2 \cos^2\omega_1 t + dE_2^2 \cos^2\omega_2 t^2 + 2dE_1 E_2 \cos\omega_1 t \cos\omega_2 t \tag{15.14}$$

Equation (15.14) can be written as

$$P = a(E_1 \cos\omega_1 t + E_2 \cos\omega_2 t) + dE_1^2\left(\frac{1+\cos 2\omega_1 t}{2}\right) + dE_2^2\left(\frac{1+\cos 2\omega_2 t}{2}\right)$$
$$+ 2dE_1 E_2[\cos(\omega_1 + \omega_2)t + \cos(\omega_1 - \omega_2)t] \tag{15.15}$$

Equation (15.15) gives the polarization produced when two coherent sources are mixed and passed through a nonlinear optical crystal. The polarization is a function of the frequencies ω_1, ω_2, $2\omega_1$, $2\omega_2$, $(\omega_1 + \omega_2)$ and $(\omega_1 - \omega_2)$. The light emerging from the crystal has two additional frequencies, $(\omega_1 + \omega_2)$ and $(\omega_1 - \omega_2)$. These additional frequencies are produced by the intensity of the laser beams when two beams are mixed together. The monochromaticity and coherence of the laser beams play no role.

The production of the sum of two frequencies ($\omega_1 + \omega_2$) is known as frequency up conversion and the production of difference between two frequencies ($\omega_1 - \omega_2$) is called frequency down conversion. The sum frequency production is shown in Figure 15.30. In Figure 15.30, two laser beams with frequencies ω_1 and ω_2 are mixed together and then passed through a proustite crystal. The frequency of the light beam emerging from the proustite crystal is found to be equal to the sum of two input frequencies ($\omega_3 = \omega_1 + \omega_2$).

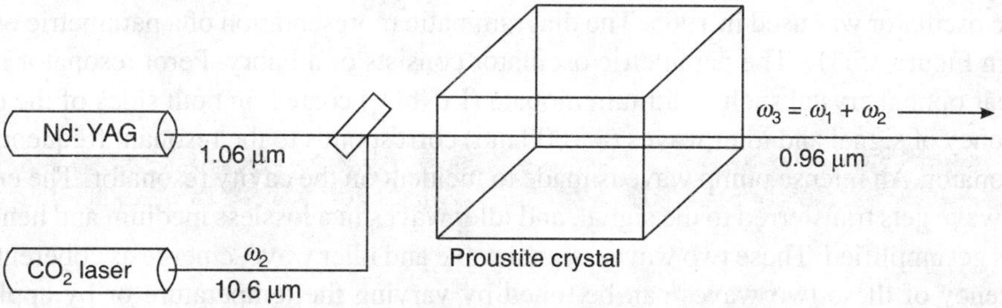

Figure 15.30 Sum frequency generation in a nonlinear crystal also called frequency up conversion.

The production of sum of two frequencies and difference between two frequencies can be explained using quantum theory of radiation. According to the quantum theory of radiation, light propagates in discrete packet of energy called photon or quantum. The sum of two frequencies ($\omega_1 + \omega_2$) is produced due to the annihilation of two photons of frequencies ω_1 and ω_2, and then they disappear. The annihilation of photons produces a new photon of frequency ω_3 (where $\omega_3 = \omega_1 + \omega_2$). The new photon of frequency ω_3 simply corresponds to a coalescing of the two original photons into a new photon. The energy of the new photon is $\hbar(\omega_1 + \omega_2)$ and hence the energy is conserved.

The difference between two frequencies ($\omega_3 = \omega_1 - \omega_2$) is produced by the interaction of the photon with frequency ω_1 with another photon of frequency ω_2. The conservation of energy and momentum requires the photons with higher frequency ω_1 vanishes, thereby it creates two new photons, one with frequency ω_2 and another with frequency ($\omega_1 - \omega_2$).

Parametric amplification

Consider three laser beams with frequencies ω_1, ω_2, and ω_3, where $\omega_3 = \omega_1 + \omega_2$ is propagating through a nonlinear optical medium. Assume ω_3 happen to be an intense beam of light than the other two waves. The energy of the wave with frequency ω_3 gets transferred to weaker signals with frequencies ω_1 and ω_2, and hence the waves with frequencies ω_1 and ω_2 get amplified. This type of amplification of waves is called parametric amplification of light.

In another way, consider an intense laser beam with frequency ω_3 and a weak laser beam with frequency ω_1 are passed through a nonlinear optical medium. The energy of the intense beam with frequency ω_3 gets transferred to the beam with frequency ω_1 and another beam with frequency ω_2, such that $\omega_2 = \omega_3 - \omega_1$ will be created so that the energy and momentum are conserved. This process is also called parametric amplification of light.

The conservation of energy and momentum can be written as:

$$\hbar\omega_3 = \hbar\omega_1 + \hbar\omega_2 \qquad (15.16)$$

and

$$\frac{\hbar\omega_3}{c} = \frac{\hbar\omega_1}{c} + \frac{\hbar\omega_2}{c} \qquad (15.17)$$

The strong wave of frequency ω_3 is called *pump light* ω_p, the weak signal of lower frequency ω_1 is called *signal light* ω_s. In the parametric amplification process, the signal wave ω_s gets amplified. The pump light (ω_p) gets converted into both signal light (ω_s) and a difference wave ω_2, called *idler light* ω_i, of frequency $\omega_2 = \omega_3 - \omega_1$ (or $\omega_i = \omega_p - \omega_s$). If the idler light ω_i beats the pump light ω_p, then the pump light gets converted into additional signal light and idler light. Thus the signal light and idler light get amplified. This process is called parametric amplification.

The parametric amplification was used in microwave frequency in 1940s. The first optical parametric oscillator was used in 1965. The diagrammatic representation of a parametric oscillator is shown in Figure 15.31. The parametric oscillator consists of a Fabry–Perot resonator made up of nonlinear optical crystal such as lithium niobate ($LiNbO_3$) coated on both sides of the crystals. The frequency of signal and idler waves ($\approx$ 1000 nm) corresponds to the resonant frequency of the cavity resonator. An intense pump wave is made to incident on the cavity resonator. The energy of the pump wave gets transferred to the signal, and idler waves in a lossless medium and hence these two waves get amplified. These two waves, signal wave and idler wave, emerge as coherent waves. The frequency of these two waves can be tuned by varying the temperature or by applying an electric field. An optical parametric oscillator is a laser-like device used to produce coherent light sources in the infrared and ultraviolet regions.

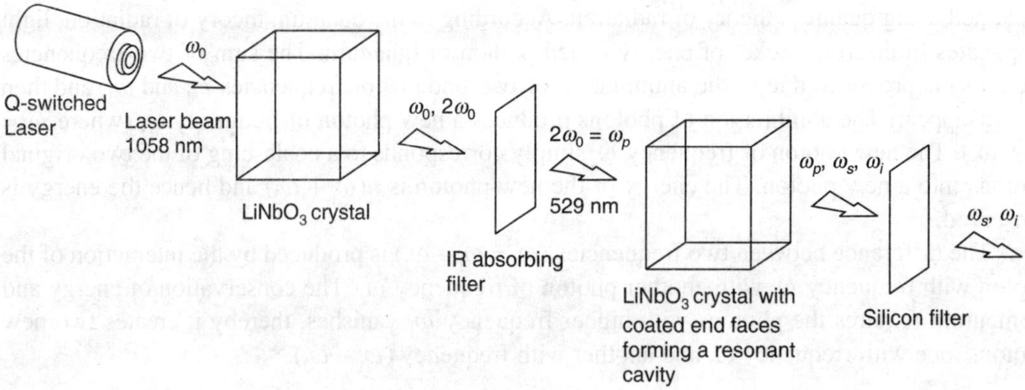

Figure 15.31 Parametric oscillator

Coherence length

Consider a wave of frequency ω is made to incident on a nonlinear material. It generates a SHG wave of frequency 2ω. If the indices of refraction of these two waves, n_ω and $n_{2\omega}$, are different, then a detailed analysis of the electromagnetic radiation in a nonlinear medium shows that the SHG wave with frequency 2ω gets amplified up to a distance,

$$l_c = \frac{1}{4}\frac{\lambda_0}{|n_\omega - n_{2\omega}|} \tag{15.18}$$

This distance is generally called coherence length.

If the difference between the indices of refraction, n_ω and $n_{2\omega}$, becomes zero, then the coherence length becomes infinity. The typical value of the coherence length of a nonlinear material varies from 10^{-5} m to 10^{-6} m.

In a nonlinear material, the energy is transferred from the primary wave to the secondary wave up to the distance l_c to $2l_c$. The energy transfer from the secondary wave to the primary wave takes place from the distance $2l_c$ to l_c. Thus the transfer of energy from primary to secondary waves and then from secondary to primary waves takes place throughout the nonlinear medium. If a perfect phase matching is obtained, then, $n_\omega = n_{2\omega}$, the energy transfer from primary to secondary takes place continuously throughout the NLO medium.

Self-focusing of light

Suppose a light beam is passed through a nonlinear medium such as gas or liquid. The third harmonic generation (THG) is produced in gases and liquids. The centre of the beam has the maximum intensity, and the intensity decreases at the periphery of the beam. The intensity distribution of the light beam is similar to a Gaussian curve (Figure 15.32). The refractive index of the medium at the centre of the beam is maximum, and it is minimum at the periphery of the beam. Due to the variation in the refractive index of the medium, different parts of the light travels with different speeds and consequently, the light contracts. The contraction of light focusses the light into a small spot as shown in Figure 15.33, and this process is called self-focussing.

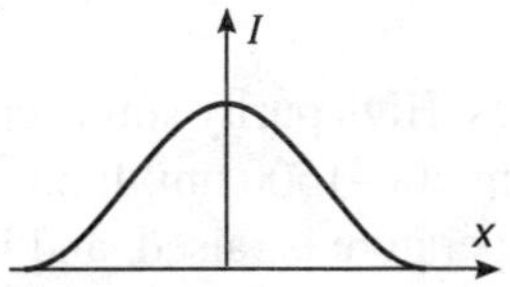

Figure 15.32 Intensity distribution of light.

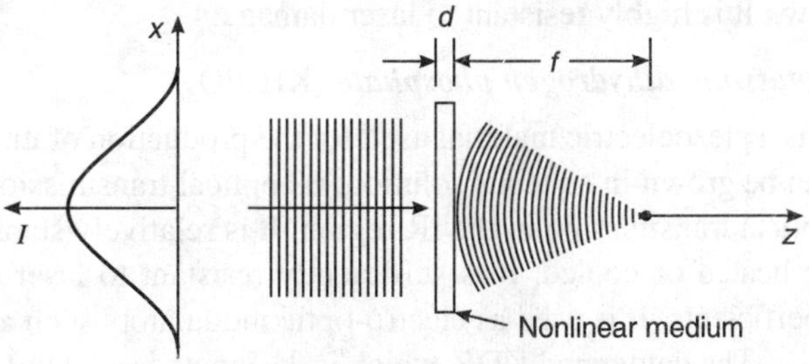

Figure 15.33 Self-focussing of light.

15.5.3 Nonlinear Materials

The nonlinear materials are classified into different categories based on the order of nonlinearity, the nonlinear interaction involved, and some other nonlinear properties. The nonlinear materials are broadly classified into (i) active or passive materials, (iii) organic or inorganic materials, and (iii) crystalline and amorphous materials. It is also classified on the basis of the order of nonlinearity. The nonlinear materials are classified as second-order nonlinear materials, third-order nonlinear materials and so on.

The materials that impose their characteristics resonance frequency onto the incident light beam are called *active* nonlinear materials. The nonlinear effects such as two-photon absorption, stimulated Raman, Rayleigh and Brillouin scattering are produced by these materials. The materials those do not impose their characteristics resonance frequencies onto the nonlinear effects are called *passive* nonlinear materials. These materials exhibit the nonlinear properties such as harmonic generation, frequency mixing, optical rectification and self-focussing of light.

Second-order NLO materials

The NLO materials used to produce new frequencies through second harmonic generation (SHG), sum and difference frequency production using optical mixing, or optical parametric oscillation (OPO) are called second order NLO materials. There are a number of NLO materials which are used to produce new frequencies, only few materials are better suited to a particular process.

First generation materials

The inorganic crystals like ammonium dihydrogen phosphate (ADP), potassium dihydrogen phosphate (KDP), *β*-barium borate (*β*-BBO), lithium niobate ($LiNbO_3$) and barium sodium niobate ($BaNaNb_5O_{15}$) are the first generation NLO materials.

Ammonium dihydrogen phosphate ($NH_4H_2PO_4$)

ADP is a piezoelectric material used as an ultrasonic transducer. It is one of the first nonlinear materials used for the production of phase-matched second harmonic generation of light. The optical transmission range of this crystal lies between 200–1200 nm. It has poor optical transmission in the IR region. It will deteriorate when heated above 100°C and upon cooling cracks will be produced. High pure optical crystal of ADP can be prepared. Its nonlinear coefficient is relatively low. It is highly resistant to laser damage.

Potassium dihydrogen phosphate (KH_2PO_4)

It is a piezoelectric material used for the production of ultrasonic waves. High purity single crystal can be grown in aqueous solution. Its optical transmission varied from 200–1500 nm. It has poor optical transmission in the IR region. It is relatively stable when temperature is raised, and it can be heated or cooled. It is sufficiently resistant to laser damage. It has relatively low nonlinear coefficients. It is used as electro-optic modulators such as Q-switches, Pockel cells, etc.

The deuterated KDP, which is designated as KD*P and called 'K-D star-P', is also used as a NLO material.

Lithium niobate ($LiNbO_3$)

The nonlinear coefficient of lithium niobate is ten times larger than that of KDP. Its optical transmission varies from 400 nm to 5 μm. It is two times efficient than KDP and SHG efficiency close to 100 per cent is possible. It has very large birefringence in the visible and near-IR region. Phase-matching can be achieved by heating the crystal for a particular wavelength of laser beam. Large size, few cm in diameter and 10-cm long, single crystal is commercially available. The single crystal is clear as water and it is insoluble in water. It is used for SHG, parametric oscillation, sum and difference frequency generation. It is also used in Pockel cells, Q-switches, and phase modulators, waveguide substrate and surface acoustic wave (SAW) wafers. It has very low damage threshold (0.1 GW/cm^2) due to its photorefractive effect.

β-barium borate (β-BaB_2O_4)

β-barium borate (*β*-BBO) has relatively large optical transmission range from UV to IR and high optical damage threshold. It has relatively low NLO coefficients. It is used to generate high power, short pulse radiation. Large, high quality single crystal is expensive. It has a broad phase matchable range from 409.6 nm to 3500 nm. It has wide transmission range 190 nm to 3500 nm. The laser damage threshold is quite high. It exhibits wide temperature bandwidth of about 55°C. It has effective SHG coefficient about 6 times greater than KDP. It is an efficient NLO material for the second, third and fourth harmonic generations using Nd:YAG lasers and the best NLO crystal for the fifth harmonic generations at 213 nm. It is also used in optical parametric amplifiers and optical parametric oscillators.

Barium sodium niobate ($BaNaNb_5O_{15}$)

It is optically transparent from 370 nm to about 5 μm. It does not suffer as much from laser damage when it is kept above room temperature. Its nonlinear coefficients are three times larger than lithium niobate. Phase-matching for SHG can be achieved using Nd:YAG (λ = 1.06 μm) around 100°C. It is used for second harmonic generation of 1.06 μm radiation and parametric oscillation.

Proustite (Ag_3AsS_3)

Proustite has the optical transmission from 600 nm to 13 μm. It is used to study the phase-matched interaction between infrared and visible region. Its nonlinear coefficients are 300 times larger than KDP. It is used for optical mixing of 10.6 μm CO_2 laser with a visible laser. Good quality proustite crystals can be grown from an aqueous solution. Phase matched second harmonic generation has been observed in proustite (Ag_3AsS_3) at 10.6 mm.

Second generation materials

The organic single crystals, organic–inorganic single crystals, chromophore-hosted polymers and Langmuir–Blodgett (LB) films belong to the second generation materials. Aniline, nitrobenzene, 4-nitroaniline, 2-nitroaniline, 3-nitroaniline, 2,4-nitroaniline, 2-chloro-4-nitroaniline, 2-bronio-4-nitroaniline, 2-methyl-4-nitroaniline (MNA) are some of the examples for second generation NLO organic molecules.

Third-order NLO materials

Semiconductors, quantum confined semiconductors (quantum wells, quantum wires and quantum dots), metal particles, organic and inorganic glasses belong to third-order NLO materials. AlInAs/GaInAs quantum wells show SHG associated intersub-band transitions. $GaAs/Al_{0.3}Ga_{0.7}As$ quantum wells have been studied for polarization-independent optical waveguide switch. The quantum dots studied for NLO properties are CdS, PbS, ZnO, ZnS, Cd_3P_2, CdSe, CuCl, CdS_XSe_{1-X}, $CdTe_XS_{1-X}$ and so on. Conductive oxides with wide band gaps such as SnO_2 and In_2O_3 showed nonlinear property such as third harmonic generation. Oxide glasses such as SiO_2, Li_2O-TiO_2-TeO_2, PbO-Bi_2O_3-GaO_3 and PbO-SiO_2 also found to have nonlinear properties.

Other materials

Some materials have shown both lasing effect and SHG, and this property is called self-frequency doubling. Nd, Mg-doped $LiNbO_3$ and $Nd_XY_{1-X}Al_3(BO_3)_4$ have shown self-frequency doubling effect with relatively low energy conversion efficiency compared to SHG materials.

15.5.4 Applications of Nonlinear Optical Materials

The nonlinear optical materials are mostly used as active optical devices. Some of the main applications of the nonlinear materials are frequency generation, optical communication, optical switching and signal processing, and optical computing.

Frequency generation

The nonlinear optical materials are used to generate light waves with new optical frequency. The second harmonic generator, optical mixing, optical parametric oscillators and Raman shifting are used to produce light waves with new frequencies. Lasers using these techniques are commercially available. Stimulated Raman oscillation is used to shift the frequency of the laser by an amount equal to the Raman mode of the molecule. The stimulated stoke Raman scattering is used to produce light wave with larger frequency than pump wave. The stimulated anti-stoke Raman scattering is used to produce light wave with shorter frequency than pump wave. Nearly 70 per cent of the pump wave is shifted using modern Raman techniques by employing the vibrational modes of hydrogen, deuterium, methane and some other gases.

Optical communication

In communication, the NLO effects are used to cancel the dispersive property of the nonlinear material and to overcome the limitations in the electronic modulation techniques. The diffraction of light is used to produce temporal and spatial soliton. The soliton propagation is used for the wavelength division multiplexing (WDM), so as to utilize the full potential capacity of the fibre. It is also possible to use polarization division multiplexing because the soliton maintains high degree of polarization over a long distance. The main advantage of the optical communication over that of electronics is because of the higher carrier frequency used that provides higher bandwidth.

Optical switching and signal processing

The optical switching devices are based on either spatial or temporal nonlinear response of a material. Let us discuss the working of a nonlinear directional coupler.

Nonlinear directional coupler

The directional couplers are used to couple the signals in the input port to the output port and hence direct the signals into different locations. The signals coming from port 1 should be coupled into the output ports 1′ and 2′, so that the signals can be available in the desired proportion between output ports 1′ and 2′. The schematic representation of a directional coupler is shown in Figure 15.34(a). The integrated optics form of a directional coupler is shown in Figure 15.34(b).

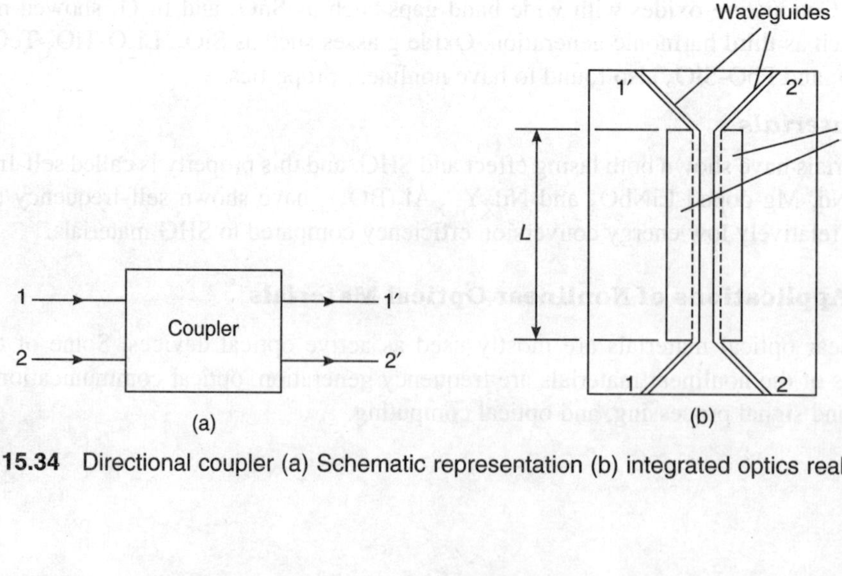

Figure 15.34 Directional coupler (a) Schematic representation (b) integrated optics realization.

It consists of waveguides 1 and 2, which are arranged in such is way that they are very close up to a distance L and hence there is a overlapping of the signals from one waveguide to other. The relative velocity of the wave propagation in the waveguides can be changed by applying a voltage between the electrodes.

If no voltage is applied between the electrodes, then the wave vectors $k_1 = k_2$, and the velocities of propagation of the two waves are identical. During that time, the exchange of power between waveguides 1 and 2 takes place. If an electric field is applied between the two electrodes, the velocity of propagation of the wave in one waveguide increases and in the other waveguide decreases. Due to different velocities of the waves, there is no transfer of signal from one waveguide to another.

A directional coupler is used as a switch. In the absence of the voltage, there is a transfer of signal from one waveguide to another, and when a voltage is applied between the two electrodes, there is no transfer of signals from one waveguide to another.

Optical computing

The electronic computers use very large scale integration of electronic transistors. Photonics is used to replace the existing electronic computers using NLO materials. The main advantages of optical computing are: high speed and higher bandwidths. The main drawback of the optical computing is the higher energy requirement and the device size limitations that is the size of the device is comparable to the wavelength of the light used.

Phase shifter

The NLO material $LiNbO_3$ is used to prepare phase shifter. A simple phase shifter is constructed by coating two electrodes on the surface of the crystal on either side of the waveguide as shown in Figure 15.35. Let d be the distance between the two electrodes, and L is the length of the electrodes.

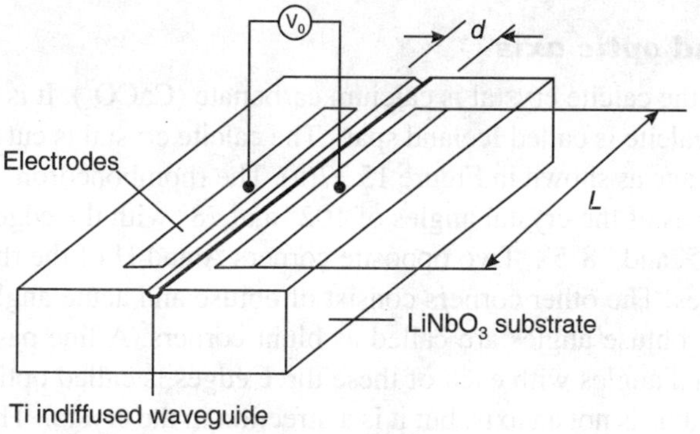

Figure 15.35 Phase shifter.

A voltage of V_0 is applied between the two electrodes. It will produce an electric field of V_0/d. The phase difference produced is given by

$$\Delta\phi = \frac{2\pi L \Delta n}{\lambda} \tag{15.19}$$

where Δn is the change in the refractive index of the medium. A voltage of 5 V, applied to the electrodes, which are separated by a distance of 5 mm, produces an electric field of 10^6 V m^{-1}.

The change in the refractive index produced by this electric field is equal to $\Delta n = 1.86 \times 10^{-4}$. A simple calculation using Eq. (15.19) gives a phase difference of π that can be produced using a light of wavelength $\lambda = 1.5$ μm and electrode length $L = 4$ mm. By varying the voltage between 0 V and 5 V, a phase modulator can also be prepared.

15.5.5 Double Refraction (Birefringence)

Consider a ray of light AB is passed through a calcite crystal. It produces two refracted rays as shown in Figure 15.36(b). This property is known as **double refraction** or **birefringence**. If the double refraction is analyzed by rotating the crystal, one image remains constant, whereas another image rotates with the rotation of the crystal as shown in Figure 15.36(a). The image that remains constant is called ordinary ray, and the another image that rotates is called extraordinary ray. The ordinary ray obeys the law of refraction, and it travels with constant speed. The refractive index of the crystal for the ordinary ray is constant. The speed of the extraordinary ray is not constant because it varies with the angle of incidence. The refractive index of the crystal for the extraordinary ray is also not constant. The materials such as calcite and quartz exhibit the double refraction.

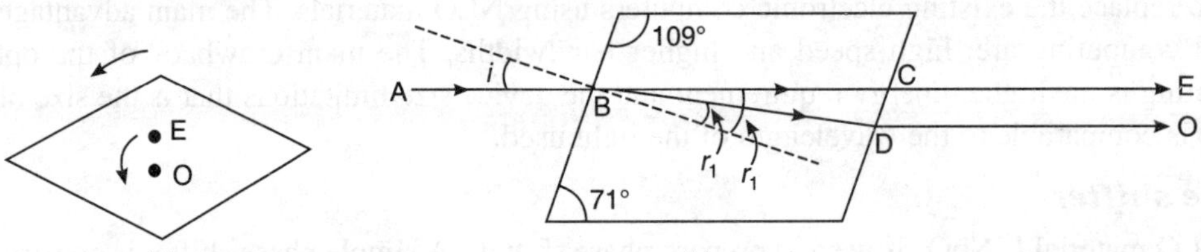

Figure 15.36 Double refraction, (a) analysis of double refraction by rotation and (b) double refraction in calcite.

Calcite crystal and optic axis

The chemical name of the calcite crystal is calcium carbonate ($CaCO_3$). It is abundantly available in Iceland. Therefore, calcite is called Iceland spar. The calcite crystal is cut into a rhombohedron shaped crystal by cleavage as shown in Figure 15.37(a). The rhombohedron-shaped calcite crystal has six faces. These faces of the crystal angles of 102° and 78° with the edges. Strictly speaking, these angles are 101°55′ and 78°5′. Two opposite corners A and H of the rhombohedron-shaped crystal are obtuse angles. The other corners consist of obtuse and acute angles. Corners A and H that are surrounded by obtuse angles are called as blunt corners. A line passing through a blunt corner and making equal angles with each of these three edges is called optic axis. It is shown in Figure 15.37(b). Optic axis is not an axis, but it is a direction in the crystal. Therefore, any line that is parallel to the optic axis is also called optic axis.

Significance of the optic axis

1. Optic axis is not an axis but it is a direction in the crystal.
2. Whenever a ray of light passes along the optic axis in a doubly refracting crystal, it does not split into two rays.
3. Along the optic axis, the ordinary and the extraordinary rays travel in the same direction with same speeds.
4. The crystal is symmetrical about the optic axis.

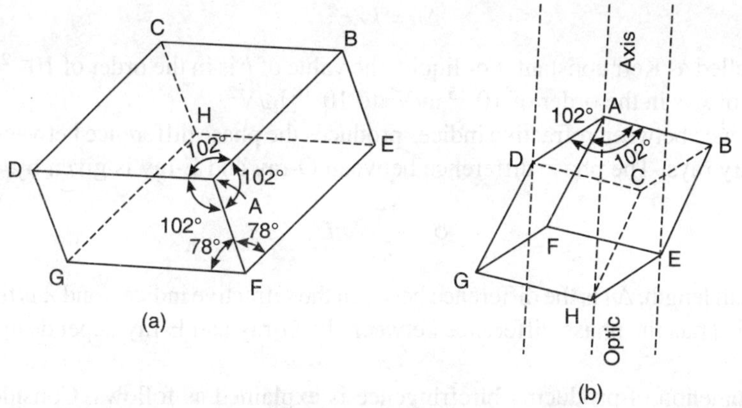

Figure 15.37 Calcite crystal (a) rhombohedron-shaped crystal and (b) optic axis.

Uniaxial and biaxial crystals

Certain materials exhibit only one optic axis. They are called uniaxial crystals. In the uniaxial crystals, the optic axis drawn from one blunt corner passes through another blunt corner. The crystals like calcite and quartz are said to be uniaxial crystals.

In certain crystals, the optic axis drawn at one blunt corner does not pass through another blunt corner of the crystal. So there are two different optic axes for such crystals. These crystals are said to be biaxial crystals. Borax and mica are the examples of the biaxial crystals.

15.5.6 Electro-optic Effect

In general, the birefringence is exhibited by anisotropic materials. But the birefringence phenomenon can also be observed in isotropic materials by applying an electric field. The isotropic medium that exhibits birefringence due to the application of electric field is called as electro-optic media. The production of birefringence in an isotropic material by applying an electric field is called electro-optic effect. There are two types of electro-optic effects, namely, Kerr effect and Pockels effect. Let us discuss about the Kerr effect in this section.

Optical Kerr effect

In certain isotropic materials such as liquid, gases and solids (typically glass), if an electric field is applied perpendicular to the direction of the light beam, the medium behaves as a uniaxial material and it exhibits birefringence. This phenomenon is called optical Kerr effect. This phenomenon was first observed by John Kerr in 1875.

Whenever an electric field is applied to the isotropic medium, an optic axis is created parallel to the electric field. The light ray passing through the medium is splitted into ordinary ray (O-ray) and extraordinary ray (E-ray). These two rays travel with different speeds and hence the medium exhibits two refractive indices: one for O-ray and another one for E-ray. The difference between the refractive indices of these two rays is given by $\Delta n = n_e - n_o$. Experimentally the difference between the refractive indices of these rays is found to be (i) directly proportional to the wavelength of the incident beam ($\Delta n \propto \lambda$) and (ii) directly proportional to the square of the applied electric field ($\Delta n \propto E^2$).

$$\therefore \qquad \Delta n = k\lambda E^2 \qquad (15.20)$$

where k is the called as Kerr constant. For liquid, the value of k is in the order of 10^{-12} m/V^2 and for gases the value of k is in the order of 10^{-15} m/V^2 to 10^{-18} m/V^2.

The difference between refractive indices produces the phase difference between the ordinary and extraordinary rays. The phase difference between O-ray and E-ray is given by

$$\phi = \frac{2\pi}{\lambda}\Delta n L \qquad (15.21)$$

where L is the path length, Δn is the difference between the refractive indices, and λ is the wavelength of the light used. Thus the phase difference between the O-ray and E-ray depends upon the length of the cell.

The phenomenon of producing birefringence is explained as follows. Consider the light is passed through polar molecules. Because of the long dipoles of the polar molecules, the medium behaves as an isotropic medium. Whenever an electric field is applied to the medium, the dipoles tend to align parallel to the applied field and hence the medium becomes anisotropic. Thus it exhibits birefringence. In the case of a non-polar medium, the electric field align the molecules parallel to the applied electric field and hence the medium becomes anisotropic. In consequence, the medium exhibits birefringence.

The cell that exhibits the optical Kerr effect is called a Kerr cell. Consider a cell of length L is filled with nitrobenzene as shown in Figure 15.38. Assume the cell is placed between a polarizer and an analyser that are kept perpendicular to each other. If an unpolarized light is passed through the polarizer, it produces polarized light. The polarized light is passed through the liquid. There is no emergent beam from the analyser indicating that no double refraction is produced. Consider an electric field of intensity E is applied to the cell, the liquid produces birefringence and it introduces some phase difference between the two refracted beams. If the electric field is applied perpendicular to the light beam at an angle of 45° to the direction of the incident polarization, then elliptically polarized light is produced. The analyser is used to analyse the emergent beam from the Kerr cell. The Kerr cell can be used as a quarter wave plate (QWP) or a half wave plate (HWP) by suitably selecting the length of the cell.

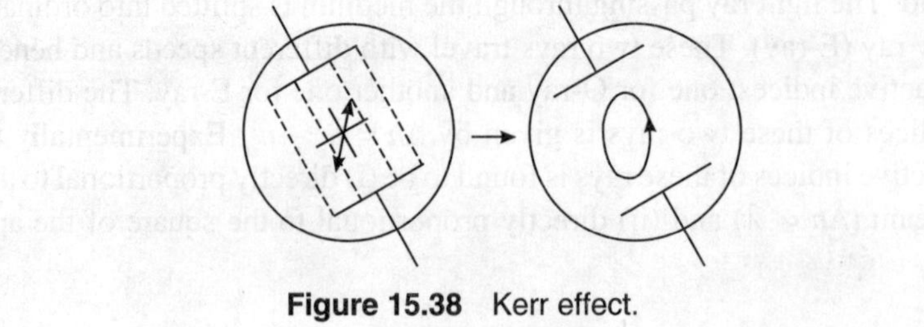

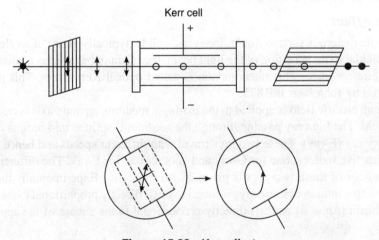

Figure 15.38 Kerr effect.

15.6 BIOMATERIALS

The materials that are used in a medical device, intended to interact with the biological systems are called biomaterials. It is used to replace a part of a function of the body in a safe, reliable, economic, and physiologically acceptable manner. A biomaterial may be any substance (other than drug), natural or synthetic, that treats, augments, or replaces any tissue, organ, and body functions. A biomaterial is used to treat many diseases, injuries in association with other therapies or procedures. It is used for the replacement of body part that has lost function (such as hip, heart), to correct abnormalities (spiral chord), to improve function (pacemaker, stent) and to assist in healing (structural, pharmaceutical effects: sutures, drug release).

15.6.1 Requirements of a Biomaterial

The exposure of materials to the body fluids provides strict restriction for an engineering material to be used as a biomaterial. A biomaterial should satisfy the following requirements:

1. A biomaterial should be biocompatible. It should not elicit an adverse effect on the body and vice-versa.
2. A biomaterial should be nontoxic, non-carcinogenic (carcinogenic means the ability or tendency to produce cancer), non-pyrogenic, non-allergenic, blood compatible and non-inflammatory.
3. A biomaterial should have good physical and mechanical properties, since biomaterial may be machined into different shapes.
4. A biomaterial may be easily machinable, mouldable, extrudable and the corrosion resistance of the material should be high.
5. A biomaterial should be readily available and should have low cost.

The uses of biomaterials are listed in Table 15.4.

Table 15.4 Uses of biomaterials

S.No.	*Organ/Tissue*	*Example*
1.	Heart	Pacemaker, artificial valve, artificial heart
2.	Eye	Contact lens, intraocular lens
3.	Ear	Artificial stapes, cochlea implant
4.	Bone	Bone plate, intramedullary rod, joint prosthesis, bone cement, bone defect repair
5.	Kidney	Dialysis machine
6.	Bladder	Catheter and stent
7.	Muscle	Sutures, muscle stimulator
8.	Circulation	Artificial blood vessels
9.	Skin	Burn dressings, artificial skin
10.	Endocrine	Encapsulated pancreatic islet cells

15.6.2 Types of Biomaterials

The above requirements eliminate most of the engineering materials to be used as biomaterials. The synthetic materials used as biomaterials are generally classified into the following categories: (i) metals, (ii) polymers, (iii) ceramics and (iv) composites.

Metals

Metals are generally used for load-bearing implants. The most common applications of metals are screws for fracture fixation plates and wires. The metals are used as artificial metallic joints for hips, knees, shoulders, ankles and so on. In addition to metallic implants, the metals are used in maxillofacial surgery, cardiovascular surgery and dental implants.

Only a small number of metals are approved as biomaterials for use in medical field. They are stainless steel, cobalt based alloys and titanium and its alloys. Medical and dental applications of some metals and alloys are given in Table 15.5.

316L stainless steel

It is a low carbon steel (<0.03% C). It has excellent corrosion resistant in the body's environment. Stainless steel with sufficient strength will be obtained using cold working. It is mostly used for the repair of fracture and other temporary applications.

Table 15.5 Medical applications of some metals and alloys

S.No.	*Metals and alloys*	*Applications*
1.	316L stainless steel	Fracture fixation, vascular stents, surgical instruments
2.	Ni-Ti alloy	Bone plates, vascular stents, orthodontic applications, pacemaker cases, artificial heart valves
3.	Gold alloys	Dental restorations
4.	Silver products	Antibacterial agents
5.	Platinum and Pt-Ir	Electrodes
6.	Ti-Al-V, Ti-Al-Nb, Ti-13Nb-13Zr, Ti-Mo-Zr-Fe	Bone and joint replacement, fracture fixation, dental implants, pacemaker encapsulation
7.	Co-Cr-Mo, Co-Ni-Cr-Mo	Bone and joint replacement, dental implants, heart valves, dental restorations
8.	Silver-tin-copper alloy	Dental amalgams

Cobalt based alloys

It has excellent corrosion resistance in the human body. It exists as castable ASTM F75 and wrought F90. The composition of F75 is 59–69.5% Co, 27–30% Cr, 5–7% Mo, with 1% Mn, 2.5%Ni, 1% Si, 0.75% Fe and 0.35% C. It is difficult to machine, therefore, prostheses equipment are prepared using casting or powder metallurgy. This alloy is used in dentistry and for the manufacture of artificial joints.

Titanium and its alloys

Commercially pure (Titanium F67) form of titanium is used as a biomaterial. It contains 0.49 per cent of oxygen. Addition of Al and V increases the strength and fatigue limit. Titanium and its alloys are used in dental implants and for hip, knee and shoulder prosthesis.

Advantages of metallic biomaterials

The metallic biomaterials have the following advantages:

1. The properties and fabrication methods of metals are well known.
2. Metals have high mechanical strength such as high stiffness, strong, high fatigue and wear resistance.
3. The joining technologies for metals are well known.

Disadvantages of metallic biomaterials

The drawbacks of the metallic biomaterials are: (i) metal ions may be toxic, and (ii) corrosion resistance of metals is very low.

Polymers

Polymers are used for a variety of applications. The polymeric materials such as acrylics, polyamides, polyesters, polyethylene, polysiloxanes and polyurethane are used as biomaterials. Some of the applications include artificial heart, kidney, liver, pancreas, bladder, bone cement, catheters, contact lenses, cornea and eye-lens replacements, external and internal ear repairs, heart valves, cardiac assist devices, implantable pumps, joint replacements, pacemaker, encapsulations, soft-tissue replacement, artificial blood vessels, artificial skin, and sutures. Medical applications of some polymeric materials are displayed in Table 15.6.

Table 15.6 Medical applications of some polymers

S.No.	*Polymers*	*Applications*
1.	Ultrahigh molecular weight polyethylene	Knee, hip and shoulder joints
2.	Silicone	Finger joints, soft-tissue replacement, ophthalmology
3.	Polylactic and polyglycolic acid, nylon	Sutures
4.	Silicone, acrylic, nylon	Tracheal tubes
5.	Acetal, polyethylene, polyurethane	Heart pacemaker
6.	Polyester, polytetrafluroethylene, PVC	Blood vessels
7.	Nylon, PVC, silicone	Gastrointestinal segments
8.	Polydimethylsiloxane, polyurethane, PVC	Facial prostheses
9.	Polymethylmethacrylate	Bone cement
10.	Hydrogels	Ophthalmology, drug-delivery systems

Advantages of polymer biomaterials

The advantages of polymers are: (i) easy fabrication, (ii) wide range of composition and properties and many ways to immobilize biomaterials/cells.

Disadvantages of polymer biomaterials

The disadvantages of polymer biomsaterials are: (i) contain leachable compounds, (ii) surface contamination, (iii) difficult to sterilize and (iv) chemical/biochemical degradation.

Ceramics

The ceramics that are used for repair and replacement of diseased and damaged parts of the musculoskeletal system are referred to as bioceramics. Ceramics are the inorganic materials and they are bonded with ionic bond. Ceramics are highly inert, hard and more resistant to degradation in many environments than metals. But they are brittle because of the ionic bond. They are widely used in orthopaedic implants, dental implants because the chemical compositions of the ceramics have some similarity with native bone. Their application in biomedicine is not extensive as metals and polymers. The poor fracture toughness of ceramics severely limits their uses in load-bearing applications. The ceramic materials used for the biomedical applications are listed in Table 15.7.

Table 15.7 Medical applications of some ceramics, glasses and composites

S.No.	*Materials*	*Applications*
Ceramics and glasses		
1.	Alumina	Joint replacement, dental implants
2.	Zirconia	Joint replacement
3.	Calcium phosphates	Bone repair and augmentation
4.	Bioactive glasses	Bone replacement
5.	Porcelain	Dental restorations
6.	Carbon	Heart valves, dental implants, percutaneous devices
Composites		
7.	BIS-GMA-quartz/silica fiber	Dental restorations
8.	PMMA-glass fibers	Dental restorations (dental cement)

Advantages of bioceramics

The advantages of bioceramics are as follows:

1. The physical properties of ceramics are similar to bone.
2. Ceramics have high compressive strength when dense and the bioreactivity of ceramics varies from low to high.
3. Ceramics are readily sterilized.

Disadvantage of bioceramics

The disadvantages of ceramics are: (i) difficult to fabricate (ii) poor fracture toughness of ceramics severely limits their uses in load-bearing applications.

Composites

The composite biomaterials are used in the field of dentistry as dental cements. Composites have unique properties and are usually stronger than any of the single materials from which they are made. Workers in this field have taken advantages of this fact and applied it to some difficult problems where tissue in-growth is necessary. The carbon–carbon and carbon-reinforced polymer components are used in bone repair and joint replacement. Composite materials with low density/weight and high strength are used in prosthetic limbs. Al_2O_3 deposited onto carbon, carbon/PTFE, Al_2O_3/PTFE, and PLA-coated carbon fibers are some of the examples for the composite materials. The medical applications of some composite materials are tabulated in Table 15.7.

15.6.3 Uses of Biomaterials

Biomaterials are used to replace hard and soft tissues that are damaged through some pathological process such as fracture, infection and cancer that cause pain, disfigurement or loss of functions. Under these conditions, the deceased tissue is removed, and it is replaced, by synthetic materials. Some of the uses of the biomaterials are discussed as under:

Orthopaedics

The most prominent applications of the biomaterials are orthopaedic implant devices. The freely moving joints such as hip, knee, shoulder, ankle and elbow are affected by osteoarthritis and rheumatoid arthritis. The weight bearing joints such as hip joint and knee joint produces considerable pain and this will affect the patient. These joints are replaced by synthetic biomaterials and thereby the pain is reduced and the mobility is restored. A variety of metals, polymers and ceramics are used for such applications.

Ophthalmics

The tissues of the eye suffer from several diseases. This results in reduced vision, and eventually blindness. Cataracts produce cloudiness of the lens. This may be replaced by synthetic (polymer) intraocular lens. Materials used for the contact lens is also considered as biomaterials.

Dental applications

The bacterially controlled diseases destroy the tooth and supporting gum tissues. The dental caries (cavities) produce extensive tooth loss. The destroyed teeth can be replaced by a variety of materials.

Cardiovascular applications

In the cardiovascular system, the problem can arise with heart valves and arteries. The heart valve may suffer structural changes that prevent the valve from fully opening and fully closing. In cardiology, the materials such as polyethylene terephthalate (PET), polytetrafluroethylene (PTFE) and polyurethanes are mostly used as synthetic vascular graft (blood vessel) materials.

Drug-delivery system

Drug-delivery system is one of the fastest growing area of implant applications. It is the process of controlled and targeted delivery of drugs. New polymeric materials are used as vehicles for drug-delivery systems.

Wound healing

The polymer and metals (e.g. stainless steel and tantalum) are used as synthetic suture materials for wound healing. Another wound healing category is fracture fixation device. Bone plates, screws, nails, rods, wires, and other devices are used as fracture fixation device. Almost all fracture fixation devices used for orthopaedic implants are made from metals, and stainless steel.

SHORT QUESTIONS

1. What are metallic glasses?
2. Mention the different types of metallic glasses.
3. Write the electronic properties of the metallic glasses.
4. Write the magnetic properties of the metallic glasses.
5. Write the superconducting property of metallic glasses.
6. Write the uses of metallic glasses as a transformer core material.
7. Write any two applications of metallic glasses.
8. What are shape memory alloys?
9. Write about the two different phases of shape memory alloys.
10. What are the types of shape memory alloys?
11. Mention the characteristics of shape memory alloys.
12. What is shape memory effect?
15. What is pseudoelasticity?
14. What are the methods used for the characterization of shape memory alloys?
15. Write about the synthesis of Ni-Ti alloy.
16. Write about the properties of Ni-Ti alloy.
17. Mention the applications of shape memory alloys.
18. What are nanomaterials?
19. What is top-down process? Give an example.
20. What is bottom-up process? Give an example.
21. What are the advantages and drawbacks of arc discharge?
22. What are the advantages and drawbacks of laser ablation techniques?
23. What are the advantages of ball milling method?
24. Mention the advantages of sol-gel process.
25. Write the advantages of chemical vapour deposition technique.
26. What are the advantages and drawbacks of electrodeposition process?
27. Mention any four properties of nanomaterials.
28. Mention the applications of nanomaterials.
29. What is meant by nonlinear optics?
30. Distinguish between linear and nonlinear properties.
31. Write any five properties of NLO materials.
32. What is meant by frequency doubling?
33. What is second harmonic generation (SHG)?
34. Write about the nonlinear charge polarization.
35. What is optical rectification?

36. What is meant by phase matching?
37. What is meant by parametric amplification?
38. What is coherence length?
39. What is self-focussing of light?
40. What are active and passive NLO materials?
41. What is frequency up conversion?
42. What is frequency down conversion?
43. Define the terms: (i) pump light, (ii) signal light and (iii) idler light.
44. What is meant by birefringence?
45. What is optic axis?
46. What are uniaxial and biaxial crystals?
47. What is Kerr effect?
48. What is a Kerr cell? What are its uses?
49. What are biomaterials?
50. What are the requirements of a biomaterial?
51. Write the names of any four metallic biomaterials.
52. Write the advantages and disadvantages of metallic biomaterials.
53. Write any four medical applications of metallic biomaterials.
54. Mention the names of any four polymeric biomaterials.
55. Write the advantages and disadvantages of polymeric biomaterials.
56. Write the medical applications of polymeric materials.
57. What are the medical applications of composite materials?
58. What are the medical applications of bioceramics?
59. Write the advantages and disadvantages of bioceramic materials.

DESCRIPTIVE TYPE QUESTIONS

1. What are metallic glasses? Describe any one method used to prepare the metallic glasses with suitable diagram.
2. Describe the piston-anvil method of preparing the metallic glasses. Write the applications of metallic glasses.
3. Describe the melt spinning method of preparing the metallic glasses. Write the properties and applications of metallic glasses.
4. What are shape memory alloys? Write about the characteristics of shape memory alloy. Mention the applications of shape memory alloys.
5. Describe the synthesis, properties and applications of Ni-Ti alloys.
6. What are nanomaterials? Describe the arc discharge method used to produce nanomaterials. What are the advantages and drawbacks of this method?

7. Describe the laser ablation technique used to produce nanomaterials. What are the advantages and drawbacks of this method?
8. Discuss the electrodeposition process used to produce nanomaterials. What are the advantages and drawbacks of this method?
9. Describe the chemical vapour deposition technique used to produce nanomaterials. What are the advantages and drawbacks of this method? Mention any four applications of nanomaterials.
10. Describe the ball milling method used to produce nanomaterials. What are the advantages and drawbacks of this method? What are the properties of nanomate.
11. Describe briefly properties, materials and applications of nonlinear optical materials.
12. What is second harmonic generation (SHG)? Derive a mathematical expression for SHG and explain. Describe an experimental method used to produce SHG.
13. What is optical mixing? Derive a mathematical expression for optical mixing and explain. Explain the sum frequency and difference frequency production using quantum principle.
14. What are NLO materials? How they are classified? Give examples.
15. Discuss about the applications of NLO materials.
16. Briefly discuss about the different types of biomaterials and their medical applications.

INDEX

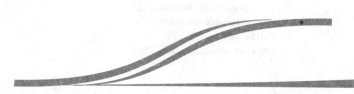